To L.R.M.

$\mathbf{D}_0$ vector of structure nodal displacements due to initial strains, global coordinates

$\mathbf{D}_A$, $\mathbf{D}_S$ respective antisymmetric and symmetric components of the **D** vector for a symmetric structure

$\mathbf{D}_c$, $\mathbf{D}_u$ subarrays of the **D** vector associated with constrained and unconstrained degrees of freedom, respectively

$\mathbf{D}_Q$ vector of relative displacements corresponding to the vector of redundant forces, **Q**

$\mathbf{D}_R$, $\mathbf{D}_L$ subarrays of the **D** vector for the right and left sides, respectively, of a symmetric structure

E modulus of elasticity

E elementary matrix

$\mathscr{f}$ array of member basic flexibility coefficients

$\overline{\mathscr{f}}$ array of member basic flexibility coefficients for the collection of all members and supports

f_c constraint basic flexibility coefficient

F array of structure flexibility coefficients, global coordinates

$\mathbf{F}_{RR}$ subarray of the **F** matrix relating **D** and **R**

$\mathbf{F}_{RQ}$ subarray of the **F** matrix relating **D** and **Q**

$\mathbf{F}_{QR}$ subarray of the **F** matrix relating $\mathbf{D}_Q$ and **R**

$\mathbf{F}_{QQ}$ subarray of the **F** matrix relating $\mathbf{D}_Q$ and **Q**

g_{ij} element of $\mathcal{G}$ array

$\mathcal{G}$ upper triangular array generated during gaussian decomposition

G modulus of rigidity

h depth of flexural member

i end of member where $x' = 0$

I moment of inertia of member cross section

I identity matrix

j end of member where $x' = L$

J member torsional geometric constant

$\mathscr{k}$ array of member basic stiffness coefficients

k array of member stiffness coefficients, global coordinates

k′ array of member stiffness coefficients, local coordinates

$\hat{\mathbf{k}}$ array of member stiffness coefficients, skew coordinates

$\overline{\mathbf{k}}$ array of stiffness coefficients for collection of all members and supports, global coordinates

$\mathbf{k}_{EE}$ substructure stiffness array relating $\mathbf{r}_E$ and $\mathbf{d}_E$

$\mathbf{k}_{EI}$ substructure stiffness array relating $\mathbf{r}_E$ and $\mathbf{d}_I$

(*Continued on next page*)

$\mathbf{k}_{IE}$ substructure stiffness array relating $\mathbf{r}_I$ and $\mathbf{d}_E$

$\mathbf{k}_{II}$ substructure stiffness array relating $\mathbf{r}_I$ and $\mathbf{d}_I$

k_c constraint basic stiffness coefficient

$\mathbf{k}_c$ constraint stiffness coefficient, global coordinates

$\mathbf{K}$ array of structure stiffness coefficients, global coordinates

$\hat{\mathbf{K}}$ array of structure stiffness coefficients, generalized coordinates or mixed coordinates

$\mathbf{K}_A$, $\mathbf{K}_S$ respective antisymmetric and symmetric components of the $\mathbf{K}$ matrix for a symmetric structure

$\mathbf{K}_{cc}$ subarray of the $\mathbf{K}$ matrix relating $\mathbf{R}_c$ and $\mathbf{D}_c$

$\mathbf{K}_{cu}$ subarray of the $\mathbf{K}$ matrix relating $\mathbf{R}_c$ and $\mathbf{D}_u$

$\mathbf{K}_{uc}$ subarray of the $\mathbf{K}$ matrix relating $\mathbf{R}_u$ and $\mathbf{D}_c$

$\mathbf{K}_{uu}$ subarray of the $\mathbf{K}$ matrix relating $\mathbf{R}_u$ and $\mathbf{D}_u$

$\mathbf{K}_{LL}$ subarray of the $\mathbf{K}$ matrix relating $\mathbf{R}_L$ and $\mathbf{D}_L$

$\mathbf{K}_{LR}$ subarray of the $\mathbf{K}$ matrix relating $\mathbf{R}_L$ and $\mathbf{D}_R$

$\mathbf{K}_{RL}$ subarray of the $\mathbf{K}$ matrix relating $\mathbf{R}_R$ and $\mathbf{D}_L$

$\mathbf{K}_{RR}$ subarray of the $\mathbf{K}$ matrix relating $\mathbf{R}_R$ and $\mathbf{D}_R$

ℓ_1, ℓ_2, ℓ_3 direction cosines of angles α_x, α_y, and α_z, respectively

ℓ_{ij} element of $\mathcal{L}$ and $\mathcal{L}^T$

$\bar{\ell}_{ij}$ element of $\bar{\mathcal{L}}$ and $\bar{\mathcal{L}}^T$

$\mathcal{L}$ upper triangular coefficient array generated during modified gaussian decomposition

$\mathcal{L}^T$ lower triangular coefficient array generated during gaussian decomposition and modified gaussian decomposition

$\bar{\mathcal{L}}$ upper triangular coefficient array generated during Choleskian decomposition

$\bar{\mathcal{L}}^T$ lower triangular coefficient array generated during Choleskian decomposition

L member length

$\mathbf{L}$ array of member force influence coefficients relating $\mathbf{r}'$ and $\mathbf{P}$

$\mathbf{L}^T$ array of member displacement influence coefficients relating $\boldsymbol{\Delta}$ and $\mathbf{d}'$

m_1, m_2, m_3 direction cosines of y' axis relative to x, y, and z axes, respectively

m_B semi-bandwidth of structure stiffness array

M member internal moment

M_i member-end basic moment, end i

M_j member-end basic moment, end j

M_c rotational constraint basic moment

(*Continued on back endpapers*)

Matrix Analysis of Structures

MATRIX ANALYSIS OF STRUCTURES

V. James Meyers

Purdue University

1817

HARPER & ROW, PUBLISHERS, New York
Cambridge, Philadelphia, San Francisco,
London, Mexico City, São Paulo, Sydney

Sponsoring Editor: Cliff Robichaud
Project Editor: Cynthia L. Indriso
Production: Delia Tedoff
Compositor: Science Press
Printer and Binder: The Maple Press Company
Art Studio: Vantage Art, Inc.

MATRIX ANALYSIS OF STRUCTURES

Library of Congress Cataloging in Publication Data

Meyers, V. James, 1933–
Matrix analysis of structures.

Includes index.
1. Structures, Theory of—Matrix methods. I. Title.
TA646.M49 1983 624.1'71 82-21308
ISBN 0-06-044438-X

Contents

Preface

There is a growing awareness throughout society of the inexorable rise of the computer. For engineers, the computer is rapidly approaching a position of dominance with respect to their technical activities. Thus, to be competitive, engineers must be capable of communicating fluently with the computer. For structural engineers, this requires an understanding of matrix structural analysis. Before studying matrix structural analysis as presented in this book, it would be desirable for students to have taken an introductory structural analysis course taught from either a matrix or a classical point of view. (Some previous study of matrix algebra would, of course, also be useful.) For course use, this book is written at a level suitable for third- or fourth-year undergraduate students or beginning graduate students. Although a classical structural analysis background is unnecessary to the understanding of the subjects presented, appropriate comparisons to classical methods are occasionally made for the benefit of students who may have such a background.

In this book, the developments of concepts are based firmly on physical reasoning. Readers are repeatedly guided to an appreciation for the physical meaning that underlies the matrix and coding operations presented. Even the matrix algebra review is offered in the context of a very idealized simple structural system rather than as a sterile set of mathematical rules.

With regard to the arrangement of subject matter, the concepts of equilibrium and compatibility are introduced first, allowing for the solution of statically determinate structures only. The range of problems that can be solved is then extended by developing stiffness and flexibility relationships. The displacement method of analysis is emphasized, with ultimate development of the direct stiffness method for automatic computation. The force method of analysis is also presented for use in an automatic computation environment. Special topics such as initial strain loadings, members of variable cross section, and shear deformation effects are relegated to a separate chapter so that the development of major ideas is uninterrupted. Thus an overall view of the displacement method is presented before adjustments for important special conditions are attempted. In line with this approach, subjects are discussed in a two-dimensional context throughout the book, with extension to three dimensions being reserved as a topic for the final chapter. Subjects related to the main theme that receive particularly effective treatment in this book are the often neglected topic of symmetry and the equally important topic of solutions of large sets of simultaneous equations.

Throughout the book, each discussion proceeds from the specific to the general. Most concepts are presented first in the context of a specific structural example and then generalized. Furthermore, the examples used are kept as simple as possible while still retaining the capacity to demonstrate the concept under discussion. These examples are reused throughout the book so that subtle differences in solution techniques are not obscured by gross differences in structures.

This book makes rather extensive use of the computer program SMIS3, which is an educational program, not a production program. Although users are relieved of much of the tedium of structural analysis calculations, they are forced to provide instructions to the computer regarding the major steps in the analysis they wish to accomplish. Thus the computer program enhances rather than obscures the readers' understanding of the analysis process. One especially useful feature of the SMIS3 program is that subroutines for certain of the important analysis steps have been written in such a way that their replacement is very convenient. Thus as the development of the subject proceeds, readers can write their own computer subroutines for particular analysis steps. The readers can thus gain the experience of transforming a structural analysis concept into FORTRAN logic without being burdened with input/output and other programming details that would otherwise consume so much of their time and effort. Proceeding in this fashion, the readers will not only learn the analysis concepts but will also have the opportunity to implement them on the computer. In the end, they will have written subroutines typical of each important step of the analysis, which should give them a reasonable insight into the functioning of a structural analysis program. Although the purpose of using the SMIS3 program is to promote a thorough understanding of matrix-computer analysis of structures, readers are also encouraged in the book to retain a healthy skepticism as to the correctness of computer results. Hand checks on computer-generated results are frequently suggested.

Because the major steps in most examples are displayed in detail independently of SMIS3, this book can also be effectively used with any other appropriate structural analysis program. It can also be used in case no computer program is available. However, it is certainly desirable for readers to be able to interact with the computer during the study of this subject if at all possible. The SMIS3 program is available at nominal reproduction, handling, and shipping cost from the author.

ACKNOWLEDGMENTS

The contributions of the many people who have helped in this effort are gratefully acknowledged. In particular, thanks are due to Dr. E. L. Wilson and Dr. A. F. Schkade for their respective permissions to use SMIS and SMIS2 as bases for SMIS3. The transformation of SMIS2 into SMIS3 is the product of the FORTRAN coding expertise as well as the creativity of thought of Messrs. William Hendrickson and Jamshid Gharib. Expert typing has been patiently supplied by Ms. Cheryl Haines, Ms. Rona Ginsburg, Ms. Molly Harrington, and Ms. Lucy Hohn. Special thanks are due to the reviewers for their helpful suggestions. The interest and encouragement shown by staff members of the School of Civil Engineering at Purdue University are sincerely appreciated. Thanks to Dr. R. H. Lee for his help in the development of the SMIS3 program and to Professor A. D. M. Lewis for permission to use his unformatted input subroutine in SMIS3. The extraordinary efforts of Dr. Atef Saleeb in proofreading the manuscript and in meticulously working out examples for the instructor's manual for this text deserve very special mention. Lastly but most importantly, thanks to my wife, Lavon, for her understanding and support.

V. JAMES MEYERS

Chapter 1
Concepts of Structural Analysis

1.1 INTRODUCTION

Structures can be classified in a variety of ways. The casual observer might first consider classifying structures according to their respective functions: buildings, bridges, ships, aircraft, towers, and so on. This basis for structural classification is in fact fundamental; all structures have some functional reasons for existence. It is the need to fulfill some function that prompts the designer to give life to a structure. Furthermore, it is the need for a safe, serviceable, feasible, and aesthetically pleasing fulfillment of a function that dictates the form, material, and manner of loading of a structure.

Once the form and material have been determined, a structure may be further classified according to either its form (e.g., an arch, truss, or suspension structure) or the material out of which it is constructed (e.g., steel, concrete, or timber). The form and material of a structure in turn dictate its behavior, which in turn dictates the character of the analytical model. Figure 1.1 illustrates schematically the relationships among the function a structure is to fulfill, the form and material of and loading on the structure, the behavior of the structure, and the analytical model of the structure.

At this point we need to discuss some of the aspects of structural behavior indicated in Fig. 1.1 and to explain their respective relationships to the form and material of the structure. A structure is linear if its response to loading, say

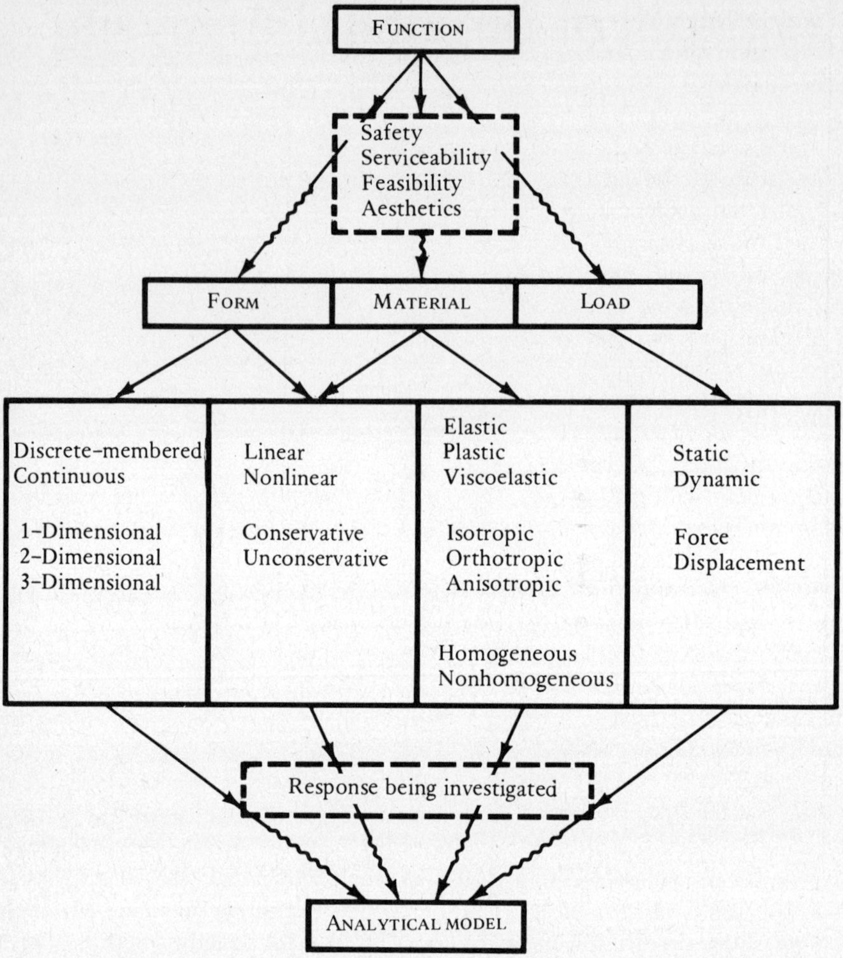

Fig. 1.1 Relationship between the function of a structure and its analytical model.

displacement at a point, is directly proportional to the magnitude of the applied load. If this proportionality does not exist, the structure is said to be nonlinear. Structural nonlinearities are of two types: (1) material nonlinearities that arise when stress is not proportional to strain, and (2) geometric nonlinearities that arise when the configuration of the structure under load is markedly changed from the unloaded configuration. (The presence of cables in a structure often leads to geometric nonlinearity because displacements can occur owing to a change in cable sag, which can be shown to be nonlinearly related to the force in the cable.) Materials, and therefore structures built from them, may be classified as elastic, plastic, or viscoelastic. Elastic materials rebound to their initial configuration when the load is removed, whereas plastic materials retain a permanent set. The deformations of viscoelastic materials depend on time and therefore load history, whereas the deformations of elastic and plastic materials do not. A structural system is unconservative or conservative depending on whether or not energy is lost from the system during a cycle of loading and unloading. Energy is generally lost if a system does not recover its initial shape after unloading owing either to plastic behavior of the material or to friction forces within or between parts of the structure.

All these behavioral aspects of the structure will have a significant influence on the nature of the analysis used in studying the structure. In addition, in developing the analytical model it will be necessary to consider whether the structural material is homogeneous or nonhomogeneous and whether it is isotropic, orthotropic, or anisotropic. (The physical properties of homogeneous materials are the same at each point; those of nonhomogeneous materials are not. The physical properties of isotropic materials are the same in all directions at a point; those of anisotropic materials are not. An orthotropic material is a special anisotropic material whose properties are different in three principal directions but whose properties in all other directions are dependent on those in the principal directions.) Other aspects of the structure, although important design considerations, will not usually have a significant impact on the analysis technique. These include brittleness, ductility, flammability, texture, color, hardness, and machinability.

Finally, the nature of the loading, which is dependent on the function of the structure, will also influence the analysis. The only truly static loading on a structure is the dead, or gravity, loading. However, if other loadings are applied gradually enough, they are called quasi-static loadings and may be considered static for analysis purposes. Whether or not the rate of loading is gradual enough depends on whether or not the time it takes to apply the load is longer than the fundamental period of vibration of the structure being analyzed. Loads usually need be treated as dynamic only if they are periodic in nature or if they are applied very suddenly. Even then, sometimes an "impact factor" is applied to an analysis with a static-loading result to account for the effect of a suddenly applied load. Loads can also be categorized as either external applied forces or internal initial distortions. Thermal loading is an example of an internal initial distortion (or initial strain) loading. (It will be shown in Chap. 2 that initial external displacements can also be considered to be initial internal distortions in support members.)

Unfortunately, the picture of structural behavior is generally not so clear as that just painted. That is, materials are not either "linear" or "nonlinear" and "elastic" or "plastic"; instead, their behavior depends on circumstances such as environment and rate of loading. The picture is further clouded in that the type of behavior that must be considered in an analysis may depend on the type of response being investigated. For example, a simpler analytical model may suffice to obtain static displacement and stress results than that which would be required for vibration or buckling results.

To clarify this picture for purposes of a rational presentation of matrix analysis of structures, we will make simplifying assumptions as to the nature of the behavior of structures. Thus we will consider only the displacement and stress response due to static loading of linear, elastic, conservative structures. We will further restrict our attention to discrete-membered structures (rigid- and pin-jointed frameworks) as opposed to continuous structures. However, it is important to recognize at the outset that the concepts that will be presented can be extended to the solution of many other classes of structural problems, including those involving dynamic response, material and geometric nonlinearities, inelasticity, instability, and continuous systems. Furthermore, the same concepts can be applied to problems from other areas of engineering, such as geotechnics, hydraulics, and heat transfer, as well as to problems outside of engineering altogether. Finally, to conserve space and time, most of our studies will deal with planar structures subjected to planar loadings in the plane of the structure. This approach will retain enough generality that the resulting analysis methods can be readily extended to three-dimensional applications.

For the sake of brevity, we will refer to our subject as "matrix structural analysis." A major feature that will be evident in matrix structural analysis is an emphasis on a *systematic* approach to the statement of the problem. Matrix notation turns out to be convenient to use in this connection because of its shorthand characteristics. Furthermore, the systematic approach together with matrix notation makes it particularly convenient to translate the statement of the problem to a computer language. Our effort will therefore be to develop matrix structural analysis as a tool for automatic computation of structural response to load.

1.2 SOLUTION METHODS

As mentioned before, the aspects of structural response to static load with which we will be mainly concerned are the internal force (and therefore stress) distribution in the structure and the displacements that the structure undergoes. These force and displacement responses constitute two infinite sets of unknowns, some portions of which must be determined in order to design the structure. Questions naturally arise as to whether we should attempt to determine displacement or force unknowns first and also whether some finite-sized subsets of each of these sets exist that are in some way fundamental. It will be shown that the consideration of either displacements or forces as the primary unknowns can lead to an effective solution method. It will be further shown that it is possible to identify a basic set of forces associated with each member (Ref. 1.1), in that not only are these forces independent of one another, but also all other forces in that member are directly dependent on this set. Thus this set of forces constitutes the minimum set that is capable of completely defining the stressed state of the member. Similarly, it will be shown that there is a corresponding basic set of displacements associated with each member (Ref. 1.1), in that not only are these displacements independent of one another, but also they are just sufficient to completely define the distorted shape of the member.

1.2.1 Force Method of Analysis

If internal forces are selected as the primary unknowns in a structure, the analysis method is referred to as the force method. It is convenient to consider the force method in the context of a pin-jointed truss structure.

A typical member of a truss is pinned at both ends and has no load applied at points intermediate to the ends. Thus the complete stress distribution in the member can be determined as soon as the axial force is known at any point in the member. A truss member therefore possesses a single basic unknown force that is not uniquely defined. That is, the basic unknown force may be taken as the axial force at any point in the member. Later we will, for the sake of consistency, always choose the axial force at a particular location to be the basic unknown, but it is not important to concern ourselves with this at present. Consider now that if the basic unknown force in each member of the truss were known, then the stress distribution in each member, and therefore the stress distribution in the entire structure, would be known. There remains the task of solving for the full set of basic forces in the structure. Consider the truss of Fig. 1.2a. The basic forces in the members must take on values such that each joint of the structure is in equilibrium. Support forces are also available to help satisfy some equilibrium requirements. Thus the full set of independent basic unknown forces in the structure consists of the sum of the basic forces in the collection of all members of the truss plus the unknown support forces. To help visualize this, it is useful to replace the supports with an appropriate collection of support members, as shown in Fig. 1.2b. Each support member is seen to possess one basic unknown force, and the total number of independent basic unknown forces in the structure consists of the sum of the basic forces in all members (including support members) of the structure. In the case of the truss of Fig. 1.2a and b, the number of basic unknown forces is exactly equal to the number of independent joint-equilbrium equations available to solve for the forces. A solution for the basic forces, and subsequently for the stress distribution in the structure, can now be effected. (Those who have studied classical structural analysis methods will recognize this solution as the "method of joints" [Ref. 1.2].)

We are left with the task of determining, if desired, the displaced configuration of the structure. In the force method, this is a backsubstitution step. We see that member strains and distortions are directly related to the member basic forces that are now known and that the displacements of the joints depend directly on the individual member distortions. For now we are interested only in discussing solution concepts. Details of carrying out solutions will be thoroughly covered in subsequent chapters.

A structure such as the truss of Fig. 1.2b is said to be statically determinate. Static determinacy implies (1) that the number of basic unknown forces is equal to the number of equilibrium equations available to solve for them, and (2) that the equations of this set are linearly independent of one another. If either condition 1 is violated (e.g., the number of unknown basic forces is less than the number of equilibrium equations) or condition 2 is violated, the structure is said to be statically unstable. The truss of Fig. 1.2c violates conditions 1 and 2, while the truss of Fig. 1.2d violates condition 2. Statically unstable structures will not be considered further in this text since most civil engineering structures either are statically stable or may be considered to be statically stable under special loadings. If structures that are intended to be statically unstable cannot be analyzed by assuming that they are statically stable under special load systems, their dynamic behavior must be considered; that is beyond the scope of this text.

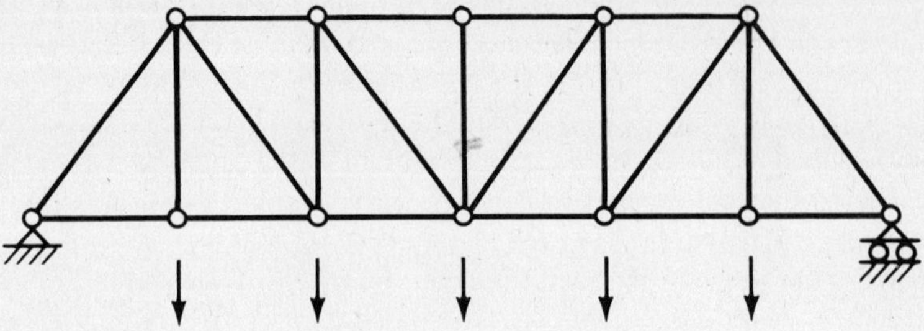

(a)

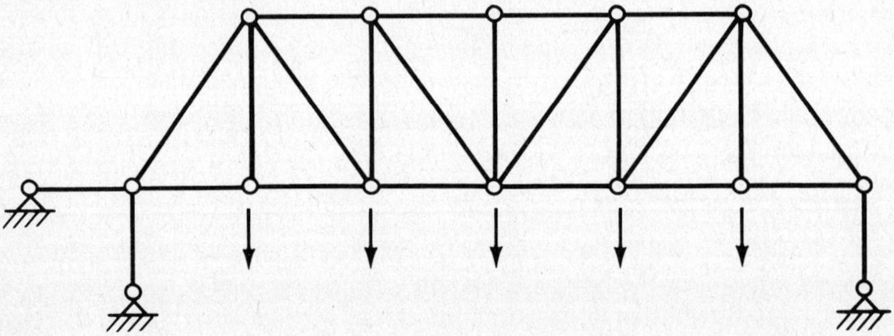

(b)

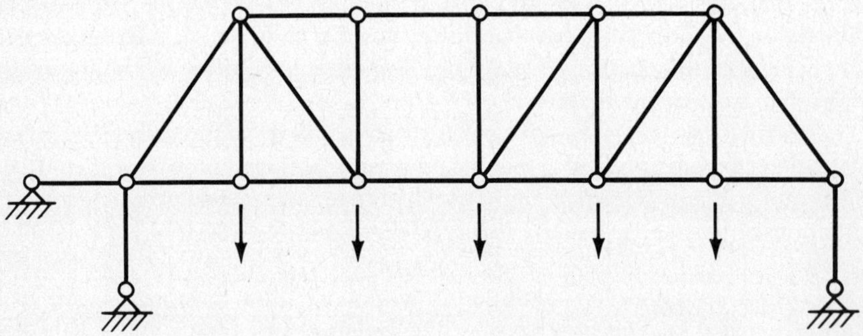

(c)

Fig. 1.2 Example truss structures: (a) statically determinate truss; (b) truss of part (a) showing support members; (c) statically unstable truss; (d) statically unstable truss; (e) statically indeterminate truss.

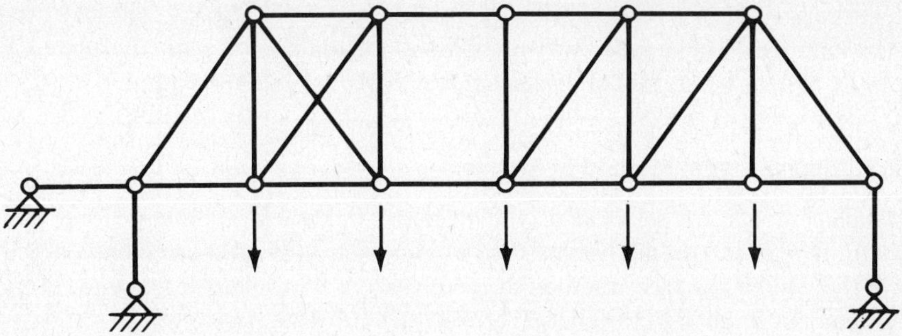

(d)

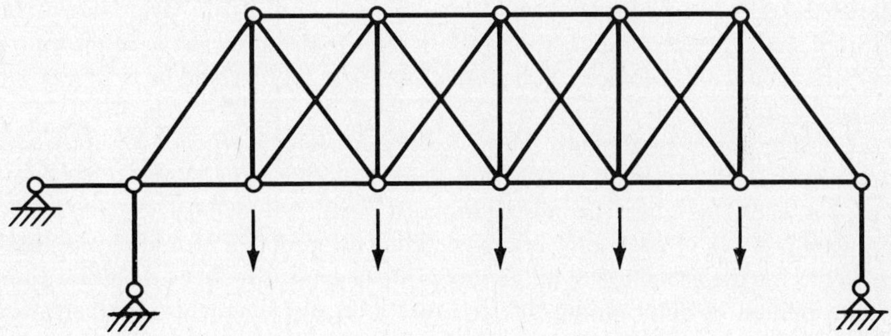

(e)

Fig. 1.2 (*continued*)

Condition 1 can also be violated when the number of basic unknown forces exceeds the number of equilibrium equations. Such structures are said to be statically indeterminate. The truss of Fig. 1.2e is statically indeterminate and statically stable. So in this case some of the basic unknown forces are not required to maintain structural equilibrium; these "extra" forces are termed redundant forces. In order to obtain a solution for the full set of basic unknown forces, it is necessary to augment the set of independent equilibrium equations with equations that are based on certain conditions that the redundant forces impose on the elastic behavior of the structure. Having solved for the full set of basic forces, we can determine the displacements by a backsubstitution process, as in the case of a statically determinate truss. Again it must be stated that in this chapter we are concerned only with solution concepts, and that the details of carrying out the solutions will be taken up in the remaining chapters.

In addition to plane trusses, we will be concerned with the matrix structural analysis of continuous beam and plane rigid-jointed frame structures. We could now outline the force method of analysis as applied to these two types of structures, identifying the basic unknown forces in each case and so on. But since the concepts will be identical with those already considered, we will leave this discussion for a more detailed treatment in later chapters. It is more useful now to move on to a consideration of displacements as primary unknowns.

1.2.2 Displacement Method of Analysis

If displacements are selected as the primary unknowns in the solution of a structural problem, the analysis method is referred to as the displacement method. As in the case of the force method, it is convenient to discuss the concepts of the displacement method in the context of truss structures.

Just as there is a single basic unknown force in a truss member, there is also a corresponding single basic unknown displacement. The truss-member basic displacement is the relative axial displacement of the ends of the member. Notice that this displacement is "basic" in the sense that it constitutes the minimum set of displacements that completely defines the distorted shape of the member. It follows that if the basic displacement for each member of a truss were known, the distorted shape of each member, and therefore the distorted shape of the entire structure, would also be known. The remaining task, then, would be to determine the basic displacements for the collection of all members of the structure.

Consider, once again, both the statically determinate truss of Fig. 1.2b and the statically indeterminate truss of Fig. 1.2e. In either case, the basic displacement for each member (including support members) will be known if the displaced position of each joint is known. Thus the problem of finding the basic unknowns for the collection of all members of the truss is reduced, in either case, to the problem of determining the structure joint displacements. Furthermore, again in either case, the number of unknown joint displacements is identical to the number of joint equilibrium equations. Thus we arrive at the somewhat surprising conclusion that a displacement method solution of a statically indeterminate structure will not be conceptually different from that of a statically determinate structure. Both solutions will involve writing the joint equilibrium equations in terms of joint displacements as primary unknowns and solving the resulting set of equations for these unknowns. Having done this, we can obtain internal stresses by backsubstitution, since the joint displacements determine each member's basic displacement and hence distortion, which in turn are directly related to the stress in the member. Once again, details of carrying out this solution will be discussed in later chapters.

It will be shown in Chaps. 4, 5, and 6 that the displacement method is generally more convenient than the force method and that most general-purpose structural analysis programs are based on it. But there are cases where it may be more desirable to use the force method.

1.3 CLASSIFICATION OF STRUCTURES

At the beginning of this chapter, we classified structures in various ways. Now our discussions of solution methods have involved the further classification of structures as to their static determinacy or indeterminacy and static stability or instability. These latter classification concepts have, of necessity, been presented rather superficially; they will be treated more fully at the end of Chap. 2. At that point we will also be in a position to consider the classification of structures according to their kinematic behavior as well as their static behavior.

1.4 COMPARISON OF CLASSICAL, MATRIX, AND APPROXIMATE METHODS OF STRUCTURAL ANALYSIS

Classical methods (Ref. 1.2) of structural analysis were intended for hand computation, and the developers of these tools took great pains to minimize the amount of calculations necessary to solve a given problem, even at the expense of making the methods somewhat unsystematic. After all, the analyst is always available for making complicated decisions when problems are being worked by hand. For example, if you have studied the slope-deflection method, you will probably recall difficulties in determining which "sidesway" unknowns were independent and how many existed in rigid-jointed frame structures. There was also the problem of choosing the proper free bodies corresponding to the sidesway unknowns. These difficulties are directly attributed to the assumption that members are inextensional, an assumption that significantly reduces the amount of calculation but complicates the analysis method conceptually. Although these difficulties are easily handled by an experienced analyst during hand computation, they make the classical method unattractive for translation to a computer code.

Thus the advent of the digital computer heralded a need for a more systematic method of analysis. As emphasized earlier, one of the hallmarks of matrix structural analysis is its systematic nature. Another descriptive term that can be applied to this method is the adjective *consistent*. By taking great care to formulate the tools of matrix structural analysis in a mathematically consistent fashion, the analyst achieves a systematic approach that is convenient for automatic computation. Every effort is made to eliminate surprises and exceptions. For example, we know from our discussion of the force and displacement methods that there is never any question about what free body to use in writing an equilibrium equation; the free body is always a joint, consistently. Further, the primary unknowns are either the basic forces in the collection of all members or the displacements of the collection of all joints. As the detailed development of these methods proceeds, you will notice this consistency and resulting systematic character at every turn.

A cautionary note should be sounded here with regard to matrix-computer analysis of structures. In some people computer results engender a fatal case of overconfidence. It is important for the analyst to maintain a critical, even skeptical, attitude toward computer results, because they can be substantially in error for a variety of reasons. One of the more common reasons, and certainly the most insidious, is misapplication of a program—that is, using a program that is based on an analytical model that does not properly reflect some important aspect of the behavior of the structure being analyzed. Thus although the results may be valid for the behavior being modeled, they may not indicate what will actually happen to the structure. This misapplication has been cited as a major factor in the recent collapse of the roof of the Hartford (Conn.) Coliseum (Ref. 1.3.). It is therefore imperative that the analyst have a proper understanding of structural behavior, both that being modeled by the program and that which can be expected in the actual structure.

Incorrect computer results can, of course, occur for other reasons, such as erroneous data or erroneous data input. In any event, it is essential that computer results be critically checked. This should be done qualitatively by examining the results for "reasonableness." The ability to make this qualitative judgment requires, again, that the analyst possess a good understanding of structural behavior. The results should also be checked quantitatively, which brings us to the subject of approximate methods of structural analysis.

Approximate methods (Ref. 1.2) involve imposing special conditions on a complex structure so that it is sufficiently simplified to allow an approximate hand result to be obtained relatively easily. And here again, the analyst must have a knowledge of structural behavior—there is no substitute. To quote Epstein and Smith (Ref. 1.3) from their paper on the Hartford catastrophe:

> No mathematical model of a structure can completely reproduce the actual behavior. The engineer is the vital link in the process of: (1) determining what approximations are reasonable for the particular structure, (2) verifying that these approximations are valid, and (3) designing the structure so that its behavior is in reasonable agreement with the model used in analyzing it.

1.5 SUMMARY

The function of a structure plays a dominating role in determining its form, material, and loading. The form, material, and loading, in turn, determine the behavior of the structure and therefore the character of the required analytical model. It is essential that the analyst understand all these relationships. It is the main purpose of this text to provide you, the potential analyst, with an understanding and insight into matrix analytical models of linear, elastic, discrete-membered structures so that you can make intelligent use of computer programs based on this systematic method of analysis.

One final comment is appropriate here. Although matrix structural analysis is more attractive to use as structures become larger and more complex, it is unnecessary and even counterproductive to present concepts of matrix structural analysis by making use of large structures as examples. Thus the examples used in succeeding chapters may seem trivial and perhaps much more easily solved by classical analysis methods. However, upon reflection you will see that the examples are just large enough to be capable of demonstrating essential ideas while requiring minimum calculations. You will also notice that the same set of examples are used repeatedly throughout the text. This format prevents the subtle differences in solution techniques from being obscured by gross differences in structures. It is hoped that, once you have mastered the concepts of matrix structural analysis in this context, you will appreciate that extension of the application of the method to larger, more meaningful examples can be easily accomplished.

REFERENCES

1.1 Meek, J. L. *Matrix Structural Analysis*. New York: McGraw-Hill, 1971, p. 86.

1.2 Norris, C. H.; Wilbur, J. B.; and Utku, S. *Elementary Structural Analysis*. New York: McGraw-Hill, 1976.

1.3 Epstein, H. I., and Smith, E. A. "Hartford Roof Failure—Can We Blame the Computer?" In ASCE (comp.), *Seventh Conference on Electronic Computation*. New York: ASCE, 1979, pp. 1–8.

Chapter 2
Equilibrium and Compatibility

2.1 INTRODUCTION

The concepts of equilibrium and compatibility are necessary for the solution of any structural problem. Taken by themselves, however, they are useful only in the study of statically determinate structures. Thus in this chapter we will be concerned with developing these concepts in the context of matrix structural analysis and applying them to the solution of certain statically determinate structures. Use of these concepts in the solution of statically indeterminate structures will be discussed in subsequent chapters. For the time being, we will consider only problems in which loads are applied at the joints between members and not within members.

2.2 TRUSS-MEMBER COORDINATE SYSTEMS

Plane trusses, being perhaps the simplest of structures, are most appropriate to be considered first in our study. To gain an understanding of the behavior of truss structures, or any structures for that matter, we must first understand the behavior of the individual members of which they are composed. It is assumed that the truss member is straight and pinned at both ends and that no loads are applied to the member between its ends. The force in the member is therefore axially oriented, and the end forces are collinear with the x' axis, as shown in Fig. 2.1a.

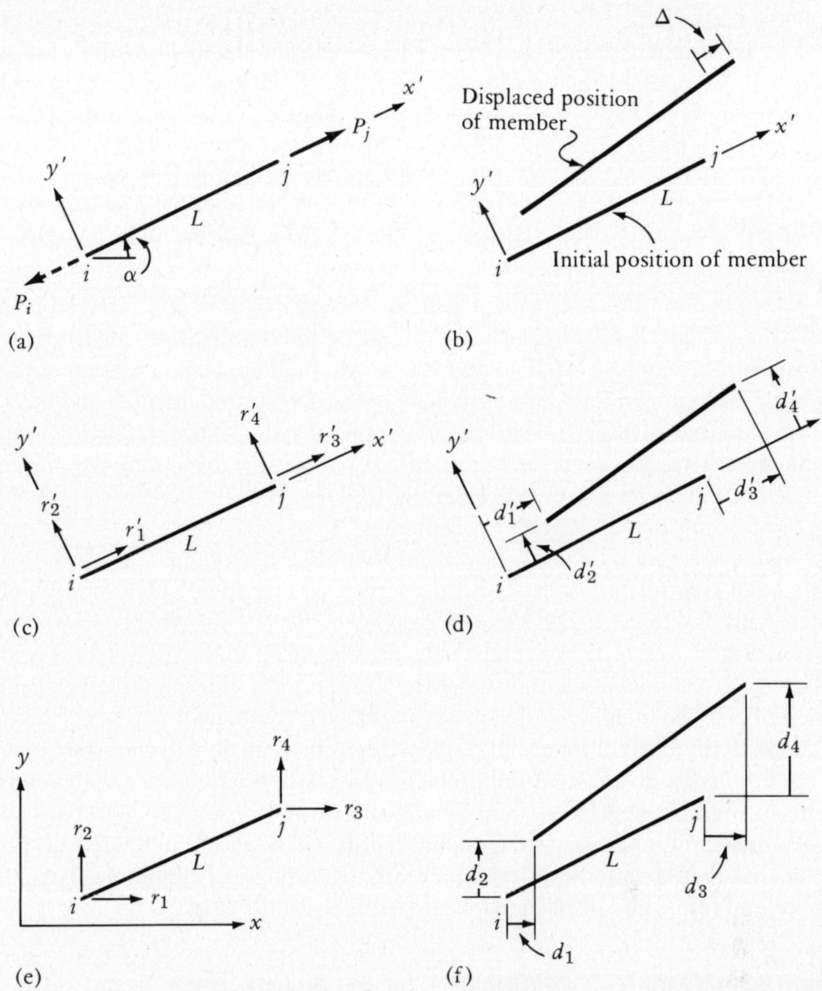

Fig. 2.1 Member-end forces and member-end displacements, plane truss member: (a) basic forces, local coordinates; (b) basic displacements, local coordinates; (c) member-end forces, local coordinates; (d) member-end displacements, local coordinates; (e) member-end forces, global coordinates; (f) member-end displacements, global coordinates.

Notice that in this figure the force at one end of the member is represented by a solid vector, and the force at the other end by a dashed vector. This representation emphasizes that only one of the forces is a "basic" force unknown. In this case P_j has arbitrarily been selected as the basic force. This force may be considered basic in that if it is known, then any other internal or external force in or on the member may be determined from equilibrium considerations only. (This single force completely defines the stressed state of the entire member.)

Corresponding to this basic force is a member elongation, or member-end relative displacement, Δ, as shown in Fig. 2.1b. This displacement is also a basic member unknown in that it is just sufficient to determine the distorted shape of the member. (Remember that we are considering the member to be loaded only at its ends.) It is important here not to confuse "distorted shape" with "displaced configuration." The basic displacement is not sufficient to locate the displaced configuration of the member in addition to defining its distorted shape. Locating this configuration would require three additional displacement components (one for each rigid body movement that the member could undergo in the plane). Thus the full set of four member-end displacements* shown in Fig. 2.1d have this capability.

The member-end displacement components of Fig. 2.1d are referred to a "local" $x'y'$ system of coordinates. This system is most convenient for studying member behavior and is "local" in that it is tied to the initial configuration of the member. (The origin of the system is at end i of the undistorted member and the x' axis runs along the member.) Member-end forces may also be referred to the local coordinate system, as shown in Fig. 2.1c. Of course, the two transverse force components will be zero in this case, but we will retain the complete set of local force components in the interest of consistency.

Finally, it is generally most convenient to study the structure as a whole in a single global coordinate system. It is therefore necessary to be able to represent member-end forces and member-end displacements in the global system, as shown in Fig. 2.1e and f. We might note, in passing, that the global force and displacement components do not have the same readily interpretable physical relationships to the member as do the basic and local components.

Since all three sets of force and displacement components will come into play in the solution of a given structural problem, the relationships among the various systems of member-end forces (and among the various systems of member-end displacements) must necessarily be established. It will be shown in the rest of this chapter that these relationships arise naturally from physical conditions that the member-end forces and displacements, respectively, must satisfy.

2.3 TRUSS-MEMBER COORDINATE TRANSFORMATIONS

Before developing truss-member coordinate transformations, a word or two on sign conventions is in order. The sign conventions for all force and displacement components have already been implied by showing all components in their positive senses in Fig. 2.1a–f. The sign convention for the basic force and displacement is very much tied to the member; that is, the basic force is positive in tension, and the basic displacement is positive if it is an elongation. On the other hand, the sign conventions for the local and global force and displacement components are referred to their respective coordinate systems; that is, they are positive when oriented in positive coordinate directions.

*The member displacements pictured in Fig. 2.1b, d, and f are greatly exaggerated for visual effect. As is the usual case in classical structural analysis, we will assume displacements are "small" in the development of the theory of matrix structural analysis. That is, the initial and distorted configurations of structures will be indistinguishable from one another to the casual observer. Thus equilibrium and compatibility equations will be written with the structure in its undistorted initial position.

2.3.1 Basic-Local Relationships

The relationship between basic and local forces may be obtained by first enforcing overall member equilibrium in Fig. 2.1a, which gives

$$P_i = P_j \tag{2.1}$$

and verifies that transverse forces at the member ends are equal to zero. Then, inspecting Fig. 2.1a and c, we may write

$$\begin{aligned} r'_1 &= -P_i \\ r'_2 &= 0 \\ r'_3 &= P_j \\ r'_4 &= 0 \end{aligned} \tag{2.2}$$

Substituting from Eq. 2.1 into 2.2 and letting $P_i = P_j = P$ gives

$$\begin{aligned} r'_1 &= -P \\ r'_2 &= 0 \\ r'_3 &= P \\ r'_4 &= 0 \end{aligned} \tag{2.3}$$

or, in matrix form,

$$\begin{pmatrix} r'_1 \\ r'_2 \\ r'_3 \\ r'_4 \end{pmatrix} = \begin{bmatrix} -1 \\ 0 \\ 1 \\ 0 \end{bmatrix} (P) \tag{2.3a}$$

or, more briefly,

$$\mathbf{r}' = \mathbf{LP} \tag{2.3b}$$

It is important to emphasize that the physical basis of Eqs. 2.3a and 2.3b is member overall equilibrium.

Comparing the displacements of Fig. 2.1b and d while recalling that all displacements are small, we see that

$$\Delta = d'_3 - d'_1 \tag{2.4}$$

or, in matrix form,

$$(\Delta) = [-1 \quad 0 \quad 1 \quad 0] \begin{pmatrix} d'_1 \\ d'_2 \\ d'_3 \\ d'_4 \end{pmatrix} \tag{2.4a}$$

or

$$\mathbf{\Delta} = \mathbf{L}^T\mathbf{d}' \tag{2.4b}$$

where $\mathbf{\Delta} = (\Delta)$, with Δ being the relative axial displacement of ends i and j; and $\mathbf{d}' = \langle d_1' \quad d_2' \quad d_3' \quad d_4' \rangle$* is the vector of member-end displacements in local coordinates. The displacement relationships are seen to be based on member kinematics.

2.3.2 Local-Global Relationships

The relationship between forces in the local and global systems may also be conveniently established. Resolving the local forces at the ends of the member of Fig. 2.1c into components in the horizontal and vertical (x and y) directions, we see that

$$\begin{aligned} r_1 &= r_1' \cos\alpha - r_2' \sin\alpha \\ r_2 &= r_1' \sin\alpha + r_2' \cos\alpha \\ r_3 &= r_3' \cos\alpha - r_4' \sin\alpha \\ r_4 &= r_3' \sin\alpha + r_4' \cos\alpha \end{aligned} \tag{2.5}$$

or, in matrix form,

$$\begin{pmatrix} r_1 \\ r_2 \\ r_3 \\ r_4 \end{pmatrix} = \begin{bmatrix} \cos\alpha & -\sin\alpha & 0 & 0 \\ \sin\alpha & \cos\alpha & 0 & 0 \\ 0 & 0 & \cos\alpha & -\sin\alpha \\ 0 & 0 & \sin\alpha & \cos\alpha \end{bmatrix} \begin{pmatrix} r_1' \\ r_2' \\ r_3' \\ r_4' \end{pmatrix} \tag{2.5a}$$

or

$$\mathbf{r} = \boldsymbol{\alpha}\mathbf{r}' \tag{2.5b}$$

Similarly, it can be shown that displacements will transform from global to local coordinates according to the relationship

$$\begin{pmatrix} d_1' \\ d_2' \\ d_3' \\ d_4' \end{pmatrix} = \begin{bmatrix} \cos\alpha & \sin\alpha & 0 & 0 \\ -\sin\alpha & \cos\alpha & 0 & 0 \\ 0 & 0 & \cos\alpha & \sin\alpha \\ 0 & 0 & -\sin\alpha & \cos\alpha \end{bmatrix} \begin{pmatrix} d_1 \\ d_2 \\ d_3 \\ d_4 \end{pmatrix} \tag{2.6}$$

or

$$\mathbf{d}' = \boldsymbol{\alpha}^T\mathbf{d} \tag{2.6a}$$

*Angle brackets, $\langle \quad \rangle$, are used in this text to indicate that the enclosed array is a column vector rather than a row vector.

Since Eq. 2.5b and 2.6a represent orthogonal transformations (Ref. 2.1),

$$\boldsymbol{\alpha\alpha}^T = \mathbf{I} \tag{2.7}$$

where **I** represents an identity matrix. It follows that

$$\mathbf{r}' = \boldsymbol{\alpha}^T\mathbf{r} \tag{2.8}$$

$$\mathbf{d} = \boldsymbol{\alpha}\mathbf{d}' \tag{2.9}$$

You should verify Eqs. 2.7, 2.8, and 2.9.

2.3.3 Basic-Global Relationships

A direct relationship between forces in the basic and global systems may now be established by substitution of Eq. 2.3b into 2.5b, giving

$$\mathbf{r} = \boldsymbol{\alpha}\mathbf{LP} \tag{2.10}$$

Writing this in detail and carrying out the matrix product $\boldsymbol{\alpha}$ **L** gives

$$\begin{pmatrix} r_1 \\ r_2 \\ r_3 \\ r_4 \end{pmatrix} = \begin{bmatrix} -\cos\alpha \\ -\sin\alpha \\ \cos\alpha \\ \sin\alpha \end{bmatrix} (P) \tag{2.10a}$$

or

$$\mathbf{r} = \mathbf{TP} \tag{2.10b}$$

thus relating basic forces **P** to global member-end forces **r** directly. Similarly, it can be shown that

$$\boldsymbol{\Delta} = \mathbf{L}^T\boldsymbol{\alpha}^T\mathbf{d} \tag{2.11}$$

or

$$(\Delta) = [-\cos\alpha \quad -\sin\alpha \quad \cos\alpha \quad \sin\alpha] \begin{pmatrix} d_1 \\ d_2 \\ d_3 \\ d_4 \end{pmatrix} \tag{2.11a}$$

or

$$\boldsymbol{\Delta} = \mathbf{T}^T\mathbf{d} \tag{2.11b}$$

thus giving basic displacements $\boldsymbol{\Delta}$ in terms of global member-end displacements **d**.

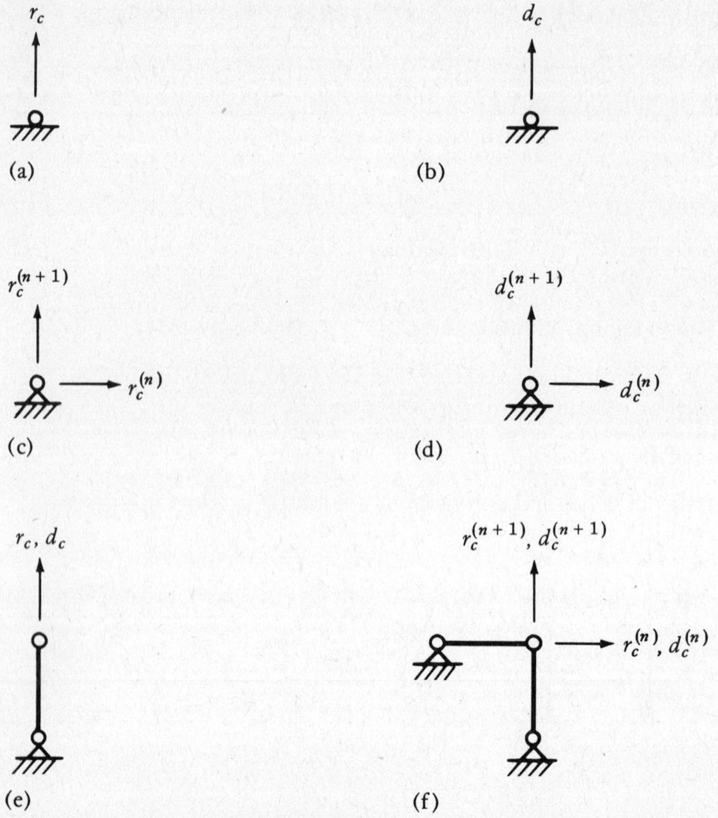

Fig. 2.2 Truss supports: (a) roller support force; (b) roller support displacements; (c) pin support forces; (d) pin support displacements; (e) roller, alternative representation; (f) pin, alternative representation.

2.4 TRUSS SUPPORTS

Supports impose one or more constraints on the behavior of a structure. In Chap. 4 it will be further observed that supports may often be considered to be infinitely rigid, thus imposing zero displacement constraints on the structure. For now, however, we need only to assume that all supports (as well as all members of the structure) have sufficient stiffness so that the small displacement assumption is fulfilled, thus allowing equations of statics and kinematics to be based on the initial undistorted geometry. Beyond this, the degree of stiffness is immaterial to our present concerns. Also for the time being, we will consider only supports whose constraints may be considered to be imposed in one or more global coordinate directions. Nonglobal constraints will be discussed in Sec. 5.4.

The global constraining force r_c induced on a roller support is shown in Fig. 2.2a. Since this force is also a basic unknown force, it may alternatively be represented by the symbol P_c, and we may write the useful identity

$$r_c = P_c \tag{2.12}$$

These forces are seen to be consistent with the force sets considered in the case of a truss member. Since the support may be nonrigid, corresponding displacements may occur. The roller-support global displacement, d_c, is shown in Fig. 2.2b. Once again we have an identity,

$$\Delta_c = d_c \tag{2.13}$$

The symbol Δ_c may be thought of as representing the basic distortion of the support (corresponding to P_c), while the symbol d_c represents the global displacement of the support (corresponding to r_c).

A pin support may be regarded as possessing two independent constraint forces and two independent constrained displacements, as shown in Fig. 2.2c and d, respectively. Once again, we could write identities of the forms of Eqs. 2.12 and 2.13. Such identities will prove useful in generating structure equilibrium and compatibility relationships in the next two sections.

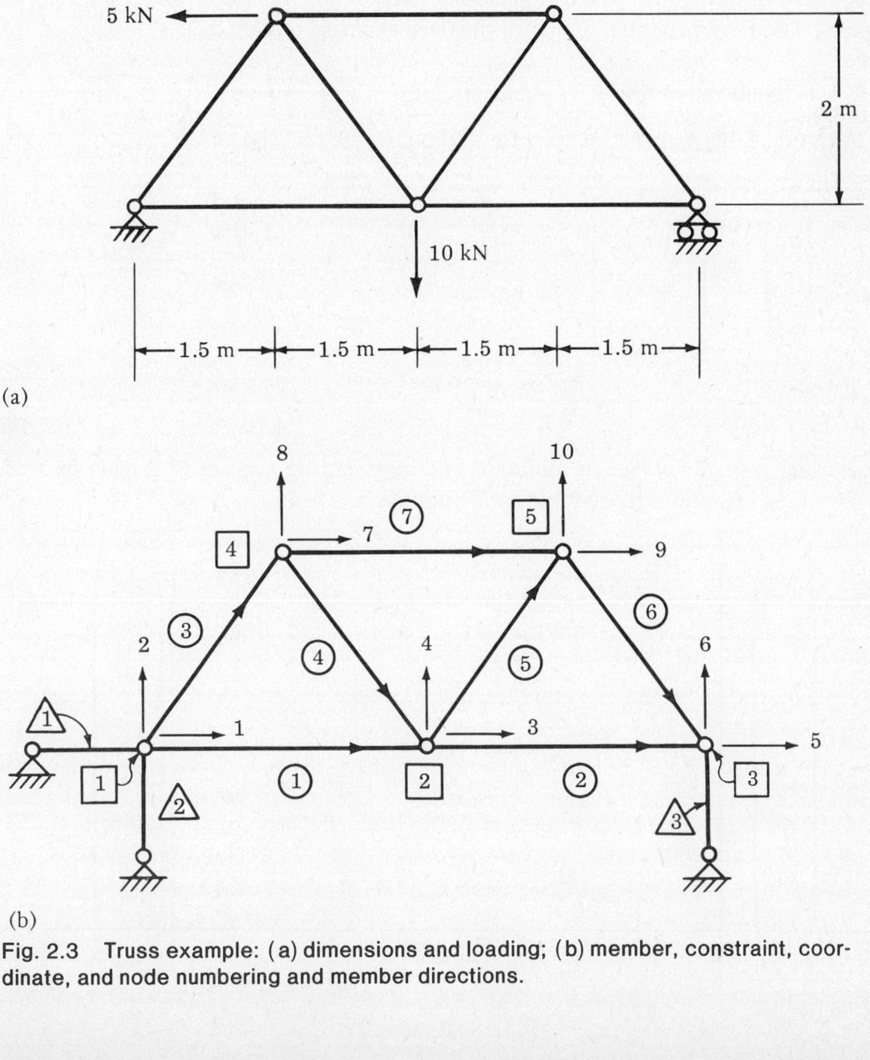

Fig. 2.3 Truss example: (a) dimensions and loading; (b) member, constraint, coordinate, and node numbering and member directions.

To better visualize the forces and displacements associated with supports, it may be helpful to represent each constraint schematically as a link, as shown in Fig. 2.2e and f for the roller and pin, respectively. The basic force P_c is then seen as the axial force in the link, whereas the force r_c is the global force at the end of the link. Correspondingly, the displacement Δ_c may be seen as the extension of the link, whereas the displacement d_c is the movement of the end of the link. This alternative representation of constraints will occasionally be used in this text (as it was in Fig. 1.2) to help clearly identify individual constraints.

2.5 SOLUTION FOR UNKNOWN INTERNAL FORCES IN STATICALLY DETERMINATE TRUSSES—EQUILIBRIUM EQUATIONS

Given the loading on a statically determinate truss, it is now possible to calculate the member-end and constraint forces by matrix structural analysis. Two methods of accomplishing this are the symbolic and coding methods, and they will be presented here in connection with the truss example of Fig. 2.3.

2.5.1 Symbolic Method: Truss Example

Equation 2.10a written for the nth member of the truss becomes

$$\begin{pmatrix} r_1 \\ r_2 \\ r_3 \\ r_4 \end{pmatrix}^{(n)} = \begin{bmatrix} -\cos \alpha_n \\ -\sin \alpha_n \\ \cos \alpha_n \\ \sin \alpha_n \end{bmatrix} (P)^{(n)} \tag{A}$$

or

$$\mathbf{r}^{(n)} = \mathbf{T}^{(n)} \mathbf{P}^{(n)} \tag{2.14}$$

Writing Eq. 2.14 for each member, and augmenting this set of equations with constraint equations (based on Eq. 2.12) gives

$$\begin{pmatrix} \mathbf{r}^{(1)} \\ \mathbf{r}^{(2)} \\ \mathbf{r}^{(3)} \\ \mathbf{r}^{(4)} \\ \mathbf{r}^{(5)} \\ \mathbf{r}^{(6)} \\ \mathbf{r}^{(7)} \\ r_c^{(1)} \\ r_c^{(2)} \\ r_c^{(3)} \end{pmatrix} = \begin{bmatrix} \mathbf{T}^{(1)} & & & & & & & & & \\ & \mathbf{T}^{(2)} & & & & & & & & \\ & & \mathbf{T}^{(3)} & & & & & & \boldsymbol{\Theta} & \\ & & & \mathbf{T}^{(4)} & & & & & & \\ & & & & \mathbf{T}^{(5)} & & & & & \\ & & & & & \mathbf{T}^{(6)} & & & & \\ & & & & & & \mathbf{T}^{(7)} & & & \\ & & \boldsymbol{\Theta} & & & & & 1 & & \\ & & & & & & & & 1 & \\ & & & & & & & & & 1 \end{bmatrix} \begin{pmatrix} \mathbf{P}^{(1)} \\ \mathbf{P}^{(2)} \\ \mathbf{P}^{(3)} \\ \mathbf{P}^{(4)} \\ \mathbf{P}^{(5)} \\ \mathbf{P}^{(6)} \\ \mathbf{P}^{(7)} \\ P_c^{(1)} \\ P_c^{(2)} \\ P_c^{(3)} \end{pmatrix} \tag{B}$$

or

$$\bar{\mathbf{r}} = \overline{\mathbf{T}}\overline{\mathbf{P}} \tag{2.15}$$

As a matter of convenience, we have adopted the convention of numbering structure members in one sequence, with constraints relegated to a second sequence. Discrimination between the two groups is accomplished with either the subscript c for constraints, as seen in the equations, or the use of special symbols, as shown in Fig. 2.3b. (This figure also establishes the convention for structure nodal and coordinate numbering used in this text.)

The vector of basic member-end and constraint forces on the right-hand sides of Eqs. B and 2.15 constitutes the complete set of independent unknown forces for the truss in question. Hence, after solving for these forces, we will be able to obtain any other desired internal structure forces in terms of these forces.

The **T** matrices for each member are

$$\mathbf{T}^{(1)} = \mathbf{T}^{(2)} = \mathbf{T}^{(7)} = \begin{bmatrix} -1 \\ 0 \\ 1 \\ 0 \end{bmatrix}$$

$$\mathbf{T}^{(3)} = \mathbf{T}^{(5)} = \begin{bmatrix} -.6 \\ -.8 \\ .6 \\ .8 \end{bmatrix}$$

$$\mathbf{T}^{(4)} = \mathbf{T}^{(6)} = \begin{bmatrix} -.6 \\ .8 \\ .6 \\ -.8 \end{bmatrix}$$

Using this information we can now write Eq. B as Eq. C as shown on page 22.

Recall now that Eq. C represents overall equilibrium of each member of the example truss. In order for the entire structure to be in equilibrium, it is further required that the member-end forces and constraint forces satisfy equilibrium of each joint (or node) of the structure. This is most conveniently accomplished by considering the free-body diagrams (FBDs) of each joint under the influence of the global member-end forces, the constraint forces, and the external applied loads also in the global system. (Notice that no decisions have to be made as to which FBDs to use.) Finally, Eq. C may be used to transform the resulting equilibrium equations so that they are in terms of the basic unknown forces.

$$
\begin{pmatrix} r_1^{(1)} \\ r_2^{(1)} \\ r_3^{(1)} \\ r_4^{(1)} \\ r_1^{(2)} \\ r_2^{(2)} \\ r_3^{(2)} \\ r_4^{(2)} \\ r_1^{(3)} \\ r_2^{(3)} \\ r_3^{(3)} \\ r_4^{(3)} \\ r_1^{(4)} \\ r_2^{(4)} \\ r_3^{(4)} \\ r_4^{(4)} \\ r_1^{(5)} \\ r_2^{(5)} \\ r_3^{(5)} \\ r_4^{(5)} \\ r_1^{(6)} \\ r_2^{(6)} \\ r_3^{(6)} \\ r_4^{(6)} \\ r_1^{(7)} \\ r_2^{(7)} \\ r_3^{(7)} \\ r_4^{(7)} \\ r_c^{(1)} \\ r_c^{(2)} \\ r_c^{(3)} \end{pmatrix}
=
\begin{bmatrix}
-1 & & & & & & & & & \\
0 & & & & & & & & & \\
1 & & & & & & & & & \\
0 & & & & & & & & & \\
 & -1 & & & & & & & & \\
 & 0 & & & & & & & & \\
 & 1 & & & & \Theta & & & & \\
 & 0 & & & & & & & & \\
 & & -.6 & & & & & & & \\
 & & -.8 & & & & & & & \\
 & & .6 & & & & & & & \\
 & & .8 & & & & & & & \\
 & & & -.6 & & & & & & \\
 & & & .8 & & & & & & \\
 & & & .6 & & & & & & \\
 & & & -.8 & & & & & & \\
 & & & & -.6 & & & & & \\
 & & & & -.8 & & & & & \\
 & & & & .6 & & & & & \\
 & & & & .8 & & & & & \\
 & & & & & -.6 & & & & \\
 & & & & & .8 & & & & \\
 & & & & & .6 & & & & \\
 & & & & & -.8 & & & & \\
 & & & & & & -1 & & & \\
 & & & & & & 0 & & & \\
 & & & & & & 1 & & & \\
 & \Theta & & & & & 0 & & & \\
 & & & & & & & 1 & & \\
 & & & & & & & & 1 & \\
 & & & & & & & & & 1
\end{bmatrix}
\begin{pmatrix} P^{(1)} \\ P^{(2)} \\ P^{(3)} \\ P^{(4)} \\ P^{(5)} \\ P^{(6)} \\ P^{(7)} \\ P_c^{(1)} \\ P_c^{(2)} \\ P_c^{(3)} \end{pmatrix}
\qquad \text{(C)}
$$

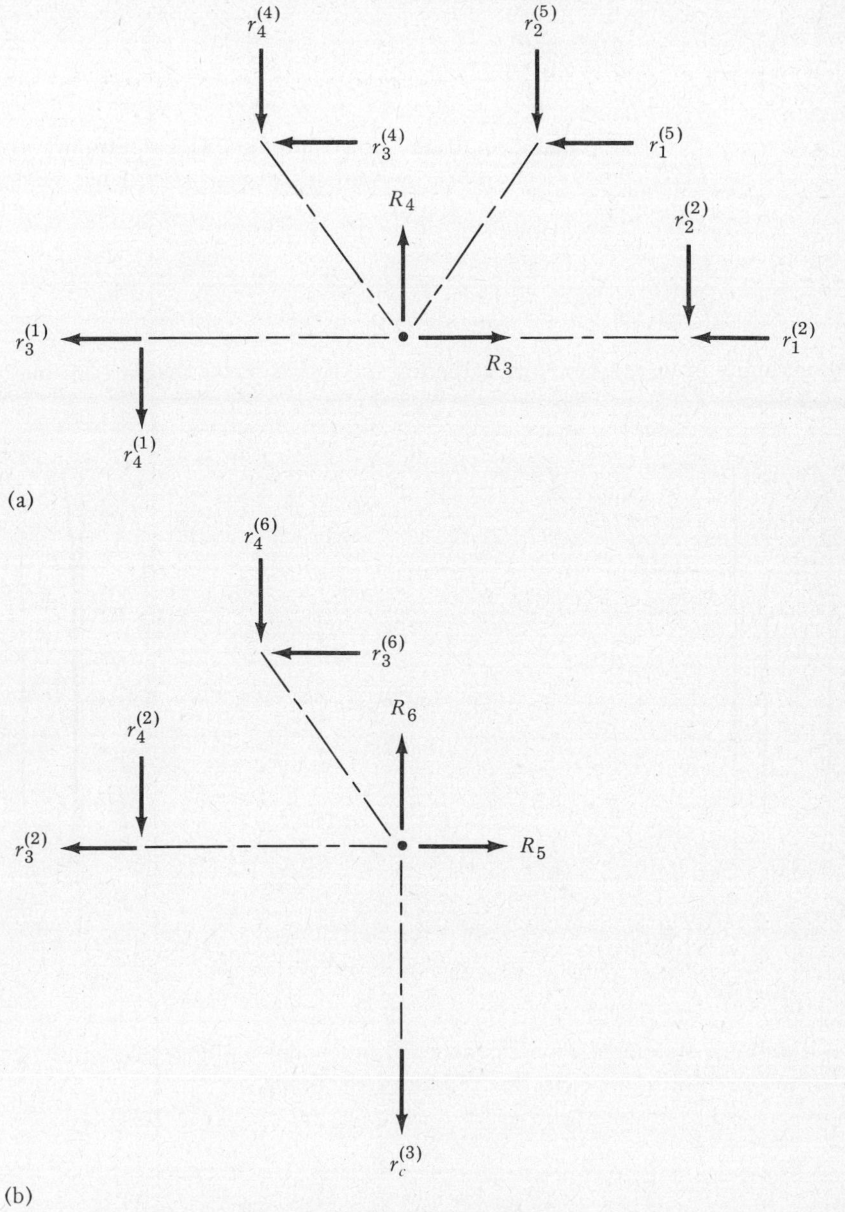

Fig. 2.4 Free body diagrams for truss: (a) node 2; (b) node 3.

Consider node 2 as an example. An FBD of this joint is shown in Fig. 2.4a. Summing forces in the x and y directions gives

$$\Sigma F_x = 0 \rightarrow R_3 - r_3^{(1)} - r_1^{(2)} - r_3^{(4)} - r_1^{(5)} = 0$$
$$\Sigma F_y = 0 \rightarrow R_4 - r_4^{(1)} - r_2^{(2)} - r_4^{(4)} - r_2^{(5)} = 0 \qquad \text{(D)}$$

or

$$
\begin{aligned}
R_3 &= r_3^{(1)} + r_1^{(2)} + r_3^{(4)} + r_1^{(5)} \\
R_4 &= r_4^{(1)} + r_2^{(2)} + r_4^{(4)} + r_2^{(5)}
\end{aligned}
\tag{E}
$$

Node 3 is not typical in that a constraint force is present. This constraint force is treated just like a member-end force. Summing forces in accordance with the FBD of Fig. 2.4b gives

$$
\begin{aligned}
R_5 &= r_3^{(2)} + r_3^{(6)} \\
R_6 &= r_4^{(2)} + r_4^{(6)} + r_c^{(3)}
\end{aligned}
\tag{F}
$$

Continuing in this fashion and gathering the results into a matrix equation, we obtain

$$
\begin{pmatrix} R_1 \\ R_2 \\ R_3 \\ R_4 \\ R_5 \\ R_6 \\ R_7 \\ R_8 \\ R_9 \\ R_{10} \end{pmatrix}
=
\left[\begin{array}{ccccccccccccccccccccccccccccccc}
1&\cdot&\cdot&\cdot&\cdot&\cdot&\cdot&\cdot&1&\cdot&\cdot&\cdot&\cdot&\cdot&\cdot&\cdot&\cdot&\cdot&\cdot&\cdot&\cdot&\cdot&\cdot&\cdot&\cdot&\cdot&\cdot&\cdot&1&\cdot&\cdot \\
\cdot&1&\cdot&\cdot&\cdot&\cdot&\cdot&\cdot&\cdot&1&\cdot&\cdot&\cdot&\cdot&\cdot&\cdot&\cdot&\cdot&\cdot&\cdot&\cdot&\cdot&\cdot&\cdot&\cdot&\cdot&\cdot&\cdot&\cdot&1&\cdot \\
\cdot&\cdot&1&\cdot&1&\cdot&\cdot&\cdot&\cdot&\cdot&\cdot&\cdot&\cdot&\cdot&1&\cdot&1&\cdot&\cdot&\cdot&\cdot&\cdot&\cdot&\cdot&\cdot&\cdot&\cdot&\cdot&\cdot&\cdot&\cdot \\
\cdot&\cdot&\cdot&1&\cdot&1&\cdot&\cdot&\cdot&\cdot&\cdot&\cdot&\cdot&\cdot&\cdot&1&\cdot&1&\cdot&\cdot&\cdot&\cdot&\cdot&\cdot&\cdot&\cdot&\cdot&\cdot&\cdot&\cdot&\cdot \\
\cdot&\cdot&\cdot&\cdot&\cdot&\cdot&1&\cdot&\cdot&\cdot&\cdot&\cdot&\cdot&\cdot&\cdot&\cdot&\cdot&\cdot&\cdot&\cdot&\cdot&\cdot&1&\cdot&\cdot&\cdot&\cdot&\cdot&\cdot&\cdot&\cdot \\
\cdot&\cdot&\cdot&\cdot&\cdot&\cdot&\cdot&1&\cdot&\cdot&\cdot&\cdot&\cdot&\cdot&\cdot&\cdot&\cdot&\cdot&\cdot&\cdot&\cdot&\cdot&\cdot&1&\cdot&\cdot&\cdot&\cdot&\cdot&\cdot&1 \\
\cdot&\cdot&\cdot&\cdot&\cdot&\cdot&\cdot&\cdot&\cdot&\cdot&1&\cdot&1&\cdot&\cdot&\cdot&\cdot&\cdot&\cdot&\cdot&\cdot&\cdot&\cdot&\cdot&1&\cdot&\cdot&\cdot&\cdot&\cdot&\cdot \\
\cdot&\cdot&\cdot&\cdot&\cdot&\cdot&\cdot&\cdot&\cdot&\cdot&\cdot&1&\cdot&1&\cdot&\cdot&\cdot&\cdot&\cdot&\cdot&\cdot&\cdot&\cdot&\cdot&\cdot&1&\cdot&\cdot&\cdot&\cdot&\cdot \\
\cdot&\cdot&\cdot&\cdot&\cdot&\cdot&\cdot&\cdot&\cdot&\cdot&\cdot&\cdot&\cdot&\cdot&\cdot&\cdot&\cdot&\cdot&1&\cdot&1&\cdot&\cdot&\cdot&\cdot&\cdot&1&\cdot&\cdot&\cdot&\cdot \\
\cdot&\cdot&\cdot&\cdot&\cdot&\cdot&\cdot&\cdot&\cdot&\cdot&\cdot&\cdot&\cdot&\cdot&\cdot&\cdot&\cdot&\cdot&\cdot&1&\cdot&1&\cdot&\cdot&\cdot&\cdot&\cdot&1&\cdot&\cdot&\cdot
\end{array}\right]
\begin{pmatrix} r_1^{(1)} \\ r_2^{(1)} \\ r_3^{(1)} \\ r_4^{(1)} \\ r_1^{(2)} \\ r_2^{(2)} \\ \vdots \\ r_c^{(1)} \\ r_c^{(2)} \\ r_c^{(3)} \end{pmatrix}
\tag{G}
$$

where the dots in the array indicate zeros. Rewriting Eq. G symbolically gives

$$\mathbf{R} = \mathbf{A}\bar{\mathbf{r}} \tag{2.16}$$

A combined statement of both member and nodal equilibrium may now be written by substituting from Eq. 2.15 into 2.16, giving

$$\mathbf{R} = \mathbf{A}\overline{\mathbf{T}}\overline{\mathbf{P}} \tag{2.17}$$

or

$$\mathbf{R} = \mathbf{a}\overline{\mathbf{P}} \tag{2.18}$$

where it is seen that $\mathbf{A}$ is 10×31 and $\overline{\mathbf{T}}$ is 31×10; therefore $\mathbf{a}$ is 10×10, and matrix Eq. 2.18 represents a set of 10 equilibrium equations relating the 10 applied external nodal forces to the 10 unknown member and constraint basic forces.

Carrying out the matrix product of Eq. 2.17 in detail, we get

$$\begin{pmatrix} R_1 \\ R_2 \\ R_3 \\ R_4 \\ R_5 \\ R_6 \\ R_7 \\ R_8 \\ R_9 \\ R_{10} \end{pmatrix} = \begin{bmatrix} -1 & 0 & -.6 & 0 & 0 & 0 & 0 & 1 & 0 & 0 \\ 0 & 0 & -.8 & 0 & 0 & 0 & 0 & 0 & 1 & 0 \\ 1 & -1 & 0 & .6 & -.6 & 0 & 0 & 0 & 0 & 0 \\ 0 & 0 & 0 & -.8 & -.8 & 0 & 0 & 0 & 0 & 0 \\ 0 & 1 & 0 & 0 & 0 & .6 & 0 & 0 & 0 & 0 \\ 0 & 0 & 0 & 0 & 0 & -.8 & 0 & 0 & 0 & 1 \\ 0 & 0 & .6 & -.6 & 0 & 0 & -1 & 0 & 0 & 0 \\ 0 & 0 & .8 & .8 & 0 & 0 & 0 & 0 & 0 & 0 \\ 0 & 0 & 0 & 0 & .6 & -.6 & 1 & 0 & 0 & 0 \\ 0 & 0 & 0 & 0 & .8 & .8 & 0 & 0 & 0 & 0 \end{bmatrix} \begin{pmatrix} P^{(1)} \\ P^{(2)} \\ P^{(3)} \\ P^{(4)} \\ P^{(5)} \\ P^{(6)} \\ P^{(7)} \\ P_c^{(1)} \\ P_c^{(2)} \\ P_c^{(3)} \end{pmatrix} \tag{H}$$

where

$$\mathbf{R} = \langle 0 \quad 0 \quad 0 \quad -10 \quad 0 \quad 0 \quad -5 \quad 0 \quad 0 \quad 0 \rangle \tag{I}$$

A solution of Eq. H with the load vector of Eq. I using the SMIS3* computer program is shown in the printout on pages 26–28, giving

$$\overline{\mathbf{P}} = \langle 0 \quad 2.5 \quad -8.33 \quad 8.33 \quad 4.17 \quad -4.17 \quad -5 \quad -5 \quad -6.67 \quad -3.33 \rangle \tag{J}$$

Symbolically, this solution takes the form

$$\overline{\mathbf{P}} = \mathbf{a}^{-1}\mathbf{R} \tag{2.19}$$

The symbolic method of building Eq. H is a very elegant and logical approach. However, because of the large numbers of zeros that must be stored and processed (see Eqs. C and G), the method is also hopelessly inefficient. Fortunately, a careful study of the method will show that the function of the nonzero terms of the **A** matrix as employed in Eq. G is simply to place the coefficients from the $\overline{\mathbf{T}}$ matrix in their proper positions in the **a** matrix. This same result can be accomplished (and nodal equilibrium enforced) by observing the correspondences of member and structure coordinates and using these correspondences to build the **a** matrix directly from the individual member **T** matrices. This concept is the basis of the coding method described in Sec. 2.5.2.

It is worth mention that solutions to sets of simultaneous equations such as Eq. H can be obtained more easily without actually carrying out the matrix inversion indicated symbolically in Eq. 2.19. Nevertheless, it will be our practice to obtain and display matrix inverses from time to time when this step serves some

*Documentation for SMIS3 is presented in App. 2.

****** *Computer Input:*

```
START
REMARK 3
SOLUTION FOR THE BASIC FORCES FOR THE EXAMPLE TRUSS

**** UNITS ARE KILONEWTONS AND METERS ****
LOAD A 10 10
-1 0 -.6 4*0 1 2*0 2*0 -.8 5*0 1 0 1 -1 0 .6 -.6
5*0 3*0 -.8 -.8 5*0 0 1 3*0 .6 4*0 5*0 -.8 3*0 1
2*0 .6 -.6 2*0 -1 3*0 2*0 .8 .8 6*0 4*0 .6 -.6 1
3*0 4*0 .8 .8 4*0
PRINT A 1
THIS IS THE A MATRIX
INVERT A
PRINT A 1
THIS IS THE MATRIX, A INVERSE
LOAD R 10 1
0 0 0 -10 0 0 -5 0 0 0
PRINT R 1
THIS IS THE LOAD VECTOR, R
MULT A R PBAR
PRINT PBAR 1
THIS IS THE VECTOR, PBAR, OF MEMBER AND CONSTRAINT BASIC FORCES
STOP
```

****** *Computer Output:*

```
START

REMARK       3

SOLUTION FOR THE BASIC FORCES FOR THE EXAMPLE TRUSS

**** UNITS ARE KILONEWTONS AND METERS ****


LOAD     A            10    10
   10 ROWS,      10 COLUMNS

PRINT    A            1


THIS IS THE A MATRIX


            1          2          3          4          5          6
 1 -1.0000E+00          0-6.0000E-01          0          0          0
 2          0          0-8.0000E-01          0          0          0
 3  1.0000E+00-1.0000E+00          0 6.0000E-01-6.0000E-01          0
 4          0          0          0-8.0000E-01-8.0000E-01          0
 5          0 1.0000E+00          0          0          0 6.0000E-01
 6          0          0          0          0          0-8.0000E-01
 7          0          0 6.0000E-01-6.0000E-01          0          0
 8          0          0 8.0000E-01 8.0000E-01          0          0
 9          0          0          0          0 6.0000E-01-6.0000E-01
10          0          0          0          0 8.0000E-01 8.0000E-01
            7          8          9         10
 1          0 1.0000E+00          0          0
 2          0          0 1.0000E+00          0
 3          0          0          0          0
 4          0          0          0          0
 5          0          0          0          0
 6          0          0          0 1.0000E+00
 7 -1.0000E+00          0          0          0
```

****** *Computer Output (continued):*

```
 8           0           0           0           0
 9  1.0000E+00           0           0           0
10           0           0           0           0


INVERT  A

PRINT   A          1

THIS IS THE MATRIX, A INVERSE


            1           2           3           4           5           6
 1          0           0 1.0000E+00-3.7500E-01 1.0000E+00           0
 2          0           0           0-3.7500E-01 1.0000E+00           0
 3          0           0           0 6.2500E-01           0           0
 4          0           0           0-6.2500E-01           0           0
 5          0           0           0-6.2500E-01           0           0
 6          0           0           0 6.2500E-01           0           0
 7          0           0           0 7.5000E-01           0           0
 8  1.0000E+00           0 1.0000E+00 4.4409E-15 1.0000E+00           0
 9          0 1.0000E+00           0 5.0000E-01           0           0
10          0           0           0 5.0000E-01           0 1.0000E+00
            7           8           9          10
 1  7.5000E-01-5.6250E-01 7.5000E-01-1.8750E-01
 2  2.5000E-01-1.8750E-01 2.5000E-01-5.6250E-01
 3  4.1667E-01 9.3750E-01 4.1667E-01 3.1250E-01
 4 -4.1667E-01 3.1250E-01-4.1667E-01-3.1250E-01
 5  4.1667E-01-3.1250E-01 4.1667E-01 3.1250E-01
 6 -4.1667E-01 3.1250E-01-4.1667E-01 9.3750E-01
 7 -5.0000E-01 3.7500E-01 5.0000E-01 3.7500E-01
 8  1.0000E+00 2.2204E-15 1.0000E+00-1.3323E-15
 9  3.3333E-01 7.5000E-01 3.3333E-01 2.5000E-01
10 -3.3333E-01 2.5000E-01-3.3333E-01 7.5000E-01


LOAD    R          10    1
   10 ROWS,     1 COLUMNS

PRINT   R          1

THIS IS THE LOAD VECTOR, R


            1
 1          0
 2          0
 3          0
 4 -1.0000E+01
 5          0
 6          0
 7 -5.0000E+00
 8          0
 9          0
10          0


MULT    A       R       PBAR

PRINT   PBAR       1

THIS IS THE VECTOR, PBAR, OF MEMBER AND CONSTRAINT BASIC FORCES
```

****** *Computer Output (continued):*

```
             1
  1 -2.8422E-14
  2  2.5000E+00
  3 -8.3333E+00
  4  8.3333E+00
  5  4.1667E+00
  6 -4.1667E+00
  7 -5.0000E+00
  8 -5.0000E+00
  9 -6.6667E+00
 10 -3.3333E+00

STOP
```

pedagogical end, as in the case of the solution displayed in Eq. 2.19. A discussion of efficient solutions of simultaneous equations is given in Chap. 7.

2.5.2 Coding Method: Truss Example

The elements in the **T** and **a** matrices may be physically interpreted as force influence coefficients. That is,

$T_m^{(n)}$ = force in mth global coordinate direction for member n per unit basic force for the member

a_{MN} = force in Mth global coordinate direction for the structure per unit Nth basic force for the structure

These two factors will obviously be the same if (1) the mth global coordinate direction for member n ≡ the Mth global coordinate direction for the structure, and (2) the basic force for member n ≡ Nth basic force for the structure. Condition 1 is fulfilled if, when the nth member is assembled into the structure, its mth coordinate coincides with the structure's Mth coordinate. Condition 2 is fulfilled for a truss if $N = n$. Consider the example truss of Fig. 2.3 again; member 3 of that truss is shown in Fig. 2.5. Figure 2.5a shows the member global coordinate numbering for member 3, while Fig. 2.5b shows the forces that would exist in these global coordinate directions owing to a unit basic force ($P^{(3)} = 1$) in member 3. These forces are, of course, seen to be precisely the elements of $\mathbf{T}^{(3)}$. Thus, for example,

$$T_4^{(3)} = .8$$

By comparing Fig. 2.5a with Fig. 2.3b, we see that the coordinate direction 4 of member 3 coincides with structure coordinate direction 8. It follows from our previous discussion that

$$a_{83} = T_4^{(3)} = .8$$

which is seen from Eq. H to be true. In other words, we could say that a_{83} relates

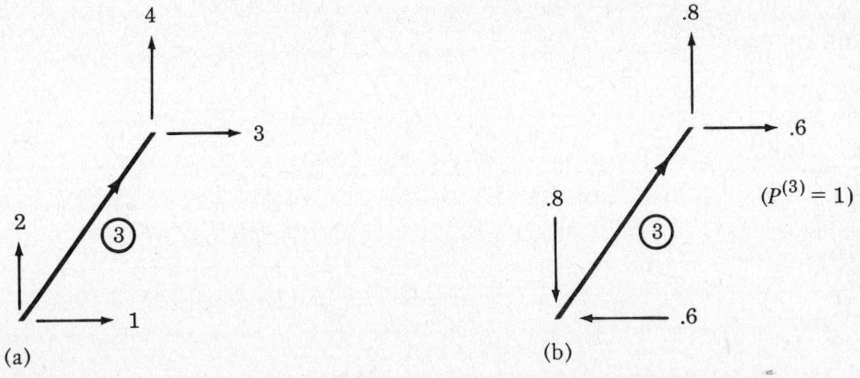

Fig. 2.5 Truss member 3: (a) member global coordinates; (b) global member-end forces.

the basic force in member 3 (which is the third basic structure force by our arrangement) to the force in structure global coordinate direction 8; and we could also say that $T_4^{(3)}$ relates the basic force in member 3 to the force in member global coordinate direction 4 (which coincides with structure coordinate direction 8). Thus the two forces are indeed the same.

The process of building the **a** matrix from the individual member **T** matrices therefore reduces to a matter of determining, by inspection, the correspondences of member and structure coordinates. Continuing, we may further observe that

$a_{73} = T_3^{(3)} = .6$ (member coordinate 3 = structure coordinate 7)
$a_{23} = T_2^{(3)} = -.8$ (member coordinate 2 = structure coordinate 2)
$a_{13} = T_1^{(3)} = -.6$ (member coordinate 1 = structure coordinate 1)
$a_{56} = T_3^{(6)} = .6$ (member coordinate 3 = structure coordinate 5)
$a_{37} = 0$ (member 7 does not have a coordinate that corresponds to structure coordinate 3)

Of course, constraint forces also contribute to structure equilibrium, and their influence on the **a** matrix may also be determined by inspection. Thus, for example,

$a_{18} = 1$ constraint 1 (1 = 8 − number of members = 8 − 7) is in structure coordinate direction 1
$a_{6,10} = 1$ constraint 3 (3 = 10 − number of members = 10 − 7) is in structure coordinate direction 6

Notice that these coefficients check those of Eq. H. We could continue in this fashion and develop the entire **a** matrix of Eq. H and subsequently solve that equation for the basic structure forces as before. It is of greater interest, however, to formalize this process to make it convenient for computer analysis.

To this end, coordinate correspondences may be tabulated as a coordinate code, as for the truss of Fig. 2.3:

Coordinate Code

Member number ↓	1	2	3	4 ← Member global coordinates
1	1	2	3	4
2	3	4	5	6
3	1	2	7	8
4	7	8	3	4
5	3	4	9	10
6	9	10	5	6
7	7	8	9	10

(Rows 1–7: Structure global coordinates)

As an aid to understanding the functioning of this code, it is helpful to list the code directly on the member **T** matrices:

$$\overset{①}{\begin{matrix}1\\2\\3\\4\end{matrix}\begin{bmatrix}-1\\0\\1\\0\end{bmatrix}}\quad\overset{②}{\begin{matrix}3\\4\\5\\6\end{matrix}\begin{bmatrix}-1\\0\\1\\0\end{bmatrix}}\quad\overset{③}{\begin{matrix}1\\2\\7\\8\end{matrix}\begin{bmatrix}-.6\\-.8\\.6\\.8\end{bmatrix}}\quad\overset{④}{\begin{matrix}7\\8\\3\\4\end{matrix}\begin{bmatrix}-.6\\.8\\.6\\-.8\end{bmatrix}}$$

$$\overset{⑤}{\begin{matrix}3\\4\\9\\10\end{matrix}\begin{bmatrix}-.6\\-.8\\.6\\.8\end{bmatrix}}\quad\overset{⑥}{\begin{matrix}9\\10\\5\\6\end{matrix}\begin{bmatrix}-.6\\.8\\.6\\-.8\end{bmatrix}}\quad\overset{⑦}{\begin{matrix}7\\8\\9\\10\end{matrix}\begin{bmatrix}-1\\0\\1\\0\end{bmatrix}}$$

It may now be seen that

$$a_{96} = T_1^{(6)} = -.6$$

$$a_{10,6} = T_2^{(6)} = .8$$

$$a_{34} = T_3^{(4)} = .6$$

and so on.

Specification of constraint information may also be formalized in a constraint list:

Constraint List

Constraint Number	Constrained Coordinate
1	1
2	2
3	6

On the basis of this constraint list, it is seen, for example, that

$$a_{6,10} = 1 \qquad (10 = 3 + \text{number of members} = 3 + 7)$$

Thus, using the member coordinate codes and the constraint list as shown, we may build the structure equilibrium equations and solve as before.

Although this coordinate coding method is commonly used in structural analysis programs, a close inspection of the arrangement that the elements of the **T** matrices assume in the **a** matrix will lead to one further improvement. Notice in particular that the elements in certain submatrices of the **T** matrices never become separated in their move to the **a** matrix. This phenomenon is a direct result of the *consistent* fashion in which both member and structure coordinates are numbered, and it leads to the concept of nodal coding. The nodal code has the advantage that data input is significantly reduced from that required for the coordinate code. The reduction will be by a factor of between 2 and 6, depending on the number of coordinates (or degrees of freedom) at each node.

The nodal code for the truss example is

Nodal Code

Member number	i	j ← Member nodes
1	1	2
2	2	3
3	1	4
4	4	2
5	2	5
6	5	3
7	4	5

(Member incidences = Structure nodes)

To be consistent, the constraint list is also written from a nodal point of view.

Constraint List

Constraint number	Structure node	Node coordinate
1	1	1
2	1	2
3	3	2

Coordinates at each node are numbered in the standard sequence; for example, for a truss, the x coordinate is 1 and the y coordinate is 2.

Rewriting the individual member **T** matrices with submatrices identified by nodal number and member number gives

$$
\overset{①}{\begin{array}{c}\boxed{1}\\ \\ \boxed{2}\\ \\ \end{array}\left[\begin{array}{c}-1\\0\\ \hline 1\\0\end{array}\right]}
\qquad
\overset{②}{\begin{array}{c}\boxed{2}\\ \\ \boxed{3}\\ \\ \end{array}\left[\begin{array}{c}-1\\0\\ \hline 1\\0\end{array}\right]}
\qquad
\overset{③}{\begin{array}{c}\boxed{1}\\ \\ \boxed{4}\\ \\ \end{array}\left[\begin{array}{c}-.6\\-.8\\ \hline .6\\.8\end{array}\right]}
\qquad
\overset{④}{\begin{array}{c}\boxed{4}\\ \\ \boxed{2}\\ \\ \end{array}\left[\begin{array}{c}-.6\\.8\\ \hline .6\\-.8\end{array}\right]}
$$

$$
\overset{⑤}{\begin{array}{c}\boxed{2}\\ \\ \boxed{5}\\ \\ \end{array}\left[\begin{array}{c}-.6\\-.8\\ \hline .6\\.8\end{array}\right]}
\qquad
\overset{⑥}{\begin{array}{c}\boxed{5}\\ \\ \boxed{3}\\ \\ \end{array}\left[\begin{array}{c}-.6\\.8\\ \hline .6\\-.8\end{array}\right]}
\qquad
\overset{⑦}{\begin{array}{c}\boxed{4}\\ \\ \boxed{5}\\ \\ \end{array}\left[\begin{array}{c}-1\\0\\ \hline 1\\0\end{array}\right]}
$$

Using the nodal code, we can now see that submatrices of the **a** matrix relating forces at various nodes of the structure to the basic forces in various members may be written as

$$\mathbf{a}_{56} = \mathbf{T}_i^{(6)} = \begin{pmatrix}-.6\\ .8\end{pmatrix}$$

$$\mathbf{a}_{24} = \mathbf{T}_j^{(4)} = \begin{pmatrix}.6\\ -.8\end{pmatrix}$$

The constraint list also makes it possible to determine certain submatrices of the **a** matrix. For example, the second row of the constraint list indicates that the force associated with constraint 2 participates in equilibrium in the second coordinate direction at node 1. Thus

$$\mathbf{a}_{19} = \begin{bmatrix}0\\1\end{bmatrix} \qquad (9 = 2 + \text{number of members} = 2 + 7)$$

To further clarify the functioning of the nodal code, the complete structure-equilibrium equations are repeated here with the code superimposed (see p. 33). Dots have been used for zeros in locations where no member contributes to the **a** matrix. A comparative study of the **T** matrices and this **a** matrix should help reinforce the concept of nodal coding.

$$
\begin{array}{c}
\\
\begin{Bmatrix} R_1 \\ R_2 \end{Bmatrix} \boxed{1} \\
\begin{Bmatrix} R_3 \\ R_4 \end{Bmatrix} \boxed{2} \\
\begin{Bmatrix} R_5 \\ R_6 \end{Bmatrix} \boxed{3} \\
\begin{Bmatrix} R_7 \\ R_8 \end{Bmatrix} \boxed{4} \\
\begin{Bmatrix} R_9 \\ R_{10} \end{Bmatrix} \boxed{5}
\end{array}
=
\begin{array}{cccccccccc}
\textcircled{1} & \textcircled{2} & \textcircled{3} & \textcircled{4} & \textcircled{5} & \textcircled{6} & \textcircled{7} & \triangle 1 & \triangle 2 & \triangle 3 \\
\end{array}
\begin{bmatrix}
-1 & . & -.6 & . & . & . & . & 1 & . & . \\
0 & . & -.8 & . & . & . & . & . & 1 & . \\
1 & -1 & . & .6 & -.6 & . & . & . & . & . \\
0 & 0 & . & -.8 & -.8 & . & . & . & . & . \\
. & 1 & . & . & . & .6 & . & . & . & . \\
. & 0 & . & . & . & -.8 & . & . & . & 1 \\
. & . & .6 & -.6 & . & . & -1 & . & . & . \\
. & . & .8 & .8 & . & . & 0 & . & . & . \\
. & . & . & . & .6 & -.6 & 1 & . & . & . \\
. & . & . & . & .8 & .8 & 0 & . & . & .
\end{bmatrix}
\begin{Bmatrix} P^{(1)} \\ P^{(2)} \\ P^{(3)} \\ P^{(4)} \\ P^{(5)} \\ P^{(6)} \\ P^{(7)} \\ P_c^{(1)} \\ P_c^{(2)} \\ P_c^{(3)} \end{Bmatrix}
\qquad \text{(K)}
$$

Finally, consider the detailed solution to Eq. K previously obtained by using SMIS3 in Sec. 2.5.1 and listed again here:

$$
\begin{Bmatrix} P^{(1)} \\ P^{(2)} \\ P^{(3)} \\ P^{(4)} \\ P^{(5)} \\ P^{(6)} \\ P^{(7)} \\ P_c^{(1)} \\ P_c^{(2)} \\ P_c^{(3)} \end{Bmatrix}
=
\begin{bmatrix}
0 & 0 & 1 & -.375 & 1 & 0 & .75 & -.5625 & .75 & -.1875 \\
0 & 0 & 0 & -.375 & 1 & 0 & .25 & -.1875 & .25 & -.5625 \\
0 & 0 & 0 & .625 & 0 & 0 & .41667 & .9375 & .41667 & .3125 \\
0 & 0 & 0 & -.625 & 0 & 0 & -.41667 & .3125 & -.41667 & -.3125 \\
0 & 0 & 0 & -.625 & 0 & 0 & .41667 & -.3125 & .41667 & .3125 \\
0 & 0 & 0 & .625 & 0 & 0 & -.41667 & .3125 & -.41667 & .9375 \\
0 & 0 & 0 & .75 & 0 & 0 & -.5 & .375 & .5 & .375 \\
1 & 0 & 1 & 0 & 1 & 0 & 1 & 0 & 1 & 0 \\
0 & 1 & 0 & .5 & 0 & 0 & .33333 & .75 & .33333 & .25 \\
0 & 0 & 0 & .5 & 0 & 1 & -.33333 & .25 & -.33333 & .75
\end{bmatrix}
\begin{Bmatrix} R_1 \\ R_2 \\ R_3 \\ R_4 \\ R_5 \\ R_6 \\ R_7 \\ R_8 \\ R_9 \\ R_{10} \end{Bmatrix}
\qquad \text{(L)}
$$

In developing the coding techniques, we utilized the concept of the elements of the **a** matrix being force influence coefficients. We can now see that the elements of $\mathbf{a}^{-1}$ constitute another set of force influence coefficients. Letting $\mathbf{a}^{-1} = \mathbf{b}$ for convenience, we have

b_{MN} = Mth basic member force per unit force in the Nth nodal coordinate direction

In fact, the elements of the **b** matrix represent the force influence coefficients encountered in standard influence-line construction. For example, the third row of the **b** matrix represents the influence on the force in member 3 of a unit load as it is applied in each global coordinate direction in turn. Suppose we elected to consider only a vertical unit load moving across the bottom chord of the truss (on a system of floor beams). We would then encounter influence coefficients $b_{32} = 0$, $b_{34} = .625$, and $b_{36} = 0$, which may be plotted as shown in Fig. 2.6a. Or, considering the influence of the same load on, say, $P_c^{(2)}$, we would get the plot of Fig. 2.6b, which is seen to be the influence line for the vertical support force at the left end of the truss. You may wish to sketch a few influence lines in this fashion (i.e., based on the **b** influence coefficients) to get a better physical understanding of solution Eq. L.

Note that the solution method discussed in this section is an application of the force method of analysis described in Sec. 1.2.1. This application is for the special case of a statically determinate structure, a structure whose internal force distribution can be determined by using equilibrium concepts only, given the applied external force loading. Extension of the application of the force method to statically indeterminate structures will be discussed in Chap. 6.

In the next section an application of the displacement method of analysis described in Sec. 1.2.2 is presented. This application is for the special case of a kinematically determinate structure, a structure whose joint displacements are completely determined by compatibility concepts, given the internal distortions present in each member. Extension of the displacement method to statically indeterminate structures will be discussed in Chaps. 4 and 5.

2.6 SOLUTION FOR UNKNOWN NODAL DISPLACEMENTS FOR STATICALLY DETERMINATE STRUCTURES—COMPATIBILITY EQUATIONS

Consider now the somewhat uncommon problem of determining the displaced position of a kinematically determinate structure, given the member (and support) basic distortions. Suppose, for example, that the truss of Fig. 2.3 undergoes a downward settlement of 10 mm at the right-hand support, and that in addition member 5 has been fabricated 5 mm too long. We want to determine the displaced configuration of this structure when subjected to no externally applied force.

From previous work we have the following kinematic relationships between member basic displacements and member-end displacements in global coordinates:

$$\boldsymbol{\Delta} = \mathbf{T}^T \mathbf{d} \tag{2.11b}$$

Writing this expression for the nth member of the truss gives

$$\boldsymbol{\Delta}^{(n)} = \mathbf{T}^{T(n)} \mathbf{d}^{(n)} \tag{2.20}$$

If we write such an expression for each member of the structure and combine these equations along with the constraint conditions (based on Eq. 2.13) into a single matrix equation, we get

$$\overline{\boldsymbol{\Delta}} = \overline{\mathbf{T}}^T \overline{\mathbf{d}} \tag{2.21}$$

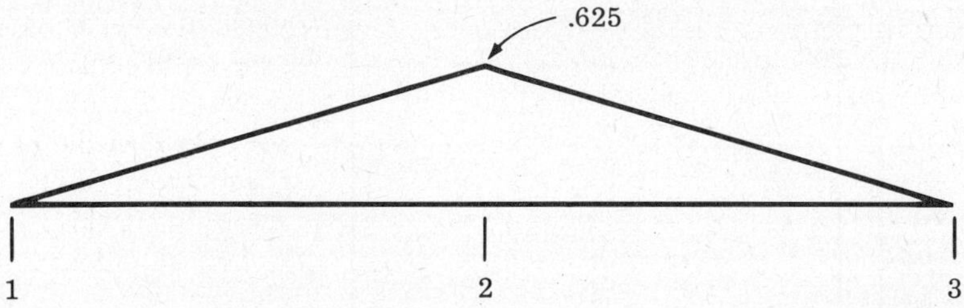

(a)

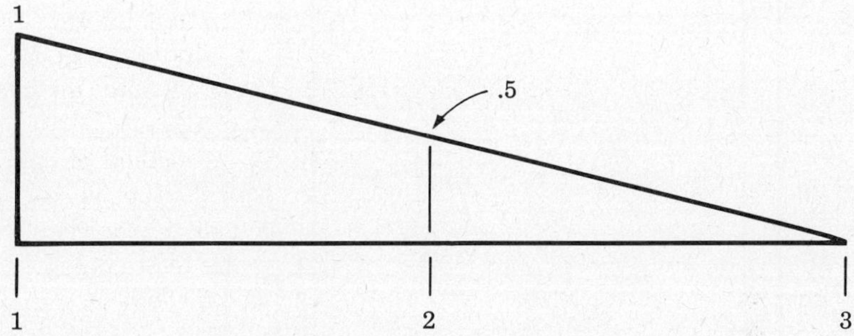

(b)

Fig. 2.6 Influence line for truss basic forces: (a) basic force, $P^{(3)}$; (b) basic force, $P_c^{(2)}$.

where matrices $\mathbf{T}^{(n)}$ and $\overline{\mathbf{T}}$ are those shown in detail on pages 21 and 22, respectively.

The member ends and supports must, of course, displace compatibly with the structure nodes to which they are attached. We can readily develop a statement of this nodal compatibility by inspecting the structure (the truss of Fig. 2.3, in this case, which is shown in an exploded view in Fig. 2.7). On inspection, we have Eq. A, displayed on page 36.

Eq. A may be represented symbolically as

$$\overline{\mathbf{d}} = \mathbf{A}^T\mathbf{D} \tag{2.22}$$

It is interesting and important to notice that this statement of nodal compatibility involves the matrix $\mathbf{A}^T$, whereas the statement of nodal equilibrium of Eq. 2.16 involved the matrix $\mathbf{A}$. (A general discussion of such dual transformations is offered in Sec. 2.12.)

Substituting from Eq. 2.22 into 2.21 gives

$$\overline{\mathbf{\Delta}} = \overline{\mathbf{T}}^T\mathbf{A}^T\mathbf{D} \tag{2.23}$$

Since we previously defined

$$\mathbf{a} = \mathbf{A}\overline{\mathbf{T}}$$

$$
\begin{pmatrix}
d_1^{(1)} \\ d_2^{(1)} \\ d_3^{(1)} \\ d_4^{(1)} \\
d_1^{(2)} \\ d_2^{(2)} \\ d_3^{(2)} \\ d_4^{(2)} \\
d_1^{(3)} \\ d_2^{(3)} \\ d_3^{(3)} \\ d_4^{(3)} \\
d_1^{(4)} \\ d_2^{(4)} \\ d_3^{(4)} \\ d_4^{(4)} \\
d_1^{(5)} \\ d_2^{(5)} \\ d_3^{(5)} \\ d_4^{(5)} \\
d_1^{(6)} \\ d_2^{(6)} \\ d_3^{(6)} \\ d_4^{(6)} \\
d_1^{(7)} \\ d_2^{(7)} \\ d_3^{(7)} \\ d_4^{(7)} \\
d_c^{(1)} \\ d_c^{(2)} \\ d_c^{(3)}
\end{pmatrix}
=
\begin{bmatrix}
1 & . & . & . & . & . & . & . & . & . \\
. & 1 & . & . & . & . & . & . & . & . \\
. & . & 1 & . & . & . & . & . & . & . \\
. & . & . & 1 & . & . & . & . & . & . \\
. & . & 1 & . & . & . & . & . & . & . \\
. & . & . & 1 & . & . & . & . & . & . \\
. & . & . & . & 1 & . & . & . & . & . \\
. & . & . & . & . & 1 & . & . & . & . \\
1 & . & . & . & . & . & . & . & . & . \\
. & 1 & . & . & . & . & . & . & . & . \\
. & . & . & . & . & . & 1 & . & . & . \\
. & . & . & . & . & . & . & 1 & . & . \\
. & . & . & . & . & . & 1 & . & . & . \\
. & . & . & . & . & . & . & 1 & . & . \\
. & . & 1 & . & . & . & . & . & . & . \\
. & . & . & 1 & . & . & . & . & . & . \\
. & . & 1 & . & . & . & . & . & . & . \\
. & . & . & 1 & . & . & . & . & . & . \\
. & . & . & . & . & . & . & . & 1 & . \\
. & . & . & . & . & . & . & . & . & 1 \\
. & . & . & . & . & . & . & . & 1 & . \\
. & . & . & . & . & . & . & . & . & 1 \\
. & . & . & . & 1 & . & . & . & . & . \\
. & . & . & . & . & 1 & . & . & . & . \\
. & . & . & . & . & . & 1 & . & . & . \\
. & . & . & . & . & . & . & 1 & . & . \\
. & . & . & . & . & . & . & . & 1 & . \\
. & . & . & . & . & . & . & . & . & 1 \\
1 & . & . & . & . & . & . & . & . & . \\
. & 1 & . & . & . & . & . & . & . & . \\
. & . & . & . & . & 1 & . & . & . & .
\end{bmatrix}
\begin{pmatrix}
D_1 \\ D_2 \\ D_3 \\ D_4 \\ D_5 \\ D_6 \\ D_7 \\ D_8 \\ D_9 \\ D_{10}
\end{pmatrix}
\qquad \text{(A)}
$$

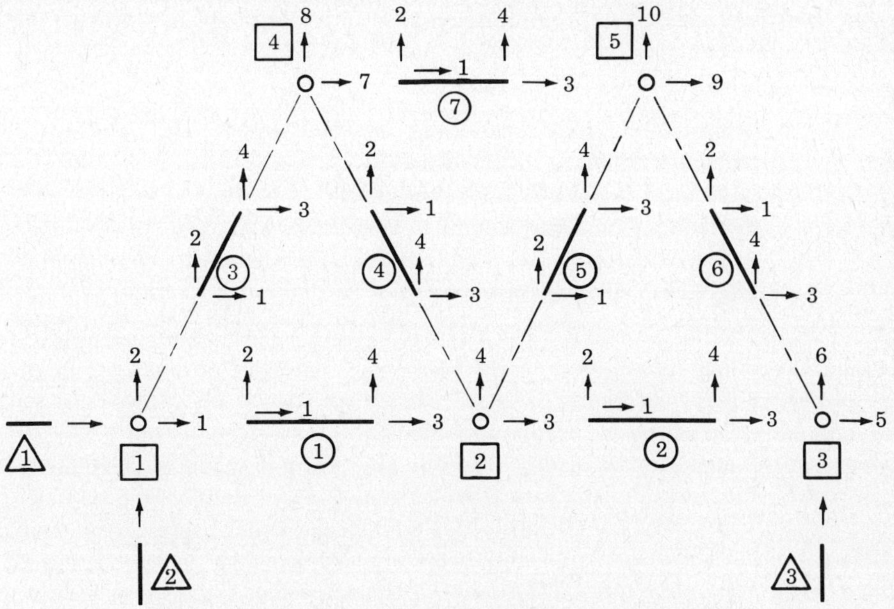

Fig. 2.7 Collection of truss members and constraints.

we may now write

$$\bar{\boldsymbol{\Delta}} = \mathbf{a}^T\mathbf{D} \tag{2.24}$$

A detailed presentation of the matrix **a** has been given in Sec. 2.5.1. Solving Eq. 2.24 gives

$$\mathbf{D} = [\mathbf{a}^T]^{-1}\,\bar{\boldsymbol{\Delta}} \tag{2.25}$$

Thus we see that a statically determinate structure is also kinematically determinate in the sense that, given a set of basic member displacements, there are just a sufficient number of compatibility relationships available to allow the structure nodal displacements to be determined.

For the truss problem, $\bar{\boldsymbol{\Delta}}$ is given (in meters) as

$$\bar{\boldsymbol{\Delta}} = \langle 0 \quad 0 \quad 0 \quad 0 \quad .005 \quad 0 \quad 0 \quad 0 \quad 0 \quad -.01\rangle \tag{B}$$

Notice that the settlement has been treated here as an initial strain in the support ($\Delta_c^{(3)} = -.01$). We will examine this concept in greater depth in Sec. 5.2.

A detailed solution to this problem using SMIS3 is displayed in the printout on pages 39–41, giving

$$\mathbf{D} = 10^{-3}\langle 0 \quad 0 \quad 0 \quad -8.125 \quad 0 \quad -10 \quad 5.4167 \quad -4.0625 \quad 5.4167 \quad -5.9375\rangle \tag{C}$$

It is apparent that the compatibility relationships may be built by coding techniques just as the equilibrium relationships were. The elements of the $\mathbf{T}^T$, $\mathbf{a}^T$, and $[\mathbf{a}^T]^{-1}$ $(= \mathbf{b}^T)$ matrices may now be interpreted as displacement influence coefficients. That is, for the truss example,

$T_m^{T^{(n)}}$ = nth member basic displacement per unit displacement in the member's mth global coordinate direction

a_{NM}^T = Nth basic structure displacement per unit displacement in the structure's Mth global coordinate direction

b_{NM}^T = displacement in the Nth structure global coordinate direction per unit Mth basic structure displacement

Once again a comparison of $T_m^{T^{(n)}}$ and a_{NM}^T may be used as a basis for developing an inspection-coding technique, and the elements of $\mathbf{a}^T$ may be used to obtain displacement influence lines.

In addition to their application to settlement and fabrication error loadings, thermal loads could also be readily considered in solving for nodal displacements of statically determinate structures. It is also obvious that external force loadings on a structure will cause member distortions and, hence, nodal displacements. But before we can solve such problems, we must develop relationships between member forces and member distortions—a procedure that is the subject of the next chapter.

Equilibrium and compatibility concepts similar to those we have considered in connection with statically determinate truss structures may also be developed for analyzing other statically determinate, discrete-membered structures. However, since statically determinate beams and rigid-jointed frameworks are generally very simple, the application of these powerful tools to the analysis of such structures is of little practical interest. It is, nevertheless, of considerable instructional value, because the member static and kinematic relationships required in this context will later be needed in the analysis of statically indeterminate structures. We will therefore now consider beams and rigid-jointed frameworks.

2.7 BEAM MEMBER

For purposes of this text, a continuous beam is taken to be a horizontal structure, a typical member of which is shown in Fig. 2.8. Axial (horizontal) forces are excluded from consideration in the case of beams. The beam member-end moments M_i and M_j are chosen as the member basic forces (Fig. 2.8a), and the corresponding basic displacements are the end rotations ϕ_i and ϕ_j (Fig. 2.8b).* As explained in Sec. 2.2, the set of basic forces is just sufficient to define the stressed state of the member, and the set of basic displacements is just sufficient to completely define the distorted shape of the member. Once again it may be noticed that these basic sets are not unique; for example, the forces M_i and V_i would work equally well as the basic force set.

*The meaning of the term *force* is generalized in this text to include moment as well as force, and the meaning of the term *displacement* is correspondingly generalized to include rotation as well as displacement.

****** *Computer Input:*

```
START
REMARK 5
SOLUTION FOR THE EXAMPLE TRUSS NODAL DISPLACEMENTS,
D, GIVEN SUPPORT SETTLEMENT AND MEMBER FABRICATION
ERROR, DELTA

**** UNITS ARE KILONEWTONS AND METERS ****
LOAD A 10 10
-1 0 -.6 4*0 1 2*0 2*0 -.8 5*0 1 0 1 -1 0 .6 -.6
5*0 3*0 -.8 -.8 5*0 0 1 3*0 .6 4*0 5*0 -.8 3*0 1
2*0 .6 -.6 2*0 -1 3*0 2*0 .8 .8 6*0 4*0 .6 -.6 1
3*0 4*0 .8 .8 4*0
TRANS A AT
PRINT AT 1
THIS THE MATRIX A-TRANSPOSE, AT
INVERT AT
PRINT AT 1
THIS IS THE INVERSE OF THE MATRIX A-TRANSPOSE
LOAD DELTA 10 1
0 0 0 0 .005 0 0 0 0 -.01
PRINT DELTA 2
THIS IS THE VECTOR, DELTA, OF MEMBER AND
SUPPORT BASIC DISTORTIONS
MULT AT DELTA D
PRINT D 1
THIS IS THE VECTOR, D, OF NODAL DISPLACEMENTS
STOP
```

****** *Computer Output:*

```
START

REMARK      5


SOLUTION FOR THE EXAMPLE TRUSS NODAL DISPLACEMENTS,
D, GIVEN SUPPORT SETTLEMENT AND MEMBER FABRICATION
ERROR, DELTA

**** UNITS ARE KILONEWTONS AND METERS ****


LOAD    A           10   10
   10 ROWS,    10 COLUMNS

TRANS   A        AT

PRINT   AT           1


THIS THE MATRIX A-TRANSPOSE, AT
```

	1	2	3	4	5	6
1	-1.0000E+00	0	1.0000E+00	0	0	0
2	0	0	-1.0000E+00	0	1.0000E+00	0
3	-6.0000E-01	-8.0000E-01	0	0	0	0
4	0	0	6.0000E-01	-8.0000E-01	0	0
5	0	0	-6.0000E-01	-8.0000E-01	0	0
6	0	0	0	0	6.0000E-01	-8.0000E-01
7	0	0	0	0	0	0
8	1.0000E+00	0	0	0	0	0
9	0	1.0000E+00	0	0	0	0
10	0	0	0	0	0	1.0000E+00

****** *Computer Output (continued):*

```
              7           8           9          10
 1            0           0           0           0
 2            0           0           0           0
 3  6.0000E-01 8.0000E-01           0           0
 4 -6.0000E-01 8.0000E-01           0           0
 5            0           0 6.0000E-01 8.0000E-01
 6            0           0-6.0000E-01 8.0000E-01
 7 -1.0000E+00           0 1.0000E+00           0
 8            0           0           0           0
 9            0           0           0           0
10            0           0           0           0
```

```
INVERT  AT

PRINT   AT          1
```

THIS IS THE INVERSE OF THE MATRIX A-TRANSPOSE

```
              1           2           3           4           5           6
 1            0           0           0           0           0           0
 2            0           0           0           0           0           0
 3  1.0000E+00           0           0           0           0           0
 4 -3.7500E-01-3.7500E-01 6.2500E-01-6.2500E-01-6.2500E-01 6.2500E-01
 5  1.0000E+00 1.0000E+00           0           0           0           0
 6            0           0           0           0           0           0
 7  7.5000E-01 2.5000E-01 4.1667E-01-4.1667E-01 4.1667E-01-4.1667E-01
 8 -5.6250E-01-1.8750E-01 9.3750E-01 3.1250E-01-3.1250E-01 3.1250E-01
 9  7.5000E-01 2.5000E-01 4.1667E-01-4.1667E-01 4.1667E-01-4.1667E-01
10 -1.8750E-01-5.6250E-01 3.1250E-01-3.1250E-01 3.1250E-01 9.3750E-01
              7           8           9          10
 1            0 1.0000E+00           0           0
 2            0           0 1.0000E+00           0
 3            0 1.0000E+00           0           0
 4  7.5000E-01 4.4409E-15 5.0000E-01 5.0000E-01
 5            0 1.0000E+00           0           0
 6            0           0           0 1.0000E+00
 7 -5.0000E-01 1.0000E+00 3.3333E-01-3.3333E-01
 8  3.7500E-01 2.2204E-15 7.5000E-01 2.5000E-01
 9  5.0000E-01 1.0000E+00 3.3333E-01-3.3333E-01
10  3.7500E-01-1.3323E-15 2.5000E-01 7.5000E-01
```

```
LOAD    DELTA        10    1
   10 ROWS,     1 COLUMNS

PRINT   DELTA        2
```

THIS IS THE VECTOR, DELTA, OF MEMBER AND
SUPPORT BASIC DISTORTIONS

```
              1
 1            0
 2            0
 3            0
 4            0
 5  5.0000E-03
 6            0
 7            0
 8            0
 9            0
10 -1.0000E-02
```

****** *Computer Output (continued):*

```
MULT    AT      DELTA   D

PRINT   D               1

THIS IS THE VECTOR, D, OF NODAL DISPLACEMENTS

              1
 1            0
 2            0
 3            0
 4 -8.1250E-03
 5            0
 6 -1.0000E-02
 7  5.4167E-03
 8 -4.0625E-03
 9  5.4167E-03
10 -5.9375E-03

STOP
```

Inspection of Fig. 2.8a and c reveals that the local and global coordinate systems will always coincide for a beam member as a consequence of the horizontal orientation of the beam. We may therefore dispense with the local member-end forces and displacements for the beam member. Thus, considering overall equilibrium of the beam member of Fig. 2.8a, we may write

$$V_i = \frac{M_i + M_j}{L}$$
$$V_j = \frac{M_i + M_j}{L} \tag{2.26}$$

Comparing the member-end forces of Fig. 2.8a and c gives

$$r_1 = V_i = \frac{M_i + M_j}{L}$$
$$r_2 = M_i$$
$$r_3 = -V_j = -\frac{M_i + M_j}{L} \tag{2.27}$$
$$r_4 = M_j$$

or, in matrix form,

$$\begin{pmatrix} r_1 \\ r_2 \\ r_3 \\ r_4 \end{pmatrix} = \begin{bmatrix} 1/L & 1/L \\ 1 & 0 \\ -1/L & -1/L \\ 0 & 1 \end{bmatrix} \begin{pmatrix} M_i \\ M_j \end{pmatrix} \tag{2.27a}$$

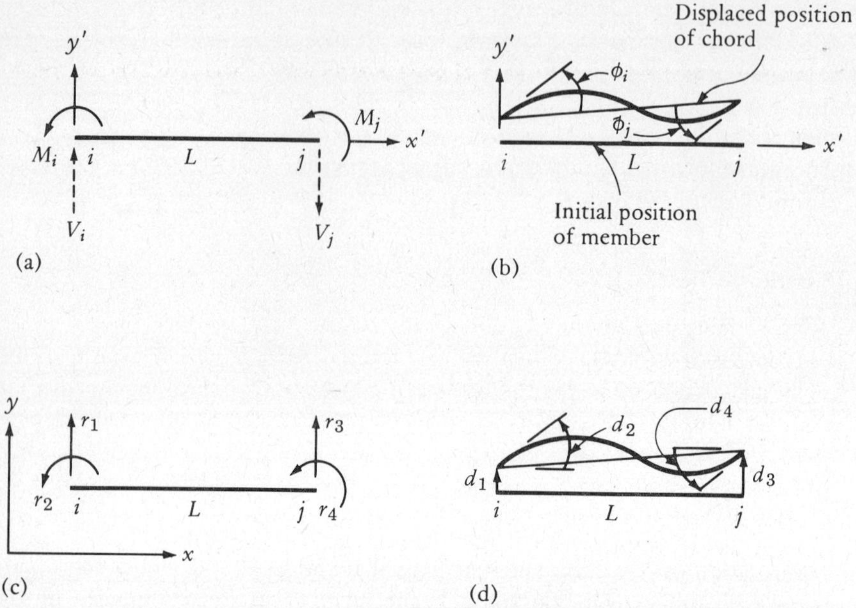

Fig. 2.8 Member-end forces and member-end displacements, beam member: (a) basic forces, local coordinates; (b) basic displacements, local coordinates; (c) member-end forces, global coordinates; (d) member-end displacements, global coordinates.

or

$$\mathbf{r} = \mathbf{TP} \tag{2.27b}$$

On comparing the displacements of Fig. 2.8b and d, we can write the following kinematic relationships:

$$\begin{aligned} \phi_i &= d_2 - \frac{d_3 - d_1}{L} \\ \phi_j &= d_4 - \frac{d_3 - d_1}{L} \end{aligned} \tag{2.28}$$

In matrix form,

$$\begin{pmatrix} \phi_i \\ \phi_j \end{pmatrix} = \begin{bmatrix} 1/L & 1 & -1/L & 0 \\ 1/L & 0 & -1/L & 1 \end{bmatrix} \begin{pmatrix} d_1 \\ d_2 \\ d_3 \\ d_4 \end{pmatrix} \tag{2.28a}$$

or

$$\boldsymbol{\Delta} = \mathbf{T}^T\mathbf{d} \tag{2.28b}$$

Although the notation used for beam members is similar to that which was used for truss members and to that which will be used for frame members, certain details are, of necessity, different. For example, r_2 represents a member-end force in the second global coordinate direction for any member, but this force is a moment in the case of a beam and a force in the x direction in the case of a truss. This situation is mentioned so that you will be on the alert to interpret the notation carefully in the context of its use.

2.8 BEAM SUPPORTS

Beam supports are commonly of three types: the roller support, the pin support, and the fixed support, shown in Fig. 2.9a, b, and c, respectively. From an analytical point of view, the roller and pin are identical when applied to the beam. The reason for this is that the horizontal constraint associated with the pin does not enter into the beam analysis, since horizontal equilibrium is automatically satisfied in the absence of axial loading. A superfluous horizontal constraint force (shown dashed) is included in the graphical representation of the beam pin support only to indicate satisfaction of horizontal static stability of the beam. The

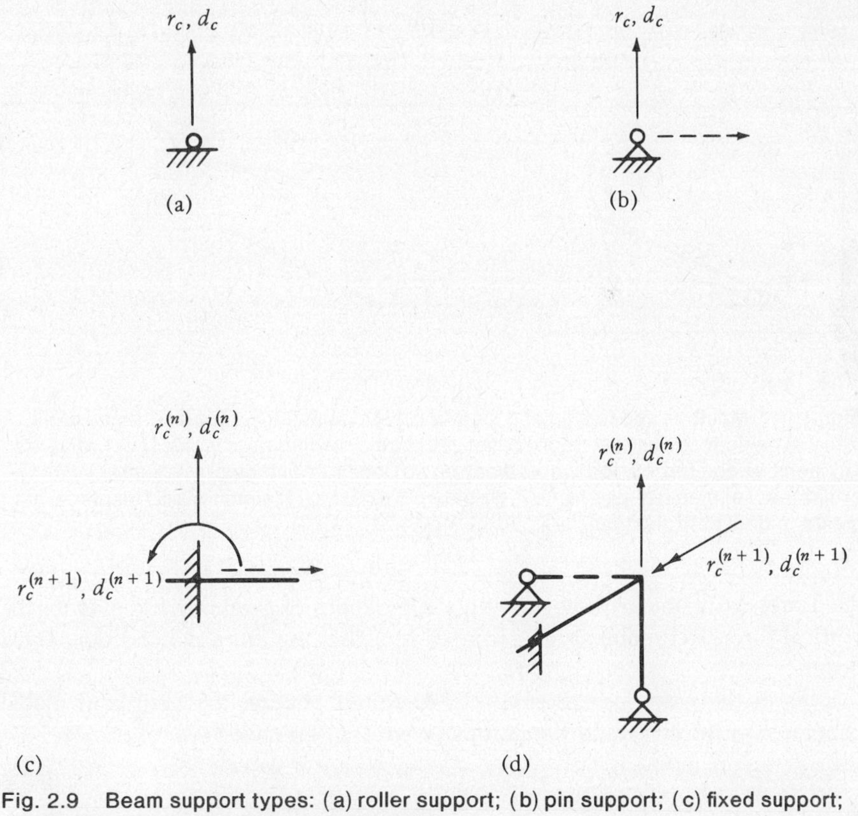

Fig. 2.9 Beam support types: (a) roller support; (b) pin support; (c) fixed support; (d) fixed support, alternative representation.

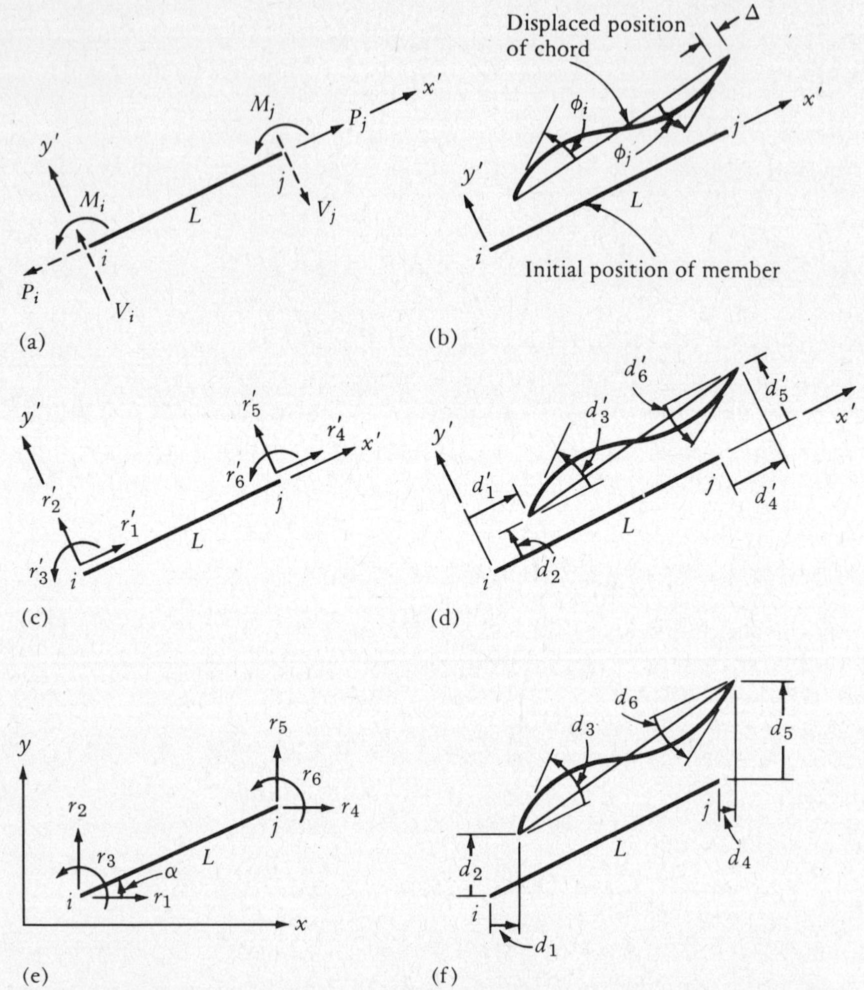

Fig. 2.10 Member-end forces and member-end displacements, plane frame member: (a) basic forces, local coordinates; (b) basic displacements, local coordinates; (c) member-end forces, local coordinates; (d) member-end displacements, local coordinates; (e) member-end forces, global coordinates; (f) member-end displacements, global coordinates.

fixed support is similarly shown with a superfluous constraining force. Only the vertical force and moment are required to satisfy equilibrium conditions for a beam.

As in the case of truss supports, identities relating the basic and global characters of the constraint forces may be written. Thus we have

$$r_c = P_c \tag{2.29}$$

for a translational constraint, and

$$r_c = M_c \tag{2.30}$$

for a rotational constraint, P_c and M_c identify the translational and rotational constraining forces, respectively, as basic unknown forces. The corresponding displacements are related as

$$\Delta_c = d_c \tag{2.31}$$

and

$$\phi_c = d_c \tag{2.32}$$

where Δ_c and ϕ_c are, respectively, the translational and rotational support distortions in basic notation. These will be zero for rigid supports, but possibly nonzero in the case of deformable supports or supports subjected to settlement.

As was the case with truss supports, the constraints associated with beam supports may also be represented by individual links. For example, Fig. 2.9d shows a collection of links represented a beam fixed support. The rotational constraint is represented with a torsional link with basic internal moment M_c and basic twisting distortion ϕ_c. The translational constraint is similar to that for the truss described in Sec. 2.4. For clarity, these alternative representations will be used from time to time later in the text.

In the interest of brevity, an application of the relationships of Secs. 2.7 and 2.8 to the analysis of a statically determinate beam is not presented here. However, the approach is very much the same as that used in the analysis of the statically determinate frame example of Sec. 2.11. Before considering that example, we must first consider the static and kinematic relationships of frame members.

2.9 FRAME MEMBER

Figure 2.10 shows a typical frame member, along with its various sets of member-end forces and displacements. Relationships among the sets of forces can be obtained, as in the cases of the truss and beam members, from considerations of member overall equilibrium. On considering the forces of Fig. 2.10a and c, we find the following relationship between the basic and local sets of forces:

$$\begin{pmatrix} r'_1 \\ r'_2 \\ r'_3 \\ r'_4 \\ r'_5 \\ r'_6 \end{pmatrix} = \begin{bmatrix} 0 & 0 & -1 \\ 1/L & 1/L & 0 \\ 1 & 0 & 0 \\ 0 & 0 & 1 \\ -1/L & -1/L & 0 \\ 0 & 1 & 0 \end{bmatrix} \begin{pmatrix} M_i \\ M_j \\ P \end{pmatrix} \tag{2.33}$$

or

$$\mathbf{r}' = \mathbf{LP} \tag{2.33a}$$

The corresponding displacement transformations may be obtained by inspection of Fig. 2.10b and d:

$$\begin{Bmatrix} \phi_i \\ \phi_j \\ \Delta \end{Bmatrix} = \begin{bmatrix} 0 & 1/L & 1 & 0 & -1/L & 0 \\ 0 & 1/L & 0 & 0 & -1/L & 1 \\ -1 & 0 & 0 & 1 & 0 & 0 \end{bmatrix} \begin{Bmatrix} d'_1 \\ d'_2 \\ d'_3 \\ d'_4 \\ d'_5 \\ d'_6 \end{Bmatrix} \tag{2.34}$$

or

$$\boldsymbol{\Delta} = \mathbf{L}^T\mathbf{d}' \tag{2.34a}$$

Comparing Fig. 2.10c and e, we see that

$$\begin{Bmatrix} r_1 \\ r_2 \\ r_3 \\ r_4 \\ r_5 \\ r_6 \end{Bmatrix} = \begin{bmatrix} \cos\alpha & -\sin\alpha & 0 & 0 & 0 & 0 \\ \sin\alpha & \cos\alpha & 0 & 0 & 0 & 0 \\ 0 & 0 & 1 & 0 & 0 & 0 \\ 0 & 0 & 0 & \cos\alpha & -\sin\alpha & 0 \\ 0 & 0 & 0 & \sin\alpha & \cos\alpha & 0 \\ 0 & 0 & 0 & 0 & 0 & 1 \end{bmatrix} \begin{Bmatrix} r'_1 \\ r'_2 \\ r'_3 \\ r'_4 \\ r'_5 \\ r'_6 \end{Bmatrix} \tag{2.35}$$

or

$$\mathbf{r} = \boldsymbol{\alpha}\mathbf{r}' \tag{2.35a}$$

Inspection of Fig. 2.10d and f gives

$$\begin{Bmatrix} d'_1 \\ d'_2 \\ d'_3 \\ d'_4 \\ d'_5 \\ d'_6 \end{Bmatrix} = \begin{bmatrix} \cos\alpha & \sin\alpha & 0 & 0 & 0 & 0 \\ -\sin\alpha & \cos\alpha & 0 & 0 & 0 & 0 \\ 0 & 0 & 1 & 0 & 0 & 0 \\ 0 & 0 & 0 & \cos\alpha & \sin\alpha & 0 \\ 0 & 0 & 0 & -\sin\alpha & \cos\alpha & 0 \\ 0 & 0 & 0 & 0 & 0 & 1 \end{bmatrix} \begin{Bmatrix} d_1 \\ d_2 \\ d_3 \\ d_4 \\ d_5 \\ d_6 \end{Bmatrix} \tag{2.36}$$

or

$$\mathbf{d}' = \boldsymbol{\alpha}^T\mathbf{d} \tag{2.36a}$$

Using Eqs. 2.33 and 2.35, we get

$$\begin{pmatrix} r_1 \\ r_2 \\ r_3 \\ r_4 \\ r_5 \\ r_6 \end{pmatrix} = \begin{bmatrix} -\sin\alpha/L & -\sin\alpha/L & \cos\alpha \\ \cos\alpha/L & \cos\alpha/L & -\sin\alpha \\ 1 & 0 & 0 \\ \sin\alpha/L & \sin\alpha/L & -\cos\alpha \\ -\cos\alpha/L & -\cos\alpha/L & \sin\alpha \\ 0 & 1 & 0 \end{bmatrix} \begin{pmatrix} M_i \\ M_j \\ P \end{pmatrix} \tag{2.37}$$

or

$$\mathbf{r} = \mathbf{TP} \tag{2.37a}$$

Combining Eqs. 2.34 and 2.36 gives

$$\begin{pmatrix} \phi_i \\ \phi_j \\ \Delta \end{pmatrix} = \begin{bmatrix} -\sin\alpha/L & \cos\alpha/L & 1 & \sin\alpha/L & -\cos\alpha/L & 0 \\ -\sin\alpha/L & \cos\alpha/L & 0 & \sin\alpha/L & -\cos\alpha/L & 1 \\ \cos\alpha & -\sin\alpha & 0 & -\cos\alpha & \sin\alpha & 0 \end{bmatrix} \begin{pmatrix} d_1 \\ d_2 \\ d_3 \\ d_4 \\ d_5 \\ d_6 \end{pmatrix} \tag{2.38}$$

or

$$\mathbf{\Delta} = \mathbf{T}^T\mathbf{d} \tag{2.38a}$$

Interestingly, the set of basic forces (displacements) of the frame member is equivalent to the collective set of basic forces (displacements) of the truss and beam members. Thus truss and beam members are, in fact, special cases of the frame member, or the frame member may be thought of as a generalized combination of a truss member and a beam member. The frame member could therefore be used to analyze trusses or beams in addition to frames, but since there are fewer basic unknowns, individually, in the truss or the beam member, it is more efficient to use either of the latter two where appropriate.

2.10 FRAME SUPPORTS

The supports for frame structures shown in Fig. 2.11 are of the same types as those shown in Fig. 2.9 for continuous beams. In the case of frame supports, however, the horizontal constraints are necessary and must be considered in the analysis. Other than this important difference, the discussion of beam supports in Sec. 2.8 applies equally well to frame supports, and you can refer to that section for further information.

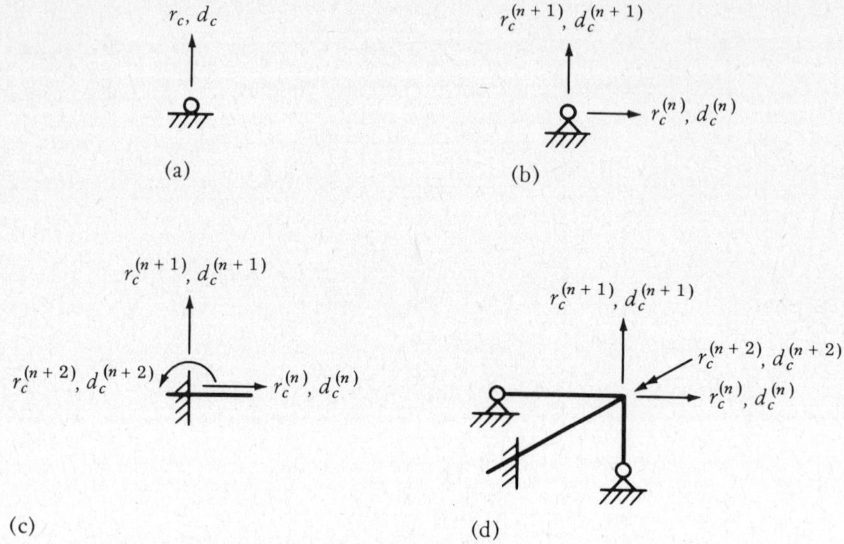

Fig. 2.11 Frame support types: (a) roller support; (b) pin support; (c) fixed support; (d) fixed support, alternative representation.

2.11 FRAME EXAMPLE*

As was mentioned earlier, statically determinate frames are generally too simple to be of innate interest for examination by the matrix force method. Even so, it is useful to consider such a frame to develop concepts that will subsequently be required in the analysis of statically indeterminate beams and frames by the force method.

2.11.1 Nodal Forces

The rigid-jointed framework of Fig. 2.12 will be analyzed using the nodal coding method. The individual member **T** matrices may be obtained by appropriately specializing Eq. 2.37:

Member 1: $\alpha_1 = 90°, L_1 = L$

$$\mathbf{T}^{(1)} = \begin{array}{c} \\ \boxed{1} \\ \\ \\ \boxed{2} \\ \\ \end{array} \overset{\textcircled{1}}{\left[\begin{array}{ccc} -1/L & -1/L & 0 \\ 0 & 0 & -1 \\ 1 & 0 & 0 \\ \hdashline 1/L & 1/L & 0 \\ 0 & 0 & 1 \\ 0 & 1 & 0 \end{array}\right]}$$

*Those interested in moving rapidly to consideration of the displacement method may wish to delay study of Sec. 2.11 until it is required as a prerequisite for understanding the force method of analysis presented in Chap. 6.

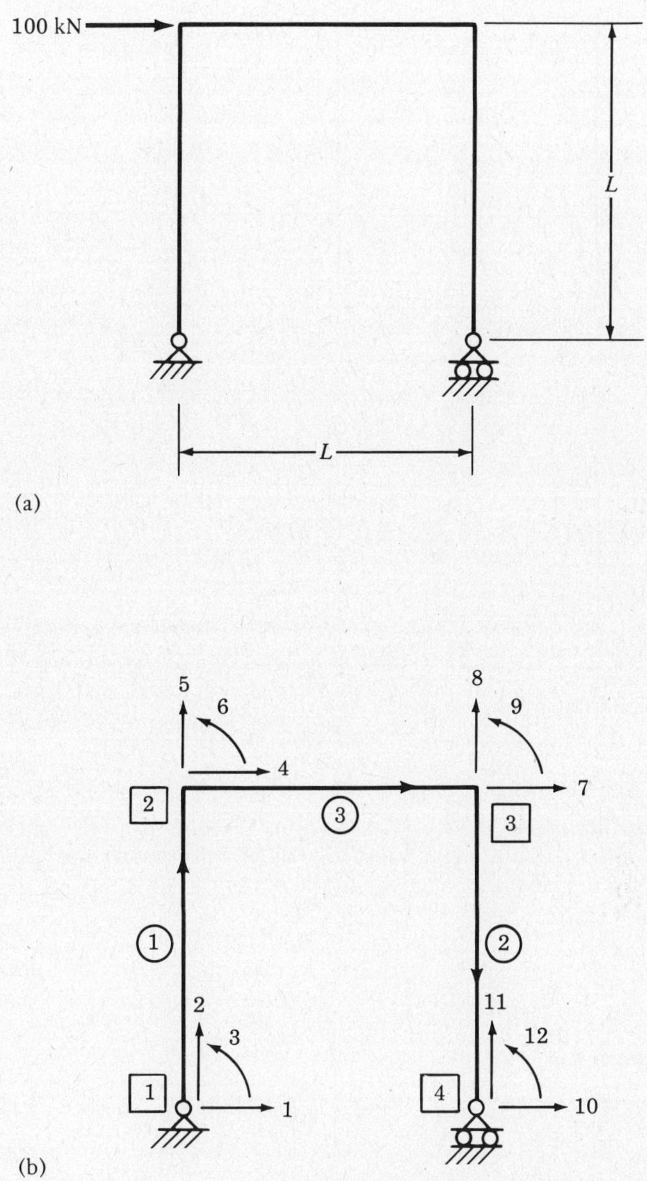

Fig. 2.12 Frame example: (a) geometry and loading; (b) numbering.

Member 2: $\alpha_2 = -90°, L_2 = L$

$$\mathbf{T}^{(2)} = \begin{array}{c} \\ \boxed{3} \\ \\ \\ \boxed{4} \\ \\ \end{array} \left[\begin{array}{ccc} & \textcircled{2} & \\ 1/L & 1/L & 0 \\ 0 & 0 & 1 \\ 1 & 0 & 0 \\ \hdashline -1/L & -1/L & 0 \\ 0 & 0 & -1 \\ 0 & 1 & 0 \end{array}\right]$$

Member 3: $\alpha_3 = 0°, L_3 = L$

$$\mathbf{T}^{(3)} = \begin{array}{c} \\ \boxed{2} \\ \\ \\ \boxed{3} \\ \\ \end{array} \left[\begin{array}{ccc} & \textcircled{3} & \\ 0 & 0 & -1 \\ 1/L & 1/L & 0 \\ 1 & 0 & 0 \\ \hdashline 0 & 0 & 1 \\ -1/L & -1/L & 0 \\ 0 & 1 & 0 \end{array}\right]$$

Note that the following nodal code has been indicated directly on the **T** matrices for convenience in assembly of the **a** matrix by hand.

Nodal Code

Member number ↓	i	j
1	1	2
2	3	4
3	2	3

Constraint information for the example frame is as follows:

Constraint List

Constraint number ↓	Node	Coordinate
1	1	1
2	1	2
3	4	2

As an example of the application of this constraint list, consider the third row of the list. This row indicates that constraint force $P_c^{(3)}$ acts in coordinate direction 2 of node 4. For a frame, the first, second, and third coordinate directions of a node have been established as the x, y, and rotational directions, respectively.

The **a** matrix may now easily be assembled as shown in Eq. A on page 51.

$$
\begin{array}{ccccccccccccc}
 & & \textcircled{1} & & & \textcircled{2} & & & \textcircled{3} & & \triangle_1 & \triangle_2 & \triangle_3
\end{array}
$$

$$
\begin{pmatrix} R_1 \\ R_2 \\ R_3 \\ R_4 \\ R_5 \\ R_6 \\ R_7 \\ R_8 \\ R_9 \\ R_{10} \\ R_{11} \\ R_{12} \end{pmatrix}
\begin{array}{c} \\ \boxed{1} \\ \\ \\ \boxed{2} \\ \\ = \\ \boxed{3} \\ \\ \\ \boxed{4} \\ \\ \end{array}
\begin{bmatrix}
-1/L & -1/L & 0 & \cdot & \cdot & \cdot & \cdot & \cdot & \cdot & 1 & \cdot & \cdot \\
0 & 0 & -1 & \cdot & \cdot & \cdot & \cdot & \cdot & \cdot & \cdot & 1 & \cdot \\
1 & 0 & 0 & \cdot & \cdot & \cdot & \cdot & \cdot & \cdot & \cdot & \cdot & \cdot \\
1/L & 1/L & 0 & \cdot & \cdot & \cdot & 0 & 0 & -1 & \cdot & \cdot & \cdot \\
0 & 0 & 1 & \cdot & \cdot & \cdot & 1/L & 1/L & 0 & \cdot & \cdot & \cdot \\
0 & 1 & 0 & \cdot & \cdot & \cdot & 1 & 0 & 0 & \cdot & \cdot & \cdot \\
\cdot & \cdot & \cdot & 1/L & 1/L & 0 & 0 & 0 & 1 & \cdot & \cdot & \cdot \\
\cdot & \cdot & \cdot & 0 & 0 & 1 & -1/L & -1/L & 0 & \cdot & \cdot & \cdot \\
\cdot & \cdot & \cdot & 1 & 0 & 0 & 0 & 1 & 0 & \cdot & \cdot & \cdot \\
\cdot & \cdot & \cdot & -1/L & -1/L & 0 & \cdot & \cdot & \cdot & \cdot & \cdot & \cdot \\
\cdot & \cdot & \cdot & 0 & 0 & -1 & \cdot & \cdot & \cdot & \cdot & \cdot & 1 \\
\cdot & \cdot & \cdot & 0 & 1 & 0 & \cdot & \cdot & \cdot & \cdot & \cdot & \cdot
\end{bmatrix}
\begin{pmatrix} M_i^{(1)} \\ M_j^{(1)} \\ P^{(1)} \\ M_i^{(2)} \\ M_j^{(2)} \\ P^{(2)} \\ M_i^{(3)} \\ M_j^{(3)} \\ P^{(3)} \\ P_c^{(1)} \\ P_c^{(2)} \\ P_c^{(3)} \end{pmatrix}
\qquad \text{(A)}
$$

where it is seen that, in this case, 3 × 3 submatrices of the individual **T** matrices that relate the three basic forces in a member to the set of three forces at a particular node are moved intact to the **a** matrix. Once again this is possible owing to our systematic coordinate and node numbering at both the member and the structure level.

Equation A may, of course, be represented symbolically as

$$\mathbf{R} = \mathbf{a}\overline{\mathbf{P}} \tag{2.18}$$

Solving symbolically gives

$$\overline{\mathbf{P}} = \mathbf{a}^{-1}\mathbf{R} \tag{2.19}$$

A computer solution to this problem using SMIS3 is shown in the next printout, with $L = 10$ m, which gives

$$\overline{\mathbf{P}} = \langle 0 \quad 1000 \quad 100 \quad 0 \quad 0 \quad -100 \quad -1000 \quad 0 \quad 0 \quad 100 \quad 100 \quad -100 \rangle \tag{B}$$

These results are seen to be in equilibrium with the applied force.

2.11.2 Distributed Loads

It is now appropriate to remove the restriction on loads within members. We will illustrate this more general problem in connection with the frame example of Fig. 2.13. This frame was previously shown under only nodal loading in Fig. 2.12, and although the loading is now changed, the member, support, nodal, and coordinate numberings shown in Fig. 2.12 will be retained in working this new example.

Solution of this example by matrix structural analysis is carried out using the principle of superposition. The total solution is broken into the two cases shown in Fig. 2.13, so that we have

$$\text{Total load} = \text{Case I load} + \text{Case II load} \tag{2.39}$$

It follows from the principle of superposition that

$$\text{Total response} = \text{Case I response} + \text{Case II response} \tag{2.40}$$

which in the case of structure basic force response may be written as

$$\overline{\mathbf{P}} = \overline{\mathbf{P}}_{\mathrm{I}} + \overline{\mathbf{P}}_{\mathrm{II}} \tag{2.41}$$

Solutions can now be obtained for Case I and Case II and superposed to obtain the total solution.

Case I

As this fixed node case involves no overall structural action but only individual, independent member actions, it is possible to consider the solution one member at

****** *Computer Input:*

```
START
REMARK 3
EXAMPLE FRAME PROBLEM...CONCENTRATED LOAD

**** UNITS ARE KILONEWTONS AND METERS ****
FORM A 12 12
LOAD T1 6 3
-.1 -.1 3*0 -1 1 2*0 .1 .1 3*0 1 0 1 0
LOAD T2 6 3
.1 .1 3*0 1 1 2*0 -.1 -.1 3*0 -1 0 1 0
LOAD T3 6 3
2*0 -1 .1 .1 0 1 4*0 1 -.1 -.1 2*0 1 0
LOAD CONSTR 1 1
1
STOSM A T1 1 1
STOSM A T2 7 4
STOSM A T3 4 7
STOSM A CONSTR 1 10
STOSM A CONSTR 2 11
STOSM A CONSTR 11 12
PRINT A 1
THIS IS THE MATRIX, A, OF STRUCTURE FORCE INFLUENCE COEFFICIENTS.
INVERT A
PRINT A 2
THIS IS THE MATRIX, A INVERSE, OF STRUCTURE
FORCE INFLUENCE COEFFICIENTS.
LOAD R 12 1
3*0 100 8*0
PRINT R 2
THIS IS THE LOAD VECTOR, R, REPRESENTING A SINGLE
CONCENTRATED HORIZONTAL NODAL LOAD.
MULT A R PBAR
PRINT PBAR 2
THIS IS THE VECTOR, PBAR, OF MEMBER AND CONSTRAINT INTERNAL BASIC
FORCES DUE TO THE CONCENTRATED LOAD OF VECTOR, R
STOP
```

****** *Computer Output:*

```
START

REMARK      3


EXAMPLE FRAME PROBLEM...CONCENTRATED LOAD

**** UNITS ARE KILONEWTONS AND METERS ****



FORM     A            12   12
   12 ROWS,     12 COLUMNS

LOAD     T1            6    3
    6 ROWS,      3 COLUMNS

LOAD     T2            6    3
    6 ROWS,      3 COLUMNS

LOAD     T3            6    3
    6 ROWS,      3 COLUMNS

LOAD     CONSTR        1    1
    1 ROWS,      1 COLUMNS

STOSM    A       T1          1    1
```

****** *Computer Output (continued):*

```
 ROW NUMBER    1,  COLUMN NUMBER    1

STOSM    A        T2            7    4
 ROW NUMBER    7,  COLUMN NUMBER    4

STOSM    A        T3            4    7
 ROW NUMBER    4,  COLUMN NUMBER    7

STOSM    A       CONSTR         1   10
 ROW NUMBER    1,  COLUMN NUMBER   10

STOSM    A       CONSTR         2   11
 ROW NUMBER    2,  COLUMN NUMBER   11

STOSM    A       CONSTR        11   12
 ROW NUMBER   11,  COLUMN NUMBER   12

PRINT    A             1


THIS IS THE MATRIX, A, OF STRUCTURE FORCE INFLUENCE COEFFICIENTS.


             1           2           3           4           5           6
 1 -1.0000E-01-1.0000E-01           0           0           0           0
 2           0           0-1.0000E+00           0           0           0
 3  1.0000E+00           0           0           0           0           0
 4  1.0000E-01 1.0000E-01           0           0           0           0
 5           0           0 1.0000E+00           0           0           0
 6           0 1.0000E+00           0           0           0           0
 7           0           0           0 1.0000E-01 1.0000E-01           0
 8           0           0           0           0           0 1.0000E+00
 9           0           0           0 1.0000E+00           0           0
10           0           0           0-1.0000E-01-1.0000E-01           0
11           0           0           0           0           0-1.0000E+00
12           0           0           0           0 1.0000E+00           0
             7           8           9          10          11          12
 1           0           0           0 1.0000E+00           0           0
 2           0           0           0           0 1.0000E+00           0
 3           0           0           0           0           0           0
 4           0           0-1.0000E+00           0           0           0
 5  1.0000E-01 1.0000E-01           0           0           0           0
 6  1.0000E+00           0           0           0           0           0
 7           0           0 1.0000E+00           0           0           0
 8 -1.0000E-01-1.0000E-01           0           0           0           0
 9           0 1.0000E+00           0           0           0           0
10           0           0           0           0           0           0
11           0           0           0           0           0 1.0000E+00
12           0           0           0           0           0           0


INVERT   A

PRINT    A             2


THIS IS THE MATRIX, A INVERSE, OF STRUCTURE
FORCE INFLUENCE COEFFICIENTS.


             1           2           3           4           5           6
 1           0           0 1.0000E+00           0           0           0
 2           0           0-1.0000E+00 1.0000E+01           0 7.1054E-15
 3           0           0-1.0000E-01 1.0000E+00 1.0000E+00-1.0000E-01
 4           0           0           0           0           0           0
 5           0           0           0           0           0           0
 6           0           0 1.0000E-01-1.0000E+00           0 1.0000E-01
 7           0           0 1.0000E+00-1.0000E+01           0 1.0000E+00
 8           0           0           0           0           0           0
 9           0           0           0           0           0           0
```

****** *Computer Output (continued):*

```
10  1.0000E+00           0 8.8818E-16 1.0000E+00           0 8.8818E-16
11           0 1.0000E+00-1.0000E-01 1.0000E+00 1.0000E+00-1.0000E-01
12           0           0 1.0000E-01-1.0000E+00           0 1.0000E-01
             7          8          9         10          11         12
 1           0          0          0          0           0          0
 2  1.0000E+01          0          0 1.0000E+01           0-7.1054E-15
 3  1.0000E+00          0-1.0000E-01-8.8818E-15           0-1.0000E-01
 4           0          0          0-1.0000E+01           0-1.0000E+00
 5           0          0          0          0           0 1.0000E+00
 6 -1.0000E+00 1.0000E+00 1.0000E-01 8.8818E-15           0 1.0000E-01
 7 -1.0000E+01          0          0-1.0000E+01           0 7.1054E-15
 8           0          0 1.0000E+00 1.0000E+01           0 1.0000E+00
 9  1.0000E+00          0          0 1.0000E+00           0-8.8818E-16
10  1.0000E+00          0          0 1.0000E+00           0-8.8818E-16
11  1.0000E+00          0-1.0000E-01-8.8818E-15           0-1.0000E-01
12 -1.0000E+00 1.0000E+00 1.0000E-01 8.8818E-15 1.0000E+00 1.0000E-01


LOAD    R           12    1
   12 ROWS,      1 COLUMNS

PRINT   R            2


THIS IS THE LOAD VECTOR, R, REPRESENTING A SINGLE
CONCENTRATED HORIZONTAL NODAL LOAD.


             1
 1           0
 2           0
 3           0
 4  1.0000E+02
 5           0
 6           0
 7           0
 8           0
 9           0
10           0
11           0
12           0


MULT    A        R        PBAR

PRINT   PBAR         2


THIS IS THE VECTOR, PBAR, OF MEMBER AND CONSTRAINT INTERNAL BASIC
FORCES DUE TO THE CONCENTRATED LOAD OF VECTOR, R


             1
 1           0
 2  1.0000E+03
 3  1.0000E+02
 4           0
 5           0
 6 -1.0000E+02
 7 -1.0000E+03
 8           0
 9           0
10  1.0000E+02
11  1.0000E+02
12 -1.0000E+02


STOP
```

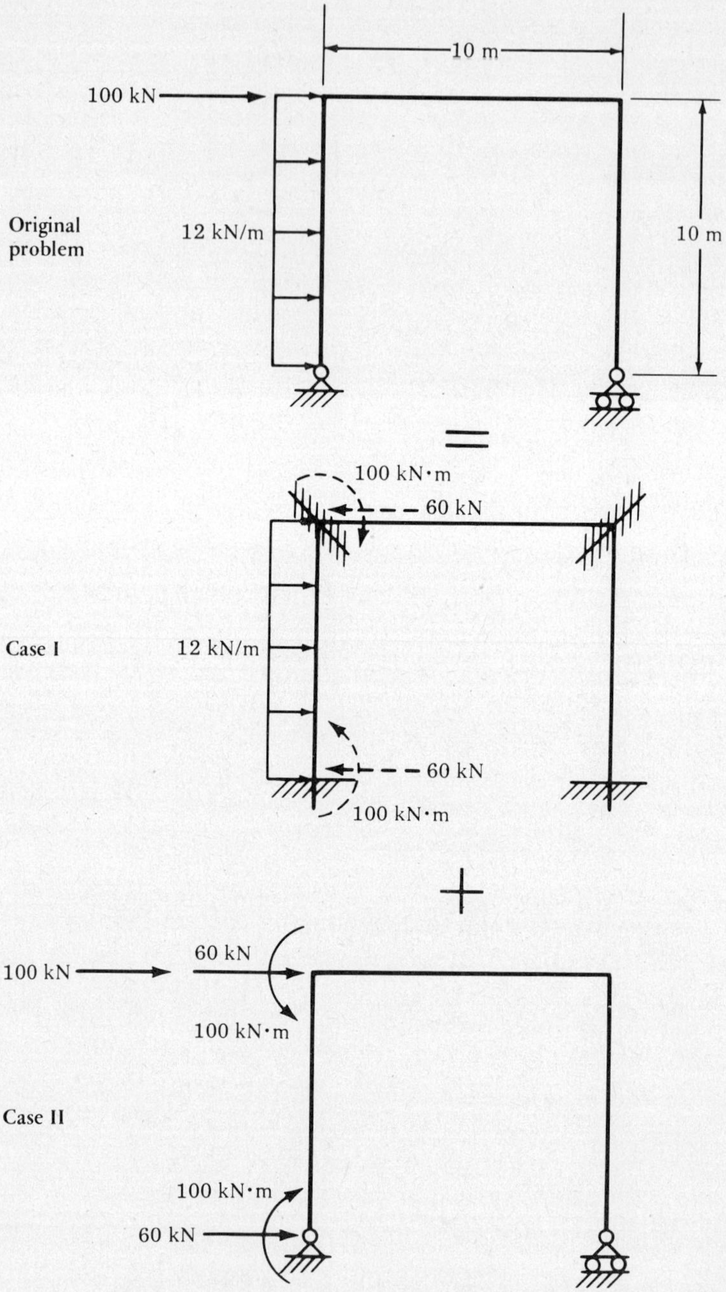

Fig. 2.13 Frame example with member loading.

a time. By inspection we get

$$\begin{aligned}
\mathbf{P}_F^{(1)} &= \langle 100 \quad -100 \quad 0 \rangle \\
\mathbf{P}_F^{(2)} &= \langle 0 \quad 0 \quad 0 \rangle \\
\mathbf{P}_F^{(3)} &= \langle 0 \quad 0 \quad 0 \rangle \\
P_{c,F}^{(1)} &= \langle 0 \rangle \\
P_{c,F}^{(2)} &= \langle 0 \rangle \\
P_{c,F}^{(3)} &= \langle 0 \rangle
\end{aligned} \tag{A}$$

where $\mathbf{P}_F^{(n)}$ and $P_{c,F}^{(n)}$ represent vectors of basic fixed-end forces for the nth member and nth constraint, respectively. Combining these results into a single vector of structure basic fixed-end forces gives for the Case I solution

$$\overline{\mathbf{P}}_F = \langle 100 \quad -100 \quad 0 \quad 0 \quad 0 \quad 0 \quad 0 \quad 0 \quad 0 \quad 0 \quad 0 \quad 0 \rangle \tag{B}$$

Thus it is seen that, in general, the Case I solution is the vector of basic fixed-end forces for the collection of all members and supports, or

$$\overline{\mathbf{P}}_I = \overline{\mathbf{P}}_F \tag{2.42}$$

Case II

Before solving Case II, we must first determine the Case II loading. By inspection of Fig. 2.13, we find that

$$\text{Total load} = \mathbf{R} + \text{distributed load} \tag{2.43}$$

and

$$\text{Case I load} = \mathbf{R}_\text{I} + \text{distributed load} \tag{2.44}$$

where $\mathbf{R}$ is a vector of applied nodal forces and $\mathbf{R}_\text{I}$ is a vector of nodal fixing forces whose components are shown dashed in Fig. 2.13 to call attention to their fictitious nature. To emphasize the fixed node nature of Case I, we let

$$\mathbf{R}_\text{I} = \mathbf{R}_F \tag{2.45}$$

and Eq. 2.44 may be rewritten as

$$\text{Case I load} = \mathbf{R}_F + \text{distributed load} \tag{2.46}$$

Solving Eq. 2.39 for the Case II load, we get

$$\text{Case II load} = \text{Total load} - \text{Case I load} \tag{2.47}$$

Substituting from Eqs. 2.43 and 2.46 into Eq. 2.47 gives

$$\text{Case II load} = \mathbf{R} + \text{distributed load} - (\mathbf{R}_F + \text{distributed load}) \tag{2.48}$$

or

$$\text{Case II load} = \mathbf{R} - \mathbf{R}_F \tag{2.48a}$$

or

$$\mathbf{R}_{II} = \mathbf{R} - \mathbf{R}_F \tag{2.48b}$$

For the example frame, inspection reveals that

$$\mathbf{R} = \langle 0 \quad 0 \quad 0 \quad 100 \quad 0 \quad 0 \quad 0 \quad 0 \quad 0 \quad 0 \quad 0 \rangle \tag{C}$$

and, by a consideration of member and nodal equilibrium for Case I, it may be shown that

$$\mathbf{R}_F = \langle -60 \quad 0 \quad 100 \quad -60 \quad 0 \quad -100 \quad 0 \quad 0 \quad 0 \quad 0 \quad 0 \quad 0 \rangle \tag{D}$$

Substituting these results into Eq. 2.48b, we get

$$\mathbf{R}_{II} = \langle 60 \quad 0 \quad -100 \quad 160 \quad 0 \quad 100 \quad 0 \quad 0 \quad 0 \quad 0 \quad 0 \quad 0 \rangle \tag{E}$$

The solution to Case II may now be obtained by using standard procedures of matrix structural analysis, since loads are applied only at the nodes. Rewriting Eq. 2.19 appropriately subscripted for this case, we have

$$\mathbf{R}_{II} = \mathbf{a}\overline{\mathbf{P}}_{II} \tag{2.49}$$

Solving Eq. 2.49 gives

$$\overline{\mathbf{P}}_{II} = \mathbf{a}^{-1}\mathbf{R}_{II} \tag{2.50}$$

Substituting from Eq. 2.48b into 2.50 yields for the Case II solution

$$\overline{\mathbf{P}}_{II} = \mathbf{a}^{-1}(\mathbf{R} - \mathbf{R}_F) \tag{2.51}$$

TOTAL SOLUTION

Finally, substituting the Case I and Case II solutions of Eq. 2.42 and 2.51 into the superposition equation, Eq. 2.41, and then rearranging yields the total solution

$$\overline{\mathbf{P}} = \mathbf{a}^{-1}(\mathbf{R} - \mathbf{R}_F) + \overline{\mathbf{P}}_F \tag{2.52}$$

The SMIS3 computer solution to this example, shown in the next printout, yields the following vector of basic forces:

$$\overline{\mathbf{P}} = \langle 0 \quad 1600 \quad 160 \quad 0 \quad 0 \quad -160 \quad -1600 \quad 0 \quad 0 \quad 220 \quad 160 \quad -160 \rangle \tag{F}$$

Inspection of these results will reveal that they completely satisfy equilibrium.

2.12 CONCLUDING REMARKS

Before we proceed to the development of member force-displacement relationships in Chap. 3, it is of interest to reflect briefly on the development of the relationships we have been discussing.

2.12.1 Load Types and Structure Types

In general, two kinds of loads come to bear on structures: external forces and internal initial displacements. It has been shown in this chapter that if a structure

****** *Computer Input:*

```
START
REMARK 3
EXAMPLE FRAME PROBLEM,CONCENTRATED LOAD + UNIFORMLY DISTRIBUTED LOAD

**** UNITS ARE KILONEWTONS AND METERS ****
FORM A 12 12
LOAD T1 6 3
-.1 -.1 3*0 -1 1 2*0 .1 .1 3*0 1 0 1 0
LOAD T2 6 3
.1 .1 3*0 1 1 2*0 -.1 -.1 3*0 -1 0 1 0
LOAD T3 6 3
2*0 -1 .1 .1 0 1 4*0 1 -.1 -.1 2*0 1 0
LOAD CONSTR 1 1
1
STOSM A T1 1 1
STOSM A T2 7 4
STOSM A T3 4 7
STOSM A CONSTR 1 10
STOSM A CONSTR 2 11
STOSM A CONSTR 11 12
PRINT A 1
THIS IS THE MATRIX, A, OF STRUCTURE FORCE INFLUENCE COEFFICIENTS.
INVERT A
PRINT A 2
THIS IS THE MATRIX, A INVERSE, OF STRUCTURE
FORCE INFLUENCE COEFFICIENTS.
LOAD PBARI 12 1
100 -100 10*0
PRINT PBARI 3
THIS IS THE VECTOR, PBARI, OF MEMBER AND CONSTRAINT INTERNAL BASIC
FORCES FOR THE FIXED-END CASE FOR A UNIFORMLY
DISTRIBUTED HORIZONTAL LOAD ON A COLUMN.
LOAD RII 12 1
60 0 -100 160 0 100 6*0
PRINT RII 3
THIS IS THE LOAD VECTOR, RII, REPRESENTING THE
NODAL LOADS MINUS THE FIXED-END FORCES FOR A UNIFORMLY
DISTRIBUTED HORIZONTAL LOAD ON A COLUMN.
MULT A RII PBARII
PRINT PBARII 2
THIS IS THE VECTOR, PBARII, OF MEMBER AND CONSTRAINT INTERNAL BASIC
FORCES DUE TO THE LOAD OF VECTOR, RII
DUPL PBARI PBAR
ADD PBAR PBARII
PRINT PBAR 3
THIS IS THE VECTOR, PBAR, OF MEMBER AND CONSTRAINT BASIC FORCES
DUE TO THE UNIFORMLY DISTRIBUTED HORIZONTAL LOAD ON A
COLUMN PLUS THE HORIZONTAL CONCENTRATED LOAD.
STOP
```

****** *Computer Output:*

```
START

REMARK      3


EXAMPLE FRAME PROBLEM,CONCENTRATED LOAD + UNIFORMLY DISTRIBUTED LOAD

**** UNITS ARE KILONEWTONS AND METERS ****


FORM    A           12   12
   12 ROWS,     12 COLUMNS
```

****** *Computer Output (continued):*

```
LOAD     T1           6    3
    6 ROWS,      3 COLUMNS

LOAD     T2           6    3
    6 ROWS,      3 COLUMNS

LOAD     T3           6    3
    6 ROWS,      3 COLUMNS

LOAD     CONSTR       1    1
    1 ROWS,      1 COLUMNS

STOSM    A       T1           1    1
 ROW NUMBER   1,  COLUMN NUMBER   1

STOSM    A       T2           7    4
 ROW NUMBER   7,  COLUMN NUMBER   4

STOSM    A       T3           4    7
 ROW NUMBER   4,  COLUMN NUMBER   7

STOSM    A       CONSTR       1   10
 ROW NUMBER   1,  COLUMN NUMBER  10

STOSM    A       CONSTR       2   11
 ROW NUMBER   2,  COLUMN NUMBER  11

STOSM    A       CONSTR      11   12
 ROW NUMBER  11,  COLUMN NUMBER  12

PRINT    A            1


THIS IS THE MATRIX, A, OF STRUCTURE FORCE INFLUENCE COEFFICIENTS.

              1           2           3           4           5           6
 1 -1.0000E-01-1.0000E-01           0           0           0           0
 2           0           0-1.0000E+00           0           0           0
 3  1.0000E+00           0           0           0           0           0
 4  1.0000E-01 1.0000E-01           0           0           0           0
 5           0           0 1.0000E+00           0           0           0
 6           0 1.0000E+00           0           0           0           0
 7           0           0           0 1.0000E-01 1.0000E-01           0
 8           0           0           0           0           0 1.0000E+00
 9           0           0           0 1.0000E+00           0           0
10           0           0           0-1.0000E-01-1.0000E-01           0
11           0           0           0           0           0-1.0000E+00
12           0           0           0           0 1.0000E+00           0
              7           8           9          10          11          12
 1           0           0           0 1.0000E+00           0           0
 2           0           0           0           0 1.0000E+00           0
 3           0           0           0           0           0           0
 4           0           0-1.0000E+00           0           0           0
 5  1.0000E-01 1.0000E-01           0           0           0           0
 6  1.0000E+00           0           0           0           0           0
 7           0           0 1.0000E+00           0           0           0
 8 -1.0000E-01-1.0000E-01           0           0           0           0
 9           0 1.0000E+00           0           0           0           0
10           0           0           0           0           0           0
11           0           0           0           0           0 1.0000E+00
12           0           0           0           0           0           0


INVERT   A

PRINT    A            2

THIS IS THE MATRIX, A INVERSE, OF STRUCTURE
```

****** *Computer Output (continued):*

FORCE INFLUENCE COEFFICIENTS.

	1	2	3	4	5	6
1	0	0	1.0000E+00	0	0	0
2	0	0	-1.0000E+00	1.0000E+01	0	7.1054E-15
3	0	0	-1.0000E-01	1.0000E+00	1.0000E+00	-1.0000E-01
4	0	0	0	0	0	0
5	0	0	0	0	0	0
6	0	0	1.0000E-01	-1.0000E+00	0	1.0000E-01
7	0	0	1.0000E+00	-1.0000E+01	0	1.0000E+00
8	0	0	0	0	0	0
9	0	0	0	0	0	0
10	1.0000E+00	0	8.8818E-16	1.0000E+00	0	8.8818E-16
11	0	1.0000E+00	-1.0000E-01	1.0000E+00	1.0000E+00	-1.0000E-01
12	0	0	1.0000E-01	-1.0000E+00	0	1.0000E-01

	7	8	9	10	11	12
1	0	0	0	0	0	0
2	1.0000E+01	0	0	1.0000E+01	0	-7.1054E-15
3	1.0000E+00	0	-1.0000E-01	-8.8818E-15	0	-1.0000E-01
4	0	0	0	-1.0000E+01	0	-1.0000E+00
5	0	0	0	0	0	1.0000E+00
6	-1.0000E+00	1.0000E+00	1.0000E-01	8.8818E-15	0	1.0000E-01
7	-1.0000E+01	0	0	-1.0000E+01	0	7.1054E-15
8	0	0	1.0000E+00	1.0000E+01	0	1.0000E+00
9	1.0000E+00	0	0	1.0000E+00	0	-8.8818E-16
10	1.0000E+00	0	0	1.0000E+00	0	-8.8818E-16
11	1.0000E+00	0	-1.0000E-01	-8.8818E-15	0	-1.0000E-01
12	-1.0000E+00	1.0000E+00	1.0000E-01	8.8818E-15	1.0000E+00	1.0000E-01

```
LOAD    PBARI       12    1
   12 ROWS,      1 COLUMNS

PRINT   PBARI        3
```

THIS IS THE VECTOR, PBARI, OF MEMBER AND CONSTRAINT INTERNAL BASIC FORCES FOR THE FIXED-END CASE FOR A UNIFORMLY DISTRIBUTED HORIZONTAL LOAD ON A COLUMN.

	1
1	1.0000E+02
2	-1.0000E+02
3	0
4	0
5	0
6	0
7	0
8	0
9	0
10	0
11	0
12	0

```
LOAD    RII         12    1
   12 ROWS,      1 COLUMNS

PRINT   RII          3
```

THIS IS THE LOAD VECTOR, RII, REPRESENTING THE NODAL LOADS MINUS THE FIXED-END FORCES FOR A UNIFORMLY DISTRIBUTED HORIZONTAL LOAD ON A COLUMN.

1

****** *Computer Output (continued):*

```
 1  6.0000E+01
 2           0
 3 -1.0000E+02
 4  1.6000E+02
 5           0
 6  1.0000E+02
 7           0
 8           0
 9           0
10           0
11           0
12           0


MULT     A        RII      PBARII

PRINT    PBARII       2


THIS IS THE VECTOR, PBARII, OF MEMBER AND CONSTRAINT INTERNAL BASIC
FORCES DUE TO THE LOAD OF VECTOR, RII


            1
 1 -1.0000E+02
 2  1.7000E+03
 3  1.6000E+02
 4           0
 5           0
 6 -1.6000E+02
 7 -1.6000E+03
 8           0
 9           0
10  2.2000E+02
11  1.6000E+02
12 -1.6000E+02


DUPL     PBARI    PBAR

ADD      PBAR     PBARII

PRINT    PBAR         3


THIS IS THE VECTOR, PBAR, OF MEMBER AND CONSTRAINT BASIC FORCES
DUE TO THE UNIFORMLY DISTRIBUTED HORIZONTAL LOAD ON A
COLUMN PLUS THE HORIZONTAL CONCENTRATED LOAD.


            1
 1           0
 2  1.6000E+03
 3  1.6000E+02
 4           0
 5           0
 6 -1.6000E+02
 7 -1.6000E+03
 8           0
 9           0
10  2.2000E+02
11  1.6000E+02
12 -1.6000E+02


STOP
```

is subjected to external applied forces and the number of basic unknown forces in the structure is equal to the number of independent equilibrium equations available, the internal forces in the structure may be determined by using equations of equilibrium only. Such structures are referred to as statically determinate structures. If the number of basic unknown forces exceeds the number of independent equilibrium equations, the internal forces cannot be obtained on the basis of statics alone, and the structure is termed statically indeterminate.

It has also been shown that if a structure is subjected to internal, initial member basic displacements and the number of unknown nodal displacements is equal to the number of independent compatibility equations available, the displaced configuration of the structure can be determined on the basis of compatibility considerations alone. Such structures may be termed kinematically determinate.* If there are more basic displacements and hence more compatibility conditions than there are nodal displacements available to satisfy them, the configuration of the structure cannot be uniquely determined by compatibility considerations alone, and the structure can be termed kinematically overdetermined.

Because the number of basic forces $\overline{\mathbf{P}}$ is always equal to the number of basic displacements $\overline{\boldsymbol{\Delta}}$, and the number of nodal equilibrium equations is always equal to the number of nodal displacements, a statically determinate structure is the same as a kinematically determinate structure, and a statically indeterminate structure is the same as a kinematically overdetermined structure. In the future we will therefore classify structures as either statically determinate or statically indeterminate.

As explained in Sec. 1.3, structures that might be classified as statically unstable are not considered in this text. But it is of interest to notice in passing that statically unstable structures (for which the number of basic unknown forces is less than the number of independent equilibrium equations) are equivalent to kinematically indeterminate structures (for which the number of unknown nodal displacements is greater than the number of independent compatibility equations). Figure 2.14 summarizes the structure and loading classifications we have been discussing. If you are interested in a more thorough discussion of structural classification, you will want to consult the literature (see Ref. 1.2).

2.12.2 Contragredient Law

During our development of the equilibrium and compatibility relationships, various corresponding pairs of force and displacement transformations have been obtained. These were

$$\mathbf{r}' = \mathbf{L}\mathbf{P} \qquad \text{and} \qquad \boldsymbol{\Delta} = \mathbf{L}^T\mathbf{d}' \qquad \text{(2.3b and 2.4b)}$$

*Caution: This definition of kinematic determinacy, which will be used throughout this text, differs from the generally accepted definition. In most texts, a kinematically determinate structure is considered to be a structure for which all nodal displacements are known.

Statically and Kinematically Determinate Structures	
External Force Loading	*Initial Displacement Loading*
Internal forces may be determined by equilibrium only.	Nodal displacements may be determined by compatibility only.
Nodal displacements can be determined by considering elastic behavior of members and compatibility relationships.	Internal forces can be determined by considering elastic behavior of members and equilibrium relationships.

Statically Indeterminate and Kinematically Overdetermined Structures

All responses to all loads require the use of equilibrium, compatibility, and elastic relationships for their determination.

Statically Overdetermined (Unstable) and Kinematically Underdetermined Structures

General static solutions are unavailable.

Fig. 2.14 Structural classification chart.

$$\mathbf{r} = \boldsymbol{\alpha}\mathbf{r}' \quad \text{and} \quad \mathbf{d}' = \boldsymbol{\alpha}^T\mathbf{d} \qquad \text{(2.5b and 2.6a)}$$

$$\mathbf{r} = \mathbf{T}\mathbf{P} \quad \text{and} \quad \boldsymbol{\Delta} = \mathbf{T}^T\mathbf{d} \qquad \text{(2.10b and 2.11b)}$$

$$\mathbf{R} = \mathbf{A}\bar{\mathbf{r}} \quad \text{and} \quad \bar{\mathbf{d}} = \mathbf{A}^T\mathbf{D} \qquad \text{(2.16 and 2.22)}$$

$$\mathbf{R} = \mathbf{a}\bar{\mathbf{P}} \quad \text{and} \quad \bar{\boldsymbol{\Delta}} = \mathbf{a}^T\mathbf{D} \qquad \text{(2.18 and 2.24)}$$

It is important to recall now that each member of each pair of these transformations was obtained independently of the other in our development. For example, Eq. 2.16 is based on nodal equilibrium, whereas Eq. 2.22 is based on nodal compatibility. Nevertheless, it is observed in each case that corresponding forces and displacements transform in a special way relative to one another. These special sets of relationships are termed contragredient relationships. The contragredient nature of any of these relationships could be established on the basis of virtual work concepts (Ref. 2.2). Consider, for example, the frame member of Fig. 2.10, which is in equilibrium under the influence of the forces at its ends. If this member is imagined to undergo some change in its configuration, then the virtual work δW done by the forces $\mathbf{r}$ of Fig. 2.10e as they move through the displacements $\delta\mathbf{d}$ will be

$$\delta W = r_1\delta d_1 + r_2\delta d_2 + r_3\delta d_3 + r_4\delta d_4 + r_5\delta d_5 + r_6\delta d_6 \tag{2.53}$$

or

$$\mathbf{W} = \delta\mathbf{d}^T\mathbf{r} \tag{2.53a}$$

This virtual work must be equal to that done by the forces $\mathbf{P}$ of Fig. 2.10a as they move through the displacements $\delta\mathbf{\Delta}$. Thus

$$\delta W = M_i\,\delta\phi_i + M_j\,\delta\phi_j + P\,\delta\Delta \tag{2.54}$$

or

$$\delta W = \delta\mathbf{\Delta}^T\,\mathbf{P} \tag{2.54a}$$

(It may be helpful to the understanding of Eq. 2.54 to notice that the virtual change in the position of the member can be broken into two parts: [1] a rigid body movement of the chord position and [2] a change in the distorted shape of the member. Since the forces on the member are in equilibrium, no virtual work is done during the rigid body movement. Furthermore, since only the basic forces undergo displacements during the distortion phase, the virtual work done may be written in terms of M_i, M_j, and P only.)

Comparing Eqs. 2.53a and 2.54a, we may write

$$\delta\mathbf{d}^T\,\mathbf{r} = \delta\mathbf{\Delta}^T\,\mathbf{P} \tag{2.55}$$

Therefore, if it is known that

$$\mathbf{\Delta} = \mathbf{T}^T\mathbf{d} \tag{2.38a}$$

it follows that

$$\delta\mathbf{\Delta} = \mathbf{T}^T\,\delta\mathbf{d} \tag{2.56}$$

and

$$\delta\mathbf{d}^T\mathbf{r} = \delta\mathbf{d}^T\,\mathbf{TP} \tag{2.57}$$

or

$$\delta\mathbf{d}^T(\mathbf{r} - \mathbf{TP}) = 0 \tag{2.57a}$$

But since the virtual displacement $\delta\mathbf{d}^T$, is arbitrary, it follows that

$$\mathbf{r} - \mathbf{TP} = \mathbf{\Theta} \tag{2.58}$$

which is seen to check Eq. 2.37a. Clearly, we could have derived the contragredient law at an early stage in our discussions. Then, after we had obtained, say, a particular force transformation, the corresponding displacement transformation would be immediately apparent. We chose not to do this, so that we could emphasize the separate physical basis of each transformation. The contragredient law does, however, offer a concept that will be useful to us in the future.

2.13 SUMMARY—EQUILIBRIUM RELATIONSHIPS

Individual Member

Member equilibrium		$\mathbf{r}' = \mathbf{LP}$	(2.3b)
Statics—resolution of forces		$\mathbf{r} = \boldsymbol{\alpha}\mathbf{r}'$	(2.5b)
Substitution of Eq. 2.3b into 2.5b		$\mathbf{r}' = \boldsymbol{\alpha}\mathbf{LP}$	(2.10)
$\boldsymbol{\alpha}\mathbf{L} = \mathbf{T}$	or	$\mathbf{r} = \mathbf{TP}$	(2.10b)

Individual Support

Identity	$r_c = P_c$	(2.12)

Collection of All Members and Supports

Based on Eqs. 2.10b and 2.12	$\bar{\mathbf{r}} = \overline{\mathbf{T}}\overline{\mathbf{P}}$	(2.15)

Structure

Nodal equilibrium		$\mathbf{R} = \mathbf{A}\bar{\mathbf{r}}$	(2.16)
Substitution of Eq. 2.15 into 2.16		$\mathbf{R} = \mathbf{A}\overline{\mathbf{T}}\overline{\mathbf{P}}$	(2.17)
$\mathbf{A}\overline{\mathbf{T}} = \mathbf{a}$	or	$\mathbf{R} = \mathbf{a}\overline{\mathbf{P}}$	(2.18)
Solution of simultaneous equations		$\overline{\mathbf{P}} = \mathbf{a}^{-1}\mathbf{R}$	(2.19)
Superposition (distributed loading)		$\overline{\mathbf{P}} = \mathbf{a}^{-1}(\mathbf{R} - \mathbf{R}_F) + \overline{\mathbf{P}}_F$	(2.52)

For statically determinate structures, the number of unknown internal forces will equal the number of nodal equilibrium equations; hence **a** will be square and nonsingular.

2.14 SUMMARY—COMPATIBILITY RELATIONSHIPS

Individual Member

Member kinematics	$\boldsymbol{\Delta} = \mathbf{L}^T\mathbf{d}'$	(2.4b)
Resolution of displacements	$\mathbf{d}' = \boldsymbol{\alpha}^T\mathbf{d}$	(2.6a)
Substitution of Eq. 2.6a into 2.4b	$\boldsymbol{\Delta} = \mathbf{L}^T\boldsymbol{\alpha}^T\mathbf{d}$	(2.11)
$\mathbf{L}^T\boldsymbol{\alpha}^T = \mathbf{T}^T$	$\boldsymbol{\Delta} = \mathbf{T}^T\mathbf{d}$	(2.11b)

Individual Support

Identity	$\Delta_c = d_c$	(2.13)

Collection of All Members and Supports

Based on Eqs. 2.11b and 2.13	$\overline{\boldsymbol{\Delta}} = \overline{\mathbf{T}}^T\bar{\mathbf{d}}$	(2.21)

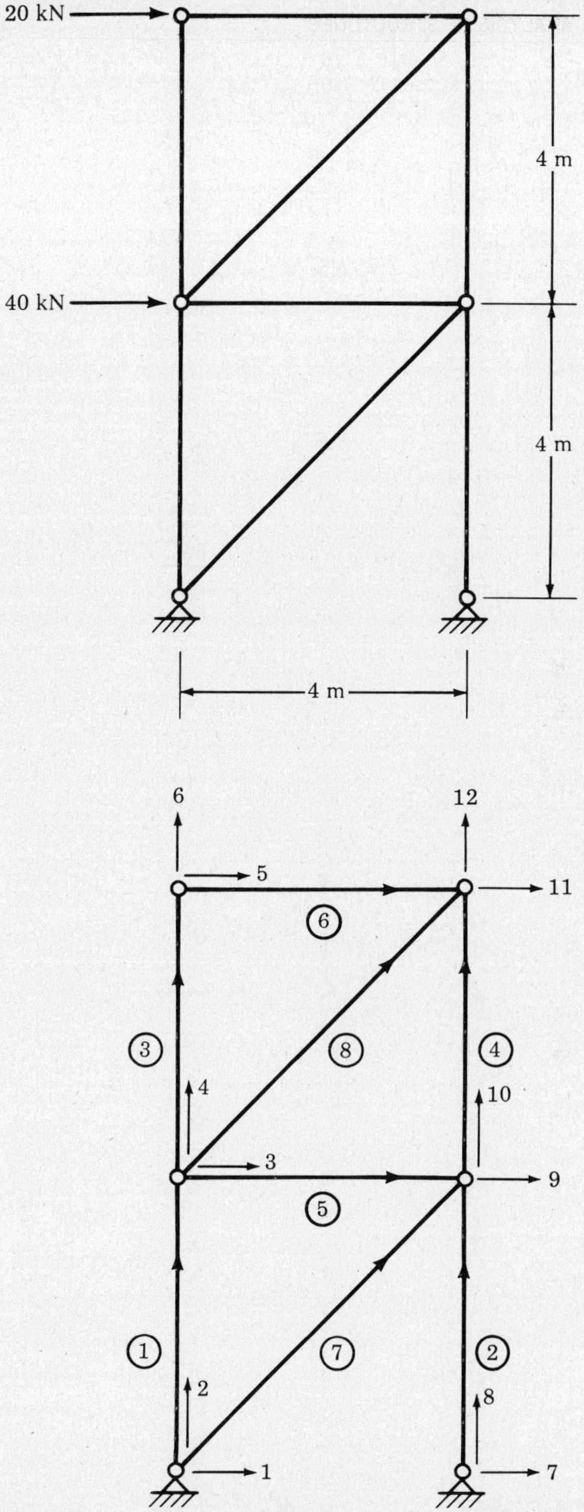

Fig. P2.2

Structure

Nodal compatibility $\qquad \bar{\mathbf{d}} = \mathbf{A}^T\mathbf{D} \qquad$ (2.22)

Substitution of Eq. 2.22 into 2.21 $\qquad \bar{\boldsymbol{\Delta}} = \bar{\mathbf{T}}^T\mathbf{A}^T\mathbf{D} \qquad$ (2.23)

$\bar{\mathbf{T}}^T\mathbf{A}^T = \mathbf{a}^T$ $\qquad$ or $\qquad \bar{\boldsymbol{\Delta}} = \mathbf{a}^T\mathbf{D} \qquad$ (2.24)

Solution of simultaneous equations $\qquad \mathbf{D} = [\mathbf{a}^T]^{-1}\bar{\boldsymbol{\Delta}} \qquad$ (2.25)

Statically determinate structures are also kinematically determinate, and the number of unknown nodal displacements will therefore equal the number of nodal compatibility conditions; hence $\mathbf{a}^T$ will be square and nonsingular.

PROBLEMS

2.1 Verify Eqs. 2.7, 2.8, and 2.9.

2.2 Solve for all member and constraint basic forces for the truss of Fig. P2.2 on page 67. Use coordinate coding to construct the **a** matrix and use SMIS3 to obtain a solution for $\bar{\mathbf{P}}$. Show coding on the $\mathbf{T}^{(n)}$ and **a** arrays. Spot-check computer results by verifying nodal equilibrium in coordinate directions 3 and 4. Also check forces in members 1, 2, and 7 by hand calculations.

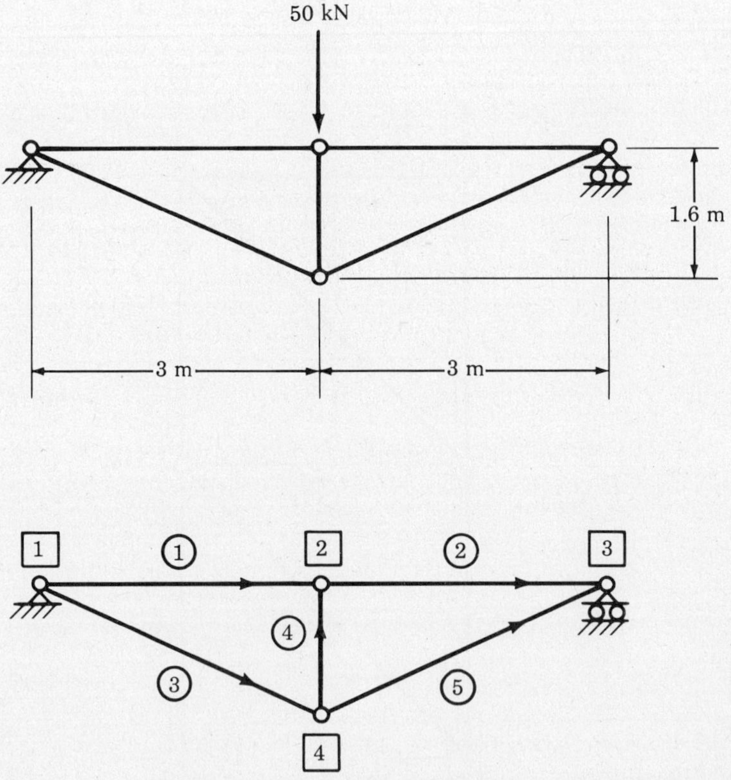

Fig. P2.3

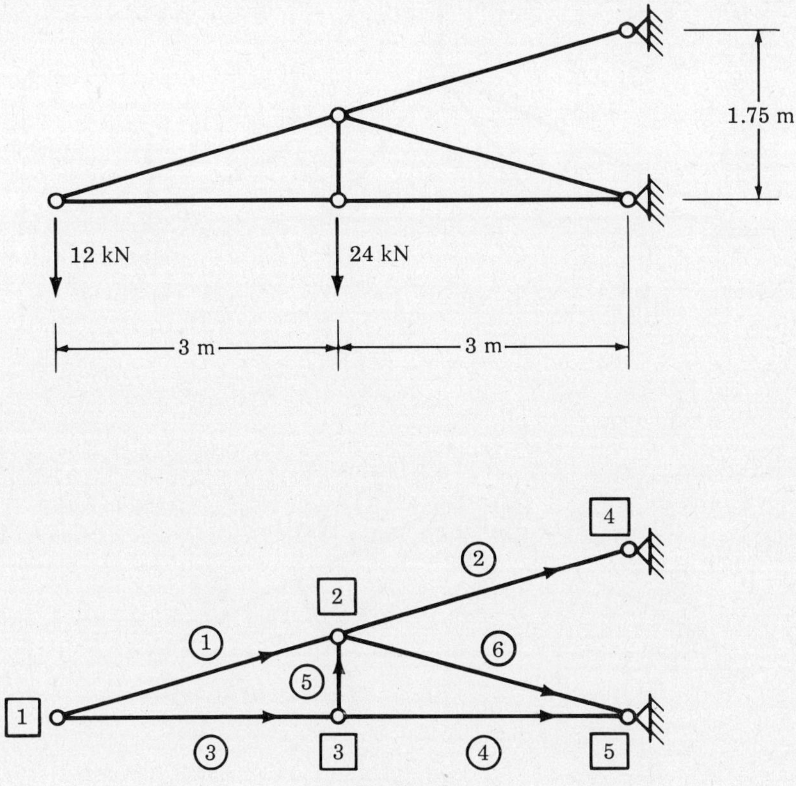

Fig. P2.4

2.3 Solve for all member and constraint basic forces for the truss of Fig. P2.3. Use nodal coding to construct the **a** matrix and use SMIS3 to obtain a solution for $\overline{\mathbf{P}}$. Show coding on the $\mathbf{T}^{(n)}$ and **a** arrays. Check nodal equilibrium at nodes 1 and 4 by hand.

2.4 Solve for all member and support basic forces for the truss of Fig. P2.4. Use coordinate coding to construct the **a** matrix and use SMIS3 to obtain a solution for $\overline{\mathbf{P}}$. Verify equilibrium in coordinate directions 3 and 4. Check forces in members 2, 4, and 6 by hand.

2.5 Trusses such as the one in Fig. P2.5 are spaced at 4 m on center and are supporting a roof subjected to a snow load of 1.5 kN/m^2 uniformly distributed over the projected horizontal area. The roof framing system is such that loads come to the trusses only at the upper chord joints. Considering snow load only, calculate the concentrated loads. Using these loads, determine the internal member and constraint forces. Use nodal coding to construct the **a** matrix and use SMIS3 to obtain a solution for $\overline{\mathbf{P}}$. Check equilibrium of nodes 1 and 4 and forces in members 2, 5, and 9 independently by hand.

2.6 Write a computer program to generate the equilibrium relationships for a plane truss. Use nodal coding. Use a library equation solver to solve the equations generated. As a check on the functioning of this program, solve Prob. 2.2, 2.3, or 2.4.

2.7 Solve for all member-end and constraint forces for the beam of Fig. P2.7. Use nodal coding to construct the **a** matrix by hand and use SMIS3 to obtain $\overline{\mathbf{P}}$. Show nodal,

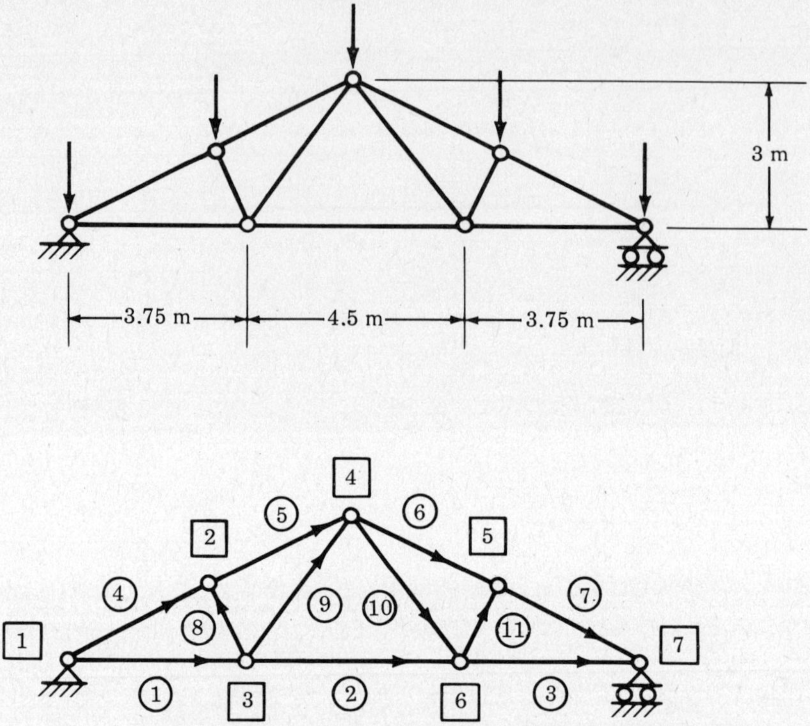

Fig. P2.5

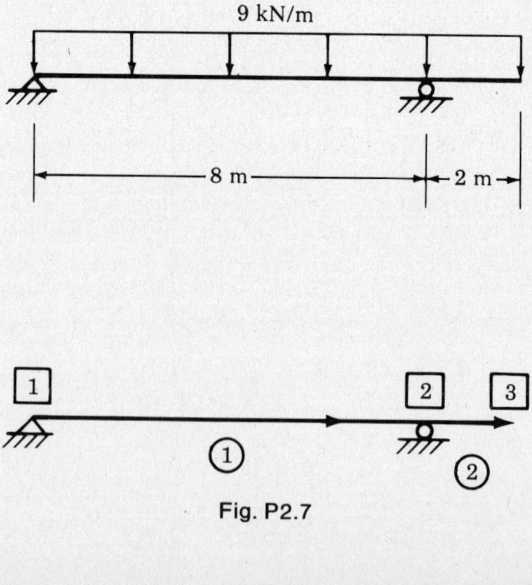

Fig. P2.7

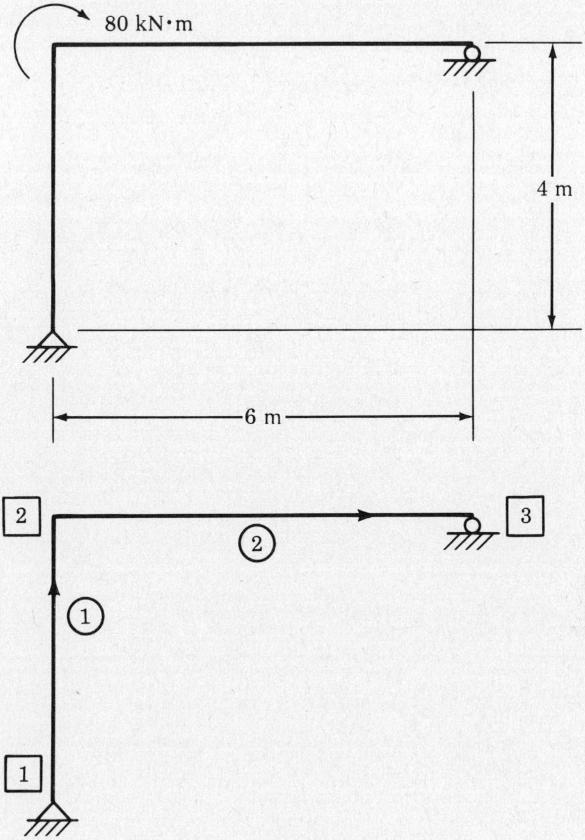

Fig. P2.9

member, and constraint numbers on the $\mathbf{T}^{(n)}$ and **a** matrices. Check equilibrium at node 2 and draw shear and moment curves.

2.8 Work Prob. 2.7 with a downward concentrated load of 30 kN applied at the right end of the beam instead of the uniform load shown in Fig. P2.7.

2.9 Solve for all member-end and constraint forces for the frame of Fig. P2.9. Use nodal coding to construct the **a** matrix by hand and use SMIS3 to solve for $\overline{\mathbf{P}}$. Check equilibrium at node 2 and draw shear and moment curves.

2.10 Work Prob. 2.9 with a concentrated load of 30 kN applied at 2 m from the left end of the girder instead of the concentrated moment shown in Fig. P2.9.

2.11 Solve for all member-end and constraint forces for the frame of Fig. P2.11. Use nodal coding to construct the **a** matrix by hand and use SMIS3 to solve for $\overline{\mathbf{P}}$. Check equilibrium at node 2 and draw shear and moment curves.

2.12 Work Prob. 2.11 with a uniformly distributed load of 6 kN/m acting downward on member 1 instead of the concentrated load shown in Fig. P2.11.

2.13 Consider the truss of Prob. 2.2 without any applied external forces. It is given that the truss is made of steel and that member 7 has become heated to a temperature of 60°C above design temperature (all other members remaining at design temperature). Determine the nodal displacements of the truss. The coefficient of thermal expansion is given as $\beta = .0000117$ (m/m)/°C.

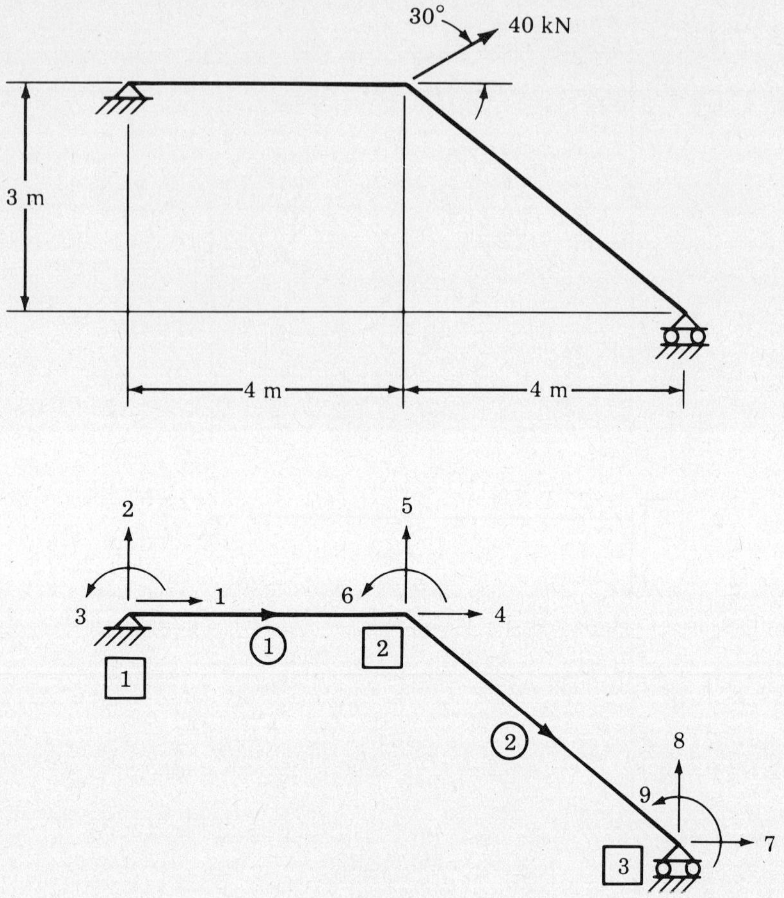

Fig. P2.11

2.14 Consider the truss of Prob. 2.4 without any applied external forces. It is given that member 6 has been accidentally fabricated .01 m too long and that the support in the direction of constraint 2 has undergone a settlement of .02 m downward. Determine the displaced position of the truss.

REFERENCES

2.1 Gere, J. M., and Weaver, W., Jr. *Matrix Algebra for Engineers*. Princeton, N.J.: Van Nostrand, 1965, pp. 61–64.

2.2 Hoff, N. J. *The Analysis of Structures*. New York: Wiley, 1956.

Chapter 3
Member Basic Force-Displacement Relationships

3.1 INTRODUCTION

Up to now we have considered only statically determinate structures in any great detail. We have further limited ourselves to problems involving either

1. the determination of internal member and support forces by enforcement of nodal equilibrium, given a certain distribution of external forces on a structure, or
2. the determination of structure nodal displacements by enforcement of nodal compatibility, given a certain distribution of member distortions and support settlements in a structure.

As stated at the end of Chap. 2, it is also of interest to be able to determine the nodal displacements for a statically determinate structure, given the applied external forces. Not only are the solutions to such problems of interest in themselves, but they can be conveniently extended to allow determination of internal forces and nodal displacements for statically indeterminate structures subjected to external applied forces and/or initial distortions. To develop such solutions, it is necessary to relate structure forces to resulting displacements. We can best accomplish this by starting at the member level, as we did before in developing the other solutions.

It is quite obvious that the basic member-end forces and corresponding member-end displacements of the frame member in Fig. 2.10 occur simultaneously and are directly related to one another. It is our present purpose to determine such relationships for truss and beam members as well as for frame members. In Secs. 3.2 to 3.5 we will develop the desired relationships using fundamental strength of materials concepts (Ref. 3.1). In Sec. 3.7 we will repeat the development, this time basing it on energy principles (Ref. 1.2). It is not essential that you study both these developments, but if you take the time and effort to do so, you will be rewarded with a greater depth of understanding of the concepts discussed. In this development it is once again convenient to initially consider forces acting only at member ends. Subsequently, the effect of forces distributed over the length of the member will be superposed on the results of this work. At this point we will consider only continuous, prismatic structural members; more general members will be discussed in Chap. 5.

3.2 TRUSS-MEMBER BASIC FLEXIBILITY AND STIFFNESS RELATIONSHIPS

Consider first of all the relationship between the basic member force and corresponding displacement for the truss member of Fig. 3.1a. Because of the nature of the loading and the uniform cross section of the member, the axial stress may be represented as

$$\sigma = \frac{P}{A} \tag{3.1}$$

Hooke's law gives

$$\sigma = E\epsilon \tag{3.2}$$

Substituting from Eq. 3.1 into 3.2 and rearranging gives

$$\epsilon = \frac{P}{EA} \tag{3.3}$$

Since this strain is constant over the length of the member, the basic member displacement that is equal to the total axial strain of the member may be written as

$$\Delta = u'_j - u'_i = \epsilon L \tag{3.4}$$

Substituting from Eq. 3.3 into 3.4 now gives

$$\Delta = \frac{L}{EA} P \tag{3.5}$$

or

$$\Delta = \mathscr{f}\mathbf{P} \tag{3.5a}$$

where the symbol $\mathscr{f}$ is used to represent the 1×1 flexibility matrix L/EA.

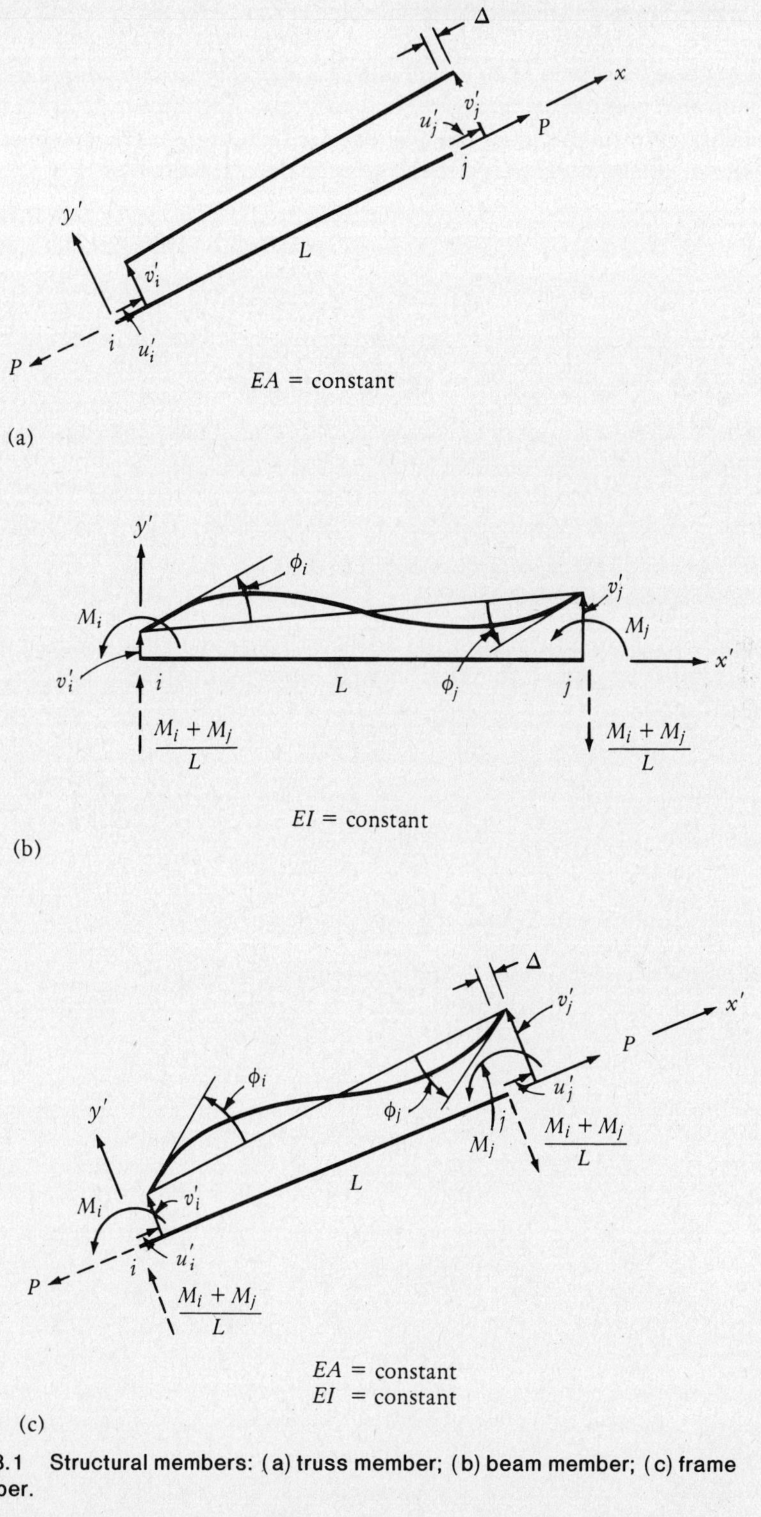

Fig. 3.1 Structural members: (a) truss member; (b) beam member; (c) frame member.

Equation 3.5 is seen to be a mathematical statement of the desired relationship between the basic member force and corresponding displacement for a truss member. Written in this form, the relationship is referred to as the member basic flexibility relationship. The inverse relationship may be written as

$$P = \frac{EA}{L}\Delta \tag{3.6}$$

or

$$\mathbf{P} = \mathbf{k}\Delta \tag{3.6a}$$

where the symbol $\mathbf{k}$ is used to represent the 1×1 member basic stiffness matrix and Eq. 3.6 is termed the member basic stiffness relationship.

3.3 BEAM-MEMBER BASIC FLEXIBILITY AND STIFFNESS RELATIONSHIPS

A similar set of flexibility and stiffness relationships may also be developed for the beam member of Fig. 3.1b. The internal moment in this member may be expressed as

$$M = -M_i + (M_i + M_j)\frac{x'}{L} \tag{3.7}$$

The moment curvature relationship may be shown to be

$$M = EI\frac{d^2v'}{dx'^2} \tag{3.8}$$

Substituting from Eq. 3.7 into 3.8 and rearranging, we obtain

$$\frac{d^2v'}{dx'^2} = -\frac{M_i}{EI} + \frac{M_i + M_j}{EI}\frac{x'}{L} \tag{3.9}$$

Integrating gives

$$\frac{dv'}{dx'} = -\frac{M_i x'}{EI} + \frac{M_i + M_j}{EI}\frac{x'^2}{2L} + C_1 \tag{3.10}$$

$$v' = -\frac{M_i x'^2}{2EI} + \frac{M_i + M_j}{EI}\frac{x'^3}{6L} + C_1 x' + C_2 \tag{3.11}$$

Enforcing boundary conditions yields

$$v'(0) = v'_i \rightarrow C_2 = v'_i$$

$$v'(L) = v'_j \rightarrow C_1 = \frac{M_i L}{3EI} - \frac{M_j L}{6EI} + \frac{v'_j - v'_i}{L}$$

Substituting these results into Eq. 3.10 and 3.11 and rearranging, we get

$$\frac{dv'}{dx'} = \frac{M_i}{EI}\left(\frac{L}{3} - x' + \frac{x'^2}{2L}\right) + \frac{M_j}{EI}\left(-\frac{L}{6} + \frac{x'^2}{2L}\right) + \frac{v'_j - v'_i}{L} \tag{3.12}$$

$$v' = \frac{M_i}{EI}\left(\frac{Lx'}{3} - \frac{x'^2}{2} + \frac{x'^3}{6L}\right) + \frac{M_j}{EI}\left(-\frac{Lx'}{3} + \frac{x'^3}{6L}\right) + \frac{v'_j - v'_i}{L}x' + v'_i \tag{3.13}$$

Evaluating Eq. 3.12 at $x' = 0$ and $x' = L$ and then rearranging, we have

$$\phi_i = \frac{dv'}{dx'}(0) - \frac{v'_j - v'_i}{L} = M_i\frac{L}{3EI} - M_j\frac{L}{6EI} \tag{3.14}$$

$$\phi_j = \frac{dv'}{dx'}(L) - \frac{v'_j - v'_i}{L} = -M_i\frac{L}{6EI} + M_j\frac{L}{3EI} \tag{3.15}$$

Rewriting Eq. 3.14 and 3.15 as a single matrix equation gives

$$\begin{pmatrix}\phi_i\\ \phi_j\end{pmatrix} = \frac{L}{6EI}\begin{bmatrix}2 & -1\\ -1 & 2\end{bmatrix}\begin{pmatrix}M_i\\ M_j\end{pmatrix} \tag{3.16}$$

or

$$\mathbf{\Delta} = \mathscr{f}\,\mathbf{P} \tag{3.16a}$$

Both Eq. 3.16 and 3.16a are seen to represent the beam-member basic flexibility relationships. The inverse of these relationships is found to be

$$\begin{pmatrix}M_i\\ M_j\end{pmatrix} = \frac{2EI}{L}\begin{bmatrix}2 & 1\\ 1 & 2\end{bmatrix}\begin{pmatrix}\phi_i\\ \phi_j\end{pmatrix} \tag{3.17}$$

or

$$\mathbf{P} = \mathscr{f}^{-1}\mathbf{\Delta} \quad \text{or} \quad \mathbf{P} = \mathscr{k}\,\mathbf{\Delta} \tag{3.17a}$$

Equations 3.17 and 3.17a are the basic stiffness relationships for the beam member. The $\mathscr{f}$ and $\mathscr{k}$ matrices are once again termed the member basic flexibility and basic stiffness matrices, respectively.

Example 3.1

Determine the rotation from the member chord that occurs at end B of the beam shown in Fig. 3.2. (This particular rotation is also the absolute rotation since the chord is stationary.)

1. Work this problem by any convenient method.
2. Check this solution using the beam flexibility relationship developed in this section.

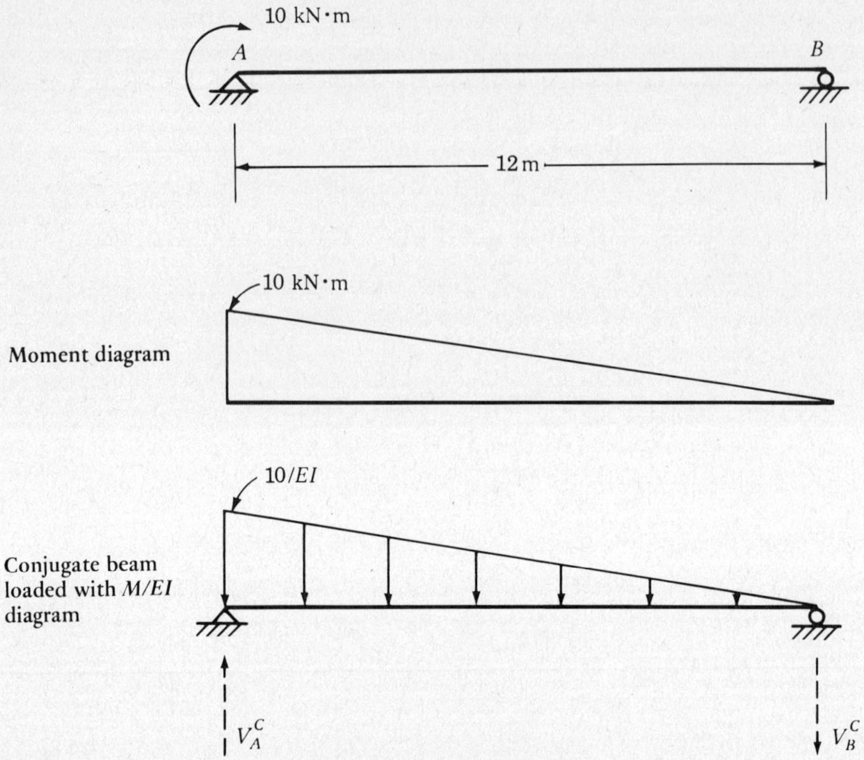

Fig. 3.2 Example beam.

SOLUTION

1. We will use the conjugate beam method (Ref. 1.2). Summing moments about point A on the conjugate beam gives

$$\frac{1}{2}12 \times \frac{10}{EI} \times 4 + 12V_B^C = 0$$

Solving yields

$$V_B^C = \phi_B = -\frac{20}{EI} \quad \text{or} \quad \phi_B = \frac{20}{EI} \text{ radians counterclockwise}$$

2. From Eq. 3.16, we obtain

$$\phi_B = \phi_j = \frac{L}{6EI}(-M_i + 2M_j) = \frac{12}{6EI}[-(-10) + 2 \times 0] = \frac{20}{EI} \text{ rad ccw}$$

3.4 FRAME-MEMBER BASIC FLEXIBILITY AND STIFFNESS RELATIONSHIPS

Figure 3.1 shows that the loading on the frame member is simply the sum of the loadings on the truss and beam members. By neglecting the interaction of axial and bending modes of member behavior, we can conclude that, by the principle of superposition, the member basic displacements in the frame member will be the sum of the basic displacements in the truss and beam members. In this connection three points should be clarified:

1. Neglecting axial and bending interaction is necessary to the application of the principle of superposition because such interaction is a nonlinear phenomenon. The subject of interaction is beyond the scope of this book (see Ref. 3.2).
2. Neglecting axial and bending interaction is allowable as long as the axial forces remain within certain limits. We will assume this to be the case.
3. Neglecting interaction is not the same as considering the frame members to be inextensional, as is often the practice in connection with classical methods of analysis (Ref. 1.2). For the time being, frame members will be considered to be extensional. The conditions under which it may be permissible to assume inextensional behavior will be examined in Sec. 5.3.3

By superposition, the frame-member basic flexibility relationships are

$$\begin{pmatrix} \phi_i \\ \phi_j \\ \Delta \end{pmatrix} = \frac{L}{6EI} \begin{bmatrix} 2 & -1 & 0 \\ -1 & 2 & 0 \\ 0 & 0 & 6I/A \end{bmatrix} \begin{pmatrix} M_i \\ M_j \\ P \end{pmatrix} \tag{3.18}$$

or

$$\boldsymbol{\Delta} = \boldsymbol{f}\,\mathbf{P} \tag{3.18a}$$

and the frame-member basic stiffness relationships are

$$\begin{pmatrix} M_i \\ M_j \\ P \end{pmatrix} = \frac{2EI}{L} \begin{bmatrix} 2 & 1 & 0 \\ 1 & 2 & 0 \\ 0 & 0 & A/2I \end{bmatrix} \begin{pmatrix} \phi_i \\ \phi_j \\ \Delta \end{pmatrix} \tag{3.19}$$

or

$$\mathbf{P} = \boldsymbol{f}^{-1}\boldsymbol{\Delta} \quad \text{or} \quad \mathbf{P} = \boldsymbol{k}\boldsymbol{\Delta} \tag{3.19a}$$

3.5 SUPPORT BASIC FORCE-DISPLACEMENT RELATIONSHIPS

Although it will later be shown that structural supports may often be considered to be completely rigid, a general approach to the solution of problems should admit the possibility of flexible supports. Furthermore, there are times when the ability to consider flexible supports will be essential.

Basic forces and displacements for truss, beam, and frame supports were defined in Secs. 2.4, 2.8, and 2.10, respectively. Each basic constraint force and its corresponding basic displacement may now be related to one another as

$$\Delta_c = f_c P_c \tag{3.20}$$

for a translational constraint and as

$$\phi_c = f_c M_c \tag{3.21}$$

for a rotational constraint. The symbol f_c represents the basic support flexibility in the direction of the constraint.

The inverse relationships may, of course, be written as

$$P_c = k_c \Delta_c \tag{3.22}$$

and

$$M_c = k_c \phi_c \tag{3.23}$$

where

$$k_c = f_c^{-1} \tag{3.24}$$

and k_c represents the basic constraint stiffness.

3.6 FLEXIBILITY AND STIFFNESS COEFFICIENT SYMMETRY CONDITIONS

To better understand the physical significance of the flexibility and stiffness relationships, consider the following definitions.

The member basic flexibility influence coefficient is defined as

f_{pq} = the pth basic displacement of a member due to a unit qth basic force on the member, all other basic forces equal to zero

The member basic stiffness influence coefficient is defined as

k_{pq} = the pth basic force on a member due to a unit qth basic displacement of the member, all other basic displacements equal to zero.

A careful scrutiny of the preceding member flexibility and stiffness relationships will bear out the truth of these definitions.

Consider now Maxwell's law of reciprocity (Ref. 3.2), which states:

> The displacement in direction p on a structure due to a unit load in direction q is equal to the displacement in direction q due to a unit load in direction p.

In view of our definition of a flexibility influence coefficient, we see that Maxwell's law requires that

$$f_{pq} = f_{qp} \tag{3.25}$$

which means that the array of flexibility coefficients must be symmetric about the main diagonal of the array. In addition, since the stiffness matrix is the inverse of the flexibility matrix and therefore the inverse of a symmetric matrix, it too must be a symmetric matrix. Thus

$$k_{pq} = k_{qp} \tag{3.26}$$

3.7 OBTAINING FLEXIBILITY AND STIFFNESS COEFFICIENTS BY ENERGY METHODS

Finally, to gain further insight into the nature of the stiffness relationships, it is worthwhile to consider alternative methods of developing these relationships. Any one of the many methods of finding deflections in structures could be used for this purpose. However, energy-based methods turn out to be particularly effective, especially in connection with the more complex finite-element type of structural member required in the analysis of continuous structures. It is therefore appropriate to use such a method to further investigate these relationships.

3.7.1 Flexibility Coefficients

Castigliano's second theorem (Ref. 3.3) is an energy method that is often used to determine flexibility coefficients. This theorem may be stated as

$$\frac{\partial U}{\partial Q} = D_Q \tag{3.27}$$

where

U = elastic strain energy stored in the structure
Q = a load on the structure
D_Q = the displacement of the structure at the point of application of load Q and in the direction of load Q

This theorem may be stated as follows: The rate of change of the elastic strain energy stored in a structure with respect to a load on the structure is equal to the displacement of the structure at the point of application of the load and in the direction of the load. Consider the application of this theorem to the frame member of Fig. 3.1c. The strain energy stored in this member can be expressed as

$$U = \int_0^L \frac{M^2 dx'}{2EI} + \int_0^L \frac{S^2\, dx'}{2EA} \tag{3.28}$$

where M and S are the member internal moment and axial force, respectively, and where shearing strain energy and hence shearing strains are neglected. This approach is consistent with our previous development of flexibility and stiffness relationships. The effect of shearing strains will be considered in Sec. 5.3.2.

From Eq. 3.7 and Fig. 3.1c, we know that

$$\begin{aligned} M &= -M_i + (M_i + M_j)\frac{x'}{L} \\ S &= P \end{aligned} \tag{3.29}$$

Substituting these expressions into Eq. 3.28 gives

$$U = \int_0^L \left[-M_i + (M_i + M_j)\frac{x'}{L}\right]^2 \frac{dx'}{2EI} + \int_0^L \frac{P^2\, dx'}{2EA} \tag{3.30}$$

Applying Castigliano's second theorem (prior to integration for convenience) gives

$$\frac{\partial U}{\partial M_i} = \int_0^L \left[-M_i + (M_i + M_j)\frac{x'}{L}\right]\left(-1 + \frac{x'}{L}\right)\frac{dx'}{EI} = \phi_i \tag{3.31}$$

Integrating gives

$$\phi_i = \left[M_i x' - (M_i + M_j)\frac{x'^2}{2L} - \frac{M_i x'^2}{2L} + (M_i + M_j)\frac{x'^3}{3L^2}\right]\Bigg|_0^L \frac{1}{EI}$$

or

$$\phi_i = \left(\frac{M_i}{3} - \frac{M_j}{6}\right)\frac{L}{EI}$$

or

$$\phi_i = \frac{L}{6EI}(2M_i - M_j) \tag{3.32}$$

Equation 3.32 is seen to verify the first of the flexibility relationships expressed in Eq. 3.18.

Consider now taking the partial derivative of Eq. 3.32 with respect to, say, M_j. This gives

$$\frac{\partial \phi_i}{\partial M_j} = -\frac{L}{6EI} \tag{3.33}$$

This derivative could be described in words as follows: The rate of change of the frame member's first basic displacement ϕ_i with respect to its second basic force M_j, all other forces held constant.

Upon investigation, this description will be seen to be a paraphrase of the definition of flexibility coefficient f_{12}. Hence

$$f_{12} = \frac{\partial \phi_i}{\partial M_j} = -\frac{L}{6EI} \tag{3.34}$$

Furthermore, since

$$\phi_i = \frac{\partial U}{\partial M_i}$$

it follows from Eq. 3.34 that

$$f_{12} = \frac{\partial^2 U}{\partial M_i\, \partial M_j} = -\frac{L}{6EI}$$

Similarly,

$$f_{11} = \frac{\partial^2 U}{\partial M_i^2} = \frac{L}{3EI}$$

$$f_{13} = \frac{\partial^2 U}{\partial M_i\, \partial P} = 0$$

and so on.

You may wish to verify the remaining basic flexibility coefficients for the frame member.

Finally, note that since

$$\frac{\partial^2 U}{\partial M_i \, \partial M_j} = \frac{\partial^2 U}{\partial M_j \, \partial M_i}$$

it follows that

$$f_{12} = f_{21}$$

or, in general,

$$f_{pq} = f_{qp}$$

which is another proof of symmetry of the member flexibility matrix and, hence, also the member stiffness matrix.

3.7.2 Stiffness Coefficients

The basic stiffness relationships may once again be obtained by inversion of the basic flexibility relationships. Alternatively, they may be obtained directly by using Castigliano's first theorem (Ref. 3.2), which may be stated as

$$\frac{\partial U}{\partial D_Q} = Q \tag{3.35}$$

where U, Q, and D_Q are the same as previously defined. This theorem may be stated in words as follows: The rate of change of the elastic strain energy stored in a structure with respect to a displacement at some point on that structure is equal to the load applied on the structure at the point of and in the direction of that displacement. Once again consider the frame member of Fig. 3.1c. Application of Castigliano's first theorem requires differentiation of the strain energy expression with respect to a displacement. To effectively carry out this operation, it is first necessary to obtain an expression for strain energy in terms of displacements. This procedure may be accomplished by appropriately transforming the strain energy expression of Eq. 3.28. Consider first of all the bending contribution to this strain energy expression:

$$U_B = \int_0^L \frac{M^2 \, dx'}{2EI} \tag{3.36}$$

Substituting from Eq. 3.8 into 3.36 and simplifying gives

$$U_B = \int_0^L \frac{EI}{2} \left(\frac{d^2 v'}{dx'^2} \right)^2 dx' \tag{3.37}$$

From Eq. 3.11 or Eq. 3.13 it is seen that the transverse displacement v' is a cubic function of the coordinate x'. This fact may be expressed as

$$v' = A_1 + A_2 x' + A_3 x'^2 + A_4 x'^3 \tag{3.38}$$

in terms of arbitrary constants $A_1, \ldots, A_4$. These constants may now be determined in terms of boundary displacements as

$$v'(0) = v'_i \rightarrow v'_i = A_1$$

$$v'(L) = v'_j \rightarrow v'_j = A_1 + A_2L + A_3L^2 + A_4L^3$$

$$\frac{dv'}{dx'}(0) = \phi_i + \frac{v'_j - v'_i}{L} \rightarrow \phi_i + \frac{v'_j - v'_i}{L} = A_2$$

$$\frac{dv'}{dx'}(L) = \phi_j + \frac{v'_j - v'_i}{L} \rightarrow \phi_j + \frac{(v'_j - v'_i)}{L} = A_2 + 2A_3L + 3A_4L^2$$

Solving the above set of equations simultaneously for $A_1, \ldots, A_4$ in terms of v'_i, v'_j, ϕ_i, and ϕ_j gives

$$A_1 = v'_i$$

$$A_2 = \phi_i + (v'_j - v'_i)/L$$

$$A_3 = -(2\phi_i + \phi_j)/L$$

$$A_4 = (\phi_i + \phi_j)/L^2$$

Backsubstituting these results into Eq. 3.38 we get

$$v' = v'_i + \left(\phi_i + \frac{v'_j - v'_i}{L}\right)x' - (2\phi_i + \phi_j)\frac{x'^2}{L} + (\phi_i + \phi_j)\frac{x'^3}{L^2} \tag{3.39}$$

Differentiating Eq. 3.39 twice with respect to x' gives

$$\frac{d^2v'}{dx'^2} = -\frac{2}{L}(2\phi_i + \phi_j) + 6(\phi_i + \phi_j)\frac{x'}{L^2} \tag{3.40}$$

Notice that curvature in the frame member is a function of member basic displacements only. Substituting this result into Eq. 3.37 gives the following expression for bending strain energy in a frame member in terms of basic member displacements:

$$U_B = \int_0^L \frac{EI}{2}\left[-\frac{2}{L}(2\phi_i + \phi_j) + 6(\phi_i + \phi_j)\frac{x'}{L^2}\right]^2 dx' \tag{3.41}$$

Consider now the axial strain energy contribution to the strain energy expression of Eq. 3.28.

$$U_A = \int_0^L \frac{S^2\, dx'}{2EA} \tag{3.42}$$

Since we know that

$$\sigma = \frac{S}{A}$$

and

$$\epsilon = \frac{\sigma}{E} = \frac{S}{EA}$$

we obtain

$$S = EA\epsilon$$

and substitution yields

$$U_A = \int_0^L \frac{EA}{2} \epsilon^2 \, dx' \tag{3.43}$$

In the case of a frame member, the axial strain is seen to be constant over the length of the member and related to the basic axial displacement as

$$\epsilon = \frac{\Delta}{L} \tag{3.44}$$

Substituting this into Eq. 3.43 gives the following expression for axial strain energy in a frame member in terms of basic member displacements:

$$U_A = \int_0^L \frac{EA}{2} \frac{\Delta^2}{L^2} dx' \tag{3.45}$$

The total strain energy in a frame member may now be written as

$$\begin{aligned} U = U_B + U_A = & \int_0^L \frac{EI}{2}\left[-\frac{2}{L}(2\phi_i + \phi_j) + 6(\phi_i + \phi_j)\frac{x'}{L^2}\right]^2 dx' \\ & + \int_0^L \frac{EA}{2}\frac{\Delta^2}{L^2} dx' \end{aligned} \tag{3.46}$$

Applying Castigliano's first theorem to find, say, the force in the direction of ϕ_i, we get

$$\begin{aligned} \frac{\partial U}{\partial \phi_i} = M_i = & \int_0^L EI\left[-\frac{2}{L}(2\phi_i + \phi_j) \right. \\ & \left. + 6(\phi_i + \phi_j)\frac{x'}{L^2}\right]\left(-\frac{4}{L} + \frac{6x'}{L^2}\right) dx' \end{aligned} \tag{3.47}$$

or, by integrating,

$$\begin{aligned} M_i = EI & \left[-\frac{2}{L}(2\phi_i + \phi_j)\left(-\frac{4x'}{L} + \frac{6x'^2}{2L^2}\right)\right. \\ & \left.\left. + 6(\phi_i + \phi_j)\left(-\frac{4x'^2}{2L^3} + \frac{6x'^3}{3L^4}\right)\right]\right|_0^L \end{aligned}$$

or

$$M_i = \frac{EI}{L}[-2(2\phi_i + \phi_j)(-1) + 6(\phi_i + \phi_j)(0)]$$

or

$$M_i = \frac{2EI}{L}(2\phi_i + \phi_j) \tag{3.48}$$

Equation 3.48 is seen to corroborate the first row of Eq. 3.19.

Consider now taking the partial derivative of Eq. 3.48 with respect to, say, ϕ_j. This gives

$$\frac{\partial M_i}{\partial \phi_j} = \frac{2EI}{L} \tag{3.49}$$

This derivative could be described in words as follows: The rate of change of basic force M_i with respect to basic displacement ϕ_j, all other basic displacements held constant. Once again this description is seen to represent a paraphrase of the definition of the stiffness coefficient k_{12}:

$$k_{12} = \frac{\partial M_i}{\partial \phi_j} = \frac{2EI}{L} \tag{3.50}$$

Furthermore, since

$$M_i = \frac{\partial U}{\partial \phi_i}$$

it follows that

$$k_{12} = \frac{\partial^2 U}{\partial \phi_i \, \partial \phi_j}$$

Similarly,

$$k_{11} = \frac{\partial^2 U}{\partial \phi_i^2}$$

$$k_{23} = \frac{\partial^2 U}{\partial \phi_j \, \partial \Delta}$$

and so on.

Finally, note that since

$$\frac{\partial^2 U}{\partial \phi_i \, \partial \phi_j} = \frac{\partial^2 U}{\partial \phi_j \, \partial \phi_i}$$

we know that

$$k_{12} = k_{21}$$

and, in general,

$$k_{pq} = k_{qp}$$

which corroborates symmetry of the member basic stiffness matrix about its main diagonal.

3.8 SUMMARY—MEMBER BASIC FLEXIBILITY AND STIFFNESS RELATIONSHIPS

It has been shown in this chapter that the basic stiffness relationship for a truss, beam, or frame member is of the form

$$\mathbf{P} = \not{k}\,\mathbf{\Delta}$$

and that the corresponding basic flexibility relationship is of the form

$$\mathbf{\Delta} = \not{f}\,\mathbf{P}$$

Before effective use can be made of these relationships, it is necessary to transform them to a global coordinate system. We will do this during the development of the displacement and force methods of analysis presented in subsequent chapters.

PROBLEMS

3.1 Determine the rotation of the right end of the beam member shown in Fig. P3.1 from its chord. (Notice that this is also the absolute rotation since the chord is stationary.)
(a) Solve the problem by writing the appropriate fourth-order differential equation and integrating.
(b) Check this result using the stiffness relationship of Eq. 3.17.

3.2 Demonstrate Maxwell's law of reciprocity by proving that the rotation in direction 2 due to a unit force in direction 1 is equal to the deflection in direction 1 due to a unit couple in direction 2 for the cantilever beam shown in Fig. P3.2. Use any convenient method for determining displacements.

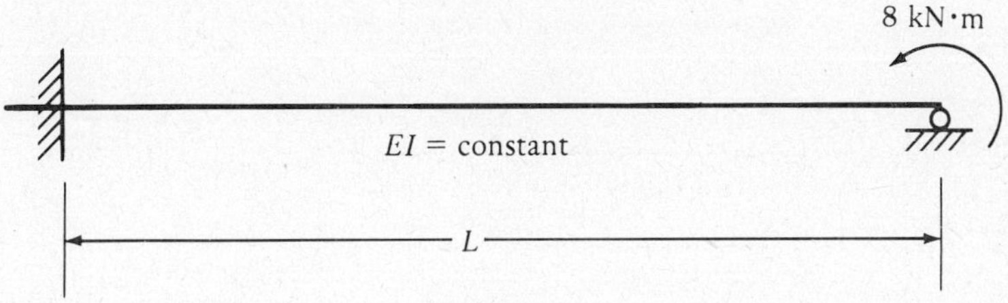

Fig. P3.1

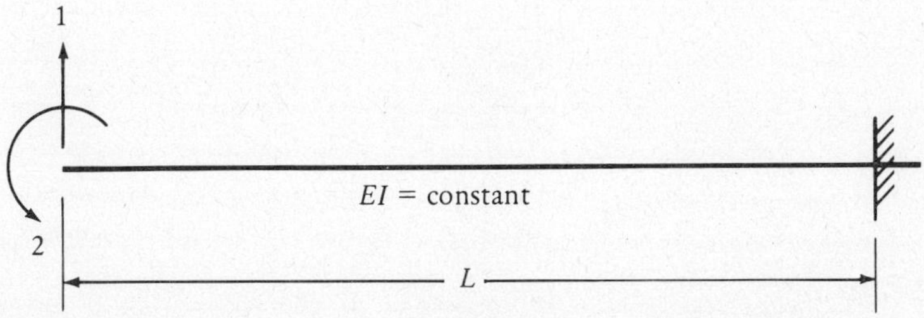

Fig. P3.2

REFERENCES

3.1 Byars, E. F., and Snyder, R. D. *Engineering Mechanics of Deformable Bodies*. 3d ed. New York: Intext, 1975.

3.2 Chajes, A. *Principles of Structural Stability Theory*. Englewood Cliffs, N.J.: Prentice-Hall, 1974.

3.3 Oden, J. T. *Mechanics of Elastic Structures*. New York: McGraw-Hill, 1967.

Chapter 4
Displacement Method

4.1 INTRODUCTION

In Chap. 3 we developed two mutually inverse relationships between basic member-end forces and corresponding displacements—the flexibility relationship and the stiffness relationship. The next step is to develop two structural analysis methods, one based on the flexibility relationship and the other based on the stiffness relationship. The two resulting methods of analysis will prove to be complementary to each other: When one is difficult to use, the other may be relatively convenient.

The structural analysis technique that is based on the stiffness relationship is often termed the displacement method—so called because the structure nodal displacements are the primary unknowns in this method. On the other hand, when the flexibility relationship is employed, the member basic end forces become the primary unknowns yielding the force method of structural analysis. Which one should be developed first? The stiffness approach will subsequently be shown to have certain inherent advantages that make it by far the more widely used method. We will therefore turn our attention in this direction first.

4.2 TRANSFORMATION OF THE MEMBER BASIC STIFFNESS RELATIONSHIPS TO GLOBAL COORDINATES

As was mentioned at the end of Chap. 3, it is convenient to transform the member basic stiffness relations to global coordinates before attempting to build the structure stiffness relationships.

4.2.1 Truss Member

The basic stiffness relationship for a truss member was expressed in Eq. 3.6 as

$$P = \frac{EA}{L}\Delta \tag{3.6}$$

But Eqs. 2.10a and 2.11a gave

$$\begin{pmatrix} r_1 \\ r_2 \\ r_3 \\ r_4 \end{pmatrix} = \begin{bmatrix} -\cos\alpha \\ -\sin\alpha \\ \cos\alpha \\ \sin\alpha \end{bmatrix} (P) \tag{2.10a}$$

and

$$[\Delta] = [-\cos\alpha \quad -\sin\alpha \quad \cos\alpha \quad \sin\alpha] \begin{pmatrix} d_1 \\ d_2 \\ d_3 \\ d_4 \end{pmatrix} \tag{2.11a}$$

Substituting from Eq. 3.6 into 2.10a and from Eq. 2.11a into that result gives

$$\begin{pmatrix} r_1 \\ r_2 \\ r_3 \\ r_4 \end{pmatrix} = \begin{bmatrix} -\cos\alpha \\ -\sin\alpha \\ \cos\alpha \\ \sin\alpha \end{bmatrix} \frac{EA}{L} [-\cos\alpha \quad -\sin\alpha \quad \cos\alpha \quad \sin\alpha] \begin{pmatrix} d_1 \\ d_2 \\ d_3 \\ d_4 \end{pmatrix} \tag{4.1}$$

or

$$\begin{pmatrix} r_1 \\ r_2 \\ r_3 \\ r_4 \end{pmatrix} = \frac{EA}{L} \begin{bmatrix} c^2 & cs & -c^2 & -cs \\ cs & s^2 & -cs & -s^2 \\ -c^2 & -cs & c^2 & cs \\ -cs & -s^2 & cs & s^2 \end{bmatrix} \begin{pmatrix} d_1 \\ d_2 \\ d_3 \\ d_4 \end{pmatrix} \tag{4.1a}$$

where

$$c = \cos\alpha$$

$$s = \sin\alpha$$

Equation 4.1a may be written symbolically as

$$\mathbf{r} = \mathbf{kd} \tag{4.1b}$$

where it is seen that $\mathbf{k} = \mathbf{T}\boldsymbol{\mathscr{k}}\mathbf{T}^T$. Equations 4.1a and 4.1b express the global stiffness relationships for a truss member.

4.2.2 Beam Member

The basic stiffness relationship for a beam member is expressed in Eq. 3.17 as

$$\begin{pmatrix} M_i \\ M_j \end{pmatrix} = \frac{2EI}{L}\begin{bmatrix} 2 & 1 \\ 1 & 2 \end{bmatrix}\begin{pmatrix} \phi_i \\ \phi_j \end{pmatrix} \tag{3.17}$$

But Eqs. 2.27a and 2.28a gave, respectively,

$$\begin{pmatrix} r_1 \\ r_2 \\ r_3 \\ r_4 \end{pmatrix} = \begin{bmatrix} 1/L & 1/L \\ 1 & 0 \\ -1/L & -1/L \\ 0 & 1 \end{bmatrix}\begin{pmatrix} M_i \\ M_j \end{pmatrix} \tag{2.27a}$$

and

$$\begin{pmatrix} \phi_i \\ \phi_j \end{pmatrix} = \begin{bmatrix} 1/L & 1 & -1/L & 0 \\ 1/L & 0 & -1/L & 1 \end{bmatrix}\begin{pmatrix} d_1 \\ d_2 \\ d_3 \\ d_4 \end{pmatrix} \tag{2.28a}$$

Using Eqs. 3.17, 2.27a, and 2.28a, we may obtain the global stiffness relationships for the beam member:

$$\begin{pmatrix} r_1 \\ r_2 \\ r_3 \\ r_4 \end{pmatrix} = \begin{bmatrix} 1/L & 1/L \\ 1 & 0 \\ -1/L & -1/L \\ 0 & 1 \end{bmatrix}\frac{2EI}{L}\begin{bmatrix} 2 & 1 \\ 1 & 2 \end{bmatrix}\begin{bmatrix} 1/L & 1 & -1/L & 0 \\ 1/L & 0 & -1/L & 1 \end{bmatrix}\begin{pmatrix} d_1 \\ d_2 \\ d_3 \\ d_4 \end{pmatrix} \tag{4.2}$$

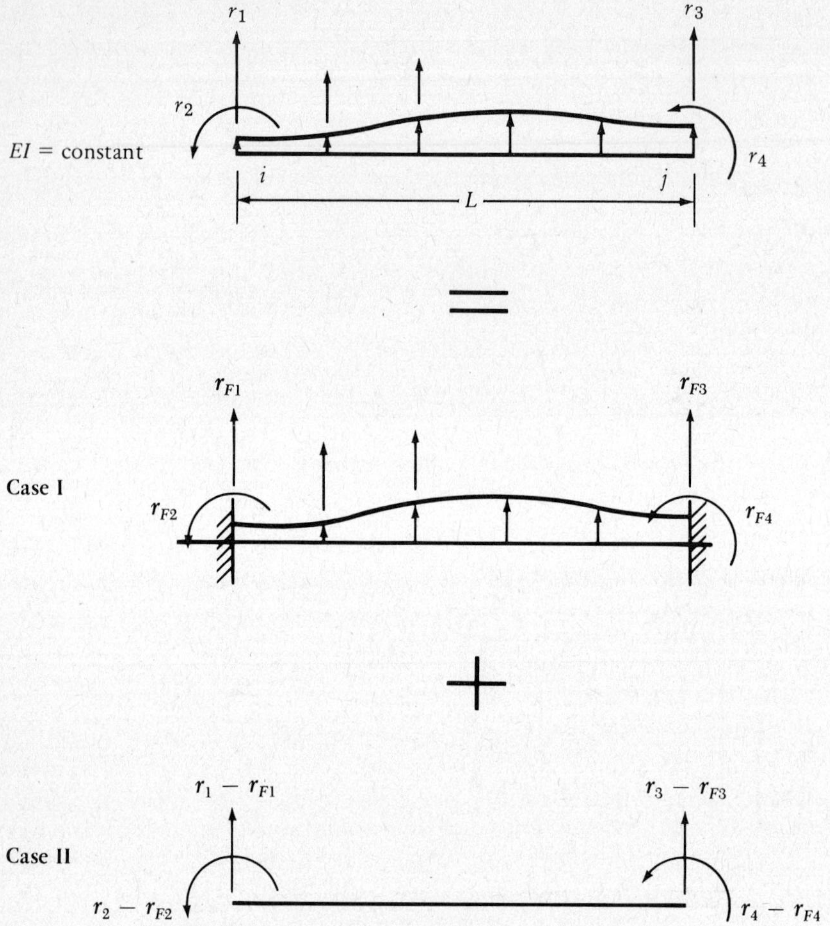

Fig. 4.1 Beam member, distributed loading.

or

$$\begin{pmatrix} r_1 \\ r_2 \\ r_3 \\ r_4 \end{pmatrix} = \frac{2EI}{L}\begin{bmatrix} 6/L^2 & 3/L & -6/L^2 & 3/L \\ 3/L & 2 & -3/L & 1 \\ -6/L^2 & -3/L & 6/L^2 & -3/L \\ 3/L & 1 & -3/L & 2 \end{bmatrix}\begin{pmatrix} d_1 \\ d_2 \\ d_3 \\ d_4 \end{pmatrix} \tag{4.2a}$$

or

$$\mathbf{r} = \mathbf{kd} \tag{4.2b}$$

Since beam members often have loads distributed along them, it is important to generalize the above stiffness relationships to account for these loads. Applying Eq. 4.2b to the Case II loading of Fig. 4.1 gives

$$\mathbf{r}_{II} = \mathbf{r} - \mathbf{r}_F = \mathbf{kd} \tag{4.3}$$

Superposing the Case I fixed-end solution, $\mathbf{r}_I = \mathbf{r}_F$, on the Case II solution shown yields

$$\mathbf{r} = \mathbf{kd} + \mathbf{r}_F \tag{4.4}$$

or

$$\begin{pmatrix} r_1 \\ r_2 \\ r_3 \\ r_4 \end{pmatrix} = \frac{2EI}{L}\begin{bmatrix} 6/L^2 & 3/L & -6/L^2 & 3/L \\ 3/L & 2 & -3/L & 1 \\ -6/L^2 & -3/L & 6/L^2 & -3/L \\ 3/L & 1 & -3/L & 2 \end{bmatrix}\begin{pmatrix} d_1 \\ d_2 \\ d_3 \\ d_4 \end{pmatrix} + \begin{pmatrix} r_{F1} \\ r_{F2} \\ r_{F3} \\ r_{F4} \end{pmatrix} \tag{4.4a}$$

where $\mathbf{r}_F$ is seen to be a vector of fixed-end forces.

4.2.3 Frame Member

In the case of the frame member, it is convenient to transform the basic stiffness relationships to global coordinates and to generalize these relationships for distributed member loadings in the following sequence of steps:

1. Transform the basic relationship to local coordinates.
2. Generalize local stiffness relationships for distributed member loadings.
3. Transform the generalized local stiffness relationships to global coordinates.

The reason for this particular sequence of steps is simply that it is most convenient to add the effect of distributed loading to the stiffness relationships in the local coordinate system.

The frame-member basic stiffness relationships were stated in Eq. 3.19 as

$$\begin{pmatrix} M_i \\ M_j \\ P \end{pmatrix} = \frac{2EI}{L}\begin{bmatrix} 2 & 1 & 0 \\ 1 & 2 & 0 \\ 0 & 0 & A/2I \end{bmatrix}\begin{pmatrix} \phi_i \\ \phi_j \\ \Delta \end{pmatrix} \tag{3.19}$$

To transform these relationships to local coordinates, it is necessary to use Eqs. 2.33 and 2.34:

$$\begin{pmatrix} r_1' \\ r_2' \\ r_3' \\ r_4' \\ r_5' \\ r_6' \end{pmatrix} = \begin{bmatrix} 0 & 0 & -1 \\ 1/L & 1/L & 0 \\ 1 & 0 & 0 \\ 0 & 0 & 1 \\ -1/L & -1/L & 0 \\ 0 & 1 & 0 \end{bmatrix}\begin{pmatrix} M_i \\ M_j \\ P \end{pmatrix} \tag{2.33}$$

$$\begin{pmatrix} \phi_i \\ \phi_j \\ \Delta \end{pmatrix} = \begin{bmatrix} 0 & 1/L & 1 & 0 & -1/L & 0 \\ 0 & 1/L & 0 & 0 & -1/L & 1 \\ -1 & 0 & 0 & 1 & 0 & 0 \end{bmatrix} \begin{pmatrix} d'_1 \\ d'_2 \\ d'_3 \\ d'_4 \\ d'_5 \\ d'_6 \end{pmatrix} \tag{2.34}$$

Combining Eqs. 3.19 with Eqs. 2.33 and 2.34 gives

$$\begin{pmatrix} r'_1 \\ r'_2 \\ r'_3 \\ r'_4 \\ r'_5 \\ r'_6 \end{pmatrix} = \begin{bmatrix} 0 & 0 & -1 \\ 1/L & 1/L & 0 \\ 1 & 0 & 0 \\ 0 & 0 & 1 \\ -1/L & -1/L & 0 \\ 0 & 1 & 0 \end{bmatrix} \frac{2EI}{L} \begin{bmatrix} 2 & 1 & 0 \\ 1 & 2 & 0 \\ 0 & 0 & A/2I \end{bmatrix} \begin{bmatrix} 0 & 1/L & 1 & 0 & -1/L & 0 \\ 0 & 1/L & 0 & 0 & -1/L & 1 \\ -1 & 0 & 0 & 1 & 0 & 0 \end{bmatrix} \begin{pmatrix} d'_1 \\ d'_2 \\ d'_3 \\ d'_4 \\ d'_5 \\ d'_6 \end{pmatrix} \tag{4.5}$$

or

$$\begin{pmatrix} r'_1 \\ r'_2 \\ r'_3 \\ r'_4 \\ r'_5 \\ r'_6 \end{pmatrix} = \frac{2EI}{L} \begin{bmatrix} A/2I & 0 & 0 & -A/2I & 0 & 0 \\ 0 & 6/L^2 & 3/L & 0 & -6/L^2 & 3/L \\ 0 & 3/L & 2 & 0 & -3/L & 1 \\ -A/2I & 0 & 0 & A/2I & 0 & 0 \\ 0 & -6/L^2 & -3/L & 0 & 6/L^2 & -3/L \\ 0 & 3/L & 1 & 0 & -3/L & 2 \end{bmatrix} \begin{pmatrix} d'_1 \\ d'_2 \\ d'_3 \\ d'_4 \\ d'_5 \\ d'_6 \end{pmatrix} \tag{4.5a}$$

or

$$\mathbf{r}' = \mathbf{k}'\mathbf{d}' \tag{4.5b}$$

These frame-member local stiffness relationships may now be generalized to include the effect of distributed member loadings by superposition. Figure 4.2 graphically portrays this superposition and emphasizes that this operation is carried out in local coordinates as a matter of convenience. Proceeding as for the beam member gives

$$\begin{pmatrix} r'_1 \\ r'_2 \\ r'_3 \\ r'_4 \\ r'_5 \\ r'_6 \end{pmatrix} = \frac{2EI}{L} \begin{bmatrix} A/2I & 0 & 0 & -A/2I & 0 & 0 \\ 0 & 6/L^2 & 3/L & 0 & -6/L^2 & 3/L \\ 0 & 3/L & 2 & 0 & -3/L & 1 \\ -A/2I & 0 & 0 & A/2I & 0 & 0 \\ 0 & -6/L^2 & -3/L & 0 & 6/L^2 & -3/L \\ 0 & 3/L & 1 & 0 & -3/L & 2 \end{bmatrix} \begin{pmatrix} d'_1 \\ d'_2 \\ d'_3 \\ d'_4 \\ d'_5 \\ d'_6 \end{pmatrix} + \begin{pmatrix} r'_{F1} \\ r'_{F2} \\ r'_{F3} \\ r'_{F4} \\ r'_{F5} \\ r'_{F6} \end{pmatrix} \tag{4.6}$$

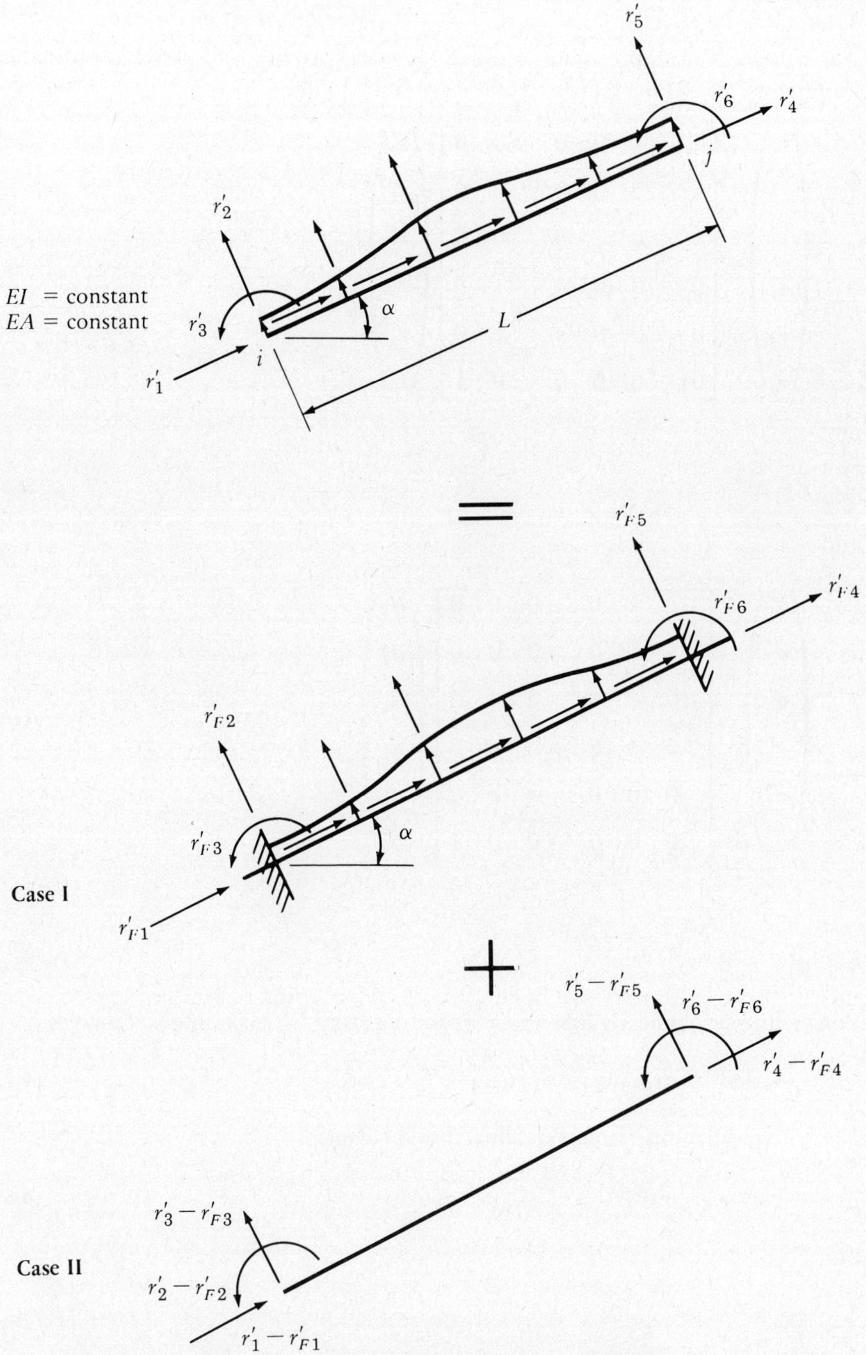

Fig. 4.2 Frame member, distributed loading.

or

$$\mathbf{r}' = \mathbf{k}'\mathbf{d}' + \mathbf{r}'_F \tag{4.6a}$$

Finally, it is appropriate to make the transformation to global coordinates. Equations 2.35, 2.35a, 2.36, and 2.36a are repeated here:

$$\begin{pmatrix} r_1 \\ r_2 \\ r_3 \\ r_4 \\ r_5 \\ r_6 \end{pmatrix} = \begin{bmatrix} c & -s & 0 & 0 & 0 & 0 \\ s & c & 0 & 0 & 0 & 0 \\ 0 & 0 & 1 & 0 & 0 & 0 \\ 0 & 0 & 0 & c & -s & 0 \\ 0 & 0 & 0 & s & c & 0 \\ 0 & 0 & 0 & 0 & 0 & 1 \end{bmatrix} \begin{pmatrix} r'_1 \\ r'_2 \\ r'_3 \\ r'_4 \\ r'_5 \\ r'_6 \end{pmatrix} \tag{2.35}$$

or

$$\mathbf{r} = \boldsymbol{\alpha}\mathbf{r}' \tag{2.35a}$$

and

$$\begin{pmatrix} d'_1 \\ d'_2 \\ d'_3 \\ d'_4 \\ d'_5 \\ d'_6 \end{pmatrix} = \begin{bmatrix} c & s & 0 & 0 & 0 & 0 \\ -s & c & 0 & 0 & 0 & 0 \\ 0 & 0 & 1 & 0 & 0 & 0 \\ 0 & 0 & 0 & c & s & 0 \\ 0 & 0 & 0 & -s & c & 0 \\ 0 & 0 & 0 & 0 & 0 & 1 \end{bmatrix} \begin{pmatrix} d_1 \\ d_2 \\ d_3 \\ d_4 \\ d_5 \\ d_6 \end{pmatrix} \tag{2.36}$$

or

$$\mathbf{d}' = \boldsymbol{\alpha}^T\mathbf{d} \tag{2.36a}$$

Proceeding symbolically first and substituting from Eq. 4.6a into 2.35a gives

$$\mathbf{r} = \boldsymbol{\alpha}\mathbf{k}'\mathbf{d}' + \boldsymbol{\alpha}\mathbf{r}'_F \tag{4.7}$$

Now, if we substitute from Eq. 2.36a into 4.7 we get

$$\mathbf{r} = \boldsymbol{\alpha}\mathbf{k}'\boldsymbol{\alpha}^T\mathbf{d} + \boldsymbol{\alpha}\mathbf{r}'_F \tag{4.8}$$

or

$$\mathbf{r} = \mathbf{k}\mathbf{d} + \mathbf{r}_F \tag{4.8a}$$

where $\mathbf{k} = \boldsymbol{\alpha}\mathbf{k}'\boldsymbol{\alpha}^T$ is displayed in Eq. 4.9.

$$\mathbf{k} = \frac{2EI}{L}\begin{bmatrix} \frac{c^2A}{2I}+\frac{s^26}{L^2} & \frac{scA}{2I}-\frac{sc6}{L^2} & \frac{-s3}{L} & \frac{-c^2A}{2I}-\frac{s^26}{L^2} & \frac{-scA}{2I}+\frac{sc6}{L^2} & \frac{-s3}{L} \\ & \frac{s^2A}{2I}+\frac{c^26}{L^2} & \frac{c3}{L} & \frac{-scA}{2I}+\frac{sc6}{L^2} & \frac{-s^2A}{2I}-\frac{c^26}{L^2} & \frac{c3}{L} \\ & & 2 & \frac{s3}{L} & \frac{-c3}{L} & 1 \\ & \text{Symmetric} & & \frac{c^2A}{2I}+\frac{s^26}{L^2} & \frac{scA}{2I}-\frac{sc6}{L^2} & \frac{s3}{L} \\ & & & & \frac{s^2A}{2I}+\frac{c^26}{L^2} & \frac{-c3}{L} \\ & & & & & 2 \end{bmatrix} \tag{4.9}$$

For the special cases of $\alpha = 0°$ and $\alpha = 90°$, this matrix reduces, respectively, to

$$\mathbf{k} = \frac{2EI}{L}\begin{bmatrix} A/2I & 0 & 0 & -A/2I & 0 & 0 \\ 0 & 6/L^2 & 3/L & 0 & -6/L^2 & 3/L \\ 0 & 3/L & 2 & 0 & -3/L & 1 \\ -A/2I & 0 & 0 & A/2I & 0 & 0 \\ 0 & -6/L^2 & -3/L & 0 & 6/L^2 & -3/L \\ 0 & 3/L & 1 & 0 & -3/L & 2 \end{bmatrix} \tag{4.9a}$$

and

$$\mathbf{k} = \frac{2EI}{L}\begin{bmatrix} 6/L^2 & 0 & -3/L & -6/L^2 & 0 & -3/L \\ 0 & A/2I & 0 & 0 & -A/2I & 0 \\ -3/L & 0 & 2 & 3/L & 0 & 1 \\ -6/L^2 & 0 & 3/L & 6/L^2 & 0 & 3/L \\ 0 & -A/2I & 0 & 0 & A/2I & 0 \\ -3/L & 0 & 1 & 3/L & 0 & 2 \end{bmatrix} \tag{4.9b}$$

Note, in passing, that each stiffness matrix that has been developed has been symmetric about its main diagonal. This symmetry follows, once again, from Maxwell's law of reciprocity, and it will continue to be evidenced at the structure level as well as at the member level.

Transformation of the basic member stiffness relationships to global coordinates for the truss, beam, and frame members is now complete.

4.2.4 Supports

We have assumed that all constraints act in global coordinate directions. Thus for translational constraints, we had

$$r_c = P_c \tag{2.29}$$

and

$$\Delta_c = d_c \tag{2.31}$$

Recalling Eq. 3.22,

$$P_c = \mathscr{k}_c \Delta_c \tag{3.22}$$

we combine the above three equations and obtain

$$r_c = \mathscr{k}_c d_c \tag{4.10}$$

or

$$r_c = k_c d_c \tag{4.10a}$$

The global support stiffness k_c is therefore seen to be identical with the basic support stiffness $\mathscr{k}_c$. This sameness is to be expected for the special case where constraints act only in global directions.

Similarly, for rotational supports, we had

$$r_c = M_c \tag{2.30}$$

$$\phi_c = d_c \tag{2.32}$$

and

$$M_c = \mathscr{k}_c \phi_c \tag{3.23}$$

Combining the above three equations gives again

$$r_c = \mathscr{k}_c d_c \tag{4.11}$$

or

$$r_c = k_c d_c \tag{4.11a}$$

Once again, global and basic support stiffnesses are seen to be identical.

For a discussion of constraints imposed in nonglobal directions, see Sec. 5.4.

4.3 ASSEMBLY AND SOLUTION OF STRUCTURE STIFFNESS RELATIONSHIPS FOR STATICALLY DETERMINATE STRUCTURES

Now that we have developed the member and support stiffness relationships in global coordinates, it is convenient to build the structure stiffness relationships. Our goal is to relate structure applied forces to structure nodal displacements so that, knowing the applied forces, we can obtain a solution for the unknown displacements. The structure stiffness relationships represent one form in which the structure force-displacement relationships can be written. (The structure flexibility relationships represent another form that will be studied subsequently.) The structure stiffness relationships may be built from the individual member and support stiffness relationships by either of two methods: (1) symbolic and (2) coding. These methods are analagous to those of the same names previously used in building the structure force influence relationships from the member force influence relationships. Furthermore, they have the same advantages and disadvantages as they did before. Since the symbolic method provides the logical mathematical framework on which the coding method is based, we will consider it first.

4.3.1 Symbolic Method

The structure stiffness relationships will first be developed for the special case of the truss structure of Fig. 4.3, which was previously shown in Fig. 2.3. Equations 4.1b and 4.10a gave the stiffness relationships symbolically for any truss member and any truss constraint, respectively, as

$$\mathbf{r} = \mathbf{k}\mathbf{d} \tag{4.1b}$$

and

$$r_c = k_c d_c \tag{4.10a}$$

To emphasize that these relationships apply to a single member, a superscript n is added, giving

$$\begin{aligned} \mathbf{r}^{(n)} &= \mathbf{k}^{(n)}\mathbf{d}^{(n)} \\ r_c^{(n)} &= k_c^{(n)} d_c^{(n)} \end{aligned} \tag{4.12}$$

Now, writing the stiffness relationships for the collection of all members and constraints in the structure as a single matrix equation, we get

$$\begin{pmatrix} \mathbf{r}^{(1)} \\ \mathbf{r}^{(2)} \\ \mathbf{r}^{(3)} \\ \mathbf{r}^{(4)} \\ \mathbf{r}^{(5)} \\ \mathbf{r}^{(6)} \\ \mathbf{r}^{(7)} \\ r_c^{(1)} \\ r_c^{(2)} \\ r_c^{(3)} \end{pmatrix} = \begin{bmatrix} \mathbf{k}^{(1)} & & & & & & & & & \\ & \mathbf{k}^{(2)} & & & & & & & & \\ & & \mathbf{k}^{(3)} & & & & \Theta & & & \\ & & & \mathbf{k}^{(4)} & & & & & & \\ & & & & \mathbf{k}^{(5)} & & & & & \\ & & & & & \mathbf{k}^{(6)} & & & & \\ & & & & & & \mathbf{k}^{(7)} & & & \\ & \Theta & & & & & & k_c^{(1)} & & \\ & & & & & & & & k_c^{(2)} & \\ & & & & & & & & & k_c^{(3)} \end{bmatrix} \begin{pmatrix} \mathbf{d}^{(1)} \\ \mathbf{d}^{(2)} \\ \mathbf{d}^{(3)} \\ \mathbf{d}^{(4)} \\ \mathbf{d}^{(5)} \\ \mathbf{d}^{(6)} \\ \mathbf{d}^{(7)} \\ d_c^{(1)} \\ d_c^{(2)} \\ d_c^{(3)} \end{pmatrix} \tag{A}$$

or

$$\bar{\mathbf{r}} = \bar{\mathbf{k}}\bar{\mathbf{d}} \tag{4.13}$$

The stiffness matrices for the individual truss members may be obtained by appropriately specializing the stiffness matrix of Eq. 4.1a. Given the member properties shown in Fig. 4.3a,

$$\mathbf{k}^{(1)} = \mathbf{k}^{(2)} = \mathbf{k}^{(7)} = \frac{EA_0}{3}\begin{bmatrix} 1 & 0 & -1 & 0 \\ 0 & 0 & 0 & 0 \\ -1 & 0 & 1 & 0 \\ 0 & 0 & 0 & 0 \end{bmatrix}$$

$$\mathbf{k}^{(3)} = \mathbf{k}^{(5)} = \frac{EA_0}{2.4}\begin{bmatrix} .36 & .48 & -.36 & -.48 \\ .48 & .64 & -.48 & -.64 \\ -.36 & -.48 & .36 & .48 \\ -.48 & -.64 & .48 & .64 \end{bmatrix} = \frac{EA_0}{3}\begin{bmatrix} .45 & .6 & -.45 & -.6 \\ .6 & .8 & -.6 & -.8 \\ -.45 & -.6 & .45 & .6 \\ -.6 & -.8 & .6 & .8 \end{bmatrix}$$

$$\mathbf{k}^{(4)} = \mathbf{k}^{(6)} = \frac{EA_0}{2.4}\begin{bmatrix} .36 & -.48 & -.36 & .48 \\ -.48 & .64 & .48 & -.64 \\ -.36 & .48 & .36 & -.48 \\ .48 & -.64 & -.48 & .64 \end{bmatrix} = \frac{EA_0}{3}\begin{bmatrix} .45 & -.6 & -.45 & .6 \\ -.6 & .8 & .6 & -.8 \\ -.45 & .6 & .45 & -.6 \\ .6 & -.8 & -.6 & .8 \end{bmatrix}$$

where, for convenience, a common factor of $EA_0/3$ has been factored as a constant multiplier of each array.

Assuming supports to be unyielding,* we may take

$$k_c^{(n)} = \infty \qquad n = 1, 2, 3$$

or, for convenience,

$$k_c^{(n)} = \frac{EA_0}{3}\infty \qquad n = 1, 2, 3$$

Substituting these arrays into Eq. A yields Eq. B, displayed on pages 102 and 103. Notice that Eq. B represents the force-displacement behaviors of all the members

*Physically, supports are generally several orders of magnitude stiffer than other members of a structure, and it is the usual practice in structural analysis to consider them as being infinitely stiff and thus capable of totally preventing displacements in the constrained coordinate directions. The concept of infinity can cause difficulties, however, in some formal mathematical proofs. Moreover, it is not possible to represent infinity numerically for computer or hand solutions. For these reasons the term *infinity* and the corresponding symbol, ∞, will often be treated in this text to represent a *very large* support stiffness, so large that other member stiffnesses are negligible by comparison and that displacements in supported directions are negligible compared with other structure nodal displacements. Other methods of considering supports in a stiffness model will be discussed in Sec. 4.5. Some of these methods will allow for exact representation of the totally rigid support.

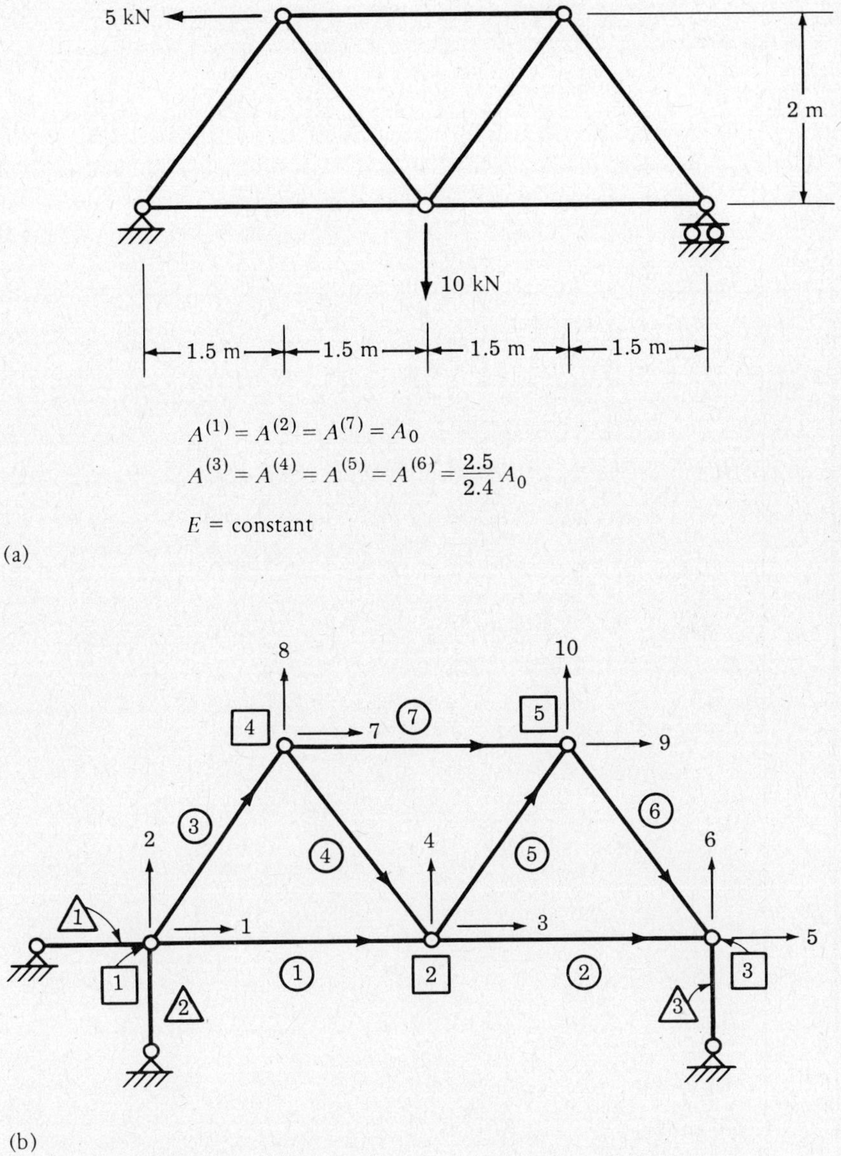

Fig. 4.3 Truss example (previously shown in Fig. 2.3): (a) dimensions and loading; (b) member, constraint, coordinate, and node numbering and member directions.

and constraints of the structure *acting independently.* The structure, however, imposes two physical restrictions on member and constraint behavior: (1) Member-end and constraint forces must satisfy equilibrium requirements at each structure node, and (2) member-end and constraint displacements must be compatible with corresponding structure nodal displacements. Enforcement of these two physical requirements—nodal equilibrium and nodal compatibility—

$$
\begin{pmatrix}
r_1^{(1)} \\ r_2^{(1)} \\ r_3^{(1)} \\ r_4^{(1)} \\
r_1^{(2)} \\ r_2^{(2)} \\ r_3^{(2)} \\ r_4^{(2)} \\
r_1^{(3)} \\ r_2^{(3)} \\ r_3^{(3)} \\ r_4^{(3)} \\
r_1^{(4)} \\ r_2^{(4)} \\ r_3^{(4)} \\ r_4^{(4)} \\
r_1^{(5)} \\ r_2^{(5)} \\ r_3^{(5)} \\ r_4^{(5)} \\
r_1^{(6)} \\ r_2^{(6)} \\ r_3^{(6)} \\ r_4^{(6)} \\
r_1^{(7)} \\ r_2^{(7)} \\ r_3^{(7)} \\ r_4^{(7)} \\
r_c^{(1)} \\ r_c^{(2)} \\ r_c^{(3)}
\end{pmatrix}
= \frac{EA_0}{3}
\left[
\begin{array}{cccccccccccccccc}
1 & 0 & -1 & 0 & \cdot & \cdot & \cdot & \cdot & \cdot & \cdot & \cdot & \cdot & \cdot & \cdot & \cdot & \cdot \\
0 & 0 & 0 & 0 & \cdot & \cdot & \cdot & \cdot & \cdot & \cdot & \cdot & \cdot & \cdot & \cdot & \cdot & \cdot \\
-1 & 0 & 1 & 0 & \cdot & \cdot & \cdot & \cdot & \cdot & \cdot & \cdot & \cdot & \cdot & \cdot & \cdot & \cdot \\
0 & 0 & 0 & 0 & \cdot & \cdot & \cdot & \cdot & \cdot & \cdot & \cdot & \cdot & \cdot & \cdot & \cdot & \cdot \\
\cdot & \cdot & \cdot & \cdot & 1 & 0 & -1 & 0 & \cdot & \cdot & \cdot & \cdot & \cdot & \cdot & \cdot & \cdot \\
\cdot & \cdot & \cdot & \cdot & 0 & 0 & 0 & 0 & \cdot & \cdot & \cdot & \cdot & \cdot & \cdot & \cdot & \cdot \\
\cdot & \cdot & \cdot & \cdot & -1 & 0 & 1 & 0 & \cdot & \cdot & \cdot & \cdot & \cdot & \cdot & \cdot & \cdot \\
\cdot & \cdot & \cdot & \cdot & 0 & 0 & 0 & 0 & \cdot & \cdot & \cdot & \cdot & \cdot & \cdot & \cdot & \cdot \\
\cdot & \cdot & \cdot & \cdot & \cdot & \cdot & \cdot & \cdot & .45 & .6 & -.45 & -.6 & \cdot & \cdot & \cdot & \cdot \\
\cdot & \cdot & \cdot & \cdot & \cdot & \cdot & \cdot & \cdot & .6 & .8 & -.6 & -.8 & \cdot & \cdot & \cdot & \cdot \\
\cdot & \cdot & \cdot & \cdot & \cdot & \cdot & \cdot & \cdot & -.45 & -.6 & .45 & .6 & \cdot & \cdot & \cdot & \cdot \\
\cdot & \cdot & \cdot & \cdot & \cdot & \cdot & \cdot & \cdot & -.6 & -.8 & .6 & .8 & \cdot & \cdot & \cdot & \cdot \\
\cdot & \cdot & \cdot & \cdot & \cdot & \cdot & \cdot & \cdot & \cdot & \cdot & \cdot & \cdot & .45 & -.6 & -.45 & .6 \\
\cdot & \cdot & \cdot & \cdot & \cdot & \cdot & \cdot & \cdot & \cdot & \cdot & \cdot & \cdot & -.6 & .8 & .6 & -.8 \\
\cdot & \cdot & \cdot & \cdot & \cdot & \cdot & \cdot & \cdot & \cdot & \cdot & \cdot & \cdot & -.45 & .6 & .45 & -.6 \\
\cdot & \cdot & \cdot & \cdot & \cdot & \cdot & \cdot & \cdot & \cdot & \cdot & \cdot & \cdot & .6 & -.8 & -.6 & .8 \\
\cdot & \cdot & \cdot & \cdot & \cdot & \cdot & \cdot & \cdot & \cdot & \cdot & \cdot & \cdot & \cdot & \cdot & \cdot & \cdot \\
\cdot & \cdot & \cdot & \cdot & \cdot & \cdot & \cdot & \cdot & \cdot & \cdot & \cdot & \cdot & \cdot & \cdot & \cdot & \cdot \\
\cdot & \cdot & \cdot & \cdot & \cdot & \cdot & \cdot & \cdot & \cdot & \cdot & \cdot & \cdot & \cdot & \cdot & \cdot & \cdot \\
\cdot & \cdot & \cdot & \cdot & \cdot & \cdot & \cdot & \cdot & \cdot & \cdot & \cdot & \cdot & \cdot & \cdot & \cdot & \cdot \\
\cdot & \cdot & \cdot & \cdot & \cdot & \cdot & \cdot & \cdot & \cdot & \cdot & \cdot & \cdot & \cdot & \cdot & \cdot & \cdot \\
\cdot & \cdot & \cdot & \cdot & \cdot & \cdot & \cdot & \cdot & \cdot & \cdot & \cdot & \cdot & \cdot & \cdot & \cdot & \cdot \\
\cdot & \cdot & \cdot & \cdot & \cdot & \cdot & \cdot & \cdot & \cdot & \cdot & \cdot & \cdot & \cdot & \cdot & \cdot & \cdot \\
\cdot & \cdot & \cdot & \cdot & \cdot & \cdot & \cdot & \cdot & \cdot & \cdot & \cdot & \cdot & \cdot & \cdot & \cdot & \cdot \\
\cdot & \cdot & \cdot & \cdot & \cdot & \cdot & \cdot & \cdot & \cdot & \cdot & \cdot & \cdot & \cdot & \cdot & \cdot & \cdot \\
\cdot & \cdot & \cdot & \cdot & \cdot & \cdot & \cdot & \cdot & \cdot & \cdot & \cdot & \cdot & \cdot & \cdot & \cdot & \cdot \\
\cdot & \cdot & \cdot & \cdot & \cdot & \cdot & \cdot & \cdot & \cdot & \cdot & \cdot & \cdot & \cdot & \cdot & \cdot & \cdot \\
\cdot & \cdot & \cdot & \cdot & \cdot & \cdot & \cdot & \cdot & \cdot & \cdot & \cdot & \cdot & \cdot & \cdot & \cdot & \cdot \\
\cdot & \cdot & \cdot & \cdot & \cdot & \cdot & \cdot & \cdot & \cdot & \cdot & \cdot & \cdot & \cdot & \cdot & \cdot & \cdot \\
\cdot & \cdot & \cdot & \cdot & \cdot & \cdot & \cdot & \cdot & \cdot & \cdot & \cdot & \cdot & \cdot & \cdot & \cdot & \cdot \\
\cdot & \cdot & \cdot & \cdot & \cdot & \cdot & \cdot & \cdot & \cdot & \cdot & \cdot & \cdot & \cdot & \cdot & \cdot & \cdot
\end{array}
\right.
$$

$$
\left.\begin{array}{ccccccccccccccc}
\cdot & \cdot & \cdot & \cdot & \cdot & \cdot & \cdot & \cdot & \cdot & \cdot & \cdot & \cdot & \cdot & \cdot & \cdot \\
\cdot & \cdot & \cdot & \cdot & \cdot & \cdot & \cdot & \cdot & \cdot & \cdot & \cdot & \cdot & \cdot & \cdot & \cdot \\
\cdot & \cdot & \cdot & \cdot & \cdot & \cdot & \cdot & \cdot & \cdot & \cdot & \cdot & \cdot & \cdot & \cdot & \cdot \\
\cdot & \cdot & \cdot & \cdot & \cdot & \cdot & \cdot & \cdot & \cdot & \cdot & \cdot & \cdot & \cdot & \cdot & \cdot \\
\cdot & \cdot & \cdot & \cdot & \cdot & \cdot & \cdot & \cdot & \cdot & \cdot & \cdot & \cdot & \cdot & \cdot & \cdot \\
\cdot & \cdot & \cdot & \cdot & \cdot & \cdot & \cdot & \cdot & \cdot & \cdot & \cdot & \cdot & \cdot & \cdot & \cdot \\
\cdot & \cdot & \cdot & \cdot & \cdot & \cdot & \cdot & \cdot & \cdot & \cdot & \cdot & \cdot & \cdot & \cdot & \cdot \\
\cdot & \cdot & \cdot & \cdot & \cdot & \cdot & \cdot & \cdot & \cdot & \cdot & \cdot & \cdot & \cdot & \cdot & \cdot \\
\cdot & \cdot & \cdot & \cdot & \cdot & \cdot & \cdot & \cdot & \cdot & \cdot & \cdot & \cdot & \cdot & \cdot & \cdot \\
\cdot & \cdot & \cdot & \cdot & \cdot & \cdot & \cdot & \cdot & \cdot & \cdot & \cdot & \cdot & \cdot & \cdot & \cdot \\
\cdot & \cdot & \cdot & \cdot & \cdot & \cdot & \cdot & \cdot & \cdot & \cdot & \cdot & \cdot & \cdot & \cdot & \cdot \\
\cdot & \cdot & \cdot & \cdot & \cdot & \cdot & \cdot & \cdot & \cdot & \cdot & \cdot & \cdot & \cdot & \cdot & \cdot \\
\cdot & \cdot & \cdot & \cdot & \cdot & \cdot & \cdot & \cdot & \cdot & \cdot & \cdot & \cdot & \cdot & \cdot & \cdot \\
\cdot & \cdot & \cdot & \cdot & \cdot & \cdot & \cdot & \cdot & \cdot & \cdot & \cdot & \cdot & \cdot & \cdot & \cdot \\
\cdot & \cdot & \cdot & \cdot & \cdot & \cdot & \cdot & \cdot & \cdot & \cdot & \cdot & \cdot & \cdot & \cdot & \cdot \\
\cdot & \cdot & \cdot & \cdot & \cdot & \cdot & \cdot & \cdot & \cdot & \cdot & \cdot & \cdot & \cdot & \cdot & \cdot \\
.45 & .6 & -.45 & -.6 & \cdot & \cdot & \cdot & \cdot & \cdot & \cdot & \cdot & \cdot & \cdot & \cdot & \cdot \\
.6 & .8 & -.6 & -.8 & \cdot & \cdot & \cdot & \cdot & \cdot & \cdot & \cdot & \cdot & \cdot & \cdot & \cdot \\
-.45 & -.6 & .45 & .6 & \cdot & \cdot & \cdot & \cdot & \cdot & \cdot & \cdot & \cdot & \cdot & \cdot & \cdot \\
-.6 & -.8 & .6 & .8 & \cdot & \cdot & \cdot & \cdot & \cdot & \cdot & \cdot & \cdot & \cdot & \cdot & \cdot \\
\cdot & \cdot & \cdot & \cdot & .45 & -.6 & -.45 & .6 & \cdot & \cdot & \cdot & \cdot & \cdot & \cdot & \cdot \\
\cdot & \cdot & \cdot & \cdot & -.6 & .8 & .6 & -.8 & \cdot & \cdot & \cdot & \cdot & \cdot & \cdot & \cdot \\
\cdot & \cdot & \cdot & \cdot & -.45 & .6 & .45 & -.6 & \cdot & \cdot & \cdot & \cdot & \cdot & \cdot & \cdot \\
\cdot & \cdot & \cdot & \cdot & .6 & -.8 & -.6 & .8 & \cdot & \cdot & \cdot & \cdot & \cdot & \cdot & \cdot \\
\cdot & \cdot & \cdot & \cdot & \cdot & \cdot & \cdot & \cdot & 1 & 0 & -1 & 0 & \cdot & \cdot & \cdot \\
\cdot & \cdot & \cdot & \cdot & \cdot & \cdot & \cdot & \cdot & 0 & 0 & 0 & 0 & \cdot & \cdot & \cdot \\
\cdot & \cdot & \cdot & \cdot & \cdot & \cdot & \cdot & \cdot & -1 & 0 & 1 & 0 & \cdot & \cdot & \cdot \\
\cdot & \cdot & \cdot & \cdot & \cdot & \cdot & \cdot & \cdot & 0 & 0 & 0 & 0 & \cdot & \cdot & \cdot \\
\cdot & \cdot & \cdot & \cdot & \cdot & \cdot & \cdot & \cdot & \cdot & \cdot & \cdot & \cdot & \infty & \cdot & \cdot \\
\cdot & \cdot & \cdot & \cdot & \cdot & \cdot & \cdot & \cdot & \cdot & \cdot & \cdot & \cdot & \cdot & \infty & \cdot \\
\cdot & \cdot & \cdot & \cdot & \cdot & \cdot & \cdot & \cdot & \cdot & \cdot & \cdot & \cdot & \cdot & \cdot & \infty
\end{array}\right]
\begin{pmatrix}
d_1^{(1)} \\ d_2^{(1)} \\ d_3^{(1)} \\ d_4^{(1)} \\
d_1^{(2)} \\ d_2^{(2)} \\ d_3^{(2)} \\ d_4^{(2)} \\
d_1^{(3)} \\ d_2^{(3)} \\ d_3^{(3)} \\ d_4^{(3)} \\
d_1^{(4)} \\ d_2^{(4)} \\ d_3^{(4)} \\ d_4^{(4)} \\
d_1^{(5)} \\ d_2^{(5)} \\ d_3^{(5)} \\ d_4^{(5)} \\
d_1^{(6)} \\ d_2^{(6)} \\ d_3^{(6)} \\ d_4^{(6)} \\
d_1^{(7)} \\ d_2^{(7)} \\ d_3^{(7)} \\ d_4^{(7)} \\
d_c^{(1)} \\ d_c^{(2)} \\ d_c^{(3)}
\end{pmatrix}
\qquad \text{(B)}
$$

knits the collection of members and constraints together and forces them to behave as a structural unit.

This result may be accomplished mathematically by developing statements of nodal equilibrium and compatibility and using these statements to transform Eq. B into a set of structural force-displacement relationships. Nodal equilibrium relationships were previously developed for this structure in Chap. 2 and listed as Eq. G on page 24, and these relationships were then represented symbolically as

$$\mathbf{R} = \mathbf{A}\bar{\mathbf{r}} \tag{2.16}$$

Nodal compatibility relationships were also previously developed for this structure and listed as Eq. A on page 36, these relationships being represented symbolically as

$$\bar{\mathbf{d}} = \mathbf{A}^T\mathbf{D} \tag{2.22}$$

Substituting from Eq. 2.22 into 4.13 and then substituting that result into Eq. 2.16 gives

$$\mathbf{R} = \mathbf{A}\bar{\mathbf{k}}\mathbf{A}^T\mathbf{D} \tag{4.14}$$

or

$$\mathbf{R} = \mathbf{K}\mathbf{D} \tag{4.14a}$$

Notice that the structure stiffness matrix $\mathbf{K}$ will be symmetric about the main diagonal because $\mathbf{A}\bar{\mathbf{k}}\mathbf{A}^T$ is a symmetric transformation of a symmetric matrix. Carrying out the operations indicated in Eq. 4.14 in detail gives

$$\begin{pmatrix} R_1 \\ R_2 \\ R_3 \\ R_4 \\ R_5 \\ R_6 \\ R_7 \\ R_8 \\ R_9 \\ R_{10} \end{pmatrix} = \frac{EA_0}{3} \begin{bmatrix} \infty & .6 & -1 & 0 & \cdot & \cdot & -.45 & -.6 & \cdot & \cdot \\ .6 & \infty & 0 & 0 & \cdot & \cdot & -.6 & -.8 & \cdot & \cdot \\ -1 & 0 & 2.9 & 0 & -1 & 0 & -.45 & .6 & -.45 & -.6 \\ 0 & 0 & 0 & 1.6 & 0 & 0 & .6 & -.8 & -.6 & -.8 \\ \cdot & \cdot & -1 & 0 & 1.45 & -.6 & \cdot & \cdot & -.45 & .6 \\ \cdot & \cdot & 0 & 0 & -.6 & \infty & \cdot & \cdot & .6 & -.8 \\ -.45 & -.6 & -.45 & .6 & \cdot & \cdot & 1.9 & 0 & -1 & 0 \\ -.6 & -.8 & .6 & -.8 & \cdot & \cdot & 0 & 1.6 & 0 & 0 \\ \cdot & \cdot & -.45 & -.6 & -.45 & .6 & -1 & 0 & 1.9 & 0 \\ \cdot & \cdot & -.6 & -.8 & .6 & -.8 & 0 & 0 & 0 & 1.6 \end{bmatrix} \begin{pmatrix} D_1 \\ D_2 \\ D_3 \\ D_4 \\ D_5 \\ D_6 \\ D_7 \\ D_8 \\ D_9 \\ D_{10} \end{pmatrix} \tag{C}$$

Solving this relationship symbolically for the structure nodal displacements gives

$$\mathbf{D} = \mathbf{K}^{-1}\mathbf{R} \tag{4.15}$$

A computer solution for these displacements with the use of SMIS3 is presented on the next printout. Rounding off results,° we get

$$\mathbf{D} = \frac{3}{EA_0}\langle -5\times 10^{-10} \quad -6.67\times 10^{-10} \quad -5\times 10^{-10} \quad -17.19 \quad 2.5 \quad -3.33\times 10^{-10} \quad .3472 \quad -8.594 \quad -4.653 \quad -9.531\rangle \tag{D}$$

The presence of the arrays **A** and $\mathbf{A}^T$ in Eq. 4.14 indicates mathematically the enforcement of nodal equilibrium and nodal compatibility in the development of the structure force-displacement relationships from the set of individual member and constraint force-displacement relationships. Physically, the nodal equilibrium and compatibility relationships are required to knit the separate members and constraints into an integrated structural unit.

For member-end displacements for the collection of all members and constraints, backsubstituting from Eq. 4.15 into 2.22 gives

$$\bar{\mathbf{d}} = \mathbf{A}^T\mathbf{K}^{-1}\mathbf{R} \tag{4.16}$$

For member-end and constraint forces for the collection of all members and constraints, backsubstituting this result into Eq. 4.13 gives

$$\bar{\mathbf{r}} = \bar{\mathbf{k}}\mathbf{A}^T\mathbf{K}^{-1}\mathbf{R} \tag{4.17}$$

For interpreting results, it is convenient to have the member-end forces in local coordinates rather than global coordinates, as in Eq. 4.17. Equation 2.8 provides a relationship between member-end forces in global coordinates and member-end forces in local coordinates as

$$\mathbf{r}' = \boldsymbol{\alpha}^T\mathbf{r} \tag{2.8}$$

To emphasize that this relationship applies to a single member, we add the superscript n and obtain

$$\mathbf{r}'^{(n)} = \boldsymbol{\alpha}^{(n)T}\mathbf{r}^{(n)} \tag{4.18}$$

Since we are presently permitting constraints to exist only in global coordinate directions, we may write

$$r_c'^{(n)} = r_c^{(n)} \tag{4.19}$$

°To accomplish this solution numerically using the computer, it is necessary to use very large numbers for ∞ in Eq. C, as mentioned earlier. In this case, inspection of the computer input and output listing on the pages that follow will show that ∞ was approximated by the number 10^{10}, where the next largest number in the coefficient array is 2.9. It is worth mention here that the presence of relatively large coefficients in a set of simultaneous equations can cause the equations to be numerically ill-conditioned to the point where an accurate solution becomes difficult, or even impossible, to achieve. Conditions under which this difficulty takes place will be discussed in Sec. 5.4, as will ways to avoid the difficulty.

Because the supports have been attributed a finite stiffness in this solution, all nodal displacements will be nonzero, as seen in Eq. D. It is true that those small displacements that occur only as a result of support distortions (D_1, D_2, D_3, and D_6 in this case) are in themselves insignificant and negligible. However, they are retained in the displacement vector for this example because they will play a significant role in the backsubstitution process that follows.

****** *Computer Input:*

```
START
REMARK 4
SOLUTION FOR THE STRUCTURE NODAL DISPLACEMENTS USING
THE STRUCTURE STIFFNESS MATRIX OF THE EXAMPLE TRUSS

**** UNITS ARE KILONEWTONS AND METERS ****
LOAD K 10 10
1E10 .6 -1 3*0 -.45 -.6 0 0 .6 1E10 4*0 -.6 -.8 0
0 -1 0 2.9 0 -1 0 -.45 .6 -.45 -.6 3*0 1.6 0 0 .6
-.8 -.6 -.8 0 0 -1 0 1.45 -.6 0 0 -.45 .6 4*0 -.6
1E10 0 0 .6 -.8 -.45 -.6 -.45 .6 0 0 1.9 0 -1 0 -.6
-.8 .6 -.8 0 0 0 1.6 0 0 0 0 -.45 -.6 -.45 .6 -1 0
1.9 0 0 0 -.6 -.8 .6 -.8 0 0 0 1.6
PRINT K 2
CONSTRAINED STRUCTURE STIFFNESS MATRIX, K, WITH
CONSTANT EA(0)/3 FACTORED OUT
SYMINV K
PRINT K 2
THIS IS THE MATRIX, K INVERSE, WITH THE CONSTANT
3/EA(0) FACTORED OUT
LOAD R 10 1
0 0 0 -10 0 0 -5 0 0 0
PRINT R 1
THIS IS THE LOAD VECTOR, R
MULT K R D
PRINT D 2
THIS IS THE VECTOR, D, OF STRUCTURE NODAL DISPLACEMENTS
WITH THE CONSTANT 3/EA(0) FACTORED OUT
STOP
```

****** *Computer Output:*

```
START

REMARK      4

SOLUTION FOR THE STRUCTURE NODAL DISPLACEMENTS USING
THE STRUCTURE STIFFNESS MATRIX OF THE EXAMPLE TRUSS

**** UNITS ARE KILONEWTONS AND METERS ****


LOAD     K            10   10
   10 ROWS,     10 COLUMNS

PRINT    K             2


CONSTRAINED STRUCTURE STIFFNESS MATRIX, K, WITH
CONSTANT EA(0)/3 FACTORED OUT


              1          2          3          4          5          6
 1   1.0000E+10 6.0000E-01-1.0000E+00          0          0          0
 2   6.0000E-01 1.0000E+10          0          0          0          0
 3  -1.0000E+00          0 2.9000E+00          0-1.0000E+00          0
 4            0          0          0 1.6000E+00          0          0
 5            0          0-1.0000E+00          0 1.4500E+00-6.0000E-01
 6            0          0          0          0-6.0000E-01 1.0000E+10
 7  -4.5000E-01-6.0000E-01-4.5000E-01 6.0000E-01          0          0
 8  -6.0000E-01-8.0000E-01 6.0000E-01-8.0000E-01          0          0
 9            0          0-4.5000E-01-6.0000E-01-4.5000E-01 6.0000E-01
10            0          0-6.0000E-01-8.0000E-01 6.0000E-01-8.0000E-01
```

****** *Computer Output (continued):*

```
           7          8          9         10
 1 -4.5000E-01-6.0000E-01          0          0
 2 -6.0000E-01-8.0000E-01          0          0
 3 -4.5000E-01 6.0000E-01-4.5000E-01-6.0000E-01
 4  6.0000E-01-8.0000E-01-6.0000E-01-8.0000E-01
 5          0          0-4.5000E-01 6.0000E-01
 6          0          0 6.0000E-01-8.0000E-01
 7  1.9000E+00          0-1.0000E+00          0
 8          0 1.6000E+00          0          0
 9 -1.0000E+00          0 1.9000E+00          0
10          0          0          0 1.6000E+00


SYMINV  K

PRINT   K              2


THIS IS THE MATRIX, K INVERSE, WITH THE CONSTANT
3/EA(0) FACTORED OUT


           1          2          3          4          5          6
 1  1.0000E-10-1.6997E-34 1.0000E-10 1.9532E-24 1.0000E-10 3.1743E-35
 2 -1.6997E-34 1.0000E-10-6.3331E-25 5.0000E-11-1.1374E-24-2.4074E-35
 3  1.0000E-10-6.3331E-25 1.0000E+00-3.7500E-01 1.0000E+00 4.6529E-25
 4  1.9532E-24 5.0000E-11-3.7500E-01 2.0938E+00-7.5000E-01 5.0000E-11
 5  1.0000E-10-1.1374E-24 1.0000E+00-7.5000E-01 2.0000E+00 6.2039E-25
 6  3.1743E-35-2.4074E-35 4.6529E-25 5.0000E-11 6.2039E-25 1.0000E-10
 7  1.0000E-10 3.3333E-11 7.5000E-01-7.5000E-01 1.0000E+00-3.3333E-11
 8  2.3586E-24 7.5000E-11-5.6250E-01 1.1875E+00-7.5000E-01 2.5000E-11
 9  1.0000E-10 3.3333E-11 7.5000E-01 1.3323E-14 1.0000E+00-3.3333E-11
10  1.2935E-24 2.5000E-11-1.8750E-01 1.1875E+00-7.5000E-01 7.5000E-11
           7          8          9         10
 1  1.0000E-10 2.3586E-24 1.0000E-10 1.2935E-24
 2  3.3333E-11 7.5000E-11 3.3333E-11 2.5000E-11
 3  7.5000E-01-5.6250E-01 7.5000E-01-1.8750E-01
 4 -7.5000E-01 1.1875E+00 1.3323E-14 1.1875E+00
 5  1.0000E+00-7.5000E-01 1.0000E+00-7.5000E-01
 6 -3.3333E-11 2.5000E-11-3.3333E-11 7.5000E-11
 7  1.4306E+00-6.5625E-01 9.3056E-01-4.6875E-01
 8 -6.5625E-01 1.4297E+00-2.8125E-01 6.6406E-01
 9  9.3056E-01-2.8125E-01 1.4306E+00-9.3750E-02
10 -4.6875E-01 6.6406E-01-9.3750E-02 1.4297E+00


LOAD     R            10     1
   10 ROWS,      1 COLUMNS

PRINT   R              1


THIS IS THE LOAD VECTOR, R


           1
 1          0
 2          0
 3          0
 4 -1.0000E+01
 5          0
 6          0
 7 -5.0000E+00
 8          0
 9          0
10          0
```

****** *Computer Output (continued):*

```
MULT    K       R       D

PRINT   D           2


THIS IS THE VECTOR, D, OF STRUCTURE NODAL DISPLACEMENTS
WITH THE CONSTANT 3/EA(0) FACTORED OUT


            1
 1 -5.0000E-10
 2 -6.6667E-10
 3 -5.0008E-10
 4 -1.7188E+01
 5  2.5000E+00
 6 -3.3333E-10
 7  3.4722E-01
 8 -8.5938E+00
 9 -4.6528E+00
10 -9.5313E+00

STOP
```

indicating that local and global constraint forces are identical. Now, writing these relationships for the collection of all members and constraints gives

$$\begin{pmatrix} \mathbf{r}'^{(1)} \\ \mathbf{r}'^{(2)} \\ \mathbf{r}'^{(3)} \\ \mathbf{r}'^{(4)} \\ \mathbf{r}'^{(5)} \\ \mathbf{r}'^{(6)} \\ \mathbf{r}'^{(7)} \\ r_c'^{(1)} \\ r_c'^{(2)} \\ r_c'^{(3)} \end{pmatrix} = \begin{bmatrix} \boldsymbol{\alpha}^{(1)T} & & & & & & & & & \\ & \boldsymbol{\alpha}^{(2)T} & & & & & & & & \\ & & \boldsymbol{\alpha}^{(3)T} & & & & & \Theta & & \\ & & & \boldsymbol{\alpha}^{(4)T} & & & & & & \\ & & & & \boldsymbol{\alpha}^{(5)T} & & & & & \\ & & & & & \boldsymbol{\alpha}^{(6)T} & & & & \\ & & & & & & \boldsymbol{\alpha}^{(7)T} & & & \\ & \Theta & & & & & & 1 & & \\ & & & & & & & & 1 & \\ & & & & & & & & & 1 \end{bmatrix} \begin{pmatrix} \mathbf{r}^{(1)} \\ \mathbf{r}^{(2)} \\ \mathbf{r}^{(3)} \\ \mathbf{r}^{(4)} \\ \mathbf{r}^{(5)} \\ \mathbf{r}^{(6)} \\ \mathbf{r}^{(7)} \\ r_c^{(1)} \\ r_c^{(2)} \\ r_c^{(3)} \end{pmatrix} \qquad \text{(E)}$$

or

$$\bar{\mathbf{r}}' = \bar{\boldsymbol{\alpha}}^T \bar{\mathbf{r}} \qquad (4.20)$$

Substituting from Eq. 4.17 into 4.20 yields

$$\bar{\mathbf{r}}' = \bar{\boldsymbol{\alpha}}^T \bar{\mathbf{k}} \mathbf{A}^T \mathbf{K}^{-1} \mathbf{R} \qquad (4.21)$$

It may be seen from Eq. 2.6 that

$$\boldsymbol{\alpha}^{(n)T} = \begin{bmatrix} \cos\alpha & \sin\alpha & 0 & 0 \\ -\sin\alpha & \cos\alpha & 0 & 0 \\ 0 & 0 & \cos\alpha & \sin\alpha \\ 0 & 0 & -\sin\alpha & \cos\alpha \end{bmatrix} \tag{F}$$

Specializing Eq. F appropriately gives

$$\boldsymbol{\alpha}^{(1)T} = \boldsymbol{\alpha}^{(2)T} = \boldsymbol{\alpha}^{(7)T} = \begin{bmatrix} 1 & 0 & 0 & 0 \\ 0 & 1 & 0 & 0 \\ 0 & 0 & 1 & 0 \\ 0 & 0 & 0 & 1 \end{bmatrix}$$

$$\boldsymbol{\alpha}^{(3)T} = \boldsymbol{\alpha}^{(5)T} = \begin{bmatrix} .6 & .8 & 0 & 0 \\ -.8 & .6 & 0 & 0 \\ 0 & 0 & .6 & .8 \\ 0 & 0 & -.8 & .6 \end{bmatrix} \qquad \boldsymbol{\alpha}^{(4)T} = \boldsymbol{\alpha}^{(6)T} = \begin{bmatrix} .6 & -.8 & 0 & 0 \\ .8 & .6 & 0 & 0 \\ 0 & 0 & .6 & -.8 \\ 0 & 0 & .8 & .6 \end{bmatrix}$$

Substituting these arrays into Eq. E produces G, displayed on pages 110 and 111 and provides a detailed $\overline{\boldsymbol{\alpha}}^T$ matrix for this problem. Carrying out the matrix products indicated on the right-hand side of Eq. 4.21 yields

$$\begin{aligned} \bar{\mathbf{r}}' = \langle 0 \;\; 0 \;\; 0 \;\; 0 \;\; -2.5 \;\; 0 \;\; 2.5 \;\; 0 \;\; 8.33 \;\; 0 \;\; -8.33 \;\; 0 \;\; -8.33 \\ 0 \;\; 8.33 \;\; 0 \;\; -4.17 \;\; 0 \;\; 4.17 \;\; 0 \;\; 4.17 \;\; 0 \;\; -4.17 \;\; 0 \;\; 5 \\ 0 \;\; -5 \;\; 0 \;\; -5 \;\; -6.67 \;\; -3.33 \rangle \end{aligned} \tag{H}$$

where the retention of the very small nonzero displacements associated with support distortions has allowed the determination of constraint forces. When multiplied by the large support stiffnesses during backsubstitution, these insignificantly small displacements produce the significant constraint force results shown. These small displacements do not significantly influence member forces, of course, because the stiffnesses of the members are relatively small. The results in Eq. H are seen to check with the basic member force results previously obtained from equilibrium considerations alone in Sec. 2.5.

Intermediate results that could be obtained in the process of developing the results of Eq. H include*

*For brevity, the small nonzero displacements associated with support distortions are taken as zero except in the constraint displacement vectors, where they will affect subsequent results.

$$
\begin{pmatrix}
r_1'^{(1)} \\ r_2'^{(1)} \\ r_3'^{(1)} \\ r_4'^{(1)} \\
r_1'^{(2)} \\ r_2'^{(2)} \\ r_3'^{(2)} \\ r_4'^{(2)} \\
r_1'^{(3)} \\ r_2'^{(3)} \\ r_3'^{(3)} \\ r_4'^{(3)} \\
r_1'^{(4)} \\ r_2'^{(4)} \\ r_3'^{(4)} \\ r_4'^{(4)} \\
r_1'^{(5)} \\ r_2'^{(5)} \\ r_3'^{(5)} \\ r_4'^{(5)} \\
r_1'^{(6)} \\ r_2'^{(6)} \\ r_3'^{(6)} \\ r_4'^{(6)} \\
r_1'^{(7)} \\ r_2'^{(7)} \\ r_3'^{(7)} \\ r_4'^{(7)} \\
r_c'^{(1)} \\ r_c'^{(2)} \\ r_c'^{(3)}
\end{pmatrix}
=
\left[
\begin{array}{cccccccccccccccc}
1 & 0 & 0 & 0 & \cdot & \cdot & \cdot & \cdot & \cdot & \cdot & \cdot & \cdot & \cdot & \cdot & \cdot & \cdot \\
0 & 1 & 0 & 0 & \cdot & \cdot & \cdot & \cdot & \cdot & \cdot & \cdot & \cdot & \cdot & \cdot & \cdot & \cdot \\
0 & 0 & 1 & 0 & \cdot & \cdot & \cdot & \cdot & \cdot & \cdot & \cdot & \cdot & \cdot & \cdot & \cdot & \cdot \\
0 & 0 & 0 & 1 & \cdot & \cdot & \cdot & \cdot & \cdot & \cdot & \cdot & \cdot & \cdot & \cdot & \cdot & \cdot \\
\cdot & \cdot & \cdot & \cdot & 1 & 0 & 0 & 0 & \cdot & \cdot & \cdot & \cdot & \cdot & \cdot & \cdot & \cdot \\
\cdot & \cdot & \cdot & \cdot & 0 & 1 & 0 & 0 & \cdot & \cdot & \cdot & \cdot & \cdot & \cdot & \cdot & \cdot \\
\cdot & \cdot & \cdot & \cdot & 0 & 0 & 1 & 0 & \cdot & \cdot & \cdot & \cdot & \cdot & \cdot & \cdot & \cdot \\
\cdot & \cdot & \cdot & \cdot & 0 & 0 & 0 & 1 & \cdot & \cdot & \cdot & \cdot & \cdot & \cdot & \cdot & \cdot \\
\cdot & \cdot & \cdot & \cdot & \cdot & \cdot & \cdot & \cdot & .6 & .8 & 0 & 0 & \cdot & \cdot & \cdot & \cdot \\
\cdot & \cdot & \cdot & \cdot & \cdot & \cdot & \cdot & \cdot & -.8 & .6 & 0 & 0 & \cdot & \cdot & \cdot & \cdot \\
\cdot & \cdot & \cdot & \cdot & \cdot & \cdot & \cdot & \cdot & 0 & 0 & .6 & .8 & \cdot & \cdot & \cdot & \cdot \\
\cdot & \cdot & \cdot & \cdot & \cdot & \cdot & \cdot & \cdot & 0 & 0 & -.8 & .6 & \cdot & \cdot & \cdot & \cdot \\
\cdot & \cdot & \cdot & \cdot & \cdot & \cdot & \cdot & \cdot & \cdot & \cdot & \cdot & \cdot & .6 & -.8 & 0 & 0 \\
\cdot & \cdot & \cdot & \cdot & \cdot & \cdot & \cdot & \cdot & \cdot & \cdot & \cdot & \cdot & .8 & .6 & 0 & 0 \\
\cdot & \cdot & \cdot & \cdot & \cdot & \cdot & \cdot & \cdot & \cdot & \cdot & \cdot & \cdot & 0 & 0 & .6 & -.8 \\
\cdot & \cdot & \cdot & \cdot & \cdot & \cdot & \cdot & \cdot & \cdot & \cdot & \cdot & \cdot & 0 & 0 & .8 & .6 \\
\cdot & \cdot & \cdot & \cdot & \cdot & \cdot & \cdot & \cdot & \cdot & \cdot & \cdot & \cdot & \cdot & \cdot & \cdot & \cdot \\
\cdot & \cdot & \cdot & \cdot & \cdot & \cdot & \cdot & \cdot & \cdot & \cdot & \cdot & \cdot & \cdot & \cdot & \cdot & \cdot \\
\cdot & \cdot & \cdot & \cdot & \cdot & \cdot & \cdot & \cdot & \cdot & \cdot & \cdot & \cdot & \cdot & \cdot & \cdot & \cdot \\
\cdot & \cdot & \cdot & \cdot & \cdot & \cdot & \cdot & \cdot & \cdot & \cdot & \cdot & \cdot & \cdot & \cdot & \cdot & \cdot \\
\cdot & \cdot & \cdot & \cdot & \cdot & \cdot & \cdot & \cdot & \cdot & \cdot & \cdot & \cdot & \cdot & \cdot & \cdot & \cdot \\
\cdot & \cdot & \cdot & \cdot & \cdot & \cdot & \cdot & \cdot & \cdot & \cdot & \cdot & \cdot & \cdot & \cdot & \cdot & \cdot \\
\cdot & \cdot & \cdot & \cdot & \cdot & \cdot & \cdot & \cdot & \cdot & \cdot & \cdot & \cdot & \cdot & \cdot & \cdot & \cdot \\
\cdot & \cdot & \cdot & \cdot & \cdot & \cdot & \cdot & \cdot & \cdot & \cdot & \cdot & \cdot & \cdot & \cdot & \cdot & \cdot \\
\cdot & \cdot & \cdot & \cdot & \cdot & \cdot & \cdot & \cdot & \cdot & \cdot & \cdot & \cdot & \cdot & \cdot & \cdot & \cdot \\
\cdot & \cdot & \cdot & \cdot & \cdot & \cdot & \cdot & \cdot & \cdot & \cdot & \cdot & \cdot & \cdot & \cdot & \cdot & \cdot \\
\cdot & \cdot & \cdot & \cdot & \cdot & \cdot & \cdot & \cdot & \cdot & \cdot & \cdot & \cdot & \cdot & \cdot & \cdot & \cdot \\
\cdot & \cdot & \cdot & \cdot & \cdot & \cdot & \cdot & \cdot & \cdot & \cdot & \cdot & \cdot & \cdot & \cdot & \cdot & \cdot \\
\cdot & \cdot & \cdot & \cdot & \cdot & \cdot & \cdot & \cdot & \cdot & \cdot & \cdot & \cdot & \cdot & \cdot & \cdot & \cdot \\
\cdot & \cdot & \cdot & \cdot & \cdot & \cdot & \cdot & \cdot & \cdot & \cdot & \cdot & \cdot & \cdot & \cdot & \cdot & \cdot \\
\cdot & \cdot & \cdot & \cdot & \cdot & \cdot & \cdot & \cdot & \cdot & \cdot & \cdot & \cdot & \cdot & \cdot & \cdot & \cdot
\end{array}
\right.
$$

$$
\left.\begin{array}{ccccccccccccccc}
\cdot & \cdot & \cdot & \cdot & \cdot & \cdot & \cdot & \cdot & \cdot & \cdot & \cdot & \cdot & \cdot & \cdot & \cdot \\
\cdot & \cdot & \cdot & \cdot & \cdot & \cdot & \cdot & \cdot & \cdot & \cdot & \cdot & \cdot & \cdot & \cdot & \cdot \\
\cdot & \cdot & \cdot & \cdot & \cdot & \cdot & \cdot & \cdot & \cdot & \cdot & \cdot & \cdot & \cdot & \cdot & \cdot \\
\cdot & \cdot & \cdot & \cdot & \cdot & \cdot & \cdot & \cdot & \cdot & \cdot & \cdot & \cdot & \cdot & \cdot & \cdot \\
\cdot & \cdot & \cdot & \cdot & \cdot & \cdot & \cdot & \cdot & \cdot & \cdot & \cdot & \cdot & \cdot & \cdot & \cdot \\
\cdot & \cdot & \cdot & \cdot & \cdot & \cdot & \cdot & \cdot & \cdot & \cdot & \cdot & \cdot & \cdot & \cdot & \cdot \\
\cdot & \cdot & \cdot & \cdot & \cdot & \cdot & \cdot & \cdot & \cdot & \cdot & \cdot & \cdot & \cdot & \cdot & \cdot \\
\cdot & \cdot & \cdot & \cdot & \cdot & \cdot & \cdot & \cdot & \cdot & \cdot & \cdot & \cdot & \cdot & \cdot & \cdot \\
\cdot & \cdot & \cdot & \cdot & \cdot & \cdot & \cdot & \cdot & \cdot & \cdot & \cdot & \cdot & \cdot & \cdot & \cdot \\
\cdot & \cdot & \cdot & \cdot & \cdot & \cdot & \cdot & \cdot & \cdot & \cdot & \cdot & \cdot & \cdot & \cdot & \cdot \\
\cdot & \cdot & \cdot & \cdot & \cdot & \cdot & \cdot & \cdot & \cdot & \cdot & \cdot & \cdot & \cdot & \cdot & \cdot \\
\cdot & \cdot & \cdot & \cdot & \cdot & \cdot & \cdot & \cdot & \cdot & \cdot & \cdot & \cdot & \cdot & \cdot & \cdot \\
\cdot & \cdot & \cdot & \cdot & \cdot & \cdot & \cdot & \cdot & \cdot & \cdot & \cdot & \cdot & \cdot & \cdot & \cdot \\
\cdot & \cdot & \cdot & \cdot & \cdot & \cdot & \cdot & \cdot & \cdot & \cdot & \cdot & \cdot & \cdot & \cdot & \cdot \\
\cdot & \cdot & \cdot & \cdot & \cdot & \cdot & \cdot & \cdot & \cdot & \cdot & \cdot & \cdot & \cdot & \cdot & \cdot \\
\cdot & \cdot & \cdot & \cdot & \cdot & \cdot & \cdot & \cdot & \cdot & \cdot & \cdot & \cdot & \cdot & \cdot & \cdot \\
.6 & .8 & 0 & 0 & \cdot & \cdot & \cdot & \cdot & \cdot & \cdot & \cdot & \cdot & \cdot & \cdot & \cdot \\
-.8 & .6 & 0 & 0 & \cdot & \cdot & \cdot & \cdot & \cdot & \cdot & \cdot & \cdot & \cdot & \cdot & \cdot \\
0 & 0 & .6 & .8 & \cdot & \cdot & \cdot & \cdot & \cdot & \cdot & \cdot & \cdot & \cdot & \cdot & \cdot \\
0 & 0 & -.8 & .6 & \cdot & \cdot & \cdot & \cdot & \cdot & \cdot & \cdot & \cdot & \cdot & \cdot & \cdot \\
\cdot & \cdot & \cdot & \cdot & .6 & -.8 & 0 & 0 & \cdot & \cdot & \cdot & \cdot & \cdot & \cdot & \cdot \\
\cdot & \cdot & \cdot & \cdot & .8 & .6 & 0 & 0 & \cdot & \cdot & \cdot & \cdot & \cdot & \cdot & \cdot \\
\cdot & \cdot & \cdot & \cdot & 0 & 0 & .6 & -.8 & \cdot & \cdot & \cdot & \cdot & \cdot & \cdot & \cdot \\
\cdot & \cdot & \cdot & \cdot & 0 & 0 & .8 & .6 & \cdot & \cdot & \cdot & \cdot & \cdot & \cdot & \cdot \\
\cdot & \cdot & \cdot & \cdot & \cdot & \cdot & \cdot & \cdot & 1 & 0 & 0 & 0 & \cdot & \cdot & \cdot \\
\cdot & \cdot & \cdot & \cdot & \cdot & \cdot & \cdot & \cdot & 0 & 1 & 0 & 0 & \cdot & \cdot & \cdot \\
\cdot & \cdot & \cdot & \cdot & \cdot & \cdot & \cdot & \cdot & 0 & 0 & 1 & 0 & \cdot & \cdot & \cdot \\
\cdot & \cdot & \cdot & \cdot & \cdot & \cdot & \cdot & \cdot & 0 & 0 & 0 & 1 & \cdot & \cdot & \cdot \\
\cdot & \cdot & \cdot & \cdot & \cdot & \cdot & \cdot & \cdot & \cdot & \cdot & \cdot & \cdot & 1 & \cdot & \cdot \\
\cdot & \cdot & \cdot & \cdot & \cdot & \cdot & \cdot & \cdot & \cdot & \cdot & \cdot & \cdot & \cdot & 1 & \cdot \\
\cdot & \cdot & \cdot & \cdot & \cdot & \cdot & \cdot & \cdot & \cdot & \cdot & \cdot & \cdot & \cdot & \cdot & 1
\end{array}\right]
\begin{pmatrix}
r_1^{(1)} \\ r_2^{(1)} \\ r_3^{(1)} \\ r_4^{(1)} \\
r_1^{(2)} \\ r_2^{(2)} \\ r_3^{(2)} \\ r_4^{(2)} \\
r_1^{(3)} \\ r_2^{(3)} \\ r_3^{(3)} \\ r_4^{(3)} \\
r_1^{(4)} \\ r_2^{(4)} \\ r_3^{(4)} \\ r_4^{(4)} \\
r_1^{(5)} \\ r_2^{(5)} \\ r_3^{(5)} \\ r_4^{(5)} \\
r_1^{(6)} \\ r_2^{(6)} \\ r_3^{(6)} \\ r_4^{(6)} \\
r_1^{(7)} \\ r_2^{(7)} \\ r_3^{(7)} \\ r_4^{(7)} \\
r_c^{(1)} \\ r_c^{(2)} \\ r_c^{(3)}
\end{pmatrix}
\qquad \text{(G)}
$$

$$\bar{\mathbf{d}} = \frac{3}{EA_0}\langle 0 \quad 0 \quad 0 \quad -17.19 \quad 0 \quad -17.19 \quad 2.5 \quad 0 \quad 0 \quad 0 \quad .3472$$
$$-8.594 \quad .3472 \quad -8.594 \quad 0 \quad -17.19 \quad 0 \quad -17.19 \quad -4.653$$
$$-9.531 \quad -4.653 \quad -9.531 \quad 2.5 \quad 0 \quad .3472 \quad -8.594 \quad -4.653$$
$$-9.531 \quad -5 \times 10^{-10} \quad -6.67 \times 10^{-10} \quad -3.33 \times 10^{-10}\rangle \tag{I}$$

and

$$\bar{\mathbf{r}} = \langle 0 \quad 0 \quad 0 \quad 0 \quad -2.5 \quad 0 \quad 2.5 \quad 0 \quad 5 \quad 6.67 \quad -5 \quad -6.67$$
$$-5 \quad 6.67 \quad 5 \quad -6.67 \quad -2.5 \quad -3.33 \quad 2.5 \quad 3.33 \quad 2.5$$
$$-3.33 \quad -2.5 \quad 3.33 \quad 5 \quad 0 \quad -5 \quad 0 \quad -5 \quad -6.67 \quad -3.33\rangle \tag{K}$$

4.3.2 Coding Method

The symbolic method is once again seen to involve the manipulation of large arrays, the majority of whose elements are zeros. Thus, although it provides an elegant presentation of the concepts of the matrix displacement method of structural analysis, its inefficiency dictates that a more effective method of building the structure force-displacement relationships be sought. A detailed inspection of the relationships of member, constraint, and structure stiffness coefficients will now be shown to lead to such a method.

Figure 4.4a and b show the coordinate numberings for the example truss structure and a typical member of a truss structure, respectively. Figure 4.4c shows the forces on the nodes of the example truss that must be present under conditions that $D_6 = 1$ and all other structure nodal displacements are zero. These forces are seen to have been designated symbolically according to the definition of a structure stiffness coefficient. That is, the force in direction 9 on the truss due to the stated unit displacement in direction 6 is K_{96}. Figure 4.4c also shows the forces induced at the ends of the individual members of the truss due to the stated displacement. These forces are represented in the notation for a member stiffness coefficient and are obviously determined according to the definition of such a stiffness coefficient. For example, member 6 undergoes a unit end displacement in direction 4 (which corresponds to structure direction 6). It follows that the force induced in, say, direction 1 on member 6 will be $k_{14}^{(6)}$. (Notice that $d_4^{(6)} = D_6$ is an enforcement of a nodal compatibility condition.) Continuing in this fashion, we could verify the representation of each force shown in Fig. 4.4c. Next, it is necessary to consider nodal equilibrium. The FBDs for each joint of the truss are shown in Fig. 4.5. Summing forces in the x and y directions on node 1 (shown in Fig. 4.5a) gives, respectively,

$$K_{16} = 0 \quad \text{and} \quad K_{26} = 0$$

This result indicates that there is no member of the truss extending from node 1 to node 3. (Structure coordinates 1 and 2 are at node 1, while structure coordinate 6 is at node 3.) There is only one way that a force can be induced in coordinate direction 1 (or 2) due to a unit displacement in direction 6 while all other structure

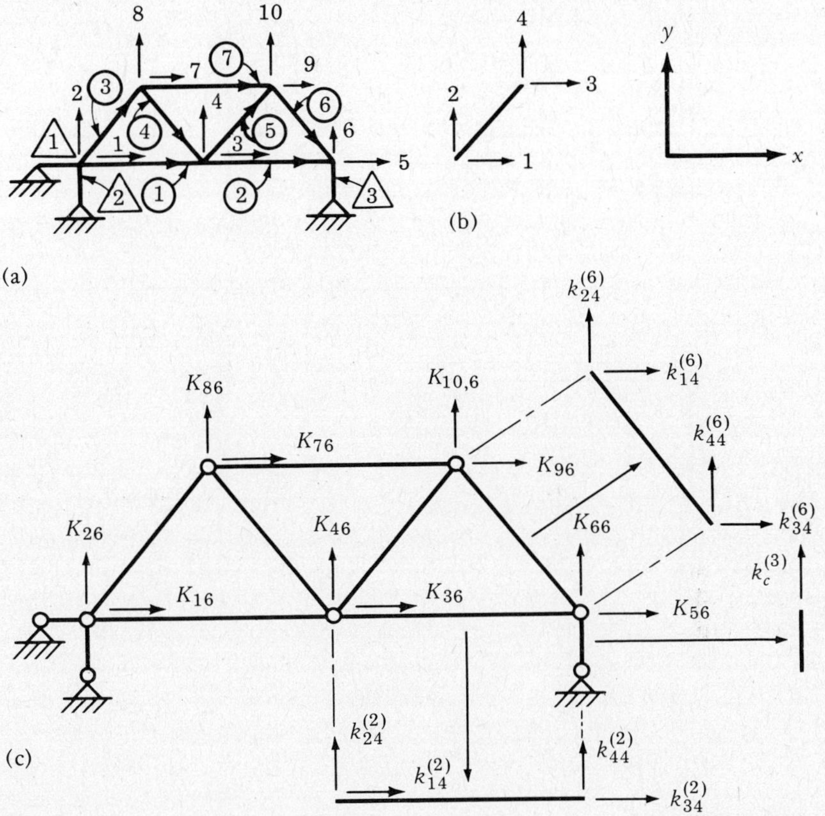

Fig. 4.4 Inspection schematic for $\mathbf{D}_6 = 1$ ($\mathbf{D}_m = 0$ for $m \neq 6$): (a) coordinate, member, and constraint numbering; (b) member coordinates; (c) structure nodal forces, member-end forces, and constraint forces.

nodal displacements are held fixed: Some *direct* physical tie must exist between coordinates 1 (or 2) and coordinate 6. This tie does not exist here. Now, summing forces in both the x and y directions on the remaining structure nodes, as shown in Fig. 4.5, gives

$$\left.\begin{aligned} K_{36} &= k_{14}^{(2)} \\ K_{46} &= k_{24}^{(2)} \end{aligned}\right\} \text{Nodes 2 and 3 tied together by member 2}$$

$$\left.\begin{aligned} K_{56} &= k_{34}^{(2)} + k_{34}^{(6)} \\ K_{66} &= k_{44}^{(2)} + k_{44}^{(6)} + k_c^{(3)} \end{aligned}\right\} \text{Node 3 attached to members 2 and 6 and to constraint 3 in direction 6}$$

$$\left.\begin{aligned} K_{76} &= 0 \\ K_{86} &= 0 \end{aligned}\right\} \text{Nodes 3 and 4 not connected}$$

$$\left.\begin{aligned} K_{96} &= k_{14}^{(6)} \\ K_{10,6} &= k_{24}^{(6)} \end{aligned}\right\} \text{Nodes 3 and 5 tied together by member 6}$$

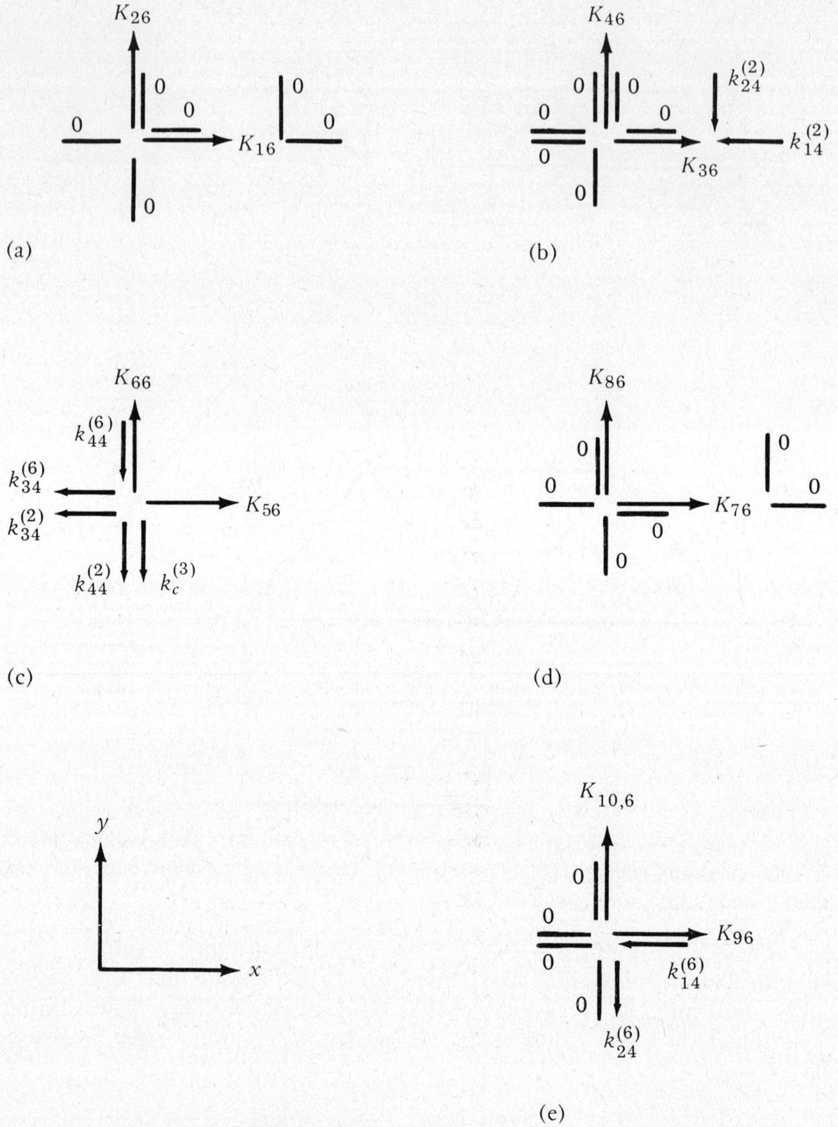

Fig. 4.5 Joint free body diagrams for example truss: (a) node 1; (b) node 2; (c) node 3; (d) node 4; (e) node 5.

Substituting into the above expressions from the member stiffness matrices on page 100 gives

$$K_{16} = 0$$

$$K_{26} = 0$$

$$K_{36} = 0$$

$$K_{46} = 0$$

$$K_{56} = 0 + \frac{EA_0}{3}(-.6) = \frac{EA_0}{3}(-.6)$$

$$K_{66} = 0 + \frac{EA_0}{3}(.8) + \frac{EA_0}{3}(\infty) = \frac{EA_0}{3}(\infty)$$

$$K_{76} = 0$$

$$K_{86} = 0$$

$$K_{96} = \frac{EA_0}{3}(.6)$$

$$K_{10,6} = \frac{EA_0}{3}(-.8)$$

These coefficients are seen to be identical with those in the sixth column of the structure stiffness matrix (shown in Eq. C on page 104).

If a unit displacement were now considered to occur in turn in each of the remaining structure coordinate directions, and the appropriate nodal compatibility and equilibrium conditions were enforced, the entire structure stiffness matrix could be developed by inspection. But it seems more appropriate at this time to use the inspection concept as a basis for developing an even more efficient method for accomplishing this task.

4.3.2.1 COORDINATE CODING

The inspection concept involves an identification of member and structure coordinate corespondences as the basis for determining the contributions of member stiffness arrays to the structure stiffness matrix. The coordinate coding method is a formalization of this concept. The coordinate code for the example truss appears to be identical with the code listed on page 30 for building the a matrix for this structure; however, its use is now somewhat different. This difference is to be expected, since the code will now be used to enforce both nodal equilibrium and nodal compatibility, whereas before (in building the a matrix) only nodal equilibrium conditions needed to be enforced. Following is the coordinate code for this example.

Coordinate Code

Member and support numbers ↓	1	2	3	4 ← Member global coordinates
1	1	2	3	4
2	3	4	5	6
3	1	2	7	8
4	7	8	3	4
5	3	4	9	10
6	9	10	5	6
7	7	8	9	10

} Corresponding structure global coordinates

As will soon be seen, the coordinate code for each member identifies the ultimate location, or "destination," of each member stiffness coefficient in the structure stiffness array. Thus, in the context of the displacement method, the coordinate code for a member is sometimes referred to as a "destination vector." We will use these two designations for a member's code interchangeably.

A formal mechanism is needed also for identifying which support stiffness coefficients contribute to which structure stiffness coefficients. This constraint information is contained in the following list.

Constraint List

Constraint number	Structure coordinate number
1	1
2	2
3	6

As an example of the use of this constraint list, the third line indicates that the third constraint is in the sixth structure coordinate direction. Thus the stiffness of the third constraint, $k_c^{(3)}$, must be added to K_{66}.

Use of the preceding destination vectors and constraint list gives

$$K_{11} = k_{11}^{(1)} + k_{11}^{(3)} + k_c^{(1)}$$

$$K_{12} = k_{12}^{(1)} + k_{12}^{(3)}$$

$$K_{13} = k_{13}^{(1)}$$

$$K_{14} = k_{14}^{(1)}$$

$$K_{15} = 0$$

$$K_{16} = 0$$

$$K_{17} = k_{13}^{(3)}$$

$$K_{18} = k_{14}^{(3)}$$

$$K_{19} = 0$$

$$K_{1,10} = 0$$

$$K_{22} = k_{22}^{(1)} + k_{22}^{(3)} + k_c^{(2)}$$

$$K_{23} = k_{23}^{(1)}$$

$$K_{24} = k_{24}^{(1)}$$

$$K_{25} = 0$$

$$K_{26} = 0$$

$$K_{27} = k_{23}^{(3)}$$

$$K_{28} = k_{24}^{(3)}$$

$$K_{29} = 0$$

$$K_{2,10} = 0$$

$$K_{33} = k_{33}^{(1)} + k_{11}^{(2)} + k_{33}^{(4)} + k_{11}^{(5)}$$

$$K_{34} = k_{34}^{(1)} + k_{12}^{(2)} + k_{34}^{(4)} + k_{12}^{(5)}$$

and so on.

Substituting into these results from the member and support stiffness matrices on page 100 gives

$$K_{11} = \frac{EA_0}{3}(1 + .45 + \infty) = \frac{EA_0}{3}(\infty)$$

$$K_{12} = \frac{EA_0}{3}(0 + .6 + 0 + 0) = \frac{EA_0}{3}(.6)$$

$$K_{13} = \frac{EA_0}{3}(-1)$$

$$K_{14} = 0$$

$$K_{22} = \frac{EA_0}{3}(0 + .8 + \infty) = \frac{EA_0}{3}(\infty)$$

$$K_{23} = 0$$

$$K_{33} = \frac{EA_0}{3}(1 + 1 + .45 + .45) = \frac{EA_0}{3}(2.9)$$

$$K_{34} = \frac{EA_0}{3}(0 + 0 + [-.6] + .6) = 0$$

and so on.

These results are seen to agree with Eq. C on page 104.

The use of such a coding technique to "directly" build the structure stiffness relationships has come to be known as the direct stiffness method. It is, of course, a displacement method. For automatic computation, a coordinate code and constraint list like the ones displayed in this section may be read into the computer and used to place the member and support stiffness coefficients properly in the structure stiffness matrix. Alternatively, a more compact nodal code may be read and used to develop the destination vectors of the coordinate code internally in the computer.

4.3.2.2 NODAL CODING

To gain an understanding of how and why the nodal coding method works, it is useful to first observe typical member stiffness arrays with their respective coordinate codes displayed thereon. Consider, for example, the stiffness arrays for members 2 and 6 for the example truss:

Member 2: Member 6:

$$\begin{array}{c} \\ 3 \\ 4 \\ 5 \\ 6 \end{array} \frac{EA_0}{3}\begin{array}{c} \begin{array}{cccc} 3 & 4 & 5 & 6 \end{array} \\ \begin{bmatrix} 1 & 0 & -1 & 0 \\ 0 & 0 & 0 & 0 \\ -1 & 0 & 1 & 0 \\ 0 & 0 & 0 & 0 \end{bmatrix} \end{array} \qquad \begin{array}{c} \\ 9 \\ 10 \\ 5 \\ 6 \end{array} \frac{EA_0}{3}\begin{array}{c} \begin{array}{cccc} 9 & 10 & 5 & 6 \end{array} \\ \begin{bmatrix} .45 & -.6 & -.45 & .6 \\ -.6 & .8 & .6 & -.8 \\ -.45 & .6 & .45 & -.6 \\ .6 & -.8 & -.6 & .8 \end{bmatrix} \end{array}$$

Because of the consistent relationship between structure coordinate numbering and structure nodal numbering, the stiffness coefficients from each quadrant of the member stiffness matrices will be added to the structure stiffness matrix as an undisturbed unit. This phenomenon is emphasized by rewriting the preceding member stiffness arrays in partitioned form along with their nodal codes:

Member 2: Member 6:

$$\begin{array}{c} \\ \boxed{2} \\ \\ \boxed{3} \end{array} \frac{EA_0}{3}\begin{array}{c} \begin{array}{cc} \boxed{2} & \boxed{3} \end{array} \\ \left[\begin{array}{cc|cc} 1 & 0 & -1 & 0 \\ 0 & 0 & 0 & 0 \\ \hline -1 & 0 & 1 & 0 \\ 0 & 0 & 0 & 0 \end{array}\right] \end{array} \qquad \begin{array}{c} \\ \boxed{5} \\ \\ \boxed{3} \end{array} \frac{EA_0}{3}\begin{array}{c} \begin{array}{cc} \boxed{5} & \boxed{3} \end{array} \\ \left[\begin{array}{cc|cc} .45 & -.6 & -.45 & .6 \\ .6 & .8 & -.6 & -.8 \\ \hline -.45 & .6 & .45 & -.6 \\ .6 & -.8 & -.6 & .8 \end{array}\right] \end{array}$$

As was the case with the force method of Chap. 2, the nodal code for each member is seen to consist simply of the member incidences for that member.

Only members 2 and 6 meet at node 3. By using the nodal code, we find their contributions to structure stiffness submatrix $\mathbf{K}_{33}$ to be

$$\mathbf{K}_{33} = \mathbf{k}_{jj}^{(2)} + \mathbf{k}_{jj}^{(6)}$$

or

$$\mathbf{K}_{33} = \frac{EA_0}{3}\begin{bmatrix} 1 & 0 \\ 0 & 0 \end{bmatrix} + \frac{EA_0}{3}\begin{bmatrix} .45 & -.6 \\ -.6 & .8 \end{bmatrix} = \frac{EA_0}{3}\begin{bmatrix} 1.45 & -.6 \\ -.6 & .8 \end{bmatrix}$$

In addition, inspection of Fig. 4.3 reveals that the third structure node is constrained in its second coordinate direction. Thus the stiffness coefficient in the 2,2 position of the $\mathbf{K}_{33}$ subarray must be augmented with $k_c^{(3)} = (EA_0/3)\infty$, giving

$$\mathbf{K}_{33} = \frac{EA_0}{3}\begin{bmatrix} 1.45 & -.6 \\ -.6 & \infty \end{bmatrix}$$

This result is seen to be the structural stiffness submatrix relating the nodal displacements at node 3 to the nodal forces at node 3, as may be observed by inspecting Eq. C in Sec. 4.3.1. We could now develop the remainder of the structure stiffness matrix in a similar manner. But instead of doing this, we will turn our attention to a class of structures that is of more interest to analyze by the displacement method—statically indeterminate structures. We will then further demonstrate the nodal coding method by using it to build the structure stiffness relationships for an example of this type of structure.

4.4 STATICALLY INDETERMINATE STRUCTURES

Consider the example truss shown in Fig. 4.6. One interesting thing you will notice as this example is analyzed is that it is no more difficult to analyze a statically indeterminate structure by the displacement method than it was to analyze a statically determinate structure by that method. This finding may be somewhat surprising, but it is consistent with the discussion of Chap. 1. And it is important, for it means that even highly redundant structures can be analyzed by this method without any penalty of added complications for added redundancies. This fact, plus the ease of building the structure force-displacement relationships by the direct stiffness method, accounts for the popularity and wide use of this method of analysis.

4.4.1 Statically Indeterminate Truss

The example truss of Fig. 4.6 is, of course, identical with that of Fig. 4.3 with the exception of added members 8 and 9. It is therefore possible to obtain the structure force-displacement relationships for the statically indeterminate truss by superposing the stiffnesses of members 8 and 9 on the structure stiffness matrix for the statically determinate truss displayed in Eq. C of Sec. 4.3.1. The member stiffness matrices for members 8 and 9 (taking E to be constant for all truss members again and taking $A^{(8)} = A^{(9)} = A_0$) are

$$\mathbf{k}^{(8)} = \frac{EA_0}{4.924}\begin{bmatrix} .835 & .371 & -.835 & -.371 \\ .371 & .165 & -.371 & -.165 \\ -.835 & -.371 & .835 & .371 \\ -.371 & -.165 & .371 & .165 \end{bmatrix}$$

$$= \frac{EA_0}{3}\begin{bmatrix} .509 & .226 & -.509 & -.226 \\ .226 & .101 & -.226 & -.101 \\ -.509 & -.226 & .509 & .226 \\ -.226 & -.101 & .226 & .101 \end{bmatrix}$$

$$\mathbf{k}^{(9)} = \frac{EA_0}{3}\begin{bmatrix} .509 & -.226 & -.509 & .226 \\ -.226 & .101 & .226 & -.101 \\ -.509 & .226 & .509 & -.226 \\ .226 & -.101 & -.226 & .101 \end{bmatrix}$$

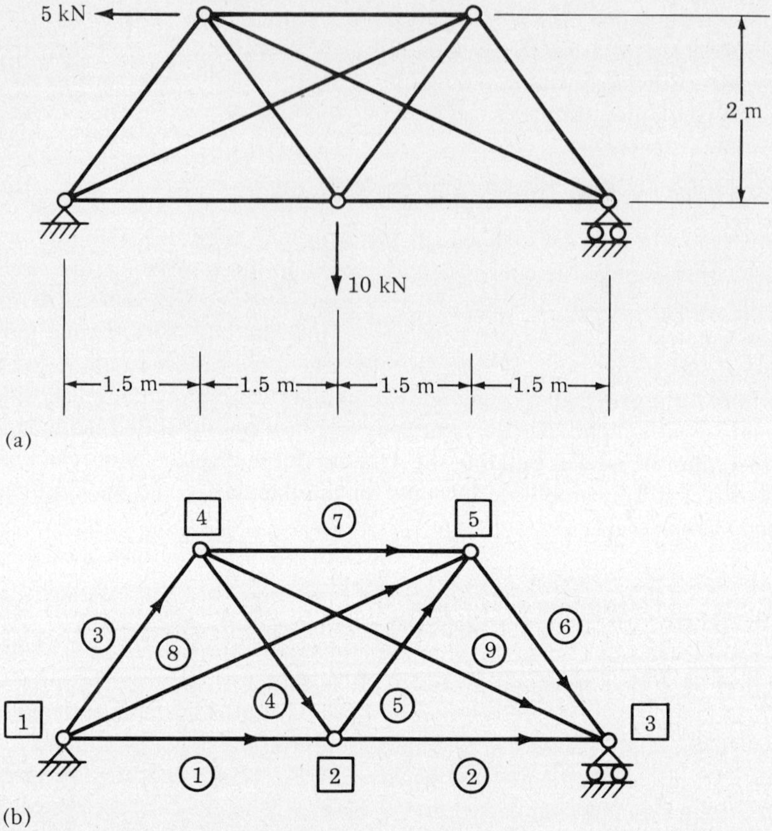

Fig. 4.6 Statically indeterminate truss: (a) dimensions and loading; (b) member and nodal numbering and member directions.

The nodal codes for all members of the statically indeterminate truss are tabulated as follows:

Nodal Code

Member numbers ↓ ; ← Member ends

Member numbers	i	j
1	1	2
2	2	3
3	1	4
4	4	2
5	2	5
6	5	3
7	4	5
8	1	5
9	4	3

} Member incidences

The codes for members 8 and 9 indicate that the submatrices $\mathbf{K}_{11}$, $\mathbf{K}_{15}$, $\mathbf{K}_{55}$, $\mathbf{K}_{51}$, $\mathbf{K}_{44}$, $\mathbf{K}_{43}$, $\mathbf{K}_{33}$, and $\mathbf{K}_{34}$ of the statically determinate structure stiffness matrix will receive added stiffness from members 8 and 9 of the statically indeterminate truss. Thus, by superposition, we obtain

$$\mathbf{K}_{11} \to \mathbf{K}_{11} + \mathbf{k}_{ii}^{(8)} = \frac{EA_0}{3}\begin{bmatrix} \infty + .509 & .6 + .226 \\ .6 + .226 & \infty + .101 \end{bmatrix} = \frac{EA_0}{3}\begin{bmatrix} \infty & .826 \\ .826 & \infty \end{bmatrix}$$

$$\mathbf{K}_{15} \to \mathbf{K}_{15} + \mathbf{k}_{ij}^{(8)} = \frac{EA_0}{3}\begin{bmatrix} 0 + (-.509) & 0 + (-.226) \\ 0 + (-.226) & 0 + (-.101) \end{bmatrix} = \frac{EA_0}{3}\begin{bmatrix} -.509 & -.226 \\ -.226 & -.101 \end{bmatrix}$$

$$\mathbf{K}_{51} = \mathbf{K}_{15}^T$$

$$\mathbf{K}_{55} \to \mathbf{K}_{55} + \mathbf{k}_{jj}^{(8)} = \frac{EA_0}{3}\begin{bmatrix} 1.9 + .509 & 0 + .226 \\ 0 + .226 & 1.6 + .101 \end{bmatrix} = \frac{EA_0}{3}\begin{bmatrix} 2.41 & .226 \\ .226 & 1.70 \end{bmatrix}$$

$$\mathbf{K}_{44} \to \mathbf{K}_{44} + \mathbf{k}_{ii}^{(9)} = \frac{EA_0}{3}\begin{bmatrix} 1.9 + .509 & 0 - .226 \\ 0 - .226 & 1.6 + .101 \end{bmatrix} = \frac{EA_0}{3}\begin{bmatrix} 2.41 & -.226 \\ -.226 & .170 \end{bmatrix}$$

$$\mathbf{K}_{43} \to \mathbf{K}_{43} + \mathbf{k}_{ij}^{(9)} = \frac{EA_0}{3}\begin{bmatrix} 0 - .509 & 0 + .226 \\ 0 + .226 & 0 - .101 \end{bmatrix} = \frac{EA_0}{3}\begin{bmatrix} -.509 & .226 \\ .226 & -.101 \end{bmatrix}$$

$$\mathbf{K}_{34} = \mathbf{K}_{43}^T$$

$$\mathbf{K}_{33} \to \mathbf{K}_{33} + \mathbf{k}_{jj}^{(9)} = \frac{EA_0}{3}\begin{bmatrix} 1.45 + .509 & -.6 - .226 \\ -.6 - .226 & \infty + .101 \end{bmatrix} = \frac{EA_0}{3}\begin{bmatrix} 1.96 & -.826 \\ -.826 & \infty \end{bmatrix}$$

The force-displacement relationships for the statically indeterminate truss may now be obtained by rewriting Eq. C (Sec. 4.3.1) with the preceding submatrix replacements. The result is

$$\begin{pmatrix} R_1 \\ R_2 \\ R_3 \\ R_4 \\ R_5 \\ R_6 \\ R_7 \\ R_8 \\ R_9 \\ R_{10} \end{pmatrix} = \frac{EA_0}{3}\begin{bmatrix} \infty & .826 & -1 & 0 & \cdot & \cdot & -.45 & -.6 & -.509 & -.226 \\ .826 & \infty & 0 & 0 & \cdot & \cdot & -.6 & -.8 & -.226 & -.101 \\ -1 & 0 & 2.9 & 0 & -1 & 0 & -.45 & .6 & -.45 & -.6 \\ 0 & 0 & 0 & 1.6 & 0 & 0 & .6 & -.8 & -.6 & -.8 \\ \cdot & \cdot & -1 & 0 & 1.96 & -.826 & -.509 & .226 & -.45 & .6 \\ \cdot & \cdot & 0 & 0 & -.826 & \infty & .226 & -.101 & .6 & -.8 \\ -.45 & -.6 & -.45 & .6 & -.509 & .226 & 2.41 & -.226 & -1 & 0 \\ -.6 & -.8 & .6 & -.8 & .226 & -.101 & -.226 & 1.70 & 0 & 0 \\ -.509 & -.226 & -.45 & -.6 & -.45 & .6 & -1 & 0 & 2.41 & .226 \\ -.226 & -.101 & -.6 & -.8 & .6 & -.8 & 0 & 0 & .226 & 1.70 \end{bmatrix}\begin{pmatrix} D_1 \\ D_2 \\ D_3 \\ D_4 \\ D_5 \\ D_6 \\ D_7 \\ D_8 \\ D_9 \\ D_{10} \end{pmatrix} \quad \text{(A)}$$

or

$$\mathbf{R} = \mathbf{KD} \tag{4.14a}$$

It is important to note that normally Eq. A would have been developed directly from the set of all member and constraint force-displacement relationships rather than by modifying the statically determinate force-displacement relationships as was just done. Our approach was used here merely as a matter of convenience, since the force-displacement relationships for a statically determinate form of the structure had already been developed in a previous example.

Solving Eq. 4.14a gives

$$\mathbf{D} = \mathbf{K}^{-1}\mathbf{R} \tag{4.15}$$

A detailed solution of Eq. A has been carried out using SMIS3 and is displayed in the next printout. The **D** vector is seen from these results to be*

$$\mathbf{D} = \frac{3}{EA_0}\langle -5\times 10^{-10} \quad -6.67\times 10^{-10} \quad 1.648 \quad -15.61 \quad 4.476 \quad -3.33\times 10^{-10} \quad 1.774 \quad -8.28 \quad -1.24 \quad -8.172\rangle \tag{B}$$

The backsubstitution process can also be expedited by use of either the coordinate code or the nodal code. For example, if we use the nodal code of page 120, we see that

$$\mathbf{d}^{(1)} = \begin{pmatrix} d_1^{(1)} \\ d_2^{(1)} \\ \hline d_3^{(1)} \\ d_4^{(1)} \end{pmatrix} = \underbrace{\begin{pmatrix} \mathbf{d}_i^{(1)} \\ \hline \mathbf{d}_j^{(1)} \end{pmatrix} = \begin{pmatrix} \mathbf{D}_1 \\ \hline \mathbf{D}_2 \end{pmatrix}}_{\text{Member incidences}} = \frac{3}{EA_0}\begin{pmatrix} -5\times 10^{-10} \\ -6.67\times 10^{-10} \\ \hline 1.648 \\ -15.61 \end{pmatrix}$$

The member incidences indicate that displacements at ends i and j of member 1 are compatible with displacements at structure nodes 1 and 2, respectively. Continuing in this fashion but without displaying all of the details, we get

$$\mathbf{d}^{(2)} = \frac{3}{EA_0}\langle 1.648 \quad -15.61 \quad 4.476 \quad -3.33\times 10^{-10}\rangle$$

$$\mathbf{d}^{(3)} = \frac{3}{EA_0}\langle -5\times 10^{-10} \quad -6.67\times 10^{-10} \quad 1.774 \quad -8.28\rangle$$

$$\mathbf{d}^{(4)} = \frac{3}{EA_0}\langle 1.774 \quad -8.28 \quad 1.648 \quad -15.61\rangle$$

$$\mathbf{d}^{(5)} = \frac{3}{EA_0}\langle 1.648 \quad -15.61 \quad -1.24 \quad -8.172\rangle$$

$$\mathbf{d}^{(6)} = \frac{3}{EA_0}\langle -1.24 \quad -8.172 \quad 4.476 \quad -3.33\times 10^{-10}\rangle$$

*See footnote on page 105.

****** *Computer Input:*

```
START
REMARK 5
SOLUTION FOR THE STRUCTURE NODAL DISPLACEMENTS
USING THE STRUCTURE STIFFNESS MATRIX OF THE
STATICALLY INDETERMINATE TRUSS

**** UNITS ARE KILONEWTONS AND METERS ****
LOAD K 10 10
 1E10 .826 -1 0 0 0 -.45 -.6 -.509 -.226 .826 1E10
0 0 0 0 -.6 -.8 -.226 -.101 -1 0 2.9 0 -1 0 -.45 .6
-.45 -.6 0 0 0 1.6 0 0 .6 -.8 -.6 -.8 0 0 -1 0 1.959
 -.826 -.509 .226 -.45 .6 0 0 0 0 -.826 1E10 .226
-.101 .6 -.8 -.45 -.6 -.45 .6 -.509 .226 2.409 -.226
-1 0 -.6 -.8 .6 -.8 .226 -.101 -.226 1.701 0 0 -.509
-.226 -.45 -.6 -.45 .6 -1 0 2.409 .226 -.226 -.101
-.6 -.8 .6 -.8  0 0 .226 1.701
PRINT K 2
THIS IS THE CONSTRAINED STRUCTURE STIFFNESS MATRIX, K,
WITH CONSTANT EA(0)/3 FACTORED OUT
LOAD R 10 1
0 0 0 -10 0 0 -5 0 0 0
PRINT R 1
THIS IS THE LOAD VECTOR, R
SOLVE K R D
PRINT D 2
THIS IS THE VECTOR, D, OF STRUCTURE NODAL DISPLACEMENTS
WITH THE CONSTANT 3/EA(0) FACTORED OUT
STOP
```

****** *Computer Output:*

```
START

REMARK       5


SOLUTION FOR THE STRUCTURE NODAL DISPLACEMENTS
USING THE STRUCTURE STIFFNESS MATRIX OF THE
STATICALLY INDETERMINATE TRUSS

**** UNITS ARE KILONEWTONS AND METERS ****



LOAD     K            10    10
   10 ROWS,     10 COLUMNS

PRINT    K             2


THIS IS THE CONSTRAINED STRUCTURE STIFFNESS MATRIX, K,
WITH CONSTANT EA(0)/3 FACTORED OUT


              1           2           3           4           5           6
 1  1.0000E+10 8.2600E-01-1.0000E+00           0           0           0
 2  8.2600E-01 1.0000E+10           0           0           0           0
 3 -1.0000E+00           0 2.9000E+00           0-1.0000E+00           0
 4           0           0           0 1.6000E+00           0           0
 5           0           0-1.0000E+00           0 1.9590E+00-8.2600E-01
 6           0           0           0           0-8.2600E-01 1.0000E+10
 7 -4.5000E-01-6.0000E-01-4.5000E-01 6.0000E-01-5.0900E-01 2.2600E-01
 8 -6.0000E-01-8.0000E-01 6.0000E-01-8.0000E-01 2.2600E-01-1.0100E-01
 9 -5.0900E-01-2.2600E-01-4.5000E-01-6.0000E-01-4.5000E-01 6.0000E-01
10 -2.2600E-01-1.0100E-01-6.0000E-01-8.0000E-01 6.0000E-01-8.0000E-01
```

****** *Computer Output (continued):*

```
              7             8            9           10
 1 -4.5000E-01-6.0000E-01-5.0900E-01-2.2600E-01
 2 -6.0000E-01-8.0000E-01-2.2600E-01-1.0100E-01
 3 -4.5000E-01 6.0000E-01-4.5000E-01-6.0000E-01
 4  6.0000E-01-8.0000E-01-6.0000E-01-8.0000E-01
 5 -5.0900E-01 2.2600E-01-4.5000E-01 6.0000E-01
 6  2.2600E-01-1.0100E-01 6.0000E-01-8.0000E-01
 7  2.4090E+00-2.2600E-01-1.0000E+00           0
 8 -2.2600E-01 1.7010E+00            0           0
 9 -1.0000E+00            0 2.4090E+00 2.2600E-01
10            0            0 2.2600E-01 1.7010E+00


LOAD     R           10    1
   10 ROWS,      1 COLUMNS

PRINT    R            1


THIS IS THE LOAD VECTOR, R


              1
 1            0
 2            0
 3            0
 4 -1.0000E+01
 5            0
 6            0
 7 -5.0000E+00
 8            0
 9            0
10            0


SOLVE    K       R       D

PRINT    D            2


THIS IS THE VECTOR, D, OF STRUCTURE NODAL DISPLACEMENTS
WITH THE CONSTANT 3/EA(0) FACTORED OUT


              1
 1 -5.0000E-10
 2 -6.6660E-10
 3  1.6484E+00
 4 -1.5607E+01
 5  4.4757E+00
 6 -3.3340E-10
 7  1.7735E+00
 8 -8.2804E+00
 9 -1.2402E+00
10 -8.1724E+00


STOP
```

$$\mathbf{d}^{(7)} = \frac{3}{EA_0}\langle 1.774 \quad -8.28 \quad -1.24 \quad -8.172\rangle$$

$$\mathbf{d}^{(8)} = \frac{3}{EA_0}\langle -5\times 10^{-10} \quad -6.67\times 10^{-10} \quad -1.24 \quad -8.172\rangle$$

$$\mathbf{d}^{(9)} = \frac{3}{EA_0}\langle 1.774 \quad -8.28 \quad 4.476 \quad -3.33\times 10^{-10}\rangle$$

From the constraint list of page 116, it is further seen that

$$d_c^{(1)} = \frac{3}{EA_0}\langle -5\times 10^{-10}\rangle$$

$$d_c^{(2)} = \frac{3}{EA_0}\langle -6.67\times 10^{-10}\rangle$$

$$d_c^{(3)} = \frac{3}{EA_0}\langle -3.33\times 10^{-10}\rangle$$

Now, using Eq. 4.12 along with the individual member stiffness matrices listed on pages 100 and 119, we can obtain the member-end forces in global coordinates. Equation 4.12 is repeated:

$$\mathbf{r}^{(n)} = \mathbf{k}^{(n)}\mathbf{d}^{(n)} \tag{4.12}$$

Substituting gives

$$\mathbf{r}^{(1)} = \begin{pmatrix} r_1^{(1)} \\ r_2^{(1)} \\ r_3^{(1)} \\ r_4^{(1)} \end{pmatrix} = \frac{EA_0}{3}\begin{bmatrix} 1 & 0 & -1 & 0 \\ 0 & 0 & 0 & 0 \\ -1 & 0 & 1 & 0 \\ 0 & 0 & 0 & 0 \end{bmatrix}\frac{3}{EA_0}\begin{pmatrix} 0 \\ 0 \\ 1.648 \\ -15.61 \end{pmatrix} = \begin{pmatrix} -1.648 \\ 0 \\ 1.648 \\ 0 \end{pmatrix}$$

Continuing in this fashion gives

$$\mathbf{r}^{(2)} = \langle -2.83 \quad 0 \quad 2.83 \quad 0\rangle$$

$$\mathbf{r}^{(3)} = \langle 4.17 \quad 5.56 \quad -4.17 \quad -5.56\rangle$$

$$\mathbf{r}^{(4)} = \langle -4.34 \quad 5.79 \quad 4.34 \quad -5.79\rangle$$

$$\mathbf{r}^{(5)} = \langle -3.16 \quad -4.22 \quad 3.16 \quad 4.22\rangle$$

$$\mathbf{r}^{(6)} = \langle 2.33 \quad -3.11 \quad -2.33 \quad 3.11\rangle$$

$$\mathbf{r}^{(7)} = \langle 3.01 \quad 0 \quad -3.01 \quad 0\rangle$$

$$\mathbf{r}^{(8)} = \langle 2.48 \quad 1.10 \quad -2.48 \quad -1.10\rangle$$

$$\mathbf{r}^{(9)} = \langle .50 \quad -.226 \quad -.50 \quad .226\rangle$$

Constraint forces may be similarly found using Eq. 4.10a as

$$r_c^{(1)} = \langle -5 \rangle$$

$$r_c^{(2)} = \langle -6.67 \rangle$$

$$r_c^{(3)} = \langle -3.33 \rangle$$

It is left to you to verify that these forces satisfy nodal equilibrium.

Finally, Eq. 4.18 may be used to obtain the member-end forces in local coordinates for stress analysis purposes.

$$\mathbf{r}'^{(n)} = \boldsymbol{\alpha}^{(n)T}\mathbf{r}^{(n)} \tag{4.18}$$

Substituting from this equation and from the arrays on page 109 gives

$$\mathbf{r}'^{(1)} = \begin{pmatrix} r_1'^{(1)} \\ r_2'^{(1)} \\ r_3'^{(1)} \\ r_4'^{(1)} \end{pmatrix} = \begin{bmatrix} 1 & 0 & 0 & 0 \\ 0 & 1 & 0 & 0 \\ 0 & 0 & 1 & 0 \\ 0 & 0 & 0 & 1 \end{bmatrix} \begin{pmatrix} -1.648 \\ 0 \\ 1.648 \\ 0 \end{pmatrix} = \begin{pmatrix} -1.648 \\ 0 \\ 1.648 \\ 0 \end{pmatrix}$$

The remaining local member-end forces may be obtained in a similar fashion.

4.4.2 Statically Indeterminate Frame

As a second example of analysis of statically indeterminate structures by the direct stiffness method, we will now consider the rigid-jointed framework of Fig. 4.7. But before carrying out a detailed direct stiffness solution, we must first generalize the symbolic approach to include distributed member loadings. The frame-member stiffness relationships in global coordinates are expressed symbolically in Eq. 4.8a as

$$\mathbf{r} = \mathbf{kd} + \mathbf{r}_F \tag{4.8a}$$

The constraint stiffness relationships are of the form of Eq. 4.11a:

$$r_c = k_c d_c \tag{4.11a}$$

To emphasize that these equations apply to the nth member and nth constraints, respectively, they may be rewritten as

$$\begin{aligned} \mathbf{r}^{(n)} &= \mathbf{k}^{(n)}\mathbf{d}^{(n)} + \mathbf{r}_F^{(n)} \\ r_c^{(n)} &= k_c^{(n)} d_c^{(n)} \end{aligned} \tag{4.22}$$

Now, writing equations of this type for all members and constraints of a frame and collecting them into a single matrix equation, we obtain

$$\bar{\mathbf{r}} = \overline{\mathbf{kd}} + \bar{\mathbf{r}}_F \tag{4.23}$$

The conditions of nodal equilibrium and compatibility may be represented, as before, as

$$\mathbf{R} = \mathbf{A}\bar{\mathbf{r}} \tag{2.16}$$

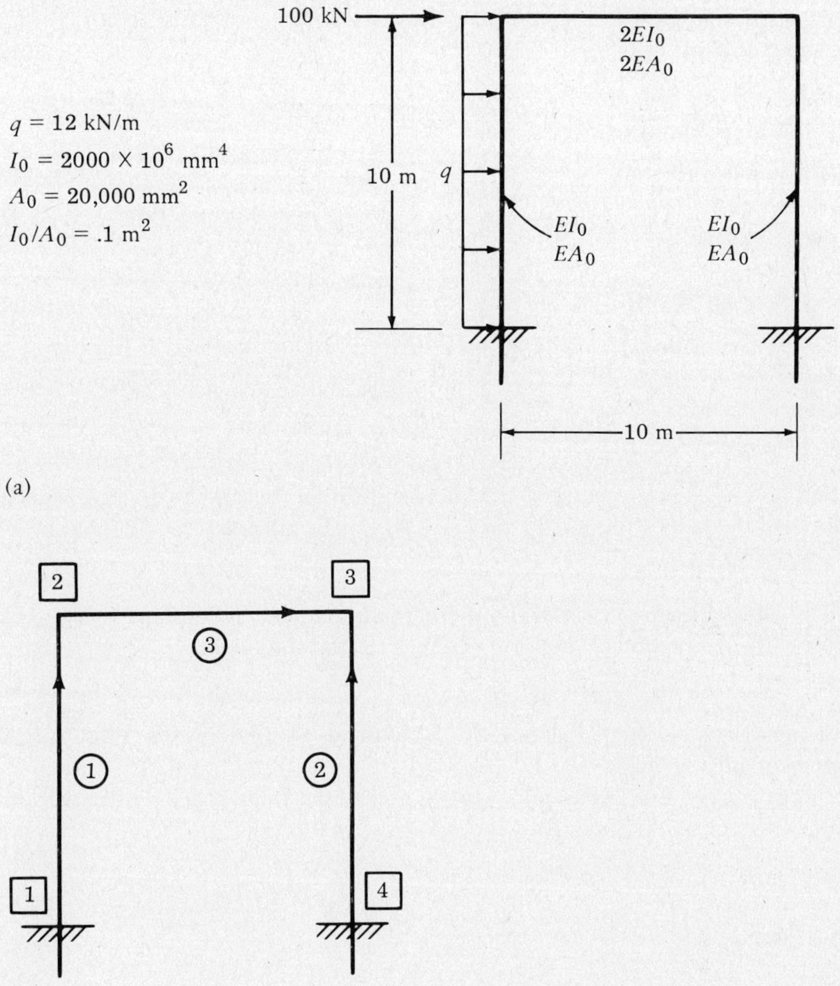

Fig. 4.7 Statically indeterminate frame: (a) dimensions and loading; (b) numbering.

and

$$\bar{\mathbf{d}} = \mathbf{A}^T\mathbf{D} \tag{2.22}$$

respectively. Substituting from Eq. 2.22 into 4.23 and then substituting that result into Eq. 2.16 gives

$$\mathbf{R} = \mathbf{A}\bar{\mathbf{k}}\mathbf{A}^T\mathbf{D} + \mathbf{A}\bar{\mathbf{r}}_F \tag{4.24}$$

or

$$\mathbf{R} = \mathbf{K}\mathbf{D} + \mathbf{R}_F \tag{4.25}$$

where

$$\mathbf{K} = \mathbf{A}\bar{\mathbf{k}}\mathbf{A}^T \quad \text{and} \quad \mathbf{R}_F = \mathbf{A}\bar{\mathbf{r}}_F \tag{4.26}$$

Solving Eq. 4.25 gives

$$\mathbf{D} = \mathbf{K}^{-1}(\mathbf{R} - \mathbf{R}_F) \tag{4.27}$$

Backsubstituting sequentially into Eqs. 2.22, 4.23, and an equation of the form of Eq. 4.20 gives, respectively,

$$\overline{\mathbf{d}} = \mathbf{A}^T\mathbf{K}^{-1}(\mathbf{R} - \mathbf{R}_F) \tag{4.28}$$

$$\overline{\mathbf{r}} = \overline{\mathbf{k}}\mathbf{A}^T\mathbf{K}^{-1}(\mathbf{R} - \mathbf{R}_F) + \overline{\mathbf{r}}_F \tag{4.29}$$

and

$$\overline{\mathbf{r}}' = \overline{\boldsymbol{\alpha}}^T\overline{\mathbf{k}}\mathbf{A}^T\mathbf{K}^{-1}(\mathbf{R} - \mathbf{R}_F) + \overline{\boldsymbol{\alpha}}^T\overline{\mathbf{r}}_F \tag{4.30}$$

It will now be demonstrated once again that the influence of the member stiffness relationships on the structure relationships may be determined by coding:

$$\mathbf{r}^{(n)} = \mathbf{k}^{(n)}\mathbf{d}^{(n)} + \mathbf{r}_F^{(n)} \xrightarrow{\text{coding}} \mathbf{R} = \mathbf{K}\mathbf{D} + \mathbf{R}_F$$

It will also be shown again that the backsubstitution process can be significantly shortened by use of coding to extract $\mathbf{d}^{(n)}$ directly from $\mathbf{D}$, or

$$\mathbf{D} \xRightarrow{\text{coding}} \mathbf{d}^{(n)}$$

which allows the direct use of the individual member relationships of Eq. 4.22 along with an equation of the form of Eq. 4.18, and thus yields

$$\mathbf{r}'^{(n)} = \boldsymbol{\alpha}^{T(n)}\mathbf{k}^{(n)}\mathbf{d}^{(n)} + \boldsymbol{\alpha}^{T(n)}\mathbf{r}_F^{(n)} \tag{4.31}$$

It is now appropriate to carry out a detailed solution of the rigid-jointed framework of Fig. 4.7.

Specializing Eqs. 4.9a and 4.9b, we obtain the following member stiffness matrices (in units of kilonewtons and meters):

$$\mathbf{k}^{(1)} = \frac{EI_0}{5} \begin{array}{c} \\ \boxed{1} \\ \\ \\ \boxed{2} \\ \\ \end{array} \left[\begin{array}{ccc|ccc} \multicolumn{3}{c}{\boxed{1}} & \multicolumn{3}{c}{\boxed{2}} \\ .06 & 0 & -.3 & -.06 & 0 & -.3 \\ 0 & 5 & 0 & 0 & -5 & 0 \\ -.3 & 0 & 2 & .3 & 0 & 1 \\ \hline -.06 & 0 & .3 & .06 & 0 & .3 \\ 0 & -5 & 0 & 0 & 5 & 0 \\ -.3 & 0 & 1 & .3 & 0 & 2 \end{array}\right]$$

$$\mathbf{k}^{(2)} = \frac{EI_0}{5} \begin{array}{c} \\ \boxed{4} \\ \\ \\ \boxed{3} \\ \\ \end{array} \left[\begin{array}{ccc|ccc} \multicolumn{3}{c}{\boxed{4}} & \multicolumn{3}{c}{\boxed{3}} \\ .06 & 0 & -.3 & -.06 & 0 & -.3 \\ 0 & 5 & 0 & 0 & -5 & 0 \\ -.3 & 0 & 2 & .3 & 0 & 1 \\ \hline -.06 & 0 & .3 & .06 & 0 & .3 \\ 0 & -5 & 0 & 0 & 5 & 0 \\ -.3 & 0 & 1 & .3 & 0 & 2 \end{array}\right]$$

$$
\mathbf{k}^{(3)} = \frac{EI_0}{5}
\begin{array}{c}
\\ \boxed{2} \\ \\ \\ \boxed{3} \\ \\
\end{array}
\left[\begin{array}{rrr|rrr}
\multicolumn{3}{c|}{\boxed{2}} & \multicolumn{3}{c}{\boxed{3}} \\
10 & 0 & 0 & -10 & 0 & 0 \\
0 & .12 & .6 & 0 & -.12 & .6 \\
0 & .6 & 4 & 0 & -.6 & 2 \\
\hline
-10 & 0 & 0 & 10 & 0 & 0 \\
0 & -.12 & -.6 & 0 & .12 & -.6 \\
0 & .6 & 2 & 0 & -.6 & 4
\end{array}\right]
$$

The fixed-end forces for member 1 may be determined by inspection as

$$
\mathbf{r}_F^{(1)} = \left\langle -\frac{qL}{2} \quad 0 \quad \frac{qL^2}{12} \quad -\frac{qL}{2} \quad 0 \quad -\frac{qL^2}{12} \right\rangle
$$

or

$$
\mathbf{r}_F^{(1)} = \langle \underbrace{-60 \quad 0 \quad 100}_{\boxed{1}} \;\Big|\; \underbrace{-60 \quad 0 \quad -100}_{\boxed{2}} \rangle
$$

As before, we can determine the influence of the supports on the structure stiffness array from a constraint list. To be consistent with the nodal code approach being used here, we also present the constraint list from a nodal point of view.

Constraint List

Constraint number	Structure node	Nodal coordinate
1	1	1
2	1	2
3	1	3
4	4	1
5	4	2
6	4	3

Consider, say, the fifth row of the constraint list; this row indicates that the fifth constraint operates in node 4's second coordinate direction. Thus an appropriate (infinite° in this case) support stiffness must be added to the second-row, second-column position of subarray $\mathbf{K}_{44}$.

Making use of our cumulated information, we may now manually assemble the structure force-displacement relationships. For convenience, the member incidences are displayed both on the member arrays shown earlier and on the structure arrays shown in Eq. A on page 130. Equation A may be represented symbolically as

$$
\mathbf{R} = \mathbf{KD} + \mathbf{R}_F \qquad (4.25)
$$

°See note on page 100.

$$\begin{pmatrix} 0 \\ 0 \\ 0 \\ 100 \\ 0 \\ 0 \\ 0 \\ 0 \\ 0 \\ 0 \\ 0 \\ 0 \end{pmatrix} = \frac{EI_0}{5} \left[\begin{array}{ccc|ccc|ccc|ccc} \infty & 0 & -.3 & -.06 & 0 & -.3 & 0 & 0 & 0 & 0 & 0 & 0 \\ 0 & \infty & 0 & 0 & -5 & 0 & 0 & 0 & 0 & 0 & 0 & 0 \\ -.6 & 0 & \infty & .3 & 0 & 1 & 0 & 0 & 0 & 0 & 0 & 0 \\ -.06 & 0 & .3 & 10.06 & 0 & .3 & -10 & 0 & 0 & 0 & 0 & 0 \\ 0 & -5 & 0 & 0 & 5.12 & .6 & 0 & -.12 & .6 & 0 & 0 & 0 \\ -.3 & 0 & 1 & .3 & .6 & 6 & 0 & -.6 & 2 & 0 & 0 & 0 \\ 0 & 0 & 0 & -10 & 0 & 0 & 10.06 & 0 & .3 & -.06 & 0 & .3 \\ 0 & 0 & 0 & 0 & -.12 & -.6 & 0 & 5.12 & -.6 & 0 & -5 & 0 \\ 0 & 0 & 0 & 0 & .6 & 2 & .3 & -.6 & 6 & -.3 & 0 & 1 \\ 0 & 0 & 0 & 0 & 0 & 0 & -.06 & 0 & -.3 & \infty & 0 & -.3 \\ 0 & 0 & 0 & 0 & 0 & 0 & 0 & -5 & 0 & 0 & \infty & 0 \\ 0 & 0 & 0 & 0 & 0 & 0 & .3 & 0 & 1 & -.3 & 0 & \infty \end{array} \right] \begin{pmatrix} D_1 \\ D_2 \\ D_3 \\ D_4 \\ D_5 \\ D_6 \\ D_7 \\ D_8 \\ D_9 \\ D_{10} \\ D_{11} \\ D_{12} \end{pmatrix} + \begin{pmatrix} -60 \\ 0 \\ 100 \\ -60 \\ 0 \\ -100 \\ 0 \\ 0 \\ 0 \\ 0 \\ 0 \\ 0 \end{pmatrix} \qquad \text{(A)}$$

(Row groups and column groups are labelled by joints 1, 2, 3, 4, each grouping three consecutive rows/columns.)

Solving symbolically gives

$$\mathbf{D} = \mathbf{K}^{-1}(\mathbf{R} - \mathbf{R}_F) \tag{4.27}$$

It is interesting to notice in passing that the total load indicated in Eq. 4.27 consists of a vector **R** of concentrated loads actually applied at the structure nodes, plus a vector $-\mathbf{R}_F$ of concentrated loads that are, in a sense, equivalent to the actual distributed loading on the structure. The concentrated loads $-\mathbf{R}_F$ are "equivalent" to the distributed loading not only statically but also in that they will cause exactly the same nodal displacements of the structure as would the distributed loading. Thus the loads of the vector $-\mathbf{R}_F$ are sometimes referred to as "equivalent nodal loads."

The solution expressed symbolically in Eq. 4.27 has been carried out in detail using the SMIS3 computer program with $\infty = 10^{10}$. This solution is shown in the printout that follows, with the nodal displacements given as*

$$\begin{aligned}\mathbf{D} = \frac{5}{EI_0}\langle 143.9 \times 10^{-10} \quad & 64.38 \times 10^{-10} \quad -541.5 \times 10^{-10} \quad 1618.255 \\ & 12.88 \quad -44.01 \quad 1610.644 \quad -12.88 \quad -68.44 \\ & 76.11 \times 10^{-10} \quad -64.38 \times 10^{-10} \quad -414.8 \times 10^{-10}\rangle\end{aligned} \tag{B}$$

Backsubstituting by nodal coding and the constraint list, gives†

$$\begin{aligned}
\mathbf{d}^{(1)} &= \frac{5}{EI_0}\langle 0 \quad 0 \quad 0 \quad 1618 \quad 12.88 \quad -44.01\rangle \\
\mathbf{d}^{(2)} &= \frac{5}{EI_0}\langle 0 \quad 0 \quad 0 \quad 1611 \quad -12.88 \quad -68.44\rangle \\
\mathbf{d}^{(3)} &= \frac{5}{EI_0}\langle 1618.255 \quad 12.88 \quad -44.01 \quad 1610.644 \quad -12.88 \quad -68.44\rangle \\
d_c^{(1)} &= \frac{5}{EI_0}\langle 143.9 \times 10^{-10}\rangle \\
d_c^{(2)} &= \frac{5}{EI_0}\langle 64.38 \times 10^{-10}\rangle \\
d_c^{(3)} &= \frac{5}{EI_0}\langle -541.5 \times 10^{-10}\rangle \\
d_c^{(4)} &= \frac{5}{EI_0}\langle 76.11 \times 10^{-10}\rangle \\
d_c^{(5)} &= \frac{5}{EI_0}\langle -64.38 \times 10^{-10}\rangle \\
d_c^{(6)} &= \frac{5}{EI_0}\langle -414.8 \times 10^{-10}\rangle
\end{aligned} \tag{C}$$

*As before, the nonzero values of the very small displacements associated with support distortions are retained for their subsequent roles in the backsubstitution process.

†See the note on page 109.

****** *Computer Input:*

```
START
REMARK 4
SOLUTION FOR THE STRUCTURE NODAL DISPLACEMENTS USING THE STRUCTURE
STIFFNESS MATRIX OF THE STATICALLY INDETERMINATE FRAME

**** UNITS ARE KILONEWTONS AND METERS ****
LOAD K 12 12
 1E10 0 -.3 -.06 0 -.3 7*0  1E10 0 0 -5 7*0 -.3 0
1E10 .3 0 1 6*0 -.06 0 .3 10.06 0 .3 -10 6*0 -5 0
0 5.12 .6 0 -.12 .6 3*0 -.3 0 1 .3 .6 6 0 -.6 2
6*0 -10 0 0 10.06 0 .3 -.06 0 .3 4*0 -.12 -.6 0
5.12 -.6 0 -5 5*0 .6 2 .3 -.6 6 -.3 0 1 6*0 -.06 0
-.3 1E10 0 -.3 7*0 -5 0 0 1E10 7*0 .3 0 1 -.3 0 1E10
PRINT K 1
CONSTRAINED STRUCTURE STIFFNESS MATRIX, K, WITH EI(0)/5 FACTORED OUT
LOAD R 12 1
3*0 100 8*0
PRINT R 1
THIS IS THE LOAD VECTOR, R
LOAD RF 12 1
-60 0 100 -60 0 -100 6*0
PRINT RF 1
THIS IS THE VECTOR OF FIXED-END FORCES, RF
SUB R RF
SOLVE K R D
PRINT D 2
THIS IS THE VECTOR OF STRUCTURE NODAL DISPLACEMENTS,
D, WITH THE CONSTANT, 5/EI(0), FACTORED OUT
STOP
```

****** *Computer Output:*

```
START

REMARK      4

SOLUTION FOR THE STRUCTURE NODAL DISPLACEMENTS USING THE STRUCTURE
STIFFNESS MATRIX OF THE STATICALLY INDETERMINATE FRAME

**** UNITS ARE KILONEWTONS AND METERS ****


LOAD    K           12   12
   12 ROWS,     12 COLUMNS

PRINT   K            1
```

CONSTRAINED STRUCTURE STIFFNESS MATRIX, K, WITH EI(0)/5 FACTORED OUT

	1	2	3	4	5	6
1	1.000000E+10	0	-3.000000E-01	-6.000000E-02	0	-3.000000E-01
2	0	1.000000E+10	0	0	-5.000000E+00	0
3	-3.000000E-01	0	1.000000E+10	3.000000E-01	0	1.000000E+00
4	-6.000000E-02	0	3.000000E-01	1.006000E+01	0	3.000000E-01
5	0	-5.000000E+00	0	0	5.120000E+00	6.000000E-01
6	-3.000000E-01	0	1.000000E+00	3.000000E-01	6.000000E-01	6.000000E+00
7	0	0	0	-1.000000E+01	0	0
8	0	0	0	0	-1.200000E-01	-6.000000E-01
9	0	0	0	0	6.000000E-01	2.000000E+00
10	0	0	0	0	0	0
11	0	0	0	0	0	0

****** *Computer Output (continued):*

12	0	0	0	0	0	0

	7	8	9	10	11	12
1	0	0	0	0	0	0
2	0	0	0	0	0	0
3	0	0	0	0	0	0
4	-1.000000E+01	0	0	0	0	0
5	0	-1.200000E-01	6.000000E-01	0	0	0
6	0	-6.000000E-01	2.000000E+00	0	0	0
7	1.006000E+01	0	3.000000E-01	-6.000000E-02	0	3.000000E-01
8	0	5.120000E+00	-6.000000E-01	0	-5.000000E+00	0
9	3.000000E-01	-6.000000E-01	6.000000E+00	-3.000000E-01	0	1.000000E+00
10	-6.000000E-02	0	-3.000000E-01	1.000000E+10	0	-3.000000E-01
11	0	-5.000000E+00	0	0	1.000000E+10	0
12	3.000000E-01	0	1.000000E+00	-3.000000E-01	0	1.000000E+10

```
LOAD    R           12    1
   12 ROWS,      1 COLUMNS

PRINT   R            1

THIS IS THE LOAD VECTOR, R
```

	1
1	0
2	0
3	0
4	1.000000E+02
5	0
6	0
7	0
8	0
9	0
10	0
11	0
12	0

```
LOAD    RF          12    1
   12 ROWS,      1 COLUMNS

PRINT   RF           1

THIS IS THE VECTOR OF FIXED-END FORCES, RF
```

	1
1	-6.000000E+01
2	0
3	1.000000E+02
4	-6.000000E+01
5	0
6	-1.000000E+02
7	0
8	0
9	0
10	0
11	0
12	0

```
SUB     R       RF

SOLVE   K       R       D
```

****** *Computer Output (continued):*

```
PRINT   D           2

THIS IS THE VECTOR OF STRUCTURE NODAL DISPLACEMENTS,
D, WITH THE CONSTANT, 5/EI(0), FACTORED OUT

            1
 1   1.438927E-08
 2   6.437768E-09
 3  -5.414678E-08
 4   1.618255E+03
 5   1.287554E+01
 6  -4.400858E+01
 7   1.610644E+03
 8  -1.287554E+01
 9  -6.843777E+01
10   7.610730E-09
11  -6.437768E-09
12  -4.147554E-08

STOP
```

Note that these displacement results are listed to four significant figures, except in two cases where seven figures are used. This type of listing is necessary because the added significant figures will be needed to obtain good force results during the following backsubstitution process. You may wish to verify this by checking the backsubstitution to obtain $\mathbf{r}^{(3)}$.

Now backsubstituting these results into Eq. 4.8a specialized for each member and Eq. 4.10a specialized for each support, in turn, gives

$$\begin{aligned}
\mathbf{r}^{(1)} &= \langle -83.88 \quad -64.40 \quad 441.4 \quad 83.88 \quad 64.40 \quad 397.4 \rangle \\
&\quad + \langle -60 \quad 0 \quad 100 \quad -60 \quad 0 \quad -100 \rangle \\
&= \langle -143.9 \quad -64.40 \quad 541.4 \quad 23.88 \quad 64.40 \quad 297.4 \rangle \\
\mathbf{r}^{(2)} &= \langle -76.13 \quad 64.40 \quad 414.9 \quad 76.13 \quad -64.40 \quad 346.4 \rangle \\
\mathbf{r}^{(3)} &= \langle 76.11 \quad -64.38 \quad -297.5 \quad -76.11 \quad 64.38 \quad -346.3 \rangle \qquad \text{(D)} \\
r_c^{(1)} &= \langle 143.9 \rangle \\
r_c^{(2)} &= \langle 64.38 \rangle \\
r_c^{(3)} &= \langle -541.5 \rangle \\
r_c^{(4)} &= \langle 76.11 \rangle \\
r_c^{(5)} &= \langle -64.38 \rangle \\
r_c^{(6)} &= \langle -414.8 \rangle
\end{aligned}$$

It is left to you to verify that these member-end and constraint forces satisfy both nodal and member equilibrium within round-off limits.

4.4.3 Statically Indeterminate Beam

The member stiffness arrays for the continuous beam of Fig. 4.8 may be obtained by specializing the stiffness array of Eq. 4.4a appropriately:

$$\mathbf{k}^{(1)} = \frac{EI_0}{12}\begin{array}{c} \\ \boxed{1} \\ \\ \boxed{2} \end{array}\left[\begin{array}{cc|cc} \multicolumn{2}{c}{\boxed{1}} & \multicolumn{2}{c}{\boxed{2}} \\ .010427 & .125 & -.010427 & .125 \\ .125 & 2 & -.125 & 1 \\ \hline -.010427 & -.125 & .010427 & -.125 \\ .125 & 1 & -.125 & 2 \end{array}\right]$$

$$\mathbf{k}^{(2)} = \frac{EI_0}{12}\begin{array}{c} \\ \boxed{2} \\ \\ \boxed{3} \end{array}\left[\begin{array}{cc|cc} \multicolumn{2}{c}{\boxed{2}} & \multicolumn{2}{c}{\boxed{3}} \\ .013333 & .2 & -.013333 & .2 \\ .2 & 4 & -.2 & 2 \\ \hline -.013333 & -.2 & .013333 & -.2 \\ .2 & 2 & -.2 & 4 \end{array}\right]$$

$$\mathbf{k}^{(3)} = \frac{EI_0}{12}\begin{array}{c} \\ \boxed{3} \\ \\ \boxed{4} \end{array}\left[\begin{array}{cc|cc} \multicolumn{2}{c}{\boxed{3}} & \multicolumn{2}{c}{\boxed{4}} \\ .010427 & .125 & -.010427 & .125 \\ .125 & 2 & -.125 & 1 \\ \hline -.010427 & -.125 & .010427 & -.125 \\ .125 & 1 & -.125 & 2 \end{array}\right]$$

where units of kilonewtons and meters are used. The fixed-end force vectors for each member are

$$\mathbf{r}_F^{(1)} = \langle \overset{\boxed{1}}{72 \quad 288} \;|\; \overset{\boxed{2}}{72 \quad -288} \rangle$$

$$\mathbf{r}_F^{(2)} = \langle \overset{\boxed{2}}{26.667 \quad 160} \;|\; \overset{\boxed{3}}{9.333 \quad -80} \rangle$$

$$\mathbf{r}_F^{(3)} = \langle \overset{\boxed{3}}{0 \quad 0} \;|\; \overset{\boxed{4}}{0 \quad 0} \rangle$$

Neglecting the foundation settlements of Fig. 4.8c for now and making use of the member incidences displayed on the preceding arrays, we can build the structure force-displacement relationships by the direct stiffness method:

$$\begin{pmatrix} 0 \\ 0 \\ 0 \\ 0 \\ 0 \\ -20 \\ 0 \\ 0 \end{pmatrix} = \frac{EI_0}{12} \begin{bmatrix} \infty & .125 & -.010427 & .125 & \cdot & \cdot & \cdot & \cdot \\ .125 & \infty & -.125 & 1 & \cdot & \cdot & \cdot & \cdot \\ -.010427 & -.125 & \infty & .075 & -.013333 & .2 & \cdot & \cdot \\ .125 & 1 & .075 & 6 & -.2 & 2 & \cdot & \cdot \\ \cdot & \cdot & -.013333 & -.2 & \infty & -.075 & -.010427 & .125 \\ \cdot & \cdot & .2 & 2 & -.075 & 6 & -.125 & 1 \\ \cdot & \cdot & \cdot & \cdot & -.010427 & -.125 & \infty & -.125 \\ \cdot & \cdot & \cdot & \cdot & .125 & 1 & -.125 & \infty \end{bmatrix} \begin{pmatrix} D_1 \\ D_2 \\ D_3 \\ D_4 \\ D_5 \\ D_6 \\ D_7 \\ D_8 \end{pmatrix} + \begin{pmatrix} 72 \\ 288 \\ 98.667 \\ -128 \\ 9.333 \\ -80 \\ 0 \\ 0 \end{pmatrix} \qquad \text{(A)}$$

(Row groups and column groups of the matrix are labelled by the boxed numbers 1, 2, 3, 4, each covering two rows/columns.)

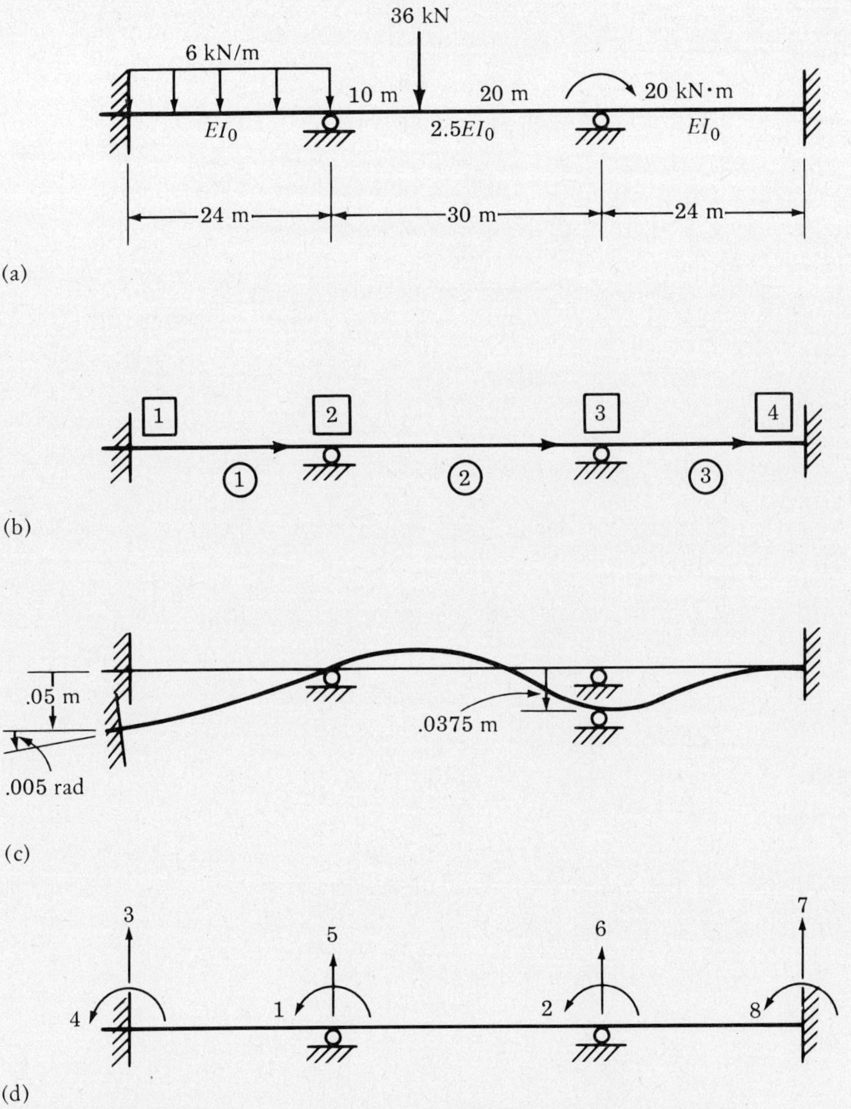

Fig. 4.8 Beam example: (a) dimensions and loading; (b) nodal and member numbering; (c) foundation settlement; (d) rearranged coordinate numbering.

or

$$\mathbf{R} = \mathbf{K}\mathbf{D} + \mathbf{R}_F \tag{4.25}$$

where the infinite support stiffnesses* have been added by inspection, since their positions in the array are so obvious. Solving gives

$$\mathbf{D} = \mathbf{K}^{-1}(\mathbf{R} - \mathbf{R}_F) \tag{4.27}$$

*To clarify the handling of terms associated with infinite support stiffnesses and corresponding small displacements, you may wish to consult the footnotes on pages 100, 109, and 131.

****** *Computer Input:*

```
START
REMARK 5
SOLUTION FOR THE STRUCTURE NODAL DISPLACEMENTS
USING THE STRUCTURE STIFFNESS MATRIX OF THE
STATICALLY INDETERMINATE BEAM

**** UNITS ARE KILONEWTONS AND METERS ****
LOAD K 8 8
 1E10 .125 -.010427 .125 4*0 .125  1E10 -.125 1 4*0
-.010427 -.125 1E10 .075 -.013333 .2 0 0 .125 1 .075
6 -.2 2 0 0 0 0 -.013333 -.2 1E10 -.075 -.010427
.125 0 0 .2 2 -.075 6 -.125 1 4*0 -.010427 -.125
1E10 -.125 4*0 .125 1 -.125 1E10
PRINT K 2
THIS IS THE CONSTRAINED STRUCTURE STIFFNESS MATRIX,
K, WITH EI(0)/12 FACTORED OUT
SYMINV K
PRINT K 1
THIS IS THE MATRIX, K INVERSE, WITH 12/EI(0) FACTORED OUT
LOAD R 8 1
5*0 -20 0 0
LOAD RF 8 1
72 288 98.666667 -128 9.333333 -80 0 0
SUB R RF
PRINT R 2
THIS IS THE VECTOR OF NODAL LOADS MINUS THE
VECTOR OF FIXED-END FORCES
MULT K R D
PRINT D 2
THIS IS THE VECTOR OF STRUCTURE NODAL DISPLACEMENTS,
D, WITH 12/EI(0) FACTORED OUT
STOP
```

****** *Computer Output:*

```
START

REMARK       5


SOLUTION FOR THE STRUCTURE NODAL DISPLACEMENTS
USING THE STRUCTURE STIFFNESS MATRIX OF THE
STATICALLY INDETERMINATE BEAM

**** UNITS ARE KILONEWTONS AND METERS ****


LOAD    K           8    8
    8 ROWS,     8 COLUMNS

PRINT   K           2


THIS IS THE CONSTRAINED STRUCTURE STIFFNESS MATRIX,
K, WITH EI(0)/12 FACTORED OUT


              1           2           3           4           5           6
 1  1.0000E+10 1.2500E-01-1.0427E-02 1.2500E-01           0           0
 2  1.2500E-01 1.0000E+10-1.2500E-01 1.0000E+00           0           0
 3 -1.0427E-02-1.2500E-01 1.0000E+10 7.5000E-02-1.3333E-02 2.0000E-01
 4  1.2500E-01 1.0000E+00 7.5000E-02 6.0000E+00-2.0000E-01 2.0000E+00
 5           0           0-1.3333E-02-2.0000E-01 1.0000E+10-7.5000E-02
 6           0           0 2.0000E-01 2.0000E+00-7.5000E-02 6.0000E+00
 7           0           0           0           0-1.0427E-02-1.2500E-01
 8           0           0           0           0 1.2500E-01 1.0000E+00
              7           8
 1           0           0
```

****** *Computer Output (continued):*

```
 2           0           0
 3           0           0
 4           0           0
 5 -1.0427E-02 1.2500E-01
 6 -1.2500E-01 1.0000E+00
 7  1.0000E+10-1.2500E-01
 8 -1.2500E-01 1.0000E+10


SYMINV  K

PRINT   K           1


THIS IS THE MATRIX, K INVERSE, WITH 12/EI(0) FACTORED OUT


             1           2           3           4           5           6
 1  1.0000E-10-1.0156E-21 1.0622E-22-2.3437E-12-4.1016E-23 7.8125E-13
 2 -1.0156E-21 1.0000E-10 1.2656E-21-1.8750E-11-3.2812E-22 6.2500E-12
 3  1.0622E-22 1.2656E-21 1.0000E-10-1.5625E-13 1.0560E-22-3.2812E-12
 4 -2.3437E-12-1.8750E-11-1.5625E-13 1.8750E-01 3.2812E-12-6.2500E-02
 5 -4.1016E-23-3.2812E-22 1.0560E-22 3.2812E-12 1.0000E-10 1.5625E-13
 6  7.8125E-13 6.2500E-12-3.2812E-12-6.2500E-02 1.5625E-13 1.8750E-01
 7  9.7656E-24 7.8125E-23-4.1016E-23-7.8125E-13 1.0622E-22 2.3437E-12
 8 -7.8125E-23-6.2500E-22 3.2812E-22 6.2500E-12-1.2656E-21-1.8750E-11
             7           8
 1  9.7656E-24-7.8125E-23
 2  7.8125E-23-6.2500E-22
 3 -4.1016E-23 3.2812E-22
 4 -7.8125E-13 6.2500E-12
 5  1.0622E-22-1.2656E-21
 6  2.3437E-12-1.8750E-11
 7  1.0000E-10 1.0156E-21
 8  1.0156E-21 1.0000E-10


LOAD    R           8    1
    8 ROWS,     1 COLUMNS

LOAD    RF          8    1
    8 ROWS,     1 COLUMNS

SUB     R       RF

PRINT   R           2


THIS IS THE VECTOR OF NODAL LOADS MINUS THE
VECTOR OF FIXED-END FORCES


             1
 1 -7.2000E+01
 2 -2.8800E+02
 3 -9.8667E+01
 4  1.2800E+02
 5 -9.3333E+00
 6  6.0000E+01
 7           0
 8           0


MULT    K       R       D

PRINT   D           2


THIS IS THE VECTOR OF STRUCTURE NODAL DISPLACEMENTS,
D, WITH 12/EI(0) FACTORED OUT
```

****** *Computer Output (continued):*

```
        1
1 -7.4531E-09
2 -3.0825E-08
3 -1.0084E-08
4  2.0250E+01
5 -5.0396E-10
6  3.2500E+00
7  4.0625E-11
8 -3.2500E-10

STOP
```

or, using SMIS3 with $\infty = 10^{10}$, as shown on the preceding printout, gives

$$\mathbf{D} = \frac{12}{EI_0}\langle -74.5 \times 10^{-10} \quad -308 \times 10^{-10} \quad -101 \times 10^{-10} \quad 20.25 \quad -5.04 \times 10^{-10} \quad 3.25 \quad .406 \times 10^{-10} \quad -3.25 \times 10^{-10}\rangle \tag{B}$$

Backsubstituting gives the following results:

$$\mathbf{d}^{(1)} = \frac{12}{EI_0}\langle 0 \quad 0 \quad 0 \quad 20.25\rangle$$

$$\mathbf{d}^{(2)} = \frac{12}{EI_0}\langle 0 \quad 20.25 \quad 0 \quad 3.25\rangle$$

$$\mathbf{d}^{(3)} = \frac{12}{EI_0}\langle 0 \quad 3.25 \quad 0 \quad 0\rangle$$

$$d_c^{(1)} = \frac{12}{EI_0}\langle -74.5 \times 10^{-10}\rangle$$

$$d_c^{(2)} = \frac{12}{EI_0}\langle -308 \times 10^{-10}\rangle \tag{C}$$

$$d_c^{(3)} = \frac{12}{EI_0}\langle -101 \times 10^{-10}\rangle$$

$$d_c^{(4)} = \frac{12}{EI_0}\langle -5.04 \times 10^{-10}\rangle$$

$$d_c^{(5)} = \frac{12}{EI_0}\langle .406 \times 10^{-10}\rangle$$

$$d_c^{(6)} = \frac{12}{EI_0}\langle -3.25 \times 10^{-10}\rangle$$

$$\mathbf{r}^{(1)} = \langle 74.53 \quad 308.25 \quad 69.47 \quad -247.5\rangle$$

$$\mathbf{r}^{(2)} = \langle 31.37 \quad 247.5 \quad 4.6 \quad -26.5\rangle \tag{D}$$

$$\mathbf{r}^{(3)} = \langle .406 \quad 6.5 \quad -.406 \quad 3.25\rangle$$

$$
\begin{aligned}
r_c^{(1)} &= \langle -74.5 \rangle \\
r_c^{(2)} &= \langle -308 \rangle \\
r_c^{(3)} &= \langle -101 \rangle \\
r_c^{(4)} &= \langle -5.04 \rangle \\
r_c^{(5)} &= \langle .406 \rangle \\
r_c^{(6)} &= \langle -3.25 \rangle
\end{aligned} \tag{D}
$$

Note that the member-end and support displacements of Eq. C satisfy nodal compatibility, while the member-end and constraint forces of Eq. D satisfy nodal equilibrium.

4.5 ALTERNATIVE FORMS OF THE STRUCTURE STIFFNESS RELATIONSHIPS

To improve solution efficiency, various alternative forms of the structure stiffness relationships are used in practice. Some of these forms are described next.

4.5.1 Modified Structure-Stiffness Relationships

One alternative involves certain modifications to the structure stiffness relationships. The rationale behind these modifications can be readily demonstrated by considering an example problem such as the continuous beam of Fig. 4.8. Once again neglecting the foundation settlements of Fig. 4.8c, we can rewrite Eq. A on page 136 for ready reference as Eq. A on page 142.

Consider the first scalar equation of this matrix equation:

$$0 = \frac{EI_0}{12}(\infty\, D_1 + .125\, D_2 - .010427\, D_3 + .125\, D_4) + 72 \tag{B}$$

Dividing this equation by $\infty\, EI_0/12$ gives*

$$0 = D_1 \tag{C}$$

It is seen therefore that Eq. C is equivalent to Eq. B and may be used in place of Eq. B in Eq. A. We may also operate on the second, third, fifth, seventh, and eighth scalar equations of matrix A in a similar fashion and obtain respective equivalent equations.

$$
\begin{aligned}
0 &= D_2 \\
0 &= D_3 \\
0 &= D_5 \\
0 &= D_7 \\
0 &= D_8
\end{aligned} \tag{D}
$$

*On the basis of our previous discussions (see the footnote on p. 100), the infinity symbol, ∞, is taken to be very large relative to other stiffness coefficients. It follows that Eq. C may be considered exact to the number of decimal places used in the solution; thus equality is indicated.

$$
\begin{pmatrix} 0 \\ 0 \\ 0 \\ 0 \\ 0 \\ -20 \\ 0 \\ 0 \end{pmatrix} = \frac{EI_0}{12} \begin{bmatrix} \infty & .125 & -.010427 & .125 & \cdot & \cdot & \cdot & \cdot \\ .125 & \infty & -.125 & 1 & \cdot & \cdot & \cdot & \cdot \\ -.010427 & -.125 & \infty & .075 & -.013333 & .2 & \cdot & \cdot \\ .125 & 1 & .075 & 6 & -.2 & 2 & \cdot & \cdot \\ \cdot & \cdot & -.013333 & -.2 & \infty & -.075 & -.010427 & .125 \\ \cdot & \cdot & .2 & 2 & -.075 & 6 & -.125 & 1 \\ \cdot & \cdot & \cdot & \cdot & -.010427 & -.125 & \infty & -.125 \\ \cdot & \cdot & \cdot & \cdot & .125 & 1 & -.125 & \infty \end{bmatrix} \begin{pmatrix} D_1 \\ D_2 \\ D_3 \\ D_4 \\ D_5 \\ D_6 \\ D_7 \\ D_8 \end{pmatrix} + \begin{pmatrix} 72 \\ 288 \\ 98.667 \\ -128 \\ 9.333 \\ -80 \\ 0 \\ 0 \end{pmatrix} \qquad \text{(A)}
$$

Finally, the fourth and sixth scalar equations may also be modified. Consider the fourth one:

$$0 = \frac{EI_0}{12}(.125D_1 + D_2 + .075D_3 + 6D_4 - .2D_5 + 2D_6) - 128 \qquad \text{(E)}$$

Substituting from Eqs. C and D yields

$$0 = \frac{EI_0}{12}(6D_4 + 2D_6) - 128 \qquad \text{(E')}$$

Similarly, the sixth scalar equation of matrix Eq. A may be written

$$-20 = \frac{EI_0}{12}(2D_4 + 6D_6) - 80 \qquad \text{(F)}$$

Rewriting Eq. A with the appropriate replacements from Eqs. C, D, E′, and F gives

$$\begin{pmatrix} 0 \\ 0 \\ 0 \\ 0 \\ 0 \\ -20 \\ 0 \\ 0 \end{pmatrix} = \begin{bmatrix} 1 & \cdot & \cdot & \cdot & \cdot & \cdot & \cdot & \cdot \\ \cdot & 1 & \cdot & \cdot & \cdot & \cdot & \cdot & \cdot \\ \cdot & \cdot & 1 & \cdot & \cdot & \cdot & \cdot & \cdot \\ \cdot & \cdot & \cdot & EI_0/2 & \cdot & EI_0/6 & \cdot & \cdot \\ \cdot & \cdot & \cdot & \cdot & 1 & \cdot & \cdot & \cdot \\ \cdot & \cdot & \cdot & EI_0/6 & \cdot & EI_0/2 & \cdot & \cdot \\ \cdot & \cdot & \cdot & \cdot & \cdot & \cdot & 1 & \cdot \\ \cdot & \cdot & \cdot & \cdot & \cdot & \cdot & \cdot & 1 \end{bmatrix} \begin{pmatrix} D_1 \\ D_2 \\ D_3 \\ D_4 \\ D_5 \\ D_6 \\ D_7 \\ D_8 \end{pmatrix} + \begin{pmatrix} 0 \\ 0 \\ 0 \\ -128 \\ 0 \\ -80 \\ 0 \\ 0 \end{pmatrix} \qquad \text{(G)}$$

A solution to Eq. G is presented in the computer printout that follows, with $E = 200{,}000$ mPa and $I_0 = 600 \times 10^6\ \text{mm}^4$, giving

$$\mathbf{D} = \langle 0 \quad 0 \quad 0 \quad .002025 \quad 0 \quad .000325 \quad 0 \quad 0 \rangle \qquad \text{(H)}$$

Factoring $12/EI_0 = 1/10{,}000\ \text{kN} \cdot \text{m}^2$ out of this vector gives

$$\mathbf{D} = \frac{12}{EI_0}\langle 0 \quad 0 \quad 0 \quad 20.25 \quad 0 \quad 3.25 \quad 0 \quad 0 \rangle \qquad \text{(I)}$$

This result is seen to be an "exact" version of the previous solution for this problem (Eq. B, page 140). The reason this solution is "exact" is, of course, that the replacement equations represent the effect of the "true" infinite stiffnesses of the supports instead of the approximate large values for support stiffnesses used previously.

Replacement Eqs. C and D used earlier in the form

$$0 = D_i \qquad \text{(J)}$$

may be rewritten as

$$D_i = 0 \qquad \text{(J')}$$

****** *Computer Input:*

```
START
REMARK 7
SOLUTION FOR THE STRUCTURE NODAL DISPLACEMENTS OF
THE BEAM EXAMPLE CONSIDERING NO INITIAL DISPLACEMENTS
AND USING THE MODIFIED STRUCTURE STIFFNESS RELATIONSHIPS

E = 2E08 KN/M**2, I(0) = .0006 M**4, EI(0)/12 = 10000 KN-M**2

**** UNITS ARE KILONEWTONS AND METERS ****
LOAD K 8 8
1 8*0 1 8*0 1 8*0 60000 0 20000 6*0 1 6*0 20000 0
60000 8*0 1 8*0 1
PRINT K 1
THIS IS THE MODIFIED STRUCTURE STIFFNESS MATRIX, K
SYMINV K
PRINT K 1
THIS IS THE MATRIX, K INVERSE.
LOAD R 8 1
0 0 0 0 0 -20 0 0
LOAD RF 8 1
0 0 0 -128 0 -80 0 0
SUB R RF
PRINT R 2
THIS IS THE VECTOR OF NODAL LOADS MINUS THE VECTOR
OF FIXED-END FORCES.
MULT K R D
PRINT D 1
THIS IS THE VECTOR, D, OF STRUCTURE NODAL DISPLACEMENTS.
STOP
```

****** *Computer Output:*

```
START

REMARK      7


SOLUTION FOR THE STRUCTURE NODAL DISPLACEMENTS OF
THE BEAM EXAMPLE CONSIDERING NO INITIAL DISPLACEMENTS
AND USING THE MODIFIED STRUCTURE STIFFNESS RELATIONSHIPS

E = 2E08 KN/M**2, I(0) = .0006 M**4, EI(0)/12 = 10000 KN-M**2

**** UNITS ARE KILONEWTONS AND METERS ****


LOAD    K           8    8
    8 ROWS,      8 COLUMNS

PRINT   K           1


THIS IS THE MODIFIED STRUCTURE STIFFNESS MATRIX, K
```

	1	2	3	4	5	6
1	1.0000E+00	0	0	0	0	0
2	0	1.0000E+00	0	0	0	0
3	0	0	1.0000E+00	0	0	0
4	0	0	0	6.0000E+04	0	2.0000E+04
5	0	0	0	0	1.0000E+00	0
6	0	0	0	2.0000E+04	0	6.0000E+04
7	0	0	0	0	0	0
8	0	0	0	0	0	0

****** *Computer Output (continued):*

```
            7           8
 1          0           0
 2          0           0
 3          0           0
 4          0           0
 5          0           0
 6          0           0
 7 1.0000E+00           0
 8          0  1.0000E+00


SYMINV  K

PRINT   K          1

THIS IS THE MATRIX, K INVERSE.

            1           2           3           4           5           6
 1 1.0000E+00           0           0           0           0           0
 2          0  1.0000E+00           0           0           0           0
 3          0           0  1.0000E+00           0           0           0
 4          0           0           0  1.8750E-05           0 -6.2500E-06
 5          0           0           0           0  1.0000E+00           0
 6          0           0 -6.2500E-06           0           0  1.8750E-05
 7          0           0           0           0           0           0
 8          0           0           0           0           0           0
            7           8
 1          0           0
 2          0           0
 3          0           0
 4          0           0
 5          0           0
 6          0           0
 7 1.0000E+00           0
 8          0  1.0000E+00


LOAD    R          8    1
    8 ROWS,     1 COLUMNS

LOAD    RF         8    1
    8 ROWS,     1 COLUMNS

SUB     R       RF

PRINT   R          2

THIS IS THE VECTOR OF NODAL LOADS MINUS THE VECTOR
OF FIXED-END FORCES.

            1
 1          0
 2          0
 3          0
 4 1.2800E+02
 5          0
 6 6.0000E+01
 7          0
 8          0


MULT    K       R       D
```

****** *Computer Output (continued):*

```
PRINT   D          1

THIS IS THE VECTOR, D, OF STRUCTURE NODAL DISPLACEMENTS.

                1
1               0
2               0
3               0
4   2.0250E-03
5               0
6   3.2500E-04
7               0
8               0

STOP
```

and be seen to represent solutions to the equations they replaced. Displacements are obviously zero in rigidly supported coordinate directions. However, in the case where some support settlement, D_{0i}, exists in the ith coordinate direction, the solution equation will be of the form

$$D_i = D_{0i} \tag{K}$$

and the equivalent equation will be*

$$D_{0i} = D_i \tag{K'}$$

Considering now the continuous beam of Fig. 4.8 subjected to support settlements in addition to the applied external loading, we see that the replacement Eqs. C and D will become

$$\begin{aligned} -.05 &= D_1 \\ .005 &= D_2 \\ 0 &= D_3 \\ -.0375 &= D_5 \\ 0 &= D_7 \\ 0 &= D_8 \end{aligned} \tag{L}$$

Replacement Eqs. E′ and F will also change. Consider Eq. E rewritten in view of Eq. L:

$$0 = \frac{EI_0}{12}[.125(-.05) + (.005) + .075(0) + 6D_4 - .2(-.0375) + 2D_6 - 128] \tag{M}$$

*Replacement Eq. K′ is written here by a rather straightforward analogy with former replacement Eq. J. A formal proof of Eq. K′ would require the writing of a more general set of original Eq. A (to include the effect of settlements). Since this is somewhat awkward and the analogy is straightforward, the formal proof will be deferred to Sec. 5.2.1.

Rearranging Eq. M gives

$$0 = \frac{EI_0}{12}(6D_4 + 2D_6) - 128 + \frac{EI_0}{12}[.125(-.05) + (.005) + .075(0) - .2(-.0375)] \qquad \text{(M')}$$

and, since $EI_0/12 = 10{,}000\ \text{kN} \cdot \text{m}^2$, Eq. M′ becomes

$$0 = \frac{EI_0}{12}(6D_4 + 2D_6) - 65.5 \qquad \text{(M'')}$$

where it is seen that the "fixed-end" effects of the foundation settlements have been lumped in with the conventional fixed-end force. As we proceed in a similar fashion, replacement Eq. F becomes

$$-20 = \frac{EI_0}{12}(2D_4 + 6D_6) - 51.875 \qquad \text{(N)}$$

Rewriting Eq. A with the new appropriate replacements from Eqs. L, M″, and N gives finally

$$\begin{pmatrix} -.05 \\ .005 \\ 0 \\ 0 \\ -.0375 \\ -20 \\ 0 \\ 0 \end{pmatrix} = \begin{bmatrix} 1 & \cdot & \cdot & \cdot & \cdot & \cdot & \cdot & \cdot \\ \cdot & 1 & \cdot & \cdot & \cdot & \cdot & \cdot & \cdot \\ \cdot & \cdot & 1 & \cdot & \cdot & \cdot & \cdot & \cdot \\ \cdot & \cdot & \cdot & EI_0/2 & \cdot & EI_0/6 & \cdot & \cdot \\ \cdot & \cdot & \cdot & \cdot & 1 & \cdot & \cdot & \cdot \\ \cdot & \cdot & \cdot & EI_0/6 & \cdot & EI_0/2 & \cdot & \cdot \\ \cdot & \cdot & \cdot & \cdot & \cdot & \cdot & 1 & \cdot \\ \cdot & \cdot & \cdot & \cdot & \cdot & \cdot & \cdot & 1 \end{bmatrix} \begin{pmatrix} D_1 \\ D_2 \\ D_3 \\ D_4 \\ D_5 \\ D_6 \\ D_7 \\ D_8 \end{pmatrix} + \begin{pmatrix} 0 \\ 0 \\ 0 \\ -65.5 \\ 0 \\ -51.875 \\ 0 \\ 0 \end{pmatrix} \qquad \text{(O)}$$

A solution to Eq. O using SMIS3 is presented in the printout that follows, giving

$$\mathbf{D} = \langle -.05 \quad .005 \quad 0 \quad .00103 \quad -.0375 \quad .000188 \quad 0 \quad 0 \rangle \qquad \text{(P)}$$

Obtaining member-end forces by backsubstitution is straightforward as before and therefore will not be carried out here.

The following general four-point rule can now be established for developing modified stiffness relationships for a structure:

1. Each row and each column of the structure stiffness matrix corresponding to a constrained structure degree of freedom should have its diagonal coefficient replaced with unity and its off-diagonal coefficients replaced with zero.
2. Initial displacements of constrained degrees of freedom should replace nodal forces at corresponding locations in the vector of applied nodal forces.

****** *Computer Input:*

```
START
REMARK 7
SOLUTION FOR THE STRUCTURE NODAL DISPLACEMENTS OF
THE BEAM EXAMPLE CONSIDERING FOUNDATION SETTLEMENT
AND USING THE MODIFIED STRUCTURE STIFFNESS RELATIONSHIPS

E = 2E08 KN/M**2, I(0) = .0006 M**4, EI(0)/12 = 10000 KN-M**2

**** UNITS ARE KILONEWTONS AND METERS ****
LOAD K 8 8
1 8*0 1 8*0 1 8*0 60000 0 20000 6*0 1 6*0 20000 0
60000 8*0 1 8*0 1
PRINT K 1
THIS IS THE MODIFIED STRUCTURE STIFFNESS MATRIX, K
LOAD R 8 1
-.05 .005 0 0 -.0375 -20 0 0
PRINT R 2
THIS IS THE VECTOR, R, OF NODAL FORCES AND INITIAL
NODAL DISPLACEMENTS DUE TO SETTLEMENT
LOAD RF 8 1
0 0 0  -65.5 0 -51.875 0 0
PRINT RF 2
THIS IS THE VECTOR OF FIXED-END FORCES, RF,
INCLUDING SUPPORT SETTLEMENT EFFECTS
SUB R RF
SOLVE K R D
PRINT D 1
THIS IS THE VECTOR OF STRUCTURE NODAL DISPLACEMENTS, D
STOP
```

****** *Computer Output:*

```
START

REMARK      7


SOLUTION FOR THE STRUCTURE NODAL DISPLACEMENTS OF
THE BEAM EXAMPLE CONSIDERING FOUNDATION SETTLEMENT
AND USING THE MODIFIED STRUCTURE STIFFNESS RELATIONSHIPS

E = 2E08 KN/M**2, I(0) = .0006 M**4, EI(0)/12 = 10000 KN-M**2

**** UNITS ARE KILONEWTONS AND METERS ****


LOAD    K             8    8
    8 ROWS,      8 COLUMNS

PRINT   K             1


THIS IS THE MODIFIED STRUCTURE STIFFNESS MATRIX, K
```

	1	2	3	4	5	6
1	1.0000E+00	0	0	0	0	0
2	0	1.0000E+00	0	0	0	0
3	0	0	1.0000E+00	0	0	0
4	0	0	0	6.0000E+04	0	2.0000E+04
5	0	0	0	0	1.0000E+00	0
6	0	0	0	2.0000E+04	0	6.0000E+04
7	0	0	0	0	0	0
8	0	0	0	0	0	0

****** *Computer Output (continued):*

```
           7           8
 1          0           0
 2          0           0
 3          0           0
 4          0           0
 5          0           0
 6          0           0
 7  1.0000E+00          0
 8          0 1.0000E+00


LOAD     R              8    1
    8 ROWS,      1 COLUMNS

PRINT    R             2


THIS IS THE VECTOR, R, OF NODAL FORCES AND INITIAL
NODAL DISPLACEMENTS DUE TO SETTLEMENT


           1
 1 -5.0000E-02
 2  5.0000E-03
 3          0
 4          0
 5 -3.7500E-02
 6 -2.0000E+01
 7          0
 8          0


LOAD     RF             8    1
    8 ROWS,      1 COLUMNS

PRINT    RF            2


THIS IS THE VECTOR OF FIXED-END FORCES, RF,
INCLUDING SUPPORT SETTLEMENT EFFECTS


           1
 1          0
 2          0
 3          0
 4 -6.5500E+01
 5          0
 6 -5.1875E+01
 7          0
 8          0


SUB      R       RF

SOLVE    K       R       D

PRINT    D             1


THIS IS THE VECTOR OF STRUCTURE NODAL DISPLACEMENTS, D


           1
 1 -5.0000E-02
 2  5.0000E-03
 3          0
```

```
****** Computer Output (continued):

 4  1.0289E-03
 5 -3.7500E-02
 6  1.8828E-04
 7           0
 8           0

STOP
```

3. The fixed-end forces of constrained degrees of freedom should be replaced by zeros.
4. The fixed-end forces of unconstrained degrees of freedom should be appropriately augmented to reflect the effect of initial displacements at constrained degrees of freedom.

The only advantage of this method of solving problems over that previously shown is that support stiffnesses need not be considered in the modified approach because the stiffness equations corresponding to supported degrees of freedom are replaced anyway. It is therefore necessary to build only an "unconstrained" structure stiffness relationship, considering the members only and not the supports. For the example beam of Fig. 4.8, we would thus get Eq. Q on page 151. The vector of nodal forces on the left of Eq. Q now contains both the unknown reaction forces (R_1, R_2, R_3, R_5, R_7, and R_8) transmitted from the supports to the structure and the known applied forces ($R_4 = 0$ and $R_6 = -20$) at unconstrained coordinate locations. Correspondingly, the vector of nodal displacements consists of known displacements ($D_1 = -.05$, $D_2 = .005$, $D_3 = 0$, $D_5 = -.0375$, $D_7 = 0$, and $D_8 = 0$) in constrained coordinate directions and of unknown displacements (D_4 and D_6) in unconstrained coordinate directions. Modification of Eq. Q as just described will, of course, eliminate the unknown reactions from the statement of the problem yielding Eq. O.

The structural analysis section of the SMIS3 computer program (see App. 2) can operate precisely according to the modified approach. The unconstrained structure stiffness relationships are first built from the individual member stiffness relationships by coordinate coding and are then modified in accordance with specified boundary (support) conditions to yield structure stiffness relationships of the form of Eq. O. These equations are finally solved for the structure nodal displacements, and backsubstitution is carried out to yield the member-end forces in local coordinates. By using the structural analysis commands of the SMIS3 program, we have a solution to the beam example of Fig. 4.8, as shown in the printout on pages 152–157.* This solution is seen to agree with that of Eq. P.

The coordinate code used in the printout may be calculated from a nodal code by using an appropriate algorithm. For a beam like that of the example, the

*In the SMIS3 computer listings, the terms *element* and *member* are used synonymously.

$$
\begin{pmatrix} R_1 \\ R_2 \\ R_3 \\ R_4 \\ R_5 \\ R_6 \\ R_7 \\ R_8 \end{pmatrix} = \frac{EI_0}{12} \begin{bmatrix} .010427 & .125 & -.010427 & .125 & \cdot & \cdot & \cdot & \cdot \\ .125 & 2 & -.125 & 1 & \cdot & \cdot & \cdot & \cdot \\ -.010427 & -.125 & .02376 & .075 & -.013333 & .2 & \cdot & \cdot \\ .125 & 1 & .075 & 6 & -.2 & 2 & \cdot & \cdot \\ \cdot & \cdot & -.013333 & -.2 & .02376 & -.075 & -.010427 & .125 \\ \cdot & \cdot & .2 & 2 & -.075 & 6 & -.125 & 1 \\ \cdot & \cdot & \cdot & \cdot & -.010427 & -.125 & .010427 & -.125 \\ \cdot & \cdot & \cdot & \cdot & .125 & 1 & -.125 & 2 \end{bmatrix} \begin{pmatrix} D_1 \\ D_2 \\ D_3 \\ D_4 \\ D_5 \\ D_6 \\ D_7 \\ D_8 \end{pmatrix} + \begin{pmatrix} 72 \\ 288 \\ 98.667 \\ -128 \\ 9.333 \\ -80 \\ 0 \\ 0 \end{pmatrix} \qquad \text{(Q)}
$$

****** *Computer Input:*

```
START
REMARK 7
SOLUTION OF BEAM EXAMPLE CONSIDERING FOUNDATION
SETTLEMENT AND USING SMIS3 STRUCTURAL ANALYSIS COMMANDS,
MODIFIED METHOD
E = 2E08 KN/M**2, I(0) = .0006 M**4, EI(0)/12 = 10000 KN-M**2
**** UNITS ARE KILONEWTONS AND METERS ****
FORM K 8 8
LOAD R 8 1
0 0 0 0 0 -20 0 0
BEAMEL K1 1 RF1
24 2E08 .0006 0 0
F 0 24 -6 -6
PRINT K1 1
THIS IS THE STIFFNESS MATRIX, K1, FOR MEMBER 1
PRINT RF1 1
THIS IS THE FIXED-END FORCE VECTOR, RF1, FOR MEMBER 1
ADDSTF  K 1
K1 C1 1 2 3 4
ADDFEF R 1
RF1 C1
PRINT K 1
THIS IS AN INTERMEDIATE STRUCTURE STIFFNESS MATRIX
PRINT R 2
THIS IS AN INTERMEDIATE STRUCTURE EQUIVALENT NODAL
LOAD VECTOR
BEAMEL K2 1 RF2
30 2E08 .0015 0 0
F 10 10 -36 -36
PRINT K2 1
THIS IS THE STIFFNESS MATRIX, K2, FOR MEMBER 2
PRINT RF2 1
THIS IS THE FIXED-END FORCE VECTOR, RF2, FOR MEMBER 2
ADDSTF K 1
K2 C2 3 4 5 6
ADDFEF R 1
RF2 C2
PRINT K 1
THIS IS AN INTERMEDIATE STRUCTURE STIFFNESS MATRIX
PRINT R 2
THIS IS AN INTERMEDIATE STRUCTURE EQUIVALENT
NODAL LOAD VECTOR
BEAMEL K3 0
24 2E08 .0006 0 0
PRINT K3 1
THIS IS THE STIFFNESS MATRIX, K3, FOR MEMBER 3
ADDSTF K 1
K3 C3 5 6 7 8
PRINT K 2
THIS IS THE COMPLETELY ASSEMBLED UNCONSTRAINED STRUCTURE
STIFFNESS MATRIX, K
BOUNDC K R 6
1 -.05
2 .005
3 0
5 -.0375
7 0
8 0
PRINT K 1
THIS IS THE MODIFIED STRUCTURE STIFFNESS MATRIX
PRINT R 1
THIS IS THE MODIFIED STRUCTURE EQUIVALENT NODAL LOAD VECTOR
SOLVE K R D
PRINT D 1
THIS IS THE VECTOR, D, OF STRUCTURE NODAL DISPLACEMENTS
BEAMFC D 3
K1 C1 RF1
K2 C2 RF2
K3 C3 0
STOP
```

****** *Computer Output:*

```
START

REMARK      7

SOLUTION OF BEAM EXAMPLE CONSIDERING FOUNDATION
SETTLEMENT AND USING SMIS3 STRUCTURAL ANALYSIS COMMANDS,
MODIFIED METHOD

E = 2E08 KN/M**2, I(0) = .0006 M**4, EI(0)/12 = 10000 KN-M**2

**** UNITS ARE KILONEWTONS AND METERS ****


FORM    K               8    8
   8 ROWS,      8 COLUMNS

LOAD    R               8    1
   8 ROWS,      1 COLUMNS

BEAMEL  K1              1       RF1
 E =  2.000E+08       I =  6.000E-04      L =  2.400E+01
 A =            0     G =             0

 LOAD TYPE = F
  DIMENSION C =             0      LOAD INTENSITY AT C = -6.000E+00
  DIMENSION D =  2.400E+01      LOAD INTENSITY AT D = -6.000E+00

PRINT   K1              1


THIS IS THE STIFFNESS MATRIX, K1, FOR MEMBER 1


              1          2          3          4
 1  1.0417E+02 1.2500E+03-1.0417E+02 1.2500E+03
 2  1.2500E+03 2.0000E+04-1.2500E+03 1.0000E+04
 3 -1.0417E+02-1.2500E+03 1.0417E+02-1.2500E+03
 4  1.2500E+03 1.0000E+04-1.2500E+03 2.0000E+04


PRINT   RF1             1


THIS IS THE FIXED-END FORCE VECTOR, RF1, FOR MEMBER 1

              1
 1  7.2000E+01
 2  2.8800E+02
 3  7.2000E+01
 4 -2.8800E+02


ADDSTF  K               1
 MEMBER K    MEMBER CODE        DEGREES OF FREEDOM
    K1           C1             1     2     3     4

ADDFEF  R               1
 MEMBER FEF    MEMBER CODE
    RF1            C1

PRINT   K               1


THIS IS AN INTERMEDIATE STRUCTURE STIFFNESS MATRIX


              1          2          3          4          5          6
```

****** *Computer Output (continued):*

```
 1  1.0417E+02 1.2500E+03-1.0417E+02 1.2500E+03          0          0
 2  1.2500E+03 2.0000E+04-1.2500E+03 1.0000E+04          0          0
 3 -1.0417E+02-1.2500E+03 1.0417E+02-1.2500E+03          0          0
 4  1.2500E+03 1.0000E+04-1.2500E+03 2.0000E+04          0          0
 5          0          0          0          0          0          0
 6          0          0          0          0          0          0
 7          0          0          0          0          0          0
 8          0          0          0          0          0          0
            7          8
 1          0          0
 2          0          0
 3          0          0
 4          0          0
 5          0          0
 6          0          0
 7          0          0
 8          0          0


PRINT   R             2


THIS IS AN INTERMEDIATE STRUCTURE EQUIVALENT NODAL
LOAD VECTOR


            1
 1 -7.2000E+01
 2 -2.8800E+02
 3 -7.2000E+01
 4  2.8800E+02
 5          0
 6 -2.0000E+01
 7          0
 8          0


BEAMEL  K2            1     RF2
 E =  2.000E+08       I =  1.500E-03      L =  3.000E+01
 A =          0       G =          0

 LOAD TYPE = F
  DIMENSION C =  1.000E+01      LOAD INTENSITY AT C = -3.600E+01
  DIMENSION D =  1.000E+01      LOAD INTENSITY AT D = -3.600E+01

PRINT   K2            1


THIS IS THE STIFFNESS MATRIX, K2, FOR MEMBER 2


            1          2          3          4
 1  1.3333E+02 2.0000E+03-1.3333E+02 2.0000E+03
 2  2.0000E+03 4.0000E+04-2.0000E+03 2.0000E+04
 3 -1.3333E+02-2.0000E+03 1.3333E+02-2.0000E+03
 4  2.0000E+03 2.0000E+04-2.0000E+03 4.0000E+04


PRINT   RF2           1


THIS IS THE FIXED-END FORCE VECTOR, RF2, FOR MEMBER 2

            1
 1  2.6667E+01
 2  1.6000E+02
 3  9.3333E+00
 4 -8.0000E+01
```

****** *Computer Output (continued):*

```
ADDSTF  K              1
 MEMBER K   MEMBER CODE         DEGREES OF FREEDOM
    K2          C2              3     4     5     6

ADDFEF  R              1
 MEMBER FEF   MEMBER CODE
    RF2           C2

PRINT   K              1


THIS IS AN INTERMEDIATE STRUCTURE STIFFNESS MATRIX


            1          2          3          4          5          6
 1  1.0417E+02 1.2500E+03-1.0417E+02 1.2500E+03          0          0
 2  1.2500E+03 2.0000E+04-1.2500E+03 1.0000E+04          0          0
 3 -1.0417E+02-1.2500E+03 2.3750E+02 7.5000E+02-1.3333E+02 2.0000E+03
 4  1.2500E+03 1.0000E+04 7.5000E+02 6.0000E+04-2.0000E+03 2.0000E+04
 5          0          0-1.3333E+02-2.0000E+03 1.3333E+02-2.0000E+03
 6          0          0 2.0000E+03 2.0000E+04-2.0000E+03 4.0000E+04
 7          0          0          0          0          0          0
 8          0          0          0          0          0          0
            7          8
 1          0          0
 2          0          0
 3          0          0
 4          0          0
 5          0          0
 6          0          0
 7          0          0
 8          0          0


PRINT   R              2


THIS IS AN INTERMEDIATE STRUCTURE EQUIVALENT
NODAL LOAD VECTOR


            1
 1 -7.2000E+01
 2 -2.8800E+02
 3 -9.8667E+01
 4  1.2800E+02
 5 -9.3333E+00
 6  6.0000E+01
 7          0
 8          0


BEAMEL  K3             0
 E =  2.000E+08        I =  6.000E-04      L =  2.400E+01
 A =          0        G =          0

PRINT   K3             1


THIS IS THE STIFFNESS MATRIX, K3, FOR MEMBER 3


            1          2          3          4
 1  1.0417E+02 1.2500E+03-1.0417E+02 1.2500E+03
 2  1.2500E+03 2.0000E+04-1.2500E+03 1.0000E+04
 3 -1.0417E+02-1.2500E+03 1.0417E+02-1.2500E+03
 4  1.2500E+03 1.0000E+04-1.2500E+03 2.0000E+04
```

****** *Computer Output (continued):*

```
ADDSTF  K           1
 MEMBER K   MEMBER CODE        DEGREES OF FREEDOM
    K3          C3             5     6     7     8

PRINT   K           2


THIS IS THE COMPLETELY ASSEMBLED UNCONSTRAINED STRUCTURE
STIFFNESS MATRIX, K


             1           2           3           4           5           6
 1  1.0417E+02 1.2500E+03-1.0417E+02 1.2500E+03           0           0
 2  1.2500E+03 2.0000E+04-1.2500E+03 1.0000E+04           0           0
 3 -1.0417E+02-1.2500E+03 2.3750E+02 7.5000E+02-1.3333E+02 2.0000E+03
 4  1.2500E+03 1.0000E+04 7.5000E+02 6.0000E+04-2.0000E+03 2.0000E+04
 5           0           0-1.3333E+02-2.0000E+03 2.3750E+02-7.5000E+02
 6           0           0 2.0000E+03 2.0000E+04-7.5000E+02 6.0000E+04
 7           0           0           0           0-1.0417E+02-1.2500E+03
 8           0           0           0           0 1.2500E+03 1.0000E+04
             7           8
 1           0           0
 2           0           0
 3           0           0
 4           0           0
 5 -1.0417E+02 1.2500E+03
 6 -1.2500E+03 1.0000E+04
 7  1.0417E+02-1.2500E+03
 8 -1.2500E+03 2.0000E+04

BOUNDC  K       R           6
 NO. OF B.C. =        6

  NEQ =      1,  DISPL =-5.000E-02
  NEQ =      2,  DISPL = 5.000E-03
  NEQ =      3,  DISPL =         0
  NEQ =      5,  DISPL =-3.750E-02
  NEQ =      7,  DISPL =         0
  NEQ =      8,  DISPL =         0

PRINT   K           1


THIS IS THE MODIFIED STRUCTURE STIFFNESS MATRIX


             1           2           3           4           5           6
 1  1.0000E+00           0           0           0           0           0
 2           0 1.0000E+00           0           0           0           0
 3           0           0 1.0000E+00           0           0           0
 4           0           0           0 6.0000E+04           0 2.0000E+04
 5           0           0           0           0 1.0000E+00           0
 6           0           0           0 2.0000E+04           0 6.0000E+04
 7           0           0           0           0           0           0
 8           0           0           0           0           0           0
             7           8
 1           0           0
 2           0           0
 3           0           0
 4           0           0
 5           0           0
 6           0           0
 7  1.0000E+00           0
 8           0 1.0000E+00

PRINT   R           1
```

****** *Computer Output (continued):*

```
THIS IS THE MODIFIED STRUCTURE EQUIVALENT NODAL LOAD VECTOR

             1
 1 -5.0000E-02
 2  5.0000E-03
 3           0
 4  6.5500E+01
 5 -3.7500E-02
 6  3.1875E+01
 7           0
 8           0


SOLVE   K       R       D

PRINT   D               1

THIS IS THE VECTOR, D, OF STRUCTURE NODAL DISPLACEMENTS

             1
 1 -5.0000E-02
 2  5.0000E-03
 3           0
 4  1.0289E-03
 5 -3.7500E-02
 6  1.8828E-04
 7           0
 8           0


BEAMFC  D               3
 MEMBER LOCAL FORCE VECTORS

 MEMBER CODE  = C1

 1  7.43278E+01
 2  3.35789E+02
 3  6.96722E+01
 4 -2.79922E+02

 MEMBER CODE  = C2

 1  3.41010E+01
 2  2.79922E+02
 3  1.89896E+00
 4  2.31094E+01

 MEMBER CODE  = C3

 1 -3.67090E+00
 2 -4.31094E+01
 3  3.67090E+00
 4 -4.49922E+01

 STOP
```

coordinate numbers and nodal numbers are related by the following algorithm:

$$\text{Translational coordinate number} = 2 \cdot (\text{node number}) - 1$$
$$\text{Rotational coordinate number} = 2 \cdot (\text{node number})$$

The advantage of using the nodal code is that it significantly reduces the amount of input data required for the computer. An example demonstrating SMIS3's built-in nodal coding capability will be worked in the next section in a slightly more complex context.

Note that since the supports do not enter into the unconstrained structure stiffness relationships, the backsubstitution process will not yield constraint forces. This problem can be circumvented by retaining (in auxiliary storage in the computer) the replaced stiffness equations of Eq. Q. Backsubstitution of structure nodal displacements into these equations will yield the unknown reaction forces. (It may also be desirable to backsubstitute into the remaining equations of Eq. Q as an equilibrium check.) This process of obtaining reaction forces will be considered in detail in connection with another alternative solution approach presented in Sec. 4.5.3.

Although the modified approach embodied in the structural analysis commands of SMIS3 offers large advantages over the previously used matrix manipulations, it is still inefficient in two ways: (1) Computer storage is wasted because a number of trivial equations of the form $D_{0i} = D_i$ are included, and (2) computer execution time is wasted because many zeros and ones associated with the above trivial equations must be processed. The modified approach is fine for small problems and has the advantage of being relatively easy to program. However, for large problems, efficiency becomes of paramount importance, and so we need other alternative solution methods.

4.5.2 Unconstrained Coordinate Code

The modifications to the structure stiffness relationships we have described are practical only because the solutions to the equations associated with the constrained coordinate directions can be obtained by inspection. This inspection leads to a simple set of replacements for and modifications to the original set of equations. The solution of the full set of modified equations then consists of a solution to a set of trivial replacement equations associated with the constrained coordinate directions (first, second, third, fifth, seventh, and eighth scaler equations of matrix Eq. O) and a solution to a set of nontrivial modified equations associated with the unconstrained coordinate directions (the fourth and sixth scaler equations of Eq. O). Further, these two sets are uncoupled after the modifications take place. It follows that, in the interest of efficiency, only the set of nontrivial equations need be generated and solved.

Several methods, differing mainly in detail, are available to generate only the set of equations associated with the unconstrained coordinate directions. One device to accomplish this task (which is very simple to program) involves the use of a modified version of the coordinate code in which only the unconstrained coordinate directions are assigned nonzero numbers, while the constrained coordinate directions are all assigned a zero. Following is the unconstrained coordinate code for the example beam problem of Fig. 4.8 (with the unconstrained coordinates at nodes 2 and 3 designated as coordinate directions 1 and 2, respectively).

Unconstrained Coordinate Code

Member number ↓	1	2	3	4 ← Member coordinates
1	0	0	0	1
2	0	1	0	2
3	0	2	0	0

(Rows: Unconstrained structure coordinate directions)

The unconstrained coordinate code is used in the following way. If a member stiffness coefficient or a fixed-end force is associated with a zero code number, it is ignored in building the structure stiffness relationships; otherwise the code operates as usual. Thus, on the basis of the unconstrained coordinate code, the constrained structure stiffness relationships for the example beam may be written as

$$\begin{pmatrix} R_1 \\ R_2 \end{pmatrix} = \begin{bmatrix} k_{44}^{(1)} + k_{22}^{(2)} & k_{24}^{(2)} \\ k_{42}^{(2)} & k_{44}^{(2)} + k_{22}^{(3)} \end{bmatrix} \begin{pmatrix} D_1 \\ D_2 \end{pmatrix} + \begin{pmatrix} r_{F4}^{(1)} + r_{F2}^{(2)} \\ r_{F4}^{(2)} + r_{F2}^{(3)} \end{pmatrix} \tag{A}$$

Or, considering the member stiffness arrays and fixed-end force vectors of page 135 as well as the nodal loading of Fig. 4.8, we may write

$$\begin{pmatrix} 0 \\ -20 \end{pmatrix} = \frac{EI_0}{12}\begin{bmatrix} 6 & 2 \\ 2 & 6 \end{bmatrix} \begin{pmatrix} D_1 \\ D_2 \end{pmatrix} + \begin{pmatrix} -128 \\ -80 \end{pmatrix} \tag{B}$$

The scalar equations of matrix Eq. B are seen to be identical with the fourth and sixth ones of Eq. G of Sec. 4.5.1. Equation B may be solved by using $EI_0/12 = 10{,}000 \text{ kN} \cdot \text{m}^2$:

$$\begin{pmatrix} D_1 \\ D_2 \end{pmatrix} = \frac{1}{10{,}000}\frac{1}{32}\begin{bmatrix} 6 & -2 \\ -2 & 6 \end{bmatrix} \begin{pmatrix} 128 \\ 60 \end{pmatrix} = \begin{pmatrix} .002025 \\ .000325 \end{pmatrix} \tag{C}$$

which is seen to agree with the solution of Eq. G of Sec. 4.5.1.

This method is more efficient than previous solution techniques in regard to both computer time and computer space requirements. Computation time is saved because (with reference to the example problem) only two simultaneous equations had to be solved (for the displacements in the directions of the unconstrained degrees of freedom) as opposed to eight previously. Although the proportion of constrained to unconstrained degrees of freedom won't generally be as high as it was in this example, a reduction in the number of simultaneous

****** *Computer Input:*

```
START
REMARK 6
SOLUTION OF CONTINOUS BEAM USING UNCONSTRAINED COORDINATE CODE.
SUPPORT SETTLEMENTS ARE NOT CONSIDERED.

E = 2E08 KN/M**2, I(0) = .0006 M**4, EI(0)/12 = 10000 KN-M**2

**** UNITS ARE KILONEWTONS AND METERS ****
FORM K 2 2
LOAD R 2 1
0 -20
PRINT R 1
THIS IS THE VECTOR, R, OF CONCENTRATED NODAL LOADS.
BEAMEL K1 1 RF1
24 2E08 .0006 0 0
F 0 24 -6 -6
ADDSTF K 1
K1 C1 0 0 0 1
PRINT K 1
THIS IS AN INTERMEDIATE STIFFNESS MATRIX, K.
ADDFEF R 1
RF1 C1
PRINT R 1
THIS IS AN INTERMEDIATE EQUIVALENT NODAL LOAD VECTOR, R.
BEAMEL K2 1 RF2
30 2E08 .0015 0 0
F 10 10 -36 -36
ADDSTF K 1
K2 C2 0 1 0 2
PRINT K 1
THIS IS AN INTERMEDIATE STIFFNESS MATRIX, K.
ADDFEF R 1
RF2 C2
PRINT R 1
THIS IS AN INTERMEDIATE EQUIVALENT NODAL LOAD VECTOR,R.
BEAMEL K3 0
24 2E08 .0006 0 0
ADDSTF K 1
K3 C3 0 2 0 0
PRINT K 1
THIS IS THE CONSTRAINED STRUCTURE STIFFNESS MATRIX,K.
PRINT R 1
THIS IS THE EQUIVALENT NODAL LOAD VECTOR,R.
SOLVE K R D
PRINT D 1
THIS IS THE VECTOR, D, OF STRUCTURE NODAL DISPLACEMENTS.
BEAMFC D 3
K1 C1 RF1
K2 C2 RF2
K3 C3 0
STOP
```

****** *Computer Output:*

```
START

REMARK      6


SOLUTION OF CONTINOUS BEAM USING UNCONSTRAINED COORDINATE CODE.
SUPPORT SETTLEMENTS ARE NOT CONSIDERED.

E = 2E08 KN/M**2, I(0) = .0006 M**4, EI(0)/12 = 10000 KN-M**2

**** UNITS ARE KILONEWTONS AND METERS ****
```

****** *Computer Output (continued):*

```
FORM    K              2    2
   2 ROWS,      2 COLUMNS

LOAD    R              2    1
   2 ROWS,      1 COLUMNS

PRINT   R              1


THIS IS THE VECTOR, R, OF CONCENTRATED NODAL LOADS.


           1
 1          0
 2 -2.0000E+01


BEAMEL  K1             1      RF1
 E =  2.000E+08        I =  6.000E-04      L =  2.400E+01
 A =           0       G =            0

 LOAD TYPE = F
  DIMENSION C =            0       LOAD INTENSITY AT C = -6.000E+00
  DIMENSION D =  2.400E+01         LOAD INTENSITY AT D = -6.000E+00

ADDSTF  K              1
 MEMBER K   MEMBER CODE          DEGREES OF FREEDOM
   K1           C1               0      0      0      1

PRINT   K              1


THIS IS AN INTERMEDIATE STIFFNESS MATRIX, K.


           1           2
 1  2.0000E+04          0
 2          0          0


ADDFEF  R              1
 MEMBER FEF   MEMBER CODE
   RF1           C1

PRINT   R              1


THIS IS AN INTERMEDIATE EQUIVALENT NODAL LOAD VECTOR, R.


           1
 1  2.8800E+02
 2 -2.0000E+01


BEAMEL  K2             1      RF2
 E =  2.000E+08        I =  1.500E-03      L =  3.000E+01
 A =           0       G =            0

 LOAD TYPE = F
  DIMENSION C =  1.000E+01         LOAD INTENSITY AT C = -3.600E+01
  DIMENSION D =  1.000E+01         LOAD INTENSITY AT D = -3.600E+01

ADDSTF  K              1
 MEMBER K   MEMBER CODE          DEGREES OF FREEDOM
```

****** *Computer Output (continued):*

```
    K2          C2              0     1     0     2

PRINT   K             1


THIS IS AN INTERMEDIATE STIFFNESS MATRIX, K.


          1           2
 1  6.0000E+04 2.0000E+04
 2  2.0000E+04 4.0000E+04


ADDFEF  R             1
 MEMBER FEF   MEMBER CODE
    RF2            C2

PRINT   R             1


THIS IS AN INTERMEDIATE EQUIVALENT NODAL LOAD VECTOR,R.


          1
 1  1.2800E+02
 2  6.0000E+01


BEAMEL  K3            0
 E =  2.000E+08       I =  6.000E-04      L =  2.400E+01
 A =           0      G =           0

ADDSTF  K             1
 MEMBER K   MEMBER CODE         DEGREES OF FREEDOM
    K3          C3              0     2     0     0

PRINT   K             1


THIS IS THE CONSTRAINED STRUCTURE STIFFNESS MATRIX,K.


          1           2
 1  6.0000E+04 2.0000E+04
 2  2.0000E+04 6.0000E+04


PRINT   R             1


THIS IS THE EQUIVALENT NODAL LOAD VECTOR,R.


          1
 1  1.2800E+02
 2  6.0000E+01


SOLVE   K       R       D

PRINT   D             1


THIS IS THE VECTOR, D, OF STRUCTURE NODAL DISPLACEMENTS.
```

****** *Computer Output (continued):*

```
            1
 1  2.0250E-03
 2  3.2500E-04


BEAMFC  D           3
 MEMBER LOCAL FORCE VECTORS


 MEMBER CODE   = C1

 1   7.45312E+01
 2   3.08250E+02
 3   6.94688E+01
 4  -2.47500E+02


 MEMBER CODE   = C2

 1   3.13667E+01
 2   2.47500E+02
 3   4.63333E+00
 4  -2.65000E+01


 MEMBER CODE   = C3

 1   4.06250E-01
 2   6.50000E+00
 3  -4.06250E-01
 4   3.25000E+00

STOP
```

equations that must be solved will still be very important, for it is this step in the overall solution of a problem that dominates time consumption. Computer core space is saved because the equations corresponding to constrained degrees of freedom are not generated.

It is seen that if a computer program for direct stiffness makes use of coordinate coding, the program can be extended to handle the unconstrained coordinate code by simply providing a single check for zeros in the code before adding member stiffness coefficients and fixed-end forces into their respective structure arrays. The SMIS3 program is capable of operating in this fashion. To demonstrate this capability, the example beam problem is reworked in the preceding printout by using SMIS3 structural analysis commands along with an unconstrained coordinate code. The results are seen to agree with the previous analyses of this beam, with support settlements not considered. (In Sec. 5.2.1 we will develop a method of considering support settlements that will allow us to apply the unconstrained coordinate code concept to the solution of problems involving support settlements.)

Further inspection of our computer solution will reveal that one additional efficiency measure has been employed. Notice that it was unnecessary to generate separate stiffness arrays for identical members 1 and 3, even though they were loaded differently. Clearly, then, a single stiffness array may be used repeatedly. This repetition makes for a significant efficiency, since structures often contain sets of many identical members. Alternate fixed-end force vectors must, of course, be generated for differently loaded members of a set even though they are physically identical.

Finally, it should be noted again that reactions cannot be obtained directly by backsubstitution, because the structure stiffness equations associated with the constrained coordinate directions are not available. But this is perhaps not terribly important, since all member-end forces may be determined by backsubstitution. Reaction forces that are of interest may then be obtained by hand, external to the computer from nodal equilibrium considerations at supported nodes. A method of solution that retains all structure stiffness equations and allows direct determination of reactions is presented next.

4.5.3 Rearranged Stiffness Relationships

The unconstrained structure stiffness relationships of Eq. Q of Sec. 4.5.1 may be constrained by substituting the known displacements in supported coordinate directions into the displacement vector. Doing this and also substituting the known applied forces into the nodal force vector, we get Eq. A on page 165. As was previously noted, this set of stiffness equations contains only two unknown displacements, those being in the unconstrained coordinate directions. It should also be noted that the two stiffness equations corresponding to the unconstrained coordinate directions involve only those two displacements as unknowns. In general, there will always be the same number of unconstrained coordinate direction stiffness equations as there are displacement unknowns in any problem. It follows that if this set of equations can be extracted from the full set of equations, they can be solved simultaneously for the unknown displacements. These displacements can then be backsubstituted into the remaining stiffness equations corresponding to the constrained structure coordinate directions to obtain the reaction force unknowns. To carry out these operations by matrix analysis, we must rearrange the structure force-displacement relationships. This rearrangement can be carried out in two steps for simplicity. First, continuing with the beam, we reorder Eq. A to place first the equations written in the unconstrained coordinate directions. Thus we get Eq. A′ on page 165. Second, for the sake of consistency, we must reorder the displacement vector in a corresponding fashion. This step requires reordering the columns of the stiffness array as well. The result is given in Eq. A″ on page 166. Notice that this last step has restored symmetry to the structure stiffness array.

We could have built the rearranged structure stiffness relationships of Eq. A″ directly by using a coordinate code based on the rearranged set of coordinates of

$$\begin{pmatrix} R_1 \\ R_2 \\ R_3 \\ 0 \\ R_5 \\ -20 \\ R_7 \\ R_8 \end{pmatrix} = \frac{EI_0}{12} \begin{bmatrix} .010427 & .125 & -.010427 & .125 & \cdot & \cdot & \cdot & \cdot \\ .125 & 2 & -.125 & 1 & \cdot & \cdot & \cdot & \cdot \\ -.010427 & -.125 & .02376 & .075 & -.013333 & .2 & \cdot & \cdot \\ .125 & 1 & .075 & 6 & -.2 & 2 & \cdot & \cdot \\ \cdot & \cdot & -.013333 & -.2 & .02376 & -.075 & -.010427 & .125 \\ \cdot & \cdot & .2 & 2 & -.075 & 6 & -.125 & 1 \\ \cdot & \cdot & \cdot & \cdot & -.010427 & -.125 & .010427 & -.125 \\ \cdot & \cdot & \cdot & \cdot & .125 & 1 & -.125 & 2 \end{bmatrix} \begin{pmatrix} -.05 \\ .005 \\ 0 \\ D_4 \\ -.0375 \\ D_6 \\ 0 \\ 0 \end{pmatrix} + \begin{pmatrix} 72 \\ 288 \\ 98.667 \\ -128 \\ 9.333 \\ -80 \\ 0 \\ 0 \end{pmatrix} \quad \text{(A)}$$

$$\begin{pmatrix} 0 \\ -20 \\ R_1 \\ R_2 \\ R_3 \\ R_5 \\ R_7 \\ R_8 \end{pmatrix} = \frac{EI_0}{12} \begin{bmatrix} .125 & 1 & .075 & 6 & -.2 & 2 & \cdot & \cdot \\ \cdot & \cdot & .2 & 2 & -.075 & 6 & -.125 & 1 \\ .010427 & .125 & -.010427 & .125 & \cdot & \cdot & \cdot & \cdot \\ .125 & 2 & -.125 & 1 & \cdot & \cdot & \cdot & \cdot \\ -.010427 & -.125 & .02376 & .075 & -.013333 & .2 & \cdot & \cdot \\ \cdot & \cdot & -.013333 & -.2 & .02376 & -.075 & -.010427 & .125 \\ \cdot & \cdot & \cdot & \cdot & -.010427 & -.125 & .010427 & -.125 \\ \cdot & \cdot & \cdot & \cdot & .125 & 1 & -.125 & 2 \end{bmatrix} \begin{pmatrix} -.05 \\ .005 \\ 0 \\ D_4 \\ -.0375 \\ D_6 \\ 0 \\ 0 \end{pmatrix} + \begin{pmatrix} -128 \\ -80 \\ 72 \\ 288 \\ 98.667 \\ 9.333 \\ 0 \\ 0 \end{pmatrix} \quad \text{(A')}$$

$$\begin{pmatrix} 0 \\ -20 \\ \hdashline R_1 \\ R_2 \\ R_3 \\ R_5 \\ R_7 \\ R_8 \end{pmatrix} = \frac{EI_0}{12} \left[\begin{array}{cc:cccccc} 6 & 2 & .125 & 1 & .075 & -.2 & \cdot & \cdot \\ 2 & 6 & \cdot & \cdot & .2 & -.075 & -.125 & 1 \\ \hdashline .125 & \cdot & .010427 & .125 & -.010427 & \cdot & \cdot & \cdot \\ 1 & \cdot & .125 & 2 & -.125 & \cdot & \cdot & \cdot \\ .075 & .2 & -.010427 & -.125 & .02376 & -.013333 & \cdot & \cdot \\ -.2 & -.075 & \cdot & \cdot & -.013333 & .02376 & .010427 & .125 \\ \cdot & -.125 & \cdot & \cdot & \cdot & -.010427 & .010427 & -.125 \\ \cdot & 1 & \cdot & \cdot & \cdot & .125 & -.125 & 2 \end{array}\right] \begin{pmatrix} D_4 \\ D_6 \\ \hdashline -.05 \\ .005 \\ 0 \\ -.0375 \\ 0 \\ 0 \end{pmatrix} + \begin{pmatrix} -128 \\ -80 \\ \hdashline 72 \\ 288 \\ 98.667 \\ 9.333 \\ 0 \\ 0 \end{pmatrix} \qquad (A'')$$

Fig. 4.8d. Such a rearranged coordinate code for the example beam is as follows:

Rearranged Coordinate Code

Member number ↓	1	2	3	4 ← Coordinate number
1	−3	−4	−5	1
2	−5	1	−6	2
3	−6	2	−7	−8

(Rows 1–3: Destination vectors)

The negative signs serve only to signal constrained coordinates allowing the arrays of Eq. A″ to be partitioned into the subarrays associated with the unconstrained and constrained degrees of freedom. Using the destination vectors of this code and the member stiffness and fixed-end force arrays of page 135, we obtain, for example,

$$K_{11} = k_{44}^{(1)} + k_{22}^{(2)} = \frac{EI_0}{12} \cdot 2 + \frac{EI_0}{12} \cdot 4 = \frac{EI_0}{12} \cdot 6$$

$$K_{12} = k_{24}^{(2)} = \frac{EI_0}{12} \cdot 2$$

$$K_{13} = k_{41}^{(1)} = \frac{EI_0}{12}(.125)$$

$$K_{55} = k_{33}^{(1)} + k_{11}^{(2)} = \frac{EI_0}{12}(.010427) + \frac{EI_0}{12}(.013333) = \frac{EI_0}{12}(.02376)$$

$$K_{56} = k_{13}^{(2)} = \frac{EI_0}{12}(-.013333)$$

$$R_{F3} = r_{F1}^{(1)} = 72$$

$$R_{F6} = r_{F3}^{(2)} + r_{F1}^{(3)} = 9.333$$

These equations are seen to check the stiffness coefficients and fixed-end forces of Eq. A″.

For added convenience and efficiency, the rearranged destination vectors of our rearranged coordinate code can be generated automatically from a constraint list and a nodal code of member incidences. Consider the constraint list for the example beam:

Constraint List

Node numbers	Constrained coordinate*	Initial displacement
1	1	−.05
1	2	.005
2	1	0
3	1	−.0375
4	1	0
4	2	0

*The nodal coordinate numbers supplied are based on the usual ordering of coordinates at a node; that is, the translational and rotational coordinates for a beam are the first and second ones, respectively.

Using this constraint list, we can obtain the rearranged structure coordinate numbering associated with each node by using the following strategy:

1. The number of constraints is counted, thus allowing the number of unconstrained coordinates, NU, to be calculated (given the total number of nodes and the number of degrees of freedom per node).
2. The constraint list is checked for each nodal coordinate, node by node. Constrained coordinates are numbered in one sequence, starting with $-\text{NU} - 1$ and incrementing by -1; unconstrained coordinates are numbered in a second sequence, starting with 1 and incrementing by 1.

Proceeding in this fashion for the example beam yields NU = 2 and the following table of rearranged coordinates:

Rearranged Coordinates

Node number	Rearranged coordinates	
1	−3	−4
2	−5	1
3	−6	2
4	−7	−8

The member incidences for the members of the example beam are listed in the following nodal code:

Nodal Code

Member number	Member incidences	
1	1	2
2	2	3
3	3	4

Matching these nodal code numbers with their corresponding coordinate numbers generated earlier using the constraint list, we get the rearranged coordinate code already shown on page 167.

The considerable efficiency gained by using the nodal code and the constraint list may not be evident from this example. To appreciate this efficiency gain, observe that the constraint list has to contain only nodes at which a constraint exists, and this would typically involve but a small fraction of the total nodes of a structure.

The information in the table of rearranged coordinates may be further used, along with the constraint list information, to generate a rearranged displacement vector. For example, the fourth row of the constraint list indicates that an initial displacement of −.0375 m is imposed in coordinate direction 1 of node 3. The third row of the table of rearranged coordinates indicates that coordinate 1 of node 3 is structure coordinate 6. Thus $D_{06} = -.0375$. Continuing in this fashion, we could fill out the entire displacement vector, with displacements in unconstrained coordinate directions being identified as unknowns D_1 and D_2.

In a similar manner, we could generate a rearranged nodal load vector from the table of rearranged coordinates and the following information on nodal loads:

Nodal Loads

Node number	Coordinate direction	Load
3	2	−20

The table of rearranged coordinates indicates that coordinate 2 of node 3 is (coincidentally, in this case) structure coordinate 2; thus $R_2 = -20$. Other forces in the nodal load vector are taken as zero by default.

Now the automatic generation of the rearranged stiffness relationships of Eq. A″ is complete. Two special structural analysis commands included in SMIS3 provide the capability of operating according to this very efficient form of the direct stiffness method. Before considering the use of these commands, however, we must first develop a detailed solution to Eq. A″.

The partitioned structure force-displacement relationships Eq. A″ may be represented symbolically as

$$\left(\begin{array}{c}\mathbf{R}_u \\ \hline \mathbf{R}_c\end{array}\right) = \left[\begin{array}{c|c}\mathbf{K}_{uu} & \mathbf{K}_{uc} \\ \hline \mathbf{K}_{cu} & \mathbf{K}_{cc}\end{array}\right]\left(\begin{array}{c}\mathbf{D}_u \\ \hline \mathbf{D}_c\end{array}\right) + \left(\begin{array}{c}\mathbf{R}_{Fu} \\ \hline \mathbf{R}_{Fc}\end{array}\right) \tag{4.32}$$

The u subscript indicates a subvector or submatrix that is related to an unconstrained degree of freedom where displacements are unknown, and the c subscript indicates a subvector or submatrix that is related to a constrained degree of freedom where reaction forces are unknown. Expanding Eq. 4.32 yields

$$\mathbf{R}_u = \mathbf{K}_{uu}\mathbf{D}_u + \mathbf{K}_{uc}\mathbf{D}_c + \mathbf{R}_{Fu} \tag{4.33}$$

and

$$\mathbf{R}_c = \mathbf{K}_{cu}\mathbf{D}_u + \mathbf{K}_{cc}\mathbf{D}_c + \mathbf{R}_{Fc} \tag{4.34}$$

Equation 4.33 may now be solved for the unknown displacements:

$$\mathbf{D}_u = -\mathbf{K}_{uu}^{-1}\mathbf{K}_{uc}\mathbf{D}_c + \mathbf{K}_{uu}^{-1}(\mathbf{R}_u - \mathbf{R}_{Fu}) \tag{4.35}$$

Backsubstituting this result into Eq. 4.34 gives, for the reactions,

$$\mathbf{R}_c = -\mathbf{K}_{cu}\mathbf{K}_{uu}^{-1}\mathbf{K}_{uc}\mathbf{D}_c + \mathbf{K}_{cu}\mathbf{K}_{uu}^{-1}(\mathbf{R}_u - \mathbf{R}_{Fu}) + \mathbf{K}_{cc}\mathbf{D}_c + \mathbf{R}_{Fc} \tag{4.36}$$

Now consider applying the general solutions of Eq. 4.35 and 4.36 to the beam example. From Eq. A″, it is seen that

$$\mathbf{R}_u = \langle\ 0 \quad -20\rangle$$

$$\mathbf{R}_{Fu} = \langle -128 \quad -80\rangle$$

$$\mathbf{D}_c = \langle -.05 \quad .005 \quad 0 \quad -.0375 \quad 0 \quad 0\rangle$$

$$\mathbf{K}_{uu} = \frac{EI_0}{12}\begin{bmatrix} 6 & 2 \\ 2 & 6 \end{bmatrix} \qquad \mathbf{K}_{uu}^{-1} = \frac{1}{32}\cdot\frac{12}{EI_0}\begin{bmatrix} 6 & -2 \\ -2 & 6 \end{bmatrix}$$

$$\mathbf{K}_{uc} = \frac{EI_0}{12}\begin{bmatrix} .125 & 1 & .075 & -.2 & \cdot & \cdot \\ \cdot & \cdot & .2 & -.075 & -.125 & 1 \end{bmatrix}$$

Substituting into Eq. 4.35 gives

$$\mathbf{D}_u = -\frac{1}{32}\frac{12}{EI_0}\begin{bmatrix} 6 & -2 \\ -2 & 6 \end{bmatrix}\frac{EI_0}{12}\begin{bmatrix} .125 & 1 & .075 & -.2 & \cdot & \cdot \\ \cdot & \cdot & .2 & -.075 & -.125 & 1 \end{bmatrix}\begin{pmatrix} -.05 \\ .005 \\ 0 \\ -.0375 \\ 0 \\ 0 \end{pmatrix}$$

$$+\frac{1}{32}\frac{12}{EI_0}\begin{bmatrix} 6 & -2 \\ -2 & 6 \end{bmatrix}\begin{pmatrix} 0 + 128 \\ -20 + 80 \end{pmatrix}$$

or

$$\mathbf{D}_u = -\frac{1}{32}\begin{bmatrix} 6 & -2 \\ -2 & 6 \end{bmatrix}\begin{pmatrix} .00625 \\ .0028125 \end{pmatrix} + \frac{1}{32}\frac{12}{EI_0}\begin{bmatrix} 6 & -2 \\ -2 & 6 \end{bmatrix}\begin{pmatrix} 128 \\ 60 \end{pmatrix}$$

Taking $EI_0/12 = 10{,}000\ \text{kN} \cdot \text{m}^2$ as before gives, upon substitution,

$$\mathbf{D}_u = -\frac{1}{32}\begin{pmatrix} .031875 \\ .004375 \end{pmatrix} + \frac{1}{32}\frac{1}{10{,}000}\begin{pmatrix} 648. \\ 104. \end{pmatrix}$$

or

$$\mathbf{D}_u = \begin{pmatrix} -.000996 \\ -.0001367 \end{pmatrix} + \frac{1}{10{,}000}\begin{pmatrix} 20.25 \\ 3.25 \end{pmatrix} = \begin{pmatrix} .001029 \\ .0001883 \end{pmatrix}$$

Displacements due to foundation settlements — Displacements due to applied loads — Total displacements

Backsubstituting these results into Eq. 4.34 gives

$$\mathbf{R}_c = \frac{EI_0}{12}\begin{bmatrix} .125 & \cdot \\ 1 & \cdot \\ .075 & .2 \\ -.2 & -.075 \\ \cdot & -.125 \\ \cdot & 1 \end{bmatrix}\begin{pmatrix} .001029 \\ .0001883 \end{pmatrix}$$

$$+\frac{EI_0}{12}\begin{bmatrix} .010427 & .125 & -.010427 & \cdot & \cdot & \cdot \\ .125 & 2 & -.125 & \cdot & \cdot & \cdot \\ -.010427 & -.125 & .02376 & -.013333 & \cdot & \cdot \\ \cdot & \cdot & -.013333 & .02376 & -.010427 & .125 \\ \cdot & \cdot & \cdot & -.010427 & .010427 & -.125 \\ \cdot & \cdot & \cdot & .125 & -.125 & 2 \end{bmatrix}$$

$$\times \begin{pmatrix} -.05 \\ .005 \\ 0 \\ -.0375 \\ 0 \\ 0 \end{pmatrix} + \begin{pmatrix} 72 \\ 288 \\ 98.667 \\ 9.333 \\ 0 \\ 0 \end{pmatrix}$$

or

$$\mathbf{R}_c = \langle 74.322 \quad 335.79 \quad 103.77 \quad -1.776 \quad 3.675 \quad -44.992 \rangle$$

Finally, the member-end displacements may be determined from the structure nodal displacements and backsubstituted into the member force-displacement relationships to obtain the member-end forces for each member. Since this step is identical with that used previously, it will not be repeated here. Once again it can be observed that this solution to the beam problem checks the previous solutions.

To conserve storage, it is likely that the stiffness relationships of Eqs. 4.33 and 4.34 would be built directly in the following respective forms:

$$(\mathbf{R}_u - \mathbf{R}_{Fu} - \mathbf{K}_{uc}\mathbf{D}_c) = \mathbf{K}_{uu}\mathbf{D}_u \tag{4.37}$$

and

$$\mathbf{R}_c = \mathbf{K}_{cu}\mathbf{D}_u + (\mathbf{K}_{cc}\mathbf{D}_c + \mathbf{R}_{Fc}) \tag{4.38}$$

Or, taking $EI_0/12 = 10{,}000 \text{ kN} \cdot \text{m}^2$, we get

$$\begin{pmatrix} 65.5 \\ 31.88 \end{pmatrix} = \frac{EI_0}{12}\begin{bmatrix} 6 & 2 \\ 2 & 6 \end{bmatrix}\mathbf{D}_u$$

which gives

$$\mathbf{D}_u = \langle .001029 \quad .0001883 \rangle$$

and

$$\mathbf{R}_c = \frac{EI_0}{12}\begin{bmatrix} .125 & \cdot \\ 1 & \cdot \\ .075 & .2 \\ -.2 & -.075 \\ \cdot & -.125 \\ \cdot & 1 \end{bmatrix}\mathbf{D}_u + \begin{pmatrix} 73.04 \\ 325.5 \\ 102.63 \\ .4268 \\ 3.908 \\ -46.88 \end{pmatrix}$$

Thus only the arrays $\mathbf{K}_{uu}$ and $\mathbf{K}_{cu}$ plus a modified $\mathbf{R}_{Fu}$ and a modified $\mathbf{R}_{Fc}$ need be generated.

Further efficiency can be gained by not generating Eq. 4.38 at all (thus sacrificing direct reaction calculation) and by not generating $\mathbf{K}_{uc}$ explicitly, but rather generating the vector resulting from the product $\mathbf{K}_{uc}\mathbf{D}_c$. This latter efficiency can be accomplished as follows.

Consider a typical coefficient from $\mathbf{K}_{uc}$, say K_{26}. From the rearranged coordinate code (p. 167), it is seen that

$$K_{26} = k_{43}^{(2)} + k_{21}^{(3)}$$

In the product $\mathbf{K}_{uc}\mathbf{D}_c$ this coefficient multiplies the displacement D_6, giving

$$[k_{43}^{(2)} + k_{21}^{(3)}]D_6$$

and, according to Eq. 4.37, this result is to be subtracted from $\mathbf{R}_{u2}$. Instead, we can subtract

$$k_{43}^{(2)}D_6 \quad \text{and} \quad k_{21}^{(3)}D_6$$

from $\mathbf{R}_{u2}$, in sequence, without forming K_{26} at all. This can be accomplished on a member-by-member basis as the structure stiffness relationships represented by Eq. 4.37 are formed.

Both of the efficiencies just described are built into the INITLZ and BLDSTF commands of SMIS3. The use of these commands is demonstrated in the printout that follows on page 174.

The three direct stiffness procedures we have described are summarized in Table 4.1. A study of this table will help to show that there are more essential similarities among the procedures than there are differences. The procedures differ mainly in regard to when constraints are imposed and what is done with the equations associated with constrained degrees of freedom. These differences have effects on both the ease of programming and the efficiency of the resulting program.

The storage and manipulation efficiencies are of particular importance when considering their use in programs that are intended for application to the solution of large problems. Solution efficiency may be further improved by other techniques.

Table 4.1 ALTERNATIVE DIRECT STIFFNESS PROCEDURES

PROCEDURE	CONSTRAINTS	EQUATIONS	SOLUTION
Modified (Sec. 4.5.1)	Imposed after stiffness equations are generated	Complete set generated in nodal order	Solve complete modified set (including trivial replacement equations associated with constrained degrees of freedom)
Unconstrained coordinate code (Sec. 4.5.2)	Imposed as stiffness equations are generated	Only equations associated with unconstrained degrees of freedom generated; in nodal order	Solve equations associated with unconstrained degrees of freedom
Rearrangement (Sec. 4.5.3)	Imposed as stiffness equations are generated	Complete set generated as two separate sets, each set in nodal order	Solve equations associated with unconstrained degrees of freedom. Backsubstitute into equations associated with constrained degrees of freedom to obtain reaction forces

Our discussions here have focused on efficiencies involved in the generation of the direct stiffness equations; in particular, we have been concerned with reducing this set of equations to a minimum size. A detailed discussion of efficient methods of solving this minimum set of simultaneous equations will be presented in Chap. 7. One further technique that can be used to reduce the size of the set of simultaneous equations, but which is applicable only to symmetric structures, is offered in the next section. One final point before proceeding to that subject: It is essential that all reasonable efforts be made to employ all manner of efficiencies in the development and use of any computer program intended for the solution of large problems.

4.6 STRUCTURAL SYMMETRY

Structural symmetry involves both the innate physical symmetry of the structure itself and the symmetry of the load applied thereon. For a structure to be classified as symmetric, one half of the structure must be a mirror image of the other half in terms of three characteristics: (1) configuration, (2) distribution of material properties, and (3) arrangement of constraints. The line that divides one half of the structure from the other half (the line along which the imaginary mirror lies) is referred to as its axis of symmetry. If the loading on one half of the

****** *Computer Input:*

```
START
REMARK 6
SOLUTION OF CONTINOUS BEAM USING INITLZ AND BLDSTF
COMMANDS. SUPPORT SETTLEMENTS ARE CONSIDERED.

E = 2E08 KN/M**2, I(0) = .0006 M**4, EI(0)/12 = 10000 KN-M**2

**** UNITS ARE KILONEWTONS AND METERS ****
INITLZ KUU RU DO 4 2 6 1 3
1 1 -.05
1 2 .005
2 1 0
3 1 -.0375
4 1 0
4 2 0
3 2 -20
C1 1 2
C2 2 3
C3 3 4
BEAMEL K1 1 RF1
24 2E08 .0006 0 0
F 0 24 -6 -6
BEAMEL K2 1 RF2
30 2E08 .0015 0 0
F 10 10 -36 -36
BLDSTF KUU RU DO 3
K1 RF1 C1
K2 RF2 C2
K1 RF3 C3
PRINT KUU 1
THIS IS THE CONSTRAINED STRUCTURE STIFFNESS MATRIX, KUU.
SOLVE KUU RU DU
PRINT DU 1
THIS IS THE VECTOR, DU, OF STRUCTURE NODAL DISPLACEMENTS.
BEAMFC DU 3
K1 C1 RF1
K2 C2 RF2
K1 C3 RF3
STOP
```

****** *Computer Output:*

```
START

REMARK      6


SOLUTION OF CONTINOUS BEAM USING INITLZ AND BLDSTF
COMMANDS. SUPPORT SETTLEMENTS ARE CONSIDERED.

E = 2E08 KN/M**2, I(0) = .0006 M**4, EI(0)/12 = 10000 KN-M**2

**** UNITS ARE KILONEWTONS AND METERS ****



INITLZ  KUU     RU      DO

 NUMBER OF NODES          =      4
 NUMBER OF DOF PER NODE   =      2
 NUMBER OF CONSTRAINTS    =      6
 NUMBER OF NODAL LOADS    =      1
 NUMBER OF ELEMENTS       =      3

 NULL    2 BY     2 STIFFNESS ARRAY, KUU    ,FORMED
```

****** *Computer Output (continued):*

```
INITIAL DISPL. VECTOR = DO        REARRANGED COORDINATE CODE
      1     -5.000E-02                      -3
      2      5.000E-03                      -4
      3              0                      -5
      4              0                       1
      5     -3.750E-02                      -6
      6              0                       2
      7              0                      -7
      8              0                      -8

NODAL LOAD VECTOR,   RU
            1
1           0
2 -2.0000E+01


MEMBER CODE          DEGREES OF FREEDOM
   C1              -3     -4     -5     1
   C2              -5      1     -6     2
   C3              -6      2     -7    -8

BEAMEL  K1            1      RF1
 E =  2.000E+08      I =  6.000E-04      L =  2.400E+01
 A =           0     G =           0

 LOAD TYPE = F
  DIMENSION C =            0     LOAD INTENSITY AT C = -6.000E+00
  DIMENSION D =  2.400E+01       LOAD INTENSITY AT D = -6.000E+00

BEAMEL  K2            1      RF2
 E =  2.000E+08      I =  1.500E-03      L =  3.000E+01
 A =           0     G =           0

 LOAD TYPE = F
  DIMENSION C =  1.000E+01       LOAD INTENSITY AT C = -3.600E+01
  DIMENSION D =  1.000E+01       LOAD INTENSITY AT D = -3.600E+01

BLDSTF  KUU      RU       DO          3

 MEMBER K   MEMBER RF   MEMBER CODE
    K1        RF1          C1
    K2        RF2          C2
    K1        RF3          C3

 STRUCTURE EQUIVALENT NODAL LOAD VECTOR,  RU
            1
 1  6.5500E+01
 2  3.1875E+01


PRINT   KUU           1

THIS IS THE CONSTRAINED STRUCTURE STIFFNESS MATRIX, KUU.


            1           2
 1  6.0000E+04 2.0000E+04
 2  2.0000E+04 6.0000E+04


SOLVE   KUU      RU       DU

PRINT   DU            1
```

****** *Computer Output (continued):*

```
THIS IS THE VECTOR, DU, OF STRUCTURE NODAL DISPLACEMENTS.


              1
 1  1.0289E-03
 2  1.8828E-04


BEAMFC  DU          3

 MEMBER LOCAL FORCE VECTORS


 MEMBER CODE   = C1

 1  7.43278E+01
 2  3.35789E+02
 3  6.96722E+01
 4 -2.79922E+02

 MEMBER CODE   = C2

 1  3.41010E+01
 2  2.79922E+02
 3  1.89896E+00
 4  2.31094E+01

 MEMBER CODE   = C3

 1 -3.67090E+00
 2 -4.31094E+01
 3  3.67090E+00
 4 -4.49922E+01

STOP
```

structure is also a mirror image about this axis of the loading on the other half, the loading is also classified as symmetric. It will be shown shortly that when a symmetric loading is applied to a symmetric structure, the structure distorts in such a way that its distorted shape remains symmetric also. In other words, with reference to the axis of symmetry, mirror-image distortions occur at corresponding points on each half of the structure. It follows that all other responses (shear, bending moment, etc.) of a symmetric structure subjected to a symmetric loading will also be symmetric in nature.

It is also possible for either the structure or the loading, or both, to be antisymmetric in nature. Since it is uncommon to encounter antisymmetric structures, that possibility will not be considered here. If the loading on one half of a symmetric structure is a reverse mirror image about the axis of symmetry of the loading on the other half, the loading is said to be antisymmetric. It will be shown that when an antisymmetric loading is applied to a symmetric structure, reverse mirror image responses occur at corresponding points on each half of the structure with reference to the axis of symmetry.

It should be noted further that a structure may possess more than one axis of symmetry. Also we will see that when an unsymmetric loading is applied to a symmetric structure, the general loading can easily be separated into its symmetric and antisymmetric components so that symmetry concepts can be employed.

To develop a feeling for the consequences of symmetric or antisymmetric loadings to the behavior of a symmetric structure, it is helpful to consider first a simple structure subjected to a variety of loadings (Fig. 4.9). Choosing coordinate axes such that one axis is collinear with the axis of symmetry of the structure makes it possible to readily determine symmetric and antisymmetric functions by inspection of their mathematical statements. Consider, for example, the expressions for the loadings shown in Fig. 4.9:

4.9a: $q(x) = q_0$ physically symmetric, mathematically even

4.9b: $q(x) = 2q_0x/L$ physically antisymmetric, mathematically odd

4.9c: $p(x) = 2p_0x/L$ physically symmetric, mathematically odd

4.9d: $p(x) = p_0$ physically antisymmetric, mathematically even

At first glance these results may seem inconsistent, since one might expect symmetric functions to be mathematically even and antisymmetric functions to be mathematically odd. Note, however, that the x axis is itself antisymmetric, whereas the y and z axes are symmetrically disposed with respect to the structure. In addition, recall that the sign convention for p loadings is referred to the x axis, whereas that for the q loadings is referred to the y axis. Thus the mathematical oddness or evenness of the expression for loading will depend on the sign convention adopted for that load. The same statement can, of course, be made for responses.

Consider the loading case of Fig. 4.9a. From elementary bending theory (Ref. 1.2) (neglecting interaction of axial and flexural modes of behavior), we may write

$$\frac{d^4v}{dx^4} = \frac{q_0}{EI}$$

Integrating gives

$$-\frac{V}{EI} = \frac{d^3v}{dx^3} = \frac{q_0x}{EI} + C_1$$

$$-\frac{M}{EI} = \frac{d^2v}{dx^2} = \frac{q_0x^2}{2EI} + C_1x + C_2$$

$$\frac{dv}{dx} = \frac{q_0x^3}{6EI} + \frac{C_1x^2}{2} + C_2x + C_3$$

$$v = \frac{q_0x^4}{24EI} + \frac{C_1x^3}{6} + \frac{C_2x^2}{2} + C_3x + C_4$$

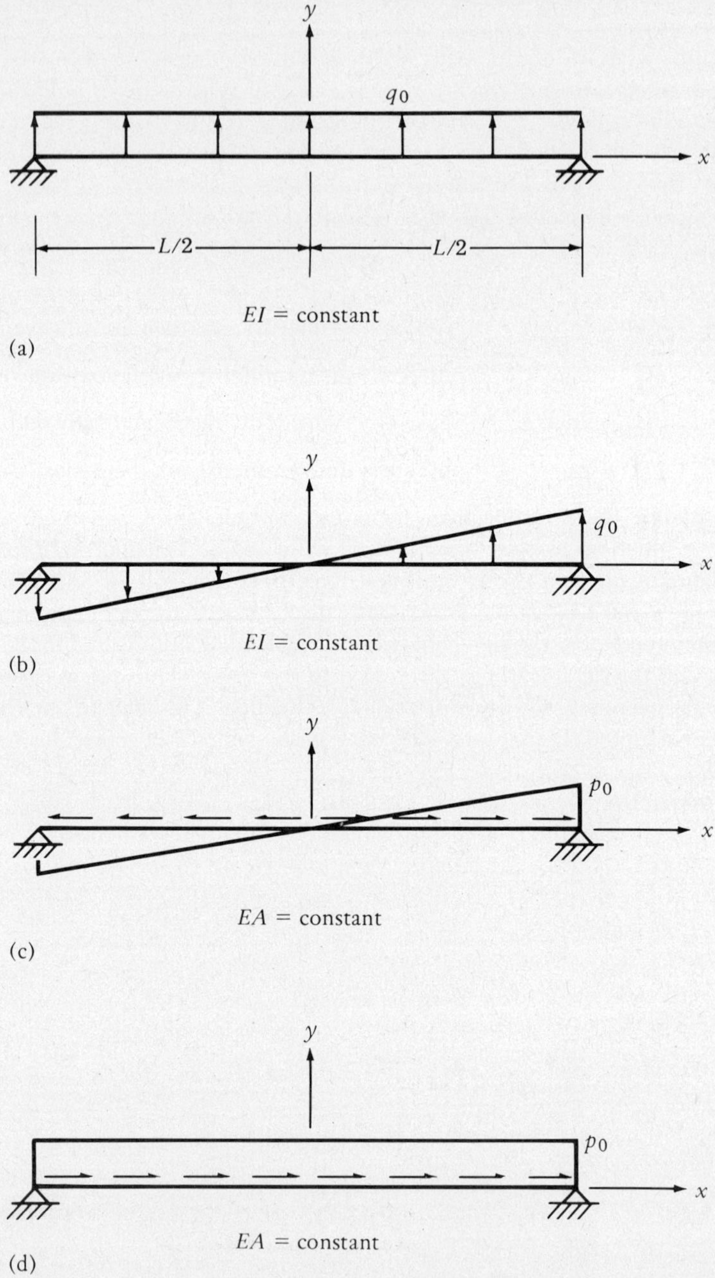

Fig. 4.9 Symmetry examples: (a) symmetric loading: *y* direction; (b) antisymmetric loading: *y* direction; (c) symmetric loading: *x* direction; (d) antisymmetric loading: *x* direction.

Enforcing boundary conditions gives

$$M\left(\pm\frac{L}{2}\right) = 0 \rightarrow \frac{d^2v}{dx^2}\left(\pm\frac{L}{2}\right) = 0 \Rightarrow \frac{q_0L^2}{8EI} \pm C_1\frac{L}{2} + C_2 = 0$$

$$\therefore C_1 = 0, \quad C_2 = -\frac{q_0L^2}{8EI}$$

$$v\left(\pm\frac{L}{2}\right) = 0 \rightarrow \frac{q_0L^4}{384EI} - \frac{q_0L^2}{16EI}\cdot\frac{L^2}{4} \pm \frac{C_3L}{2} + C_4 = 0$$

$$\therefore C_3 = 0, \quad C_4 = \frac{5qL^4}{384EI}$$

Backsubstituting, we see that where

$q(x) = q_0$ mathematically even

we find

$V = -q_0x$ mathematically odd

$M = -\dfrac{q_0}{2}\left(x^2 - \dfrac{L^2}{4}\right)$ mathematically even

$\dfrac{dv}{dx} = \dfrac{q_0}{EI}\left(\dfrac{x^3}{6} - \dfrac{xL^2}{8}\right)$ mathematically odd

and

$v = \dfrac{q_0}{EI}\left(\dfrac{x^4}{24} - \dfrac{x^2L^2}{16} + \dfrac{5L^4}{384}\right)$ mathematically even

Thus although all the preceding functions are physically symmetric, some are mathematically even and some are odd. A little experience with results of this type will make it possible to anticipate which result to expect. You may wish to carry out a detailed study of the remaining examples of Fig. 4.9 to gain more confidence in these symmetry concepts.

The continuous beam of Fig. 4.8 has an axis of symmetry but is subjected to an unsymmetric loading. It is seen from Fig. 4.10a, however, that the unsymmetric loading may be broken into its symmetric and antisymmetric components. Thus the general applied loads may be represented as

$$\begin{aligned} R_4 &= R_{S4} + R_{A4} \\ R_6 &= R_{S6} + R_{A6} = -R_{S4} + R_{A4} \end{aligned} \qquad \text{(A)}$$

General (unsymmetrical) loading

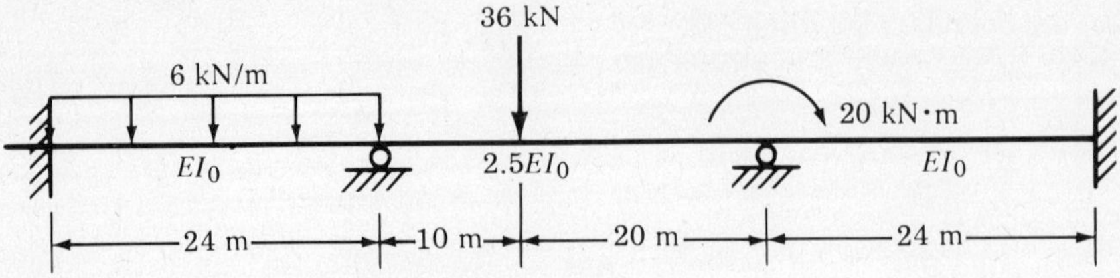

$\mathbf{D}_o = <-.05\ .005\ 0\ 0\ -.0375\ 0\ 0\ 0>$

=

Symmetrical loading

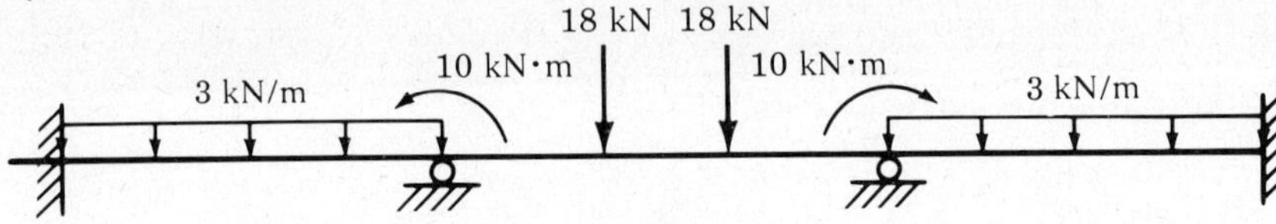

$\mathbf{D}_o = <-.025\ .0025\ -.01875\ 0\ -.01875\ 0\ -.025\ -.0025>$

+

Antisymmetrical loading

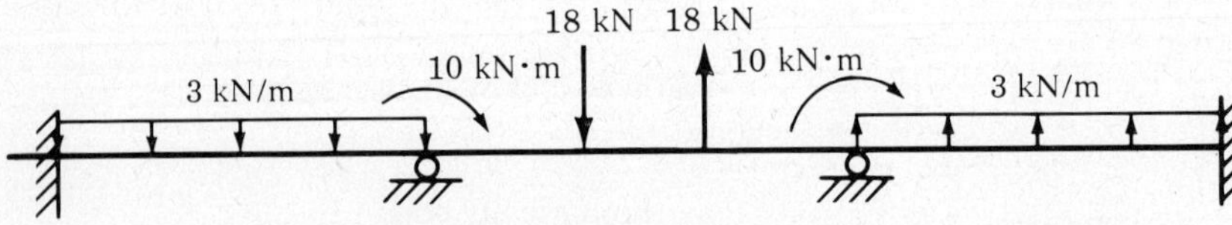

$\mathbf{D}_o = <-.025\ .0025\ .01875\ 0\ -.01875\ 0\ .025\ .0025>$

(a)

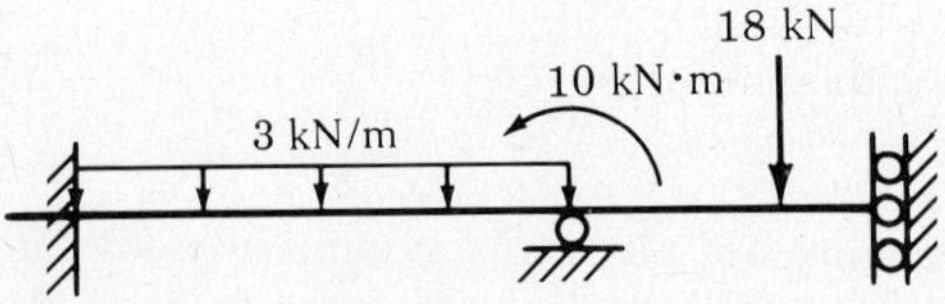

$\mathbf{D}_o = <-.025\ .0025\ -.01875\ 0\ 0\ 0>$

(b)

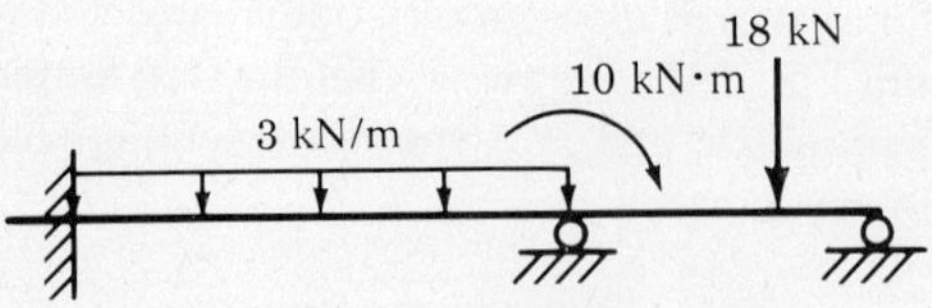

$\mathbf{D}_o = <-.025\ .0025\ .01875\ 0\ 0\ 0>$

(c)

Fig. 4.10 Symmetrical beam: (a) symmetry example; (b) left half of symmetrical component; (c) left half of antisymmetrical component.

where the S and A subscripts denote symmetric and antisymmetric components, respectively. Similarly, the responding reactions will be

$$
\begin{aligned}
R_1 &= R_{S1} + R_{A1} \\
R_2 &= R_{S2} + R_{A2} \\
R_3 &= R_{S3} + R_{A3} \\
R_5 &= R_{S5} + R_{A5} = R_{S3} - R_{A3} \\
R_7 &= R_{S7} + R_{A7} = R_{S1} - R_{A1} \\
R_3 &= R_{S8} + R_{A8} = -R_{S2} + R_{A2}
\end{aligned}
\tag{B}
$$

where it is seen that $\mathbf{R}$ can be written entirely in terms of, say, left-hand side symmetric and antisymmetric force components. Solving Eqs. A and B gives

$$
\begin{aligned}
R_{S1} &= \frac{R_1 + R_7}{2} & R_{A1} &= \frac{R_1 - R_7}{2} \\
R_{S2} &= \frac{R_2 - R_8}{2} & R_{A2} &= \frac{R_2 + R_8}{2} \\
R_{S3} &= \frac{R_3 + R_5}{2} & R_{A3} &= \frac{R_3 - R_5}{2} \\
R_{S4} &= \frac{R_4 - R_6}{2} & R_{A4} &= \frac{R_4 + R_6}{2}
\end{aligned}
\tag{C}
$$

or, in matrix form,

$$
\begin{pmatrix} R_{S1} \\ R_{S2} \\ R_{S3} \\ R_{S4} \\ \hdashline R_{A1} \\ R_{A2} \\ R_{A3} \\ R_{A4} \end{pmatrix}
= \frac{1}{2}
\left[\begin{array}{cccc:cccc}
1 & \cdot & \cdot & \cdot & \cdot & \cdot & 1 & \cdot \\
\cdot & 1 & \cdot & \cdot & \cdot & \cdot & \cdot & -1 \\
\cdot & \cdot & 1 & \cdot & 1 & \cdot & \cdot & \cdot \\
\cdot & \cdot & \cdot & 1 & \cdot & -1 & \cdot & \cdot \\
\hdashline
1 & \cdot & \cdot & \cdot & \cdot & \cdot & -1 & \cdot \\
\cdot & 1 & \cdot & \cdot & \cdot & \cdot & \cdot & 1 \\
\cdot & \cdot & 1 & \cdot & -1 & \cdot & \cdot & \cdot \\
\cdot & \cdot & \cdot & 1 & \cdot & 1 & \cdot & \cdot
\end{array}\right]
\begin{pmatrix} R_1 \\ R_2 \\ R_3 \\ R_4 \\ \hdashline R_5 \\ R_6 \\ R_7 \\ R_8 \end{pmatrix}
\tag{C'}
$$

or

$$
\begin{pmatrix} \mathbf{R}_S \\ \hdashline \mathbf{R}_A \end{pmatrix}
= \frac{1}{2}
\left[\begin{array}{c:c} \mathbf{I} & \mathbf{E} \\ \hdashline \mathbf{I} & -\mathbf{E} \end{array}\right]
\begin{pmatrix} \mathbf{R}_L \\ \hdashline \mathbf{R}_R \end{pmatrix}
\tag{C''}
$$

where the L and R subscripts indicate structure left-hand side and right-hand side forces, respectively; and the **I** and **E** arrays are identity and elementary arrays, as discussed in App. 1.

Similarly, it is seen that both the imposed and responding displacements for the general unsymmetric case may be represented in terms of the symmetric and antisymmetric displacement components as

$$\begin{aligned}
D_1 &= D_{S1} + D_{A1} \\
D_2 &= D_{S2} + D_{A2} \\
D_3 &= D_{S3} + D_{A3} \\
D_4 &= D_{S4} + D_{A4} \\
D_5 &= D_{S5} + D_{A5} = D_{S3} - D_{A3} \\
D_6 &= D_{S6} + D_{A6} = -D_{S4} + D_{A4} \\
D_7 &= D_{S7} + D_{A7} = D_{S1} - D_{A1} \\
D_8 &= D_{S8} + D_{A8} = -D_{S2} + D_{A2}
\end{aligned} \tag{D}$$

or, in matrix form,

$$\begin{pmatrix} D_1 \\ D_2 \\ D_3 \\ D_4 \\ \hline D_5 \\ D_6 \\ D_7 \\ D_8 \end{pmatrix} = \left[\begin{array}{cccc|cccc} 1 & \cdot & \cdot & \cdot & 1 & \cdot & \cdot & \cdot \\ \cdot & 1 & \cdot & \cdot & \cdot & 1 & \cdot & \cdot \\ \cdot & \cdot & 1 & \cdot & \cdot & \cdot & 1 & \cdot \\ \cdot & \cdot & \cdot & 1 & \cdot & \cdot & \cdot & 1 \\ \hline \cdot & \cdot & 1 & \cdot & \cdot & \cdot & -1 & \cdot \\ \cdot & \cdot & \cdot & -1 & \cdot & \cdot & \cdot & 1 \\ 1 & \cdot & \cdot & \cdot & -1 & \cdot & \cdot & \cdot \\ \cdot & -1 & \cdot & \cdot & \cdot & 1 & \cdot & \cdot \end{array}\right] \begin{pmatrix} D_{S1} \\ D_{S2} \\ D_{S3} \\ D_{S4} \\ \hline D_{A1} \\ D_{A2} \\ D_{A3} \\ D_{A4} \end{pmatrix} \tag{D'}$$

or

$$\begin{pmatrix} \mathbf{D}_L \\ \hline \mathbf{D}_R \end{pmatrix} = \left[\begin{array}{c|c} \mathbf{I} & \mathbf{I} \\ \hline \mathbf{E} & -\mathbf{E} \end{array}\right] \begin{pmatrix} \mathbf{D}_S \\ \hline \mathbf{D}_A \end{pmatrix} \tag{D''}$$

In Eq. C″ the factor of $\frac{1}{2}$, which seems to contradict the contragredient law, is present because the vectors $\langle \mathbf{R}_S \mid \mathbf{R}_A \rangle$ and $\langle \mathbf{D}_S \mid \mathbf{D}_A \rangle$ represent the action of only "half" of the total structure. Equation D may be solved for $D_{S1}, \ldots, D_{S4}$ and

$D_{A1}, \ldots, D_{A4}$ in terms of $D_1, \ldots, D_8$, giving

$$
\begin{aligned}
D_{S1} &= \frac{D_1 + D_7}{2} & D_{A1} &= \frac{D_1 - D_7}{2} \\
D_{S2} &= \frac{D_2 - D_8}{2} & D_{A2} &= \frac{D_2 + D_8}{2} \\
D_{S3} &= \frac{D_3 + D_5}{2} & D_{A3} &= \frac{D_3 - D_5}{2} \\
D_{S4} &= \frac{D_4 - D_6}{2} & D_{A4} &= \frac{D_4 + D_6}{2}
\end{aligned}
\tag{E}
$$

Equation E, along with Eq. C, is useful for determining symmetric and antisymmetric loadings, given the general loading. Thus, for example,

$$R_{S4} = \frac{R_4 - R_6}{2} = \frac{0 - (-20)}{2} = 10 \text{ kN} \cdot \text{m}$$

and

$$D_{A2} = \frac{D_2 + D_8}{2} = \frac{.05 + 0}{2} = .0025 \text{ m}$$

Rewriting Eq. Q of Sec. 4.5.1 in general symbolic terms gives

$$
\begin{pmatrix} R_1 \\ R_2 \\ R_3 \\ R_4 \\ R_5 \\ R_6 \\ R_7 \\ R_8 \end{pmatrix}
=
\begin{bmatrix}
K_{11} & K_{12} & K_{13} & K_{14} & K_{15} & K_{16} & K_{17} & K_{18} \\
K_{21} & K_{22} & K_{23} & K_{24} & K_{25} & K_{26} & K_{27} & K_{28} \\
K_{31} & K_{32} & K_{33} & K_{34} & K_{35} & K_{36} & K_{37} & K_{38} \\
K_{41} & K_{42} & K_{43} & K_{44} & K_{45} & K_{46} & K_{47} & K_{48} \\
K_{51} & K_{52} & K_{53} & K_{54} & K_{55} & K_{56} & K_{57} & K_{58} \\
K_{61} & K_{62} & K_{63} & K_{64} & K_{65} & K_{66} & K_{67} & K_{68} \\
K_{71} & K_{72} & K_{73} & K_{74} & K_{75} & K_{76} & K_{77} & K_{78} \\
K_{81} & K_{82} & K_{83} & K_{84} & K_{85} & K_{86} & K_{87} & K_{88}
\end{bmatrix}
\begin{pmatrix} D_1 \\ D_2 \\ D_3 \\ D_4 \\ D_5 \\ D_6 \\ D_7 \\ D_8 \end{pmatrix}
+
\begin{pmatrix} R_{F1} \\ R_{F2} \\ R_{F3} \\ R_{F4} \\ R_{F5} \\ R_{F6} \\ R_{F7} \\ R_{F8} \end{pmatrix}
\tag{F}
$$

or

$$
\begin{pmatrix} \mathbf{R}_L \\ \hline \mathbf{R}_R \end{pmatrix}
=
\left[\begin{array}{c|c} \mathbf{K}_{LL} & \mathbf{K}_{LR} \\ \hline \mathbf{K}_{RL} & \mathbf{K}_{RR} \end{array}\right]
\begin{pmatrix} \mathbf{D}_L \\ \hline \mathbf{D}_R \end{pmatrix}
+
\begin{pmatrix} \mathbf{R}_{FL} \\ \hline \mathbf{R}_{FR} \end{pmatrix}
\tag{F'}
$$

where the arrays in Eq. F′ have been partitioned appropriately for the convenience of subsequent operations. As previously stated, the subscripts L and R on the subarrays may be interpreted as *l*eft and *r*ight, respectively; thus $\mathbf{K}_{RL}$, for example, relates the forces on the right side of the structure to the displacements on the left side. Using Eqs. C″, D″, and F′, we obtain

$$\left(\begin{array}{c}\mathbf{R}_S\\ \hline \mathbf{R}_A\end{array}\right)=\frac{1}{2}\left[\begin{array}{c|c}\mathbf{I}&\mathbf{E}\\ \hline \mathbf{I}&-\mathbf{E}\end{array}\right]\left[\begin{array}{c|c}\mathbf{K}_{LL}&\mathbf{K}_{LR}\\ \hline \mathbf{K}_{RL}&\mathbf{K}_{RR}\end{array}\right]\left[\begin{array}{c|c}\mathbf{I}&\mathbf{I}\\ \hline \mathbf{E}&-\mathbf{E}\end{array}\right]\left(\begin{array}{c}\mathbf{D}_S\\ \hline \mathbf{D}_A\end{array}\right)+\frac{1}{2}\left[\begin{array}{c|c}\mathbf{I}&\mathbf{E}\\ \hline \mathbf{I}&-\mathbf{E}\end{array}\right]\left(\begin{array}{c}\mathbf{R}_{FL}\\ \hline \mathbf{R}_{FR}\end{array}\right) \tag{G}$$

Carrying out the operations of Eq. G in detail will show that

$$\left(\begin{array}{c}\mathbf{R}_S\\ \hline \mathbf{R}_A\end{array}\right)=\left[\begin{array}{c|c}\mathbf{K}_{SS}&\mathbf{\Theta}\\ \hline \mathbf{\Theta}&\mathbf{K}_{AA}\end{array}\right]\left(\begin{array}{c}\mathbf{D}_S\\ \hline \mathbf{D}_A\end{array}\right)+\left(\begin{array}{c}\mathbf{R}_{FS}\\ \hline \mathbf{R}_{FA}\end{array}\right) \tag{H}$$

or

$$\begin{aligned}\mathbf{R}_S&=\mathbf{K}_{SS}\,\mathbf{D}_S+\mathbf{R}_{FS}\\ \mathbf{R}_A&=\mathbf{K}_{AA}\,\mathbf{D}_A+\mathbf{R}_{FA}\end{aligned} \tag{I}$$

Thus the general statement is broken into a symmetric and an antisymmetric statement, each being half the size of the original statement. You may wish to verify Eq. H in detail for the example beam. In somewhat more detail, the first equation of Eq. I may be shown as Eq. J on page 185. If we further specialize Eq. J for this example by making use of Eq. A of Sec. 4.4.3, we get

$$\begin{pmatrix}R_{S1}\\ R_{S2}\\ R_{S3}\\ 10\end{pmatrix}=\frac{EI_0}{12}\begin{bmatrix}.010427&.125&-.010427&.125\\ .125&2&-.125&1\\ -.010427&-.125&.010427&-.125\\ .125&1&-.125&4\end{bmatrix}\begin{pmatrix}-.025\\ .0025\\ -.01875\\ D_{S4}\end{pmatrix}+\begin{pmatrix}36\\ 144\\ 54\\ -24\end{pmatrix} \tag{J′}$$

Solving the last scalar equation of matrix Eq. J′, with $EI_0/12 = 10{,}000\ \text{kN}\cdot\text{m}^2$ as before, gives

$$D_{S4}=\frac{1}{4}\cdot\frac{1}{10{,}000}(10+24)-\frac{1}{4}(.001719)=.0004203\ \text{m} \tag{K}$$

Backsubstituting into the remaining three equations of Eq. J′ yields

$$\langle R_{S1}\quad R_{S2}\quad R_{S3}\rangle=\langle 39.0\quad 190.4\quad 51.0\rangle \tag{L}$$

The antisymmetric component of the total solution may now be obtained by considering the second equation of Eq. I. Rewriting in somewhat more detail, we get Eq. M on page 185. Further specializing Eq. M for this example, we find that

$$\begin{pmatrix}R_{A1}\\ R_{A2}\\ R_{A3}\\ -10\end{pmatrix}=\frac{EI_0}{12}\begin{bmatrix}.010427&.125&-.010427&.125\\ .125&2&-.125&1\\ -.010427&-.125&.037093&.275\\ .125&1&.275&8\end{bmatrix}\begin{pmatrix}-.025\\ .0025\\ .01875\\ D_{A4}\end{pmatrix}+\begin{pmatrix}36\\ 144\\ 44.667\\ -104\end{pmatrix} \tag{M′}$$

$$
\begin{pmatrix} R_{S1} \\ R_{S2} \\ R_{S3} \\ R_{S4} \end{pmatrix} = \frac{1}{2}\begin{bmatrix}
(K_{11} + K_{17} + K_{71} + K_{77}) & (K_{12} - K_{18} + K_{72} - K_{78}) & (K_{13} + K_{15} + K_{73} + K_{75}) & (K_{14} - K_{16} + K_{74} - K_{76}) \\
(K_{21} + K_{27} - K_{81} - K_{87}) & (K_{22} - K_{28} - K_{82} + K_{88}) & (K_{23} + K_{25} - K_{83} - K_{85}) & (K_{24} - K_{26} - K_{84} + K_{86}) \\
(K_{31} + K_{37} + K_{51} + K_{57}) & (K_{32} - K_{38} + K_{52} - K_{58}) & (K_{33} + K_{35} + K_{53} + K_{55}) & (K_{34} - K_{36} + K_{54} - K_{56}) \\
(K_{41} + K_{47} - K_{61} - K_{67}) & (K_{42} - K_{48} - K_{62} + K_{68}) & (K_{43} + K_{45} - K_{63} - K_{65}) & (K_{44} - K_{46} - K_{64} + K_{66})
\end{bmatrix} \begin{pmatrix} D_{S1} \\ D_{S2} \\ D_{S3} \\ D_{S4} \end{pmatrix} + \frac{1}{2}\begin{pmatrix} R_{F1} + R_{F7} \\ R_{F2} - R_{F8} \\ R_{F3} + R_{F5} \\ R_{F4} - R_{F6} \end{pmatrix} \quad \text{(J)}
$$

$$
\begin{pmatrix} R_{A1} \\ R_{A2} \\ R_{A3} \\ R_{A4} \end{pmatrix} = \frac{1}{2}\begin{bmatrix}
(K_{11} - K_{17} - K_{71} + K_{77}) & (K_{12} + K_{18} - K_{72} - K_{78}) & (K_{13} - K_{15} - K_{73} + K_{75}) & (K_{14} + K_{16} - K_{74} - K_{76}) \\
(K_{21} - K_{27} + K_{81} - K_{87}) & (K_{22} + K_{28} + K_{82} + K_{88}) & (K_{23} - K_{25} + K_{83} - K_{85}) & (K_{24} + K_{26} + K_{84} + K_{86}) \\
(K_{31} - K_{37} - K_{51} + K_{57}) & (K_{32} + K_{38} - K_{52} - K_{58}) & (K_{33} - K_{35} - K_{53} + K_{55}) & (K_{34} + K_{36} - K_{54} - K_{56}) \\
(K_{41} - K_{47} + K_{61} - K_{67}) & (K_{42} + K_{48} + K_{62} + K_{68}) & (K_{43} - K_{45} + K_{63} - K_{65}) & (K_{44} + K_{46} + K_{64} + K_{66})
\end{bmatrix} \begin{pmatrix} D_{A1} \\ D_{A2} \\ D_{A3} \\ D_{A4} \end{pmatrix} + \frac{1}{2}\begin{pmatrix} R_{F1} - R_{F7} \\ R_{F2} + R_{F8} \\ R_{F3} - R_{F5} \\ R_{F4} + R_{F6} \end{pmatrix} \quad \text{(M)}
$$

Solving gives

$$D_{A4} = \frac{1}{8}\frac{1}{10{,}000}(-10 + 104) - \frac{1}{8}(.000453125) = .0006086 \text{ m} \tag{N}$$

and backsubstituting yields

$$\langle R_{A1} \quad R_{A2} \quad R_{A3} \rangle = \langle 35.3 \quad 145.4 \quad 52.77 \rangle \tag{O}$$

The general unsymmetric solution may now be obtained by backsubstituting the symmetric and antisymmetric solution results into Eqs. A, B, and D. For example,

$$D_4 = D_{S4} + D_{A4} = .0004203 + .0006086 = .001029 \text{ m}$$

$$D_6 = -D_{S4} + D_{A4} = -.0004203 + .0006086 = .000188 \text{ m}$$

which are seen to check the results previously obtained on page 147. Continuing in this fashion we could obtain all the structure nodal displacements as well as the nodal forces. Backsubstitution of the displacement results into the individual member stiffness relationships would then yield the member-end forces in the usual fashion.

The example we have used demonstrates that by taking advantage of conditions of symmetry and antisymmetry, we can reduce a given problem to two problems, each "roughly" half the size of the original—"roughly" because there may be certain degrees of freedom in the original problem that will respond only to the symmetric load component and others that will respond only to the antisymmetric component. This condition arises for a frame structure when there is a structure node on the axis of symmetry. It may be seen that at such a node both the rotational degree of freedom and the displacement degree of freedom normal to the axis of symmetry will respond only to an antisymmetric load, whereas the displacement degree of freedom parallel to the axis of symmetry will respond only to a symmetric load.

It also is evident that if the loading is either purely symmetric or purely antisymmetric, the use of symmetry conditions reduces the general problem to a single problem of about half the original size. Returning to the more general loading, however, we must ask, "Is there any advantage to working two problems of one-half size as opposed to a single full-size problem?" The answer is yes, for the following important reasons:

1. The solution time for a general set of simultaneous equations is proportional to the third power of the number of equations. That is, time to solve n equations $= Cn^3$ and time to solve 2 sets of $\frac{n}{2}$ equations $= 2C(\frac{n}{2})^3 = \frac{1}{4}Cn^3$. Thus considerable computation time can be saved.*

*This time saving will be reduced or may even disappear for stiffness equations having a well-banded coefficient array (see Chap. 7).

2. The storage requirements may be cut about in half, because each problem will have about half the number of degrees of freedom as the original.
3. Input requirements will also be cut about in half.

To realize the second and third of these advantages, we must develop a technique to build Eqs. J′ and M′ directly rather than by operating on the entire set of stiffness force-displacement relations (Eq. F) as was done earlier. Although it would be possible (and perhaps desirable) to develop a coding technique to accomplish this goal, most structural analysis computer programs do not offer this capability. To use these programs and at the same time take advantage of the efficiencies associated with the concepts of symmetry and antisymmetry, it will be necessary to develop an alternative approach to such problems.

An alternative that accomplishes the desired goals with only a small sacrifice in problem-size reduction compared to that of the earlier approach involves the enforcement of symmetry or antisymmetry conditions at the outset to physically reduce the problem to "half" its original size. Thus the example of Fig. 4.10b exactly represents the left half of the symmetrically loaded example of Fig. 4.10a. This reduction is accomplished because, under symmetric loading, no shearing force is transmitted and no rotation occurs at the center of the symmetrically loaded beam shown in Fig. 4.10a. These symmetry conditions in the whole beam are enforced by boundary conditions imposed on the left half of the beam, as shown in Fig. 4.10b. In a similar way, the left half of the antisymmetric example of Fig. 4.10a may be reduced to the example of Fig. 4.10c. The solutions to the examples of Fig. 4.10b and c may now be carried out by using any structural analysis program. Solutions using the SMIS3 program, presented in the next printout on pages 189–192, are seen to correspond with the solutions to the examples of Fig. 4.10a.

It is worth noticing that although the examples of Fig. 4.10b and c show extremely well how the alternative approach works, they are very misleading with respect to demonstrating the efficiency of the approach. For example, each of these "reduced" problems has two free degrees of freedom, just as each of the original problems did. It will therefore still be necessary to solve two simultaneous equations and to use as much computer time and storage to solve each of these reduced problems as it took to solve the original problem (Fig. 4.10a). The question is, why? The answer is twofold:

1. Since there was no structural node on the axis of symmetry of the original structure, a node had to be added at this point when the structure was cut in half. Thus although one degree of freedom was eliminated, one was added too.
2. Although nodes will always have to be added if they do not naturally occur on the line of symmetry of a structure, the effect of adding them will not usually be so disastrous to the desired efficiencies. This is true because the number of free degrees of freedom, n, in the original structure will usually be much larger than the number of new degrees of freedom, m, that must be added. Thus, $n/2 + m \ll n$ and $(n/2 + m)^3 \lll n^3$.

Examples that demonstrate greater efficiency are shown in Fig. 4.11.

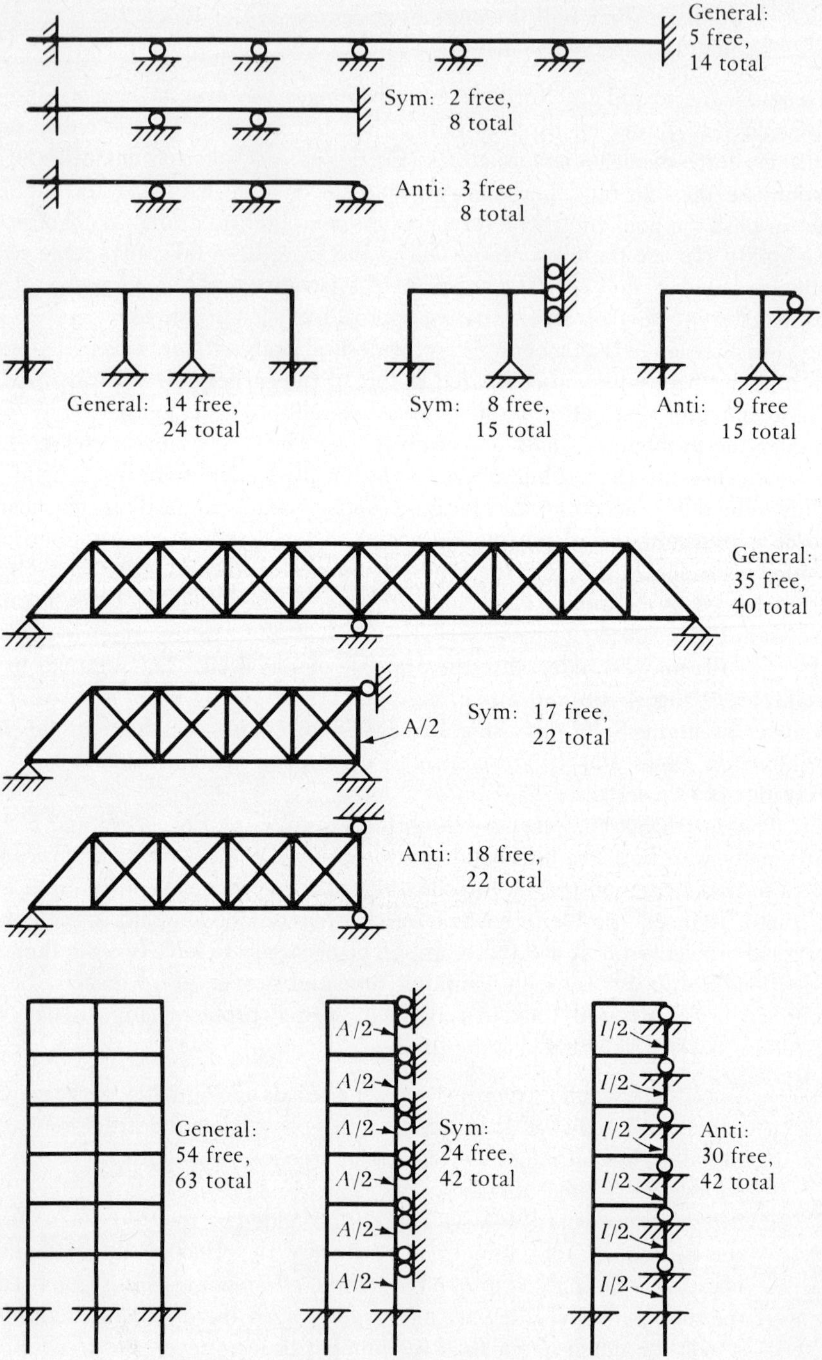

Fig. 4.11 Symmetry and antisymmetry examples.

****** *Computer Input:*

```
START
REMARK 3
SOLUTION OF THREE SPAN BEAM, SYMMETRIC COMPONENT

**** UNITS ARE KILONEWTONS AND METERS ****
INITLZ KUU RU DO 3 2 4 1 2
1 1 -.025
1 2 .0025
2 1 -.01875
3 2 0
2 2 10
C1 1 2
C2 2 3
BEAMEL K1 1 RF1
24 2E08 .0006 0 0
F 0 24 -3 -3
BEAMEL K2 1 RF2
15 2E08 .0015 0 0
F 10 10 -18 -18
BLDSTF KUU RU DO 2
K1 RF1 C1
K2 RF2 C2
SOLVE KUU RU DUS
PRINT DUS 2
THIS IS THE VECTOR, DUS, WHICH IS THE SYMMETRIC
COMPONENT OF THE NODAL DISPLACEMENT VECTOR
BEAMFC DUS 2
K1 C1 RF1
K2 C2 RF2
STOP
```

****** *Computer Output:*

```
START

REMARK      3


SOLUTION OF THREE SPAN BEAM, SYMMETRIC COMPONENT

**** UNITS ARE KILONEWTONS AND METERS ****



INITLZ  KUU     RU      DO

 NUMBER OF NODES          =     3
 NUMBER OF DOF PER NODE =     2
 NUMBER OF CONSTRAINTS  =     4
 NUMBER OF NODAL LOADS  =     1
 NUMBER OF ELEMENTS     =     2

 NULL    2 BY     2 STIFFNESS ARRAY, KUU    ,FORMED

 INITIAL DISPL. VECTOR = DO        REARRANGED COORDINATE CODE
       1     -2.500E-02                            -3
       2      2.500E-03                            -4
       3     -1.875E-02                            -5
       4              0                             1
       5              0                             2
       6              0                            -6

 NODAL LOAD VECTOR,  RU
               1
 1  1.0000E+01
 2           0
```

****** *Computer Output (continued):*

```
 MEMBER CODE        DEGREES OF FREEDOM
    C1            -3     -4     -5      1
    C2            -5      1      2     -6

BEAMEL  K1           1      RF1
 E =  2.000E+08      I =  6.000E-04      L =  2.400E+01
 A =          0      G =          0

 LOAD TYPE = F
  DIMENSION C =              0     LOAD INTENSITY AT C = -3.000E+00
  DIMENSION D =  2.400E+01         LOAD INTENSITY AT D = -3.000E+00

BEAMEL  K2           1      RF2
 E =  2.000E+08      I =  1.500E-03      L =  1.500E+01
 A =          0      G =          0

 LOAD TYPE = F
  DIMENSION C =  1.000E+01         LOAD INTENSITY AT C = -1.800E+01
  DIMENSION D =  1.000E+01         LOAD INTENSITY AT D = -1.800E+01

BLDSTF  KUU      RU       DO          2

 MEMBER K   MEMBER RF   MEMBER CODE
    K1         RF1          C1
    K2         RF2          C2

 STRUCTURE EQUIVALENT NODAL LOAD VECTOR,  RU
            1
 1  2.6681E+02
 2 -3.3333E+01


SOLVE   KUU      RU       DUS

PRINT   DUS           2


THIS IS THE VECTOR, DUS, WHICH IS THE SYMMETRIC
COMPONENT OF THE NODAL DISPLACEMENT VECTOR


            1
 1  4.2031E-04
 2 -2.8098E-02


BEAMFC  DUS           2

 MEMBER LOCAL FORCE VECTORS


 MEMBER CODE   = C1

 1  3.83993E+01
 2  1.90391E+02
 3  3.30007E+01
 4 -1.18406E+02


 MEMBER CODE   = C2

 1  1.80000E+01
 2  1.28406E+02
 3  2.27374E-13
 4  5.15938E+01
STOP
```

****** *Computer Input:*

```
START
REMARK 3
SOLUTION OF THREE SPAN BEAM, ANTISYMMETRIC COMPONENT

**** UNITS ARE KILONEWTONS AND METERS ****
INITLZ KUU RU DO 3 2 4 1 2
1 1 -.025
1 2 .0025
2 1 .01875
3 1 0
2 2 -10
C1 1 2
C2 2 3
BEAMEL K1 1 RF1
24 2E08 .0006 0 0
F 0 24 -3 -3
BEAMEL K2 1 RF2
15 2E08 .0015 0 0
F 10 10 -18 -18
BLDSTF KUU RU DO 2
K1 RF1 C1
K2 RF2 C2
SOLVE KUU RU DUA
PRINT DUA 2
THIS IS THE VECTOR,DUA,WHICH IS THE ANTISYMMETRIC
COMPONENT OF THE NODAL DISPLACEMENT VECTOR
BEAMFC DUA 2
K1 C1 RF1
K2 C2 RF2
STOP
```

****** *Computer Output:*

```
START

REMARK      3


SOLUTION OF THREE SPAN BEAM, ANTISYMMETRIC COMPONENT

**** UNITS ARE KILONEWTONS AND METERS ****



INITLZ  KUU      RU       DO

 NUMBER OF NODES          =       3
 NUMBER OF DOF PER NODE =       2
 NUMBER OF CONSTRAINTS   =       4
 NUMBER OF NODAL LOADS   =       1
 NUMBER OF ELEMENTS      =       2

 NULL    2 BY     2 STIFFNESS ARRAY, KUU    ,FORMED

 INITIAL DISPL. VECTOR = DO        REARRANGED COORDINATE CODE
       1     -2.500E-02                          -3
       2      2.500E-03                          -4
       3      1.875E-02                          -5
       4              0                           1
       5              0                          -6
       6              0                           2

 NODAL LOAD VECTOR,   RU
              1
 1 -1.0000E+01
 2           0
```

****** *Computer Output (continued):*

```
 MEMBER CODE        DEGREES OF FREEDOM
    C1             -3     -4     -5     1
    C2             -5      1     -6     2

BEAMEL  K1              1      RF1
 E =  2.000E+08      I =  6.000E-04      L =  2.400E+01
 A =            0    G =            0

 LOAD TYPE = F
  DIMENSION C =             0     LOAD INTENSITY AT C = -3.000E+00
  DIMENSION D =  2.400E+01        LOAD INTENSITY AT D = -3.000E+00

BEAMEL  K2              1      RF2
 E =  2.000E+08      I =  1.500E-03      L =  1.500E+01
 A =            0    G =            0

 LOAD TYPE = F
  DIMENSION C =  1.000E+01        LOAD INTENSITY AT C = -1.800E+01
  DIMENSION D =  1.000E+01        LOAD INTENSITY AT D = -1.800E+01

BLDSTF  KUU      RU       DO          2

 MEMBER K   MEMBER RF   MEMBER CODE
    K1         RF1          C1
    K2         RF2          C2

 STRUCTURE EQUIVALENT NODAL LOAD VECTOR,  RU
           1
 1 -6.3125E+00
 2 -1.1000E+02


SOLVE   KUU      RU       DUA

PRINT   DUA           2


THIS IS THE VECTOR,DUA,WHICH IS THE ANTISYMMETRIC
COMPONENT OF THE NODAL DISPLACEMENT VECTOR


           1
 1  6.0859E-04
 2 -1.6793E-03


BEAMFC  DUA           2

 MEMBER LOCAL FORCE VECTORS


 MEMBER CODE  = C1

 1  3.53285E+01
 2  1.45398E+02
 3  3.66715E+01
 4 -1.61516E+02


 MEMBER CODE  = C2

 1  1.61010E+01
 2  1.51516E+02
 3  1.89896E+00
 4  9.09495E-13
STOP
```

4.7 SUMMARY—DISPLACEMENT METHOD

Individual Member

Hooke's law, internal member equilibrium and compatability	$\mathbf{P} = \boldsymbol{k}\boldsymbol{\Delta}$	(3.6a, etc.)
Member overall equilibrium	$\mathbf{r} = \mathbf{TP}$	(2.10b)
Member deformed geometry	$\boldsymbol{\Delta} = \mathbf{T}^T\mathbf{d}$	(2.11b)
Combining Eq. 3.6a with Eqs. 2.10b and 2.11b	$\mathbf{r} = \mathbf{T}\boldsymbol{k}\mathbf{T}^T\mathbf{d}$	
$\mathbf{T}\boldsymbol{k}\mathbf{T}^T = \mathbf{k}$ or	$\mathbf{r} = \mathbf{kd}$	(4.1b, etc.)
Superposition	$\mathbf{r} = \mathbf{kd} + \mathbf{r}_F$	(4.4, etc.)

Individual Support

Support stiffness	$r_c = k_c d_c$	(4.10a, etc.)

Collection of All Members and Supports

Based on Eqs. 4.4 and 4.10a	$\bar{\mathbf{r}} = \bar{\mathbf{k}}\bar{\mathbf{d}} + \bar{\mathbf{r}}_F$	(4.23)

Structure

Nodal equilibrium	$\mathbf{R} = \mathbf{A}\bar{\mathbf{r}}$	(2.16)
Nodal compatibility	$\bar{\mathbf{d}} = \mathbf{A}^T\mathbf{D}$	(2.22)
Combining Eq. 4.23 with Eqs. 2.16 and 2.22	$\mathbf{R} = \mathbf{A}\bar{\mathbf{k}}\mathbf{A}^T\mathbf{D} + \mathbf{A}\bar{\mathbf{r}}_F$	(4.24)
$\mathbf{A}\bar{\mathbf{k}}\mathbf{A}^T = \mathbf{K}$ and $\mathbf{A}\bar{\mathbf{r}}_F = \mathbf{R}_F$ or	$\mathbf{R} = \mathbf{KD} + \mathbf{R}_F$	(4.25)
Solution possible for any stable structure	$\mathbf{D} = \mathbf{K}^{-1}(\mathbf{R} - \mathbf{R}_F)$	(4.27)
Backsubstitution from Eq. 4.27 into 2.22	$\bar{\mathbf{d}} = \mathbf{A}^T\mathbf{K}^{-1}(\mathbf{R} - \mathbf{R}_F)$	(4.28)
Backsubstitution from Eq. 4.28 into 4.23	$\bar{\mathbf{r}} = \bar{\mathbf{k}}\mathbf{A}^T\mathbf{K}^{-1}(\mathbf{R} - \mathbf{R}_F) + \bar{\mathbf{r}}_F$	(4.29)
Backsubstitution from Eq. 4.29 into 4.20	$\bar{\mathbf{r}}' = \bar{\boldsymbol{\alpha}}^T\bar{\mathbf{k}}\mathbf{A}^T\mathbf{K}^{-1}(\mathbf{R} - \mathbf{R}_F) + \bar{\boldsymbol{\alpha}}^T\bar{\mathbf{r}}_F$	(4.30)

PROBLEMS

4.1 Solve for the nodal displacements of the example truss of Fig. P4.1. Use the direct stiffness method and coordinate coding. Take $EA = EA_0$ as a constant and take $EA_0/4$ as a factor. Determine local end forces in members 1, 2, and 7, and check against the results previously obtained in Prob. 2.2. If Prob. 2.2 was not worked, check the results by hand calculations.

4.2 Solve for the nodal displacements of the example truss of Fig. P4.2. Use the direct stiffness method and coordinate coding. Take $EA = EA_0$ as a constant and take $EA_0/3$

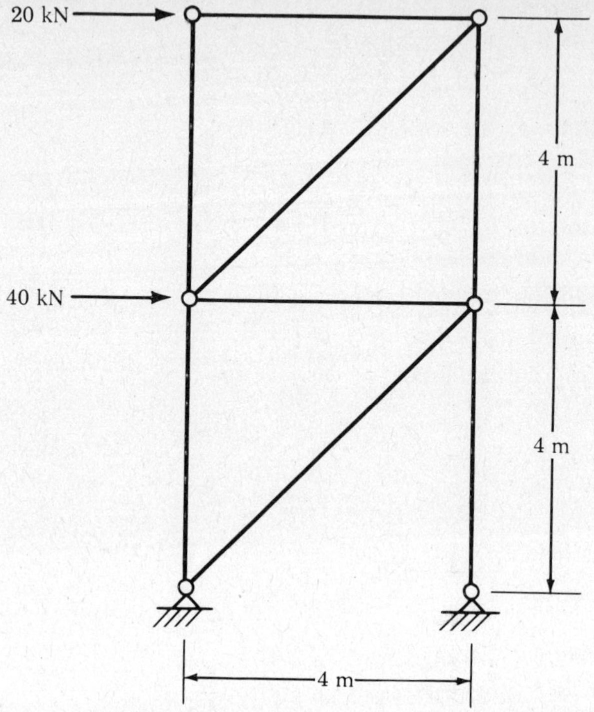

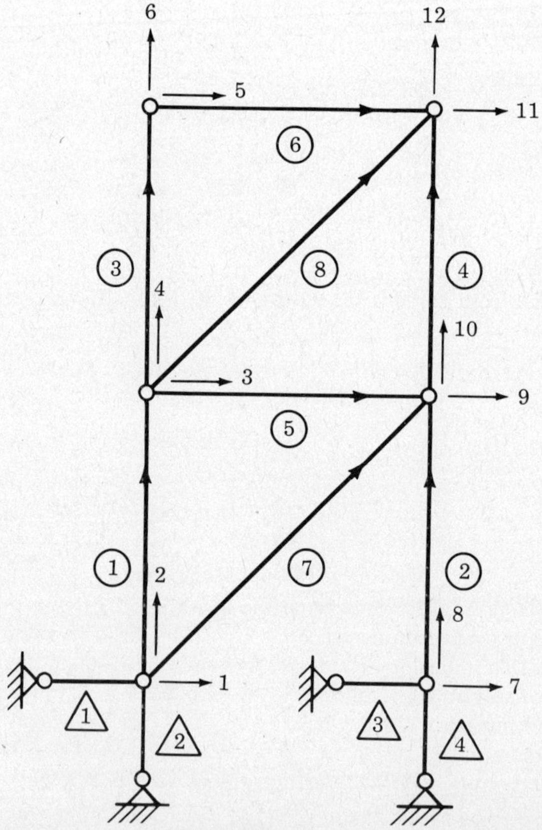

Fig. P4.1

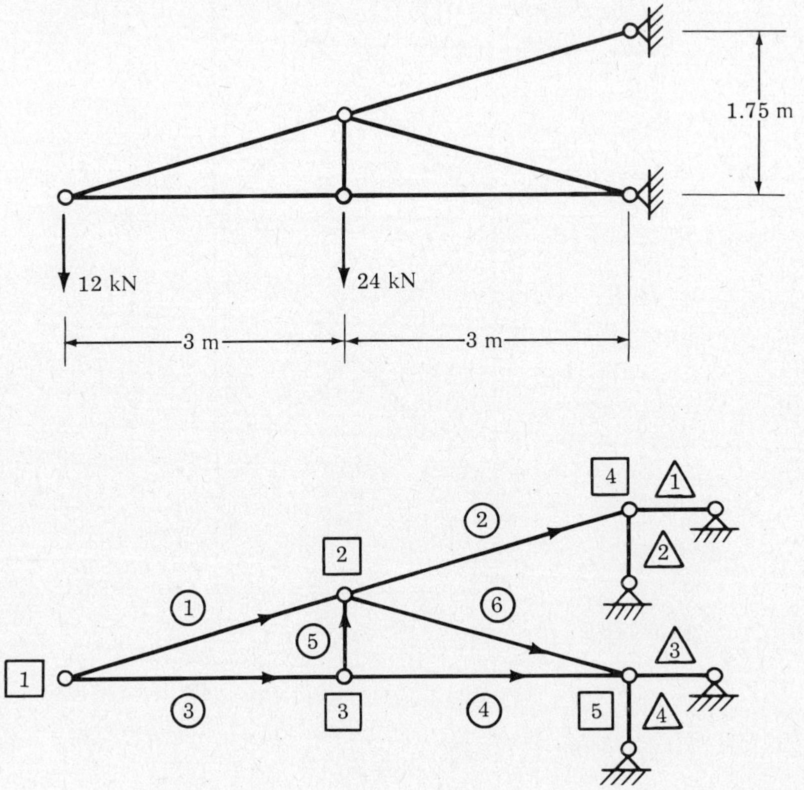

Fig. P4.2

as a factor. Determine local end forces in members 2, 4 and 6, and check against the results of Prob. 2.4 and/or by hand calculations.

4.3 Solve for the nodal displacements of the example truss of Fig. P4.3 by the direct stiffness method. Use nodal coding showing the nodal code on $\mathbf{k}^{(n)}$ and $\mathbf{K}$ arrays. Take $EA = EA_0$ as a constant, and take $EA_0/4$ as a factor. Backsubstitute displacement results to determine the local end forces in members 3, 6 and 10, and check equilibrium at node 3.

4.4 Solve for the nodal displacements of the example truss of Fig. P4.4 by the direct stiffness method. Use nodal coding showing the nodal code on $\mathbf{k}^{(n)}$ and $\mathbf{K}$ arrays. Take $EA = EA_0$ as a constant, and take $EA_0/3$ as a factor. Backsubstitute displacement results to determine the local end forces in members 3, 4, 5 and 7, and check equilibrium at node 3.

4.5 Solve for the nodal displacements of the example frame of Fig. P4.5 by the direct stiffness method. Use nodal coding showing the code on the $\mathbf{k}^{(n)}$, $\mathbf{K}$, $\mathbf{r}_F^{(n)}$, and $\mathbf{R}_F$ arrays. Take $EI_0/2$ as a factor with $I_0/A_0 = .005\ \text{m}^2$. Determine member-end forces in local coordinates and draw shear and moment curves.

4.6 Solve for the nodal displacements of the example frame of Fig. P4.6 by the direct stiffness method. Use nodal coding showing the code on the $\mathbf{k}^{(n)}$, $\mathbf{K}$, $\mathbf{r}_F^{(n)}$, and $\mathbf{R}_F$ arrays. Take $EI = EI_0$ and $EA = EA_0$ both as constants with $I_0/A_0 = .006452\ \text{m}^2$. Take $EI_0/2$

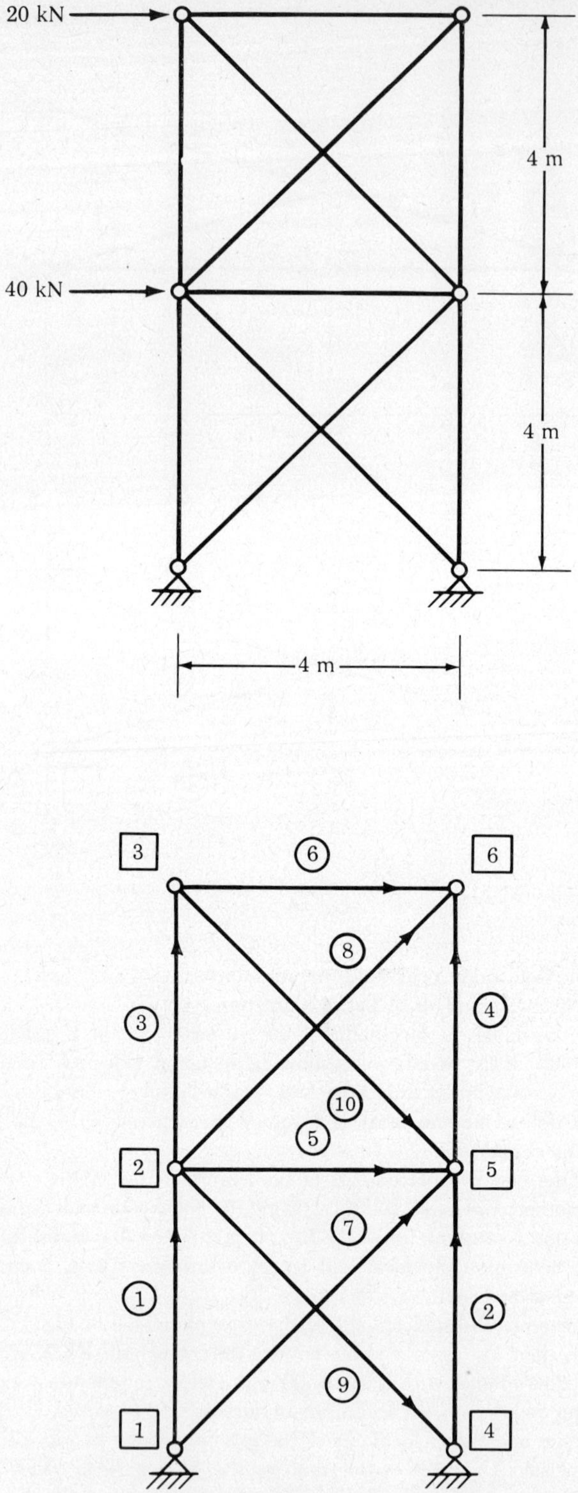

Fig. P4.3

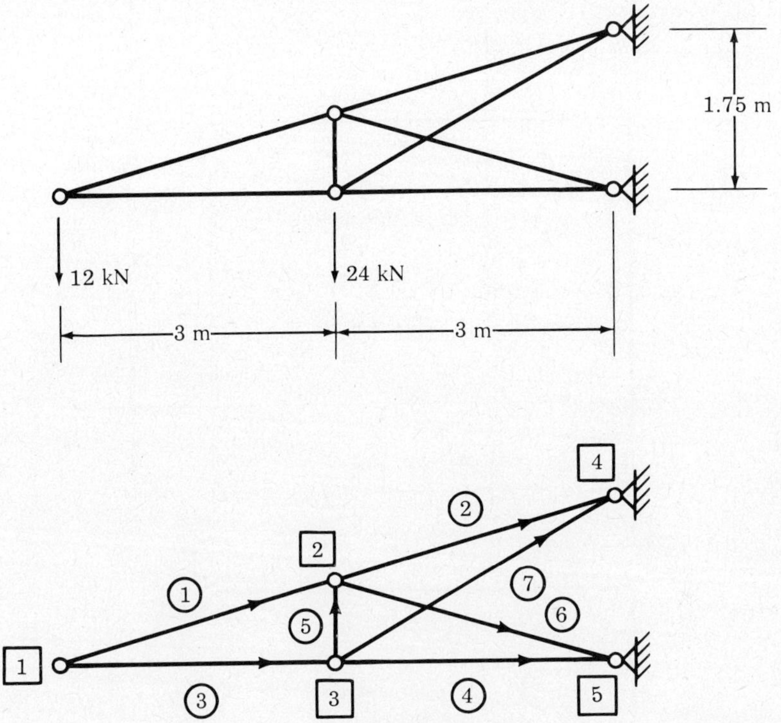

Fig. P4.4

as a factor. Determine member-end forces in local coordinates and draw shear and moment curves.

4.7 Solve for the nodal displacements for the example beam of Fig. P4.7 by the direct stiffness method. Use nodal coding showing the code on the $\mathbf{k}^{(n)}$, $\mathbf{K}$, $\mathbf{r}_F^{(n)}$, and $\mathbf{R}_F$ arrays. Take EI_0 as a factor. Determine member-end forces and draw shear and moment curves.

4.8 Solve for the nodal displacements of the example truss of Fig. P4.8 by the six variations of the direct stiffness method in the following list, taking $EA = EA_0$ as a constant and as a factor in the solution.

(a) Nodal coding: Generate equations by hand and solve using SMIS3.
(b) Modified method: Modify equations generated in part a and solve by SMIS3.
(c) SMIS3 structural commands: Use a complete coordinate code.
(d) Unconstrained coordinate code: Generate and solve the equations by hand.
(e) SMIS3 structural analysis commands: Use an unconstrained coordinate code.
(f) Rearrangement: Use the INITLZ and BLDSTF commands of SMIS3.

4.9 Work Prob. 4.3, but instead of using nodal coding, use (a) the SMIS3 structural analysis commands and the unconstrained coordinate code, and (b) the INITLZ and BLDSTF commands.

4.10 (a) Work Prob. 4.5, but instead of using nodal coding, use the SMIS3 INITLZ and BLDSTF structural analysis commands.
(b) Repeat part a, adding a foundation settlement at the left-hand support of .015 m downward and taking $E = 2.10^8$ kN/m^2 and $A_0 = .003$ m^2.

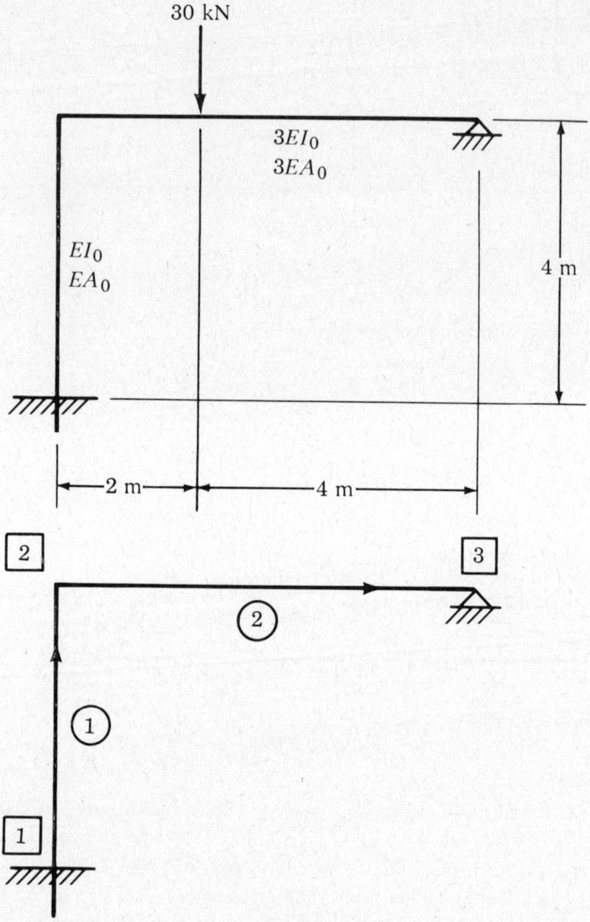

Fig. P4.5

4.11 Work Prob. 4.6, but instead of using nodal coding, use the SMIS3 INITLZ and BLDSTF commands.

4.12 Using symmetry, solve for member-end forces due to a dead load plus snow load of 2.5 kN/m^2 uniformly distributed over the horizontal projected area of the roof supported by the gable frame of Fig. P4.12. The gable frames are spaced at 10 m on centers. Use SMIS3 structural analysis commands.

4.13 Solve for the nodal displacements and member-end forces of the structure of Fig. P4.13. Use SMIS3 structural analysis commands, taking $EI_{\text{columns}} = EI_0$, $EA_{\text{columns}} = EA_0$, $EI_{\text{girders}} = 4EI_0$, and $EA_{\text{girders}} = 4EA_0$. Let $I_0/A_0 = .006\ m^2$. Use symmetry.

The SMIS3 program has been written to allow certain subroutines to be conveniently replaced by user-generated subroutines. Problems 4.14 to 4.16 make use of this feature. To successfully accomplish these replacements, study the appropriate sections of the documentation presented in Sec. A2.6 of App. 2.

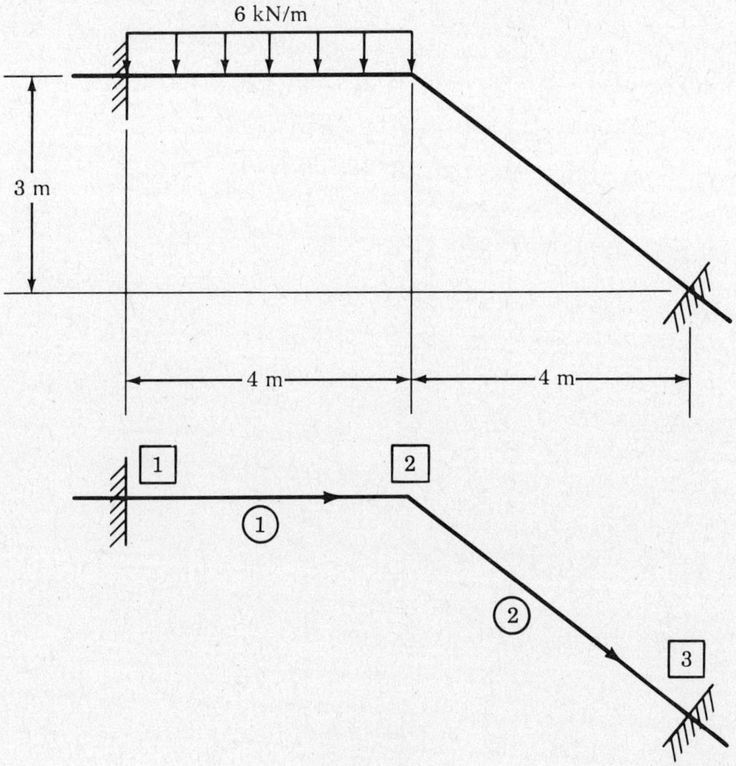

Fig. P4.6

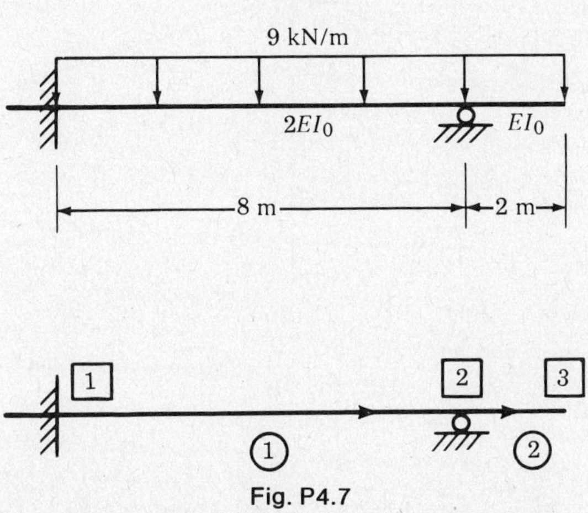

Fig. P4.7

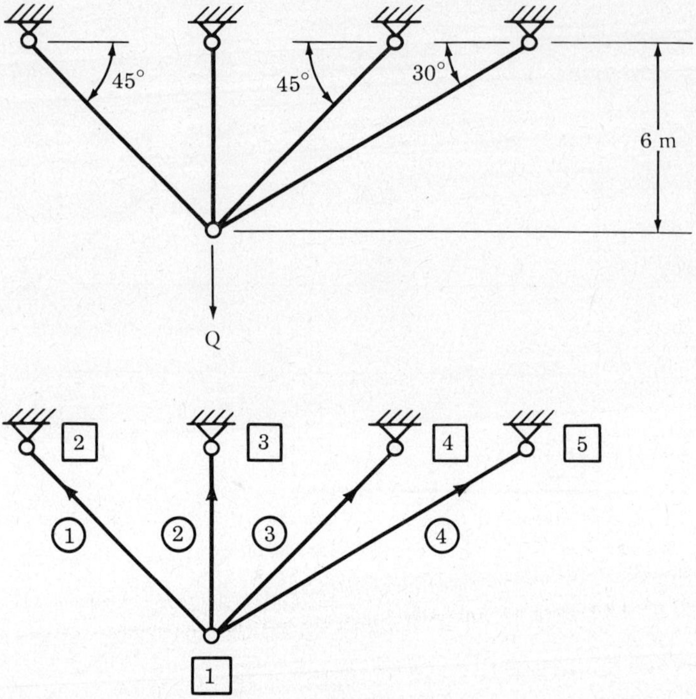

Fig. P4.8

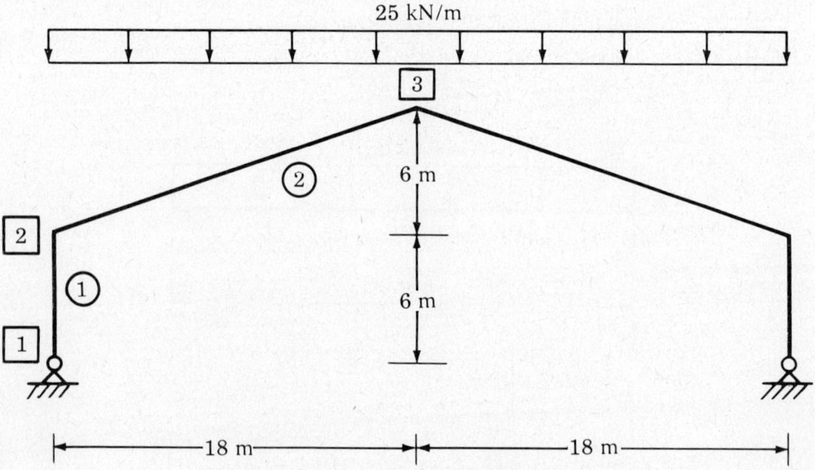

$E = 2 \times 10^8 \text{ kN/m}^2$

$I = .0024 \text{ m}^4$

$A = .025 \text{ m}^2$

Fig. P4.12

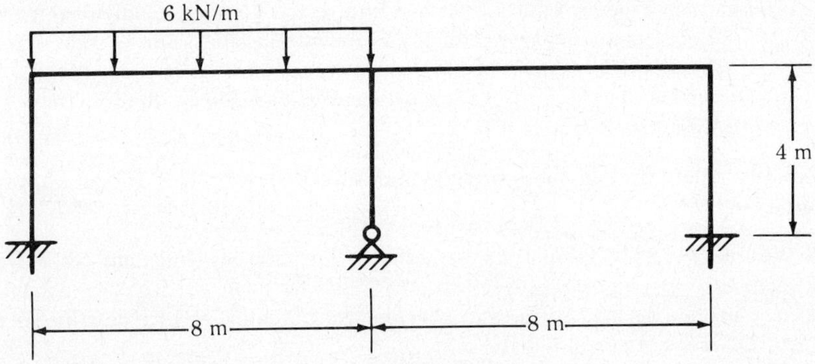

Fig. P4.13

4.14 Write FORTRAN subroutines to perform the following functions for a truss member:

(a) Generate the basic stiffness array $\mathscr{k}$. Replace subroutine TRBKF described in part C2 of Sec. A2.6.

(b) Generate the basic-local transformation array **L**. Replace subroutine TRL described in part C4 of Sec. A2.6.

(c) Generate the local-global transformation array $\boldsymbol{\alpha}$. Replace subroutine TRA described in part C3 of Sec. A2.6.

(d) Transform the basic stiffness array $\mathscr{k}$ into the global stiffness array **k**. Replace subroutine FORMKT described in part C5 of Sec. A2.6.

(e) Build the structure stiffness matrix **K** from the individual member stiffness matrices **k**, using a complete coordinate code. Replace subroutine ADDST described in part G1 of Sec. A2.6.

(f) Modify the **K** and **R** arrays in accordance with constraint conditions. Replace subroutine BOUDC described in part G3 of Sec. A2.6.

(g) Determine the vector $\mathbf{r}'$ of local member-end forces. Replace subroutine FRTRF described in part H2 of Sec. A2.6.

Check the functioning of these replacement subroutines by working an appropriate example.

4.15 Write FORTRAN subroutines to perform the following functions for a beam member:

(a) Generate the basic stiffness array $\mathscr{k}$. Replace subroutine BEBKF described in paragraph B2 of Sec. A2.6.

(b) Generate the basic-global transformation array **T**. Replace subroutine BMT described in part B3 of Sec. A2.6.

(c) Transform the basic stiffness array $\mathscr{k}$ into the global stiffness array **k**. Replace subroutine FORMK described in part B4 of Sec. A2.6.

(d) Generate the vector $\mathbf{r}_F$ of global fixed-end forces due to a transverse load distributed uniformly over the entire length of the member. Replace subroutine BMFEF described in part B6 of Sec. A2.6.

(e) Build the structure stiffness matrix **K** from the individual member stiffness matrices **k**, using an unconstrained coordinate code. Replace subroutine ADDST described in part G1 of Sec. A2.6.

(f) Build the structure equivalent nodal load vector **R** from the individual member fixed-end force vectors $\mathbf{r}_F$, using an unconstrained coordinate code. Replace subroutine ADDFE described in part G2 of Sec. A2.6.
(g) Determine the vector **r** of global member-end forces. Replace subroutine BEAMF described in part H1 of Sec. A2.6.

Check the functioning of these replacement subroutines by working an appropriate example.

4.16 Write FORTRAN subroutines to perform the following functions for a frame member:
(a) Generate the basic stiffness array $\boldsymbol{k}$. Replace subroutine FRBKF described in part D2 of Sec. A2.6.
(b) Generate the basic-local transformation array **L**. Replace subroutine FRL described in part D4 of Sec. A2.6.
(c) Generate the local-global transformation array $\boldsymbol{\alpha}$. Replace subroutine FRA described in part D3 of Sec. A2.6.
(d) Transform the basic stiffness array $\boldsymbol{k}$ into the global stiffness array **k**. Replace subroutine FORMKT described in part D5 of Sec. A2.6.
(e) Generate the vector $\mathbf{r}'_F$, of local fixed-end forces due to a concentrated load applied in an arbitrary direction at an arbitrary position along the member. Replace subroutine FRFEF described in part D7 of Sec. A2.6.
(f) Transform the local fixed-end force vector $\mathbf{r}'_F$ into the global fixed-end forced vector $\mathbf{r}_F$. Replace subroutine FORMRF described in part D9 of Sec. A2.6.
(g) Build the structure stiffness relationships on the basis of a nodal code and a constraint list. Replace subroutines ADDST, ADDFE, INITLD, INITLR, INITLC, REARRD, and FEFINT described in part G of Sec. A2.6.
(h) Determine the vector $\mathbf{r}'$ of local fixed-end forces. Replace subroutine FRTRF described in part H2 of Sec A2.6.

Check the functioning of these replacement subroutines by working an appropriate example.

Chapter 5
Displacement Method—Special Considerations

5.1 INTRODUCTION

The subjects to be discussed in this chapter fall under the heading of "special considerations," not because they are infrequent or unusual in occurrence but rather because their inclusion in the previous chapter might have interrupted the flow of the general development of the displacement method of analysis. Included in this chapter is a variety of subjects ranging from "special" loadings to "special" internal member conditions to "special" supports.

5.2 INITIALLY STRAINED MEMBERS

Initial strains are defined as strains that are not directly associated with a corresponding stress. Initial strains in members generally result from one of two causes: either (1) fabrication error (e.g., member made the wrong length) or (2) thermal loading. If the member is unconstrained, there will be a set of initial member-end displacements associated with these initial strains. But there will be no initial member-end forces associated with the initial displacements for an unconstrained member. For a member constrained to act as part of a structure, member-end forces will therefore be proportional to the difference between the total member-end displacements and the initial member-end displacements. Or, modifying the general member force-displacement relationships (see Eq. 4.4 or

4.8a, for example), yields

$$\mathbf{r} = \mathbf{k}(\mathbf{d} - \mathbf{d}_0) + \mathbf{r}_F \tag{5.1}$$

or

$$\mathbf{r} = \mathbf{kd} - \mathbf{kd}_0 + \mathbf{r}_F \tag{5.1a}$$

or

$$\mathbf{r} = \mathbf{kd} + \mathbf{r}_{F0} + \mathbf{r}_F \tag{5.1b}$$

where

$$\mathbf{r}_{F0} = -\mathbf{kd}_0 \tag{5.2}$$

and where $\mathbf{d}_0$ is a vector of initial member-end displacements for the member. It has been shown in Eqs. 4.1, 4.2, 4.5, and 4.8 that the member global stiffness matrix $\mathbf{k}$ for a truss, beam, or frame member is of the form

$$\mathbf{k} = \mathbf{T}\boldsymbol{k}T^T \tag{5.3}$$

or

$$\mathbf{k} = \boldsymbol{\alpha}\mathbf{L}\boldsymbol{k}\mathbf{L}^T\boldsymbol{\alpha}^T \tag{5.4}$$

Substituting from Eq. 5.4 into 5.2 gives

$$\mathbf{r}_{F0} = -\boldsymbol{\alpha}\mathbf{L}\boldsymbol{k}\mathbf{L}^T\boldsymbol{\alpha}^T\mathbf{d}_0 \tag{5.5}$$

but from Eq. 2.12 it is seen that

$$\boldsymbol{\Delta}_0 = \mathbf{L}^T\boldsymbol{\alpha}^T\mathbf{d}_0 \tag{5.6}$$

Substituting into Eq. 5.5 gives

$$\mathbf{r}_{F0} = -\boldsymbol{\alpha}\mathbf{L}\boldsymbol{k}\boldsymbol{\Delta}_0 \tag{5.7}$$

5.2.1 Fabrication Error

Frame Member

The most common fabrication error is an error in member length. For this case,

$$\boldsymbol{\Delta}_0 = \langle 0 \quad 0 \quad \Delta_0 \rangle \tag{5.8}$$

Substituting from Eqs. 2.37, 3.19, and 5.8 into Eq. 5.7 gives

$$\begin{pmatrix} r_{F01} \\ r_{F02} \\ r_{F03} \\ r_{F04} \\ r_{F05} \\ r_{F06} \end{pmatrix} = - \begin{bmatrix} -\sin\alpha/L & -\sin\alpha/L & -\cos\alpha \\ \cos\alpha/L & \cos\alpha/L & -\sin\alpha \\ 1 & 0 & 0 \\ \sin\alpha/L & \sin\alpha/L & \cos\alpha \\ -\cos\alpha/L & -\cos\alpha/L & \sin\alpha \\ 0 & 1 & 0 \end{bmatrix} \frac{2EI}{L} \begin{bmatrix} 2 & 1 & 0 \\ 1 & 2 & 0 \\ 0 & 0 & A/2I \end{bmatrix} \begin{pmatrix} 0 \\ 0 \\ \Delta_0 \end{pmatrix} \tag{5.9}$$

or

$$\mathbf{r}_{F0} = \frac{EA\,\Delta_0}{L}\langle \cos\alpha \quad \sin\alpha \quad 0 \quad -\cos\alpha \quad -\sin\alpha \quad 0\rangle \tag{5.9a}$$

Truss Member

Equation 5.9a can be specialized for a truss:

$$\mathbf{r}_{F0} = \frac{EA\,\Delta_0}{L}\langle \cos\alpha \quad \sin\alpha \quad -\cos\alpha \quad -\sin\alpha\rangle \tag{5.10}$$

This fabrication error will not affect a beam member.

In Chap. 2 the truss of Fig. 5.1 was subjected to initial distortions consisting of a 10-mm downward settlement of the right-hand support and a fabrication error that resulted in member 5 being 5 mm too long. Since this truss is statically determinate, no internal forces were induced due to the initial distortions, and the nodal displacements were determined by using nodal compatibility conditions only. It is interesting now to consider this example again, this time using the stiffness method just developed. Member properties for this truss are shown in Fig. 4.3. Given that $A^{(5)} = (2.5/2.4)A_0$, it is seen from Eq. 5.10 that

$$\mathbf{r}_{F0}^{(5)} = \frac{(2.5/2.4)EA_0(.005)}{2.5}\langle .6 \quad .8 \quad -.6 \quad -.8\rangle \tag{A}$$

or

$$\mathbf{r}_{F0}^{(5)} = \frac{EA_0}{3}\langle .00375 \quad .005 \quad -.00375 \quad -.005\rangle \tag{A'}$$

It follows from coding that the structure fixed-end force vector is

$$\mathbf{R}_F = \mathbf{R}_{F0} = \frac{EA_0}{3}\langle 0 \quad 0 \quad .00375 \quad .005 \quad 0 \quad 0 \quad 0 \quad 0 \quad -.00375 \quad -.005\rangle \tag{B}$$

In addition, the support settlement gives an initial displacement vector:

$$\mathbf{D}_0 = \langle 0 \quad 0 \quad 0 \quad 0 \quad 0 \quad -.01 \quad 0 \quad 0 \quad 0 \quad 0\rangle \tag{C}$$

A SMIS3 computer solution to this problem is presented in the next printout. (This solution makes use of the complete coordinate code so that the degrees of freedom will coincide with those previously used in the solution of this example.) Notice that the internally generated vectors of member fixed-end forces and structure equivalent nodal loads are consistent with the corresponding vectors shown in Eqs. A′ and B. The vector of nodal displacements,

$$\mathbf{D} = \langle 0 \quad 0 \quad 0 \quad -.008125 \quad 0 \quad -.01 \quad .0054167 \quad -.0040625 \quad .0054167 \quad -.0059375\rangle \tag{D}$$

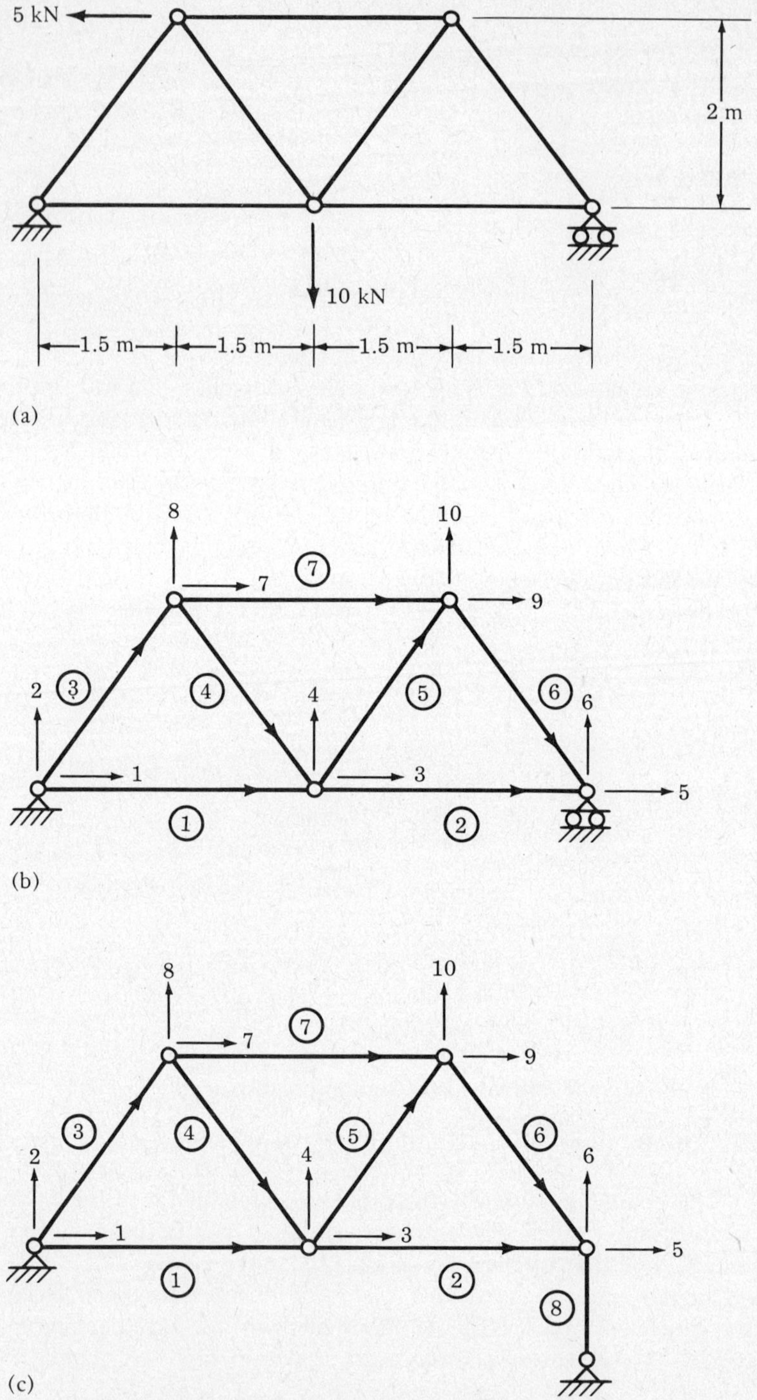

Fig. 5.1 Truss of Fig. 2.3: (a) dimensions and loading; (b) member and coordinate numbering and member directions; (c) alternate right-hand support.

****** *Computer Input:*

```
START
REMARK 7
STATICALLY DETERMINATE TRUSS SUBJECTED TO SUPPORT SETTLEMENT AND
FABRICATION ERROR. FABRICATION ERROR IS TREATED AS AN INITIAL STRAIN.
SUPPORT SETTLEMENT IS TREATED AS AN INITIAL DISPLACEMENT.
VALUES OF E AND A ARE TAKEN SUCH THAT EA(0)/3 = 1 AND MAY BE
CONSIDERED AS HAVING BEEN FACTORED FROM THE SOLUTION.

**** UNITS ARE KILONEWTONS AND METERS****
FORM K 10 10
FORM R 10 1
TRUSEL A1 K1
3 0 3 1
TRUSEL A3 K3
1.5 2 3 2.5/2.4
TRUSLD 1 RF5
I .005
PRINT RF5 2
THIS IS THE FIXED-END FORCE VECTOR,RF5, FOR MEMBER 5 DUE TO
THE INITIAL STRAIN IN MEMBER 5, EA(0)/3 IS FACTORED OUT.
TRUSEL A4 K4
1.5 -2 3 2.5/2.4
ADDSTF K 7
K1 C1 1 2 3 4
K1 C2 3 4 5 6
K3 C3 1 2 7 8
K4 C4 7 8 3 4
K3 C5 3 4 9 10
K4 C6 9 10 5 6
K1 C7 7 8 9 10
ADDFEF R 1
RF5 C5
BOUNDC K R 3
1 0
2 0
6 -.01
PRINT K 1
THIS IS THE STRUCTURE STIFFNESS MATRIX,K.
PRINT R 2
THIS IS THE STRUCTURE EQUIVALENT NODAL LOAD VECTOR,R,
WITH EA(0)/3 FACTORED OUT.
SOLVE K R D
PRINT D 1
THIS IS THE VECTOR OF NODAL DISPLACEMENTS,D.
REMARK 1
FORCES HAVE EA(0)/3 FACTORED OUT
TRUSFC D 7
A1 K1 C1 0
A1 K1 C2 0
A3 K3 C3 0
A4 K4 C4 0
A3 K3 C5 RF5
A4 K4 C6 0
A1 K1 C7 0
STOP
```

****** *Computer Output:*

```
START

  REMARK    7

STATICALLY DETERMINATE TRUSS SUBJECTED TO SUPPORT SETTLEMENT AND
FABRICATION ERROR. FABRICATION ERROR IS TREATED AS AN INITIAL STRAIN.
SUPPORT SETTLEMENT IS TREATED AS AN INITIAL DISPLACEMENT.
VALUES OF E AND A ARE TAKEN SUCH THAT EA(0)/3 = 1 AND MAY BE
```

****** *Computer Output (continued):*

```
CONSIDERED AS HAVING BEEN FACTORED FROM THE SOLUTION.

**** UNITS ARE KILONEWTONS AND METERS****

  FORM     K          10   10
   10 ROWS,    10 COLUMNS

  FORM     R          10    1
   10 ROWS,     1 COLUMNS

  TRUSEL   A1       K1
 DX=    .300E+01      DY=    .000E+00      L =    .300E+01
 E =    .300E+01      A =    .100E+01

  TRUSEL   A3       K3
 DX=    .150E+01      DY=    .200E+01      L =    .250E+01
 E =    .300E+01      A =    .104E+01

  TRUSLD    1         RF5

 LOAD TYPE =          I
  CHANGE IN LENGTH =    .500E-02

  PRINT   RF5         2

THIS IS THE FIXED-END FORCE VECTOR,RF5, FOR MEMBER 5 DUE TO
THE INITIAL STRAIN IN MEMBER 5, EA(0)/3 IS FACTORED OUT.

           1
 1    .3750E-02
 2    .5000E-02
 3   -.3750E-02
 4   -.5000E-02


  TRUSEL   A4       K4
 DX=    .150E+01      DY=   -.200E+01      L =    .250E+01
 E =    .300E+01      A =    .104E+01

  ADDSTF   K          7
 MEMBER K   MEMBER CODE        DEGREES OF FREEDOM
      K1          C1           1     2     3     4
      K1          C2           3     4     5     6
      K3          C3           1     2     7     8
      K4          C4           7     8     3     4
      K3          C5           3     4     9    10
      K4          C6           9    10     5     6
      K1          C7           7     8     9    10

  ADDFEF   R          1
 MEMBER FEF    MEMBER CODE
      RF5            C5

  BOUNDC  K        R         3
 NO. OF B.C. =        3

  NEQ =      1,   DISPL =   .000E+00
  NEQ =      2,   DISPL =   .000E+00
  NEQ =      6,   DISPL = -.100E-01

  PRINT   K           1

THIS IS THE STRUCTURE STIFFNESS MATRIX,K.

           1            2            3            4            5            6
 1   .1000E+01   .0000E+00   .0000E+00   .0000E+00   .0000E+00   .0000E+00
 2   .0000E+00   .1000E+01   .0000E+00   .0000E+00   .0000E+00   .0000E+00
 3   .0000E+00   .0000E+00   .2900E+01   .0000E+00  -.1000E+01   .0000E+00
```

****** *Computer Output (continued):*

```
  4    .0000E+00  .0000E+00  .0000E+00  .1600E+01  .0000E+00  .0000E+00
  5    .0000E+00  .0000E+00 -.1000E+01  .0000E+00  .1450E+01  .0000E+00
  6    .0000E+00  .0000E+00  .0000E+00  .0000E+00  .0000E+00  .1000E+01
  7    .0000E+00  .0000E+00 -.4500E+00  .6000E+00  .0000E+00  .0000E+00
  8    .0000E+00  .0000E+00  .6000E+00 -.8000E+00  .0000E+00  .0000E+00
  9    .0000E+00  .0000E+00 -.4500E+00 -.6000E+00 -.4500E+00  .0000E+00
 10    .0000E+00  .0000E+00 -.6000E+00 -.8000E+00  .6000E+00  .0000E+00
               7          8          9         10
  1    .0000E+00  .0000E+00  .0000E+00  .0000E+00
  2    .0000E+00  .0000E+00  .0000E+00  .0000E+00
  3   -.4500E+00  .6000E+00 -.4500E+00 -.6000E+00
  4    .6000E+00 -.8000E+00 -.6000E+00 -.8000E+00
  5    .0000E+00  .0000E+00 -.4500E+00  .6000E+00
  6    .0000E+00  .0000E+00  .0000E+00  .0000E+00
  7    .1900E+01  .0000E+00 -.1000E+01  .0000E+00
  8    .0000E+00  .1600E+01  .0000E+00  .0000E+00
  9   -.1000E+01  .0000E+00  .1900E+01  .0000E+00
 10    .0000E+00  .0000E+00  .0000E+00  .1600E+01


   PRINT    R           2

 THIS IS THE STRUCTURE EQUIVALENT NODAL LOAD VECTOR,R,
 WITH EA(0)/3 FACTORED OUT.

               1
  1    .0000E+00
  2    .0000E+00
  3   -.3750E-02
  4   -.5000E-02
  5   -.6000E-02
  6   -.1000E-01
  7    .0000E+00
  8    .0000E+00
  9    .9750E-02
 10   -.3000E-02


   SOLVE    K        R        D

   PRINT    D           1

 THIS IS THE VECTOR OF NODAL DISPLACEMENTS,D.

               1
  1    .0000E+00
  2    .0000E+00
  3   -.1746E-09
  4   -.8125E-02
  5    .6985E-09
  6   -.1000E-01
  7    .5417E-02
  8   -.4063E-02
  9    .5417E-02
 10   -.5938E-02


   REMARK    1

 FORCES HAVE EA(0)/3 FACTORED OUT


   TRUSFC  D          7

  MEMBER LOCAL FORCE VECTORS


  MEMBER CODE  =   C1
```

****** *Computer Output (continued):*

```
 1    .17462E-09
 2    .00000E+00
 3   -.17462E-09
 4    .00000E+00

 MEMBER CODE  =    C2

 1   -.87311E-09
 2    .00000E+00
 3    .87311E-09
 4    .00000E+00

 MEMBER CODE  =    C3

 1   -.46566E-09
 2   -.80715E-10
 3    .46566E-09
 4    .80715E-10

 MEMBER CODE  =    C4

 1    .93132E-09
 2   -.80715E-10
 3   -.93132E-09
 4    .80715E-10

 MEMBER CODE  =    C5

 1   -.93132E-09
 2    .29181E-09
 3    .93132E-09
 4   -.29181E-09

 MEMBER CODE  =    C6

 1    .18626E-08
 2   -.80714E-10
 3   -.18626E-08
 4    .80714E-10

 MEMBER CODE  =    C7

 1    .00000E+00
 2    .00000E+00
 3    .00000E+00
 4    .00000E+00
STOP
```

is seen to agree with the previous displacement result shown in Eq. C on page 37. Notice also that the internal forces induced are shown to be zero, as expected for a statically determinate structure.

It is interesting to consider an alternative method of accounting for the support settlement in this example. A support settlement may be regarded as an initial strain in the support. Modeling the support with a truss member as shown in

Fig. 5.1c and using Eq. 5.10, with $EA = (EA_0/3) \cdot \infty$, gives

$$r_{F0}^{(8)} = \frac{EA_0}{3} \cdot \infty(-.01)\langle 0 \quad 1 \quad 0 \quad -1\rangle \tag{E}$$

or

$$\mathbf{r}_{F0}^{(8)} = \frac{EA_0}{3}\langle 0 \quad -.01\infty \quad 0 \quad .01\infty\rangle \tag{E'}$$

For the structure fixed-end force vector, coding this result into Eq. B gives

$$\mathbf{R}_F = \mathbf{R}_{F0} = \frac{EA_0}{3}\langle 0 \quad 0 \quad .00375 \quad .005 \quad 0 \quad .01\infty \quad 0 \quad 0 \quad -.00375 \quad -.005\rangle \tag{F}$$

An SMIS3 solution that uses this latter approach is presented in the next printout, and it gives virtually identical results to those just listed.

Once again the solution was carried out using the complete coordinate code and the modified method to maintain the same coordinate numbering as before. It is important to notice, however, that any of the variations of the direct stiffness method presented in our discussions could be conveniently used when employing this initial-strain method of considering nonzero constraints. This possibility was mentioned in Sec. 4.5 and is now seen to be true.

Furthermore, this latter approach for considering the effects of specified nonzero nodal displacements is more fundamental in concept than the former method. In fact, the former method, which was developed in a somewhat superficial fashion in Sec. 4.5, can now be proved on the basis of the concept of initially strained supports. Consider, for this purpose, the stiffness equation corresponding to the sixth coordinate direction for the example truss. From Eq. C of Sec. 4.3.1 and from Eq. F in the present section, this equation is seen to be

$$R_6 = \frac{EA_0}{3}(-.6D_5 + \infty D_6 + .6D_9 - .8D_{10}) + \frac{EA_0}{3}(.01)\infty \tag{G}$$

where the loading is due to initially strained members 5 and 8, as described before. Dividing Eq. G by $(EA_0/3)\infty$ gives*

$$0 = D_6 + .01 \tag{G'}$$

or

$$-.01 = D_6 \tag{G''}$$

which may be written as

$$D_{06} = D_6 \tag{G'''}$$

This result proves the legitimacy of using replacement equations such as Eq. G‴ in place of original stiffness equations, as discussed in Sec. 4.5.

*See footnotes on pages 100 and 141 for a discussion of the term *infinity* and its symbol, ∞.

****** *Computer Input:*

```
START
REMARK 7
STATICALLY DETERMINATE TRUSS SUBJECTED TO SUPPORT SETTLEMENT
AND FABRICATION ERROR. BOTH SUPPORT SETTLEMENT AND
FABRICATION ERROR ARE TREATED AS INITIAL STRAINS.
VALUES OF E AND A ARE TAKEN SUCH THAT EA(0)/3 = 1 AND MAY BE
CONSIDERED AS HAVING BEEN FACTORED FROM THE SOLUTION.

**** UNITS ARE KILONEWTONS AND METERS****
FORM K 10 10
FORM R 10 1
TRUSEL A1 K1
3 0 3 1
TRUSEL A3 K3
1.5 2 3 2.5/2.4
TRUSLD 1 RF5
I .005
PRINT RF5 2
THIS IS THE FIXED-END FORCE VECTOR,RF5, FOR MEMBER 5 DUE TO
THE INITIAL STRAIN IN MEMBER 5, EA(0)/3 IS FACTORED OUT.
TRUSEL A4 K4
1.5 -2 3 2.5/2.4
TRUSEL A8 K8
0 1 3 1E10
TRUSLD 1 RF8
I -.01
PRINT RF8 2
THIS IS THE FIXED-END FORCE VECTOR,RF8, FOR MEMBER 8 DUE TO
THE INITIAL STRAIN IN MEMBER 8, EA(0)/3 IS FACTORED OUT.
ADDSTF K 8
K1 C1 1 2 3 4
K1 C2 3 4 5 6
K3 C3 1 2 7 8
K4 C4 7 8 3 4
K3 C5 3 4 9 10
K4 C6 9 10 5 6
K1 C7 7 8 9 10
K8 C8 0 0 5 6
ADDFEF R 2
RF5 C5
RF8 C8
PRINT R 2
THIS IS THE STRUCTURE EQUIVALENT NODAL LOAD VECTOR,R,
WITH EA(0)/3 FACTORED OUT.
BOUNDC K R 2
1 0
2 0
PRINT K 1
THIS IS THE STRUCTURE STIFFNESS MATRIX,K.
SOLVE K R D
PRINT D 1
THIS IS THE VECTOR OF NODAL DISPLACEMENTS,D.
REMARK 1
FORCES HAVE EA(0)/3 FACTORED OUT
TRUSFC D 8
A1 K1 C1 0
A1 K1 C2 0
A3 K3 C3 0
A4 K4 C4 0
A3 K3 C5 RF5
A4 K4 C6 0
A1 K1 C7 0
A8 K8 C8 RF8
STOP
```

****** *Computer Output:*

```
START

 REMARK    7

STATICALLY DETERMINATE TRUSS SUBJECTED TO SUPPORT SETTLEMENT
AND FABRICATION ERROR. BOTH SUPPORT SETTLEMENT AND              ND
FABRICATION ERROR ARE TREATED AS INITIAL STRAINS.
VALUES OF E AND A ARE TAKEN SUCH THAT EA(0)/3 = 1 AND MAY BE
CONSIDERED AS HAVING BEEN FACTORED FROM THE SOLUTION.

**** UNITS ARE KILONEWTONS AND METERS****


 FORM     K          10   10
  10 ROWS,     10 COLUMNS

 FORM     R          10    1
  10 ROWS,      1 COLUMNS

 TRUSEL  A1        K1
DX=   .300E+01      DY=   .000E+00      L =   .300E+01
E =   .300E+01      A =   .100E+01

 TRUSEL  A3        K3
DX=   .150E+01      DY=   .200E+01      L =   .250E+01
E =   .300E+01      A =   .104E+01

 TRUSLD     1        RF5

LOAD TYPE =       I
 CHANGE IN LENGTH =   .500E-02

 PRINT   RF5         2

THIS IS THE FIXED-END FORCE VECTOR,RF5, FOR MEMBER 5 DUE TO
THE INITIAL STRAIN IN MEMBER 5, EA(0)/3 IS FACTORED OUT.

           1
 1   .3750E-02
 2   .5000E-02
 3  -.3750E-02
 4  -.5000E-02


 TRUSEL  A4        K4
DX=   .150E+01      DY=  -.200E+01      L =   .250E+01
E =   .300E+01      A =   .104E+01

 TRUSEL  A8        K8
DX=   .000E+00      DY=   .100E+01      L =   .100E+01
E =   .300E+01      A =   .100E+11

 TRUSLD     1        RF8

LOAD TYPE =       I
 CHANGE IN LENGTH =  -.100E-01

 PRINT   RF8         2

THIS IS THE FIXED-END FORCE VECTOR,RF8, FOR MEMBER 8 DUE TO
THE INITIAL STRAIN IN MEMBER 8, EA(0)/3 IS FACTORED OUT.

           1
 1   .0000E+00
 2  -.3000E+09
 3   .0000E+00
 4   .3000E+09
```

****** *Computer Output (continued):*

```
 ADDSTF  K          8
MEMBER K   MEMBER CODE        DEGREES OF FREEDOM
     K1         C1            1     2     3     4
     K1         C2            3     4     5     6
     K3         C3            1     2     7     8
     K4         C4            7     8     3     4
     K3         C5            3     4     9    10
     K4         C6            9    10     5     6
     K1         C7            7     8     9    10
     K8         C8            0     0     5     6

 PRINT   K          1

THIS IS THE STRUCTURE STIFFNESS MATRIX,K.

            1          2          3          4          5          6
 1   .1450E+01  .6000E+00 -.1000E+01  .0000E+00  .0000E+00  .0000E+00
 2   .6000E+00  .8000E+00  .0000E+00  .0000E+00  .0000E+00  .0000E+00
 3  -.1000E+01  .0000E+00  .2900E+01  .0000E+00 -.1000E+01  .0000E+00
 4   .0000E+00  .0000E+00  .0000E+00  .1600E+01  .0000E+00  .0000E+00
 5   .0000E+00  .0000E+00 -.1000E+01  .0000E+00  .1450E+01 -.6000E+00
 6   .0000E+00  .0000E+00  .0000E+00  .0000E+00 -.6000E+00  .3000E+11
 7  -.4500E+00 -.6000E+00 -.4500E+00  .6000E+00  .0000E+00  .0000E+00
 8  -.6000E+00 -.8000E+00  .6000E+00 -.8000E+00  .0000E+00  .0000E+00
 9   .0000E+00  .0000E+00 -.4500E+00 -.6000E+00 -.4500E+00  .6000E+00
10   .0000E+00  .0000E+00 -.6000E+00 -.8000E+00  .6000E+00 -.8000E+00
            7          8          9         10
 1  -.4500E+00 -.6000E+00  .0000E+00  .0000E+00
 2  -.6000E+00 -.8000E+00  .0000E+00  .0000E+00
 3  -.4500E+00  .6000E+00 -.4500E+00 -.6000E+00
 4   .6000E+00 -.8000E+00 -.6000E+00 -.8000E+00
 5   .0000E+00  .0000E+00 -.4500E+00  .6000E+00
 6   .0000E+00  .0000E+00  .6000E+00 -.8000E+00
 7   .1900E+01  .0000E+00 -.1000E+01  .0000E+00
 8   .0000E+00  .1600E+01  .0000E+00  .0000E+00
 9  -.1000E+01  .0000E+00  .1900E+01  .0000E+00
10   .0000E+00  .0000E+00  .0000E+00  .1600E+01


 ADDFEF  R          2
MEMBER FEF   MEMBER CODE
     RF5          C5
     RF8          C8

 PRINT   R          2

THIS IS THE STRUCTURE EQUIVALENT NODAL LOAD VECTOR,R,
WITH EA(0)/3 FACTORED OUT.

            1
 1   .0000E+00
 2   .0000E+00
 3  -.3750E-02
 4  -.5000E-02
 5   .0000E+00
 6  -.3000E+09
 7   .0000E+00
 8   .0000E+00
 9   .3750E-02
10   .5000E-02


 BOUNDC  K          R          2
NO. OF B.C. =       2

 NEQ =      1,  DISPL =  .000E+00
 NEQ =      2,  DISPL =  .000E+00
```

****** *Computer Output (continued):*

```
 SOLVE   K       R       D

 PRINT   D           1

THIS IS THE VECTOR OF NODAL DISPLACEMENTS,D.

             1
 1    .0000E+00
 2    .0000E+00
 3   -.5821E-09
 4   -.8125E-02
 5   -.6985E-09
 6   -.1000E-01
 7    .5417E-02
 8   -.4063E-02
 9    .5417E-02
10   -.5938E-02


 REMARK    1

FORCES HAVE EA(0)/3 FACTORED OUT


 TRUSFC  D          8

 MEMBER LOCAL FORCE VECTORS


 MEMBER CODE   =   C1

 1    .58208E-09
 2    .00000E+00
 3   -.58208E-09
 4    .00000E+00

 MEMBER CODE   =   C2

 1    .11642E-09
 2    .00000E+00
 3   -.11642E-09
 4    .00000E+00

 MEMBER CODE   =   C3

 1    .00000E+00
 2   -.80715E-10
 3    .00000E+00
 4    .80715E-10

 MEMBER CODE   =   C4

 1    .00000E+00
 2   -.80715E-10
 3    .00000E+00
 4    .80715E-10

 MEMBER CODE   =   C5

 1    .00000E+00
 2    .29181E-09
 3    .00000E+00
 4   -.29181E-09
```

****** *Computer Output (continued):*

```
 MEMBER CODE  =    C6

 1    .27940E-08
 2   -.80715E-10
 3   -.27940E-08
 4    .80715E-10


 MEMBER CODE  =    C7

 1    .46566E-09
 2    .00000E+00
 3   -.46566E-09
 4    .00000E+00


 MEMBER CODE  =    C8

 1    .00000E+00
 2    .00000E+00
 3    .00000E+00
 4    .00000E+00
STOP
```

One final point of importance before leaving this example! The initial-strain approach to considering support settlement has the further advantage that it allows the effects of specified nonzero nodal displacements to be determined even when using direct stiffness method computer programs that contain no special provision for considering such displacements. This determination may be accomplished by loading the fixed-end force vector of Eq. F into the computer as a vector of equivalent nodal loads while modeling the support with a member of appropriate stiffness. Problem 5.1 at the end of this chapter offers the reader an opportunity to verify this approach.

Our example of a statically determinate truss subjected to initial strains was worked by the displacement method mainly for comparison with previous results of another method of analysis. It is, of course, of more interest to apply the displacement method to statically indeterminate structures. A statically indeterminate form of the example truss is shown in Fig. 5.2. An SMIS3 solution for this truss subjected to the same initial strains as was the statically determinate truss is shown in the next printout on pages 219–223. The nodal displacement vector for this example is

$$\mathbf{D} = \langle 0 \;\; 0 \;\; -.000508 \;\; -.008125 \;\; 0 \;\; -.01 \;\; .004316 \;\; -.003660 \;\; .004316 \;\; -.006340 \rangle \tag{H}$$

As expected, the addition of the two redundant members causes the nodal displacements for the statically indeterminate truss to be somewhat different from those of the statically determinate truss. Further, for the statically indeter-

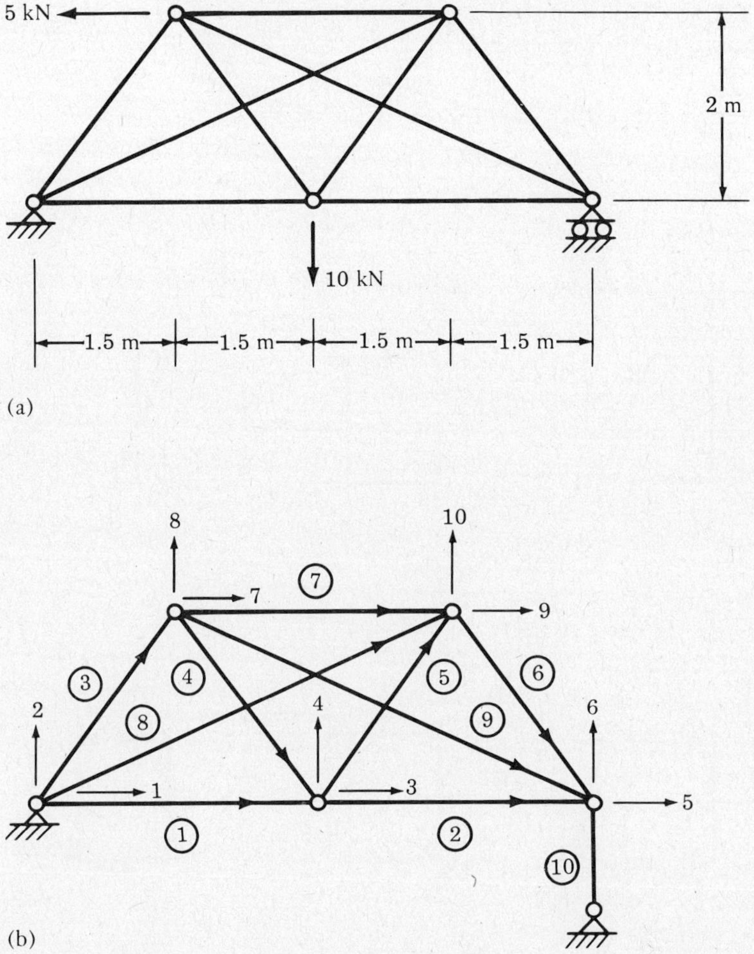

Fig. 5.2 Truss of Fig. 4.6: (a) dimensions and loading; (b) member and coordinate numbering and member directions.

minate truss, initially strained members produce internal forces in the structure, as may be observed from the SMIS3 solution.

5.2.2 Thermal Loading

Frame Member

The temperature variation over a member is assumed to be constant with x' and linear with y', as shown in Fig. 5.3a and b. Letting

t_1 = temperature change on $+y'$ side of member
t_2 = temperature change on $-y'$ side of member

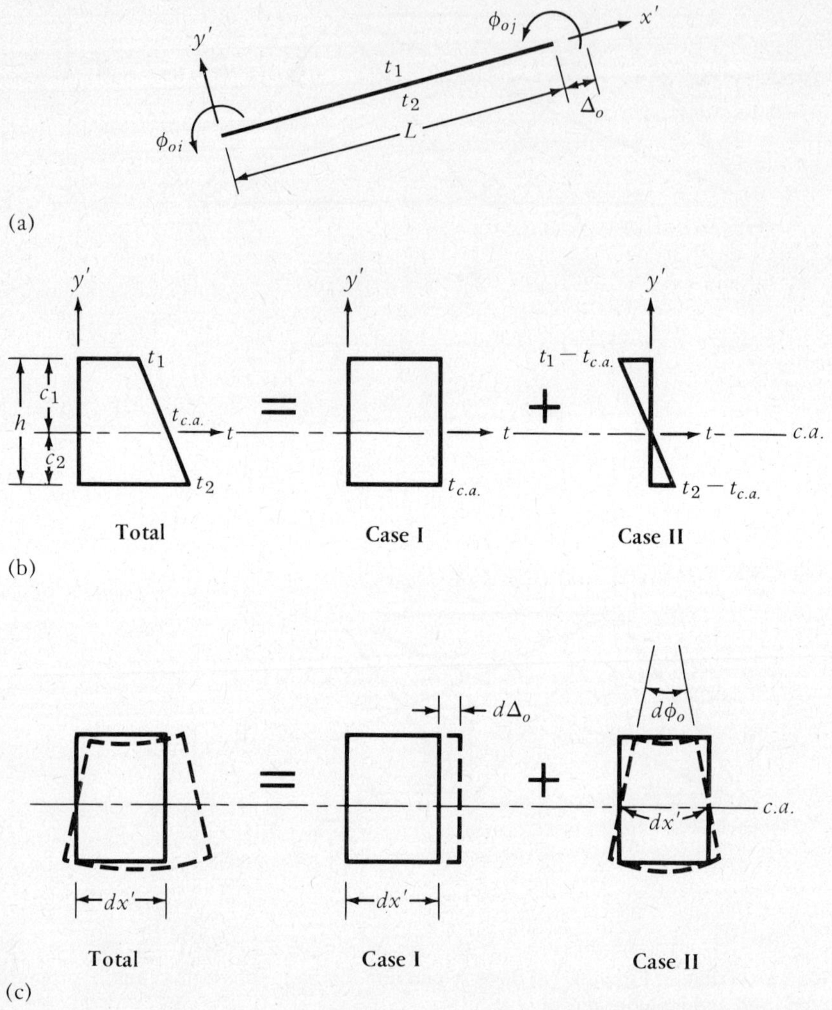

Fig. 5.3 Thermal loading, frame member: (a) frame member under thermal loading; (b) thermal load distribution; (c) distortion of infinitesimal element.

and referring to Fig. 5.3b, we can calculate the temperature change at the centroidal axis (*c.a.*) of the member to be

$$t_{c.a.} = t_1 + \frac{c_1}{c_1 + c_2}(t_2 - t_1) \tag{5.11}$$

It will be found to be convenient to consider the given temperature distribution through the member thickness to be the sum of two distributions, as shown in Fig. 5.3b. It is seen that the Case I temperature distribution results in a pure extension of a typical infinitesimal element of the member and that the Case II distribution causes a pure change in curvature of the element, as shown in Fig.

****** *Computer Input:*

```
START
REMARK 7
STATICALLY INDETERMINATE TRUSS SUBJECTED TO SUPPORT SETTLEMENT
AND FABRICATION ERROR. BOTH SUPPORT SETTLEMENT AND
FABRICATION ERROR ARE TREATED AS INITIAL STRAINS.
VALUES OF E AND A ARE TAKEN SUCH THAT EA(0)/3 = 1 AND MAY BE
CONSIDERED AS HAVING BEEN FACTORED FROM THE SOLUTION.


**** UNITS ARE KILONEWTONS AND METERS****
FORM K 10 10
FORM R 10 1
TRUSEL A1 K1
3 0 3 1
TRUSEL A3 K3
1.5 2 3 2.5/2.4
TRUSLD 1 RF5
I .005
TRUSEL A4 K4
1.5 -2 3 2.5/2.4
TRUSEL A8 K8
4.5 2 3 1
TRUSEL A9 K9
4.5 -2 3 1
TRUSEL A10 K10
0 1 3 1E10
TRUSLD 1 RF10
I -.01
ADDSTF K 10
K1 C1 1 2 3 4
K1 C2 3 4 5 6
K3 C3 1 2 7 8
K4 C4 7 8 3 4
K3 C5 3 4 9 10
K4 C6 9 10 5 6
K1 C7 7 8 9 10
K8 C8 1 2 9 10
K9 C9 7 8 5 6
K10 C10 0 0 5 6
ADDFEF R 2
RF5 C5
RF10 C10
PRINT R 1
THIS IS THE STRUCTURE EQUIVALENT NODAL LOAD VECTOR,R.
BOUNDC K R 2
1 0
2 0
SOLVE K R D
PRINT D 1
THIS IS THE VECTOR OF NODAL DISPLACEMENTS,D.
REMARK 1
FORCES HAVE EA(0)/3 FACTORED OUT
TRUSFC D 10
A1 K1 C1 0
A1 K1 C2 0
A3 K3 C3 0
A4 K4 C4 0
A3 K3 C5 RF5
A4 K4 C6 0
A1 K1 C7 0
A8 K8 C8 0
A9 K9 C9 0
A10 K10 C10 RF10
STOP
```

****** *Computer Output:*

```
START

  REMARK    7

STATICALLY INDETERMINATE TRUSS SUBJECTED TO SUPPORT SETTLEMENT
AND FABRICATION ERROR. BOTH SUPPORT SETTLEMENT AND
FABRICATION ERROR ARE TREATED AS INITIAL STRAINS.
VALUES OF E AND A ARE TAKEN SUCH THAT EA(0)/3 = 1 AND MAY BE
CONSIDERED AS HAVING BEEN FACTORED FROM THE SOLUTION.

**** UNITS ARE KILONEWTONS AND METERS****


  FORM    K          10   10
   10 ROWS,     10 COLUMNS

  FORM    R          10    1
   10 ROWS,      1 COLUMNS

  TRUSEL  A1       K1
 DX=   .300E+01      DY=   .000E+00      L =   .300E+01
 E =   .300E+01      A =   .100E+01

  TRUSEL  A3       K3
 DX=   .150E+01      DY=   .200E+01      L =   .250E+01
 E =   .300E+01      A =   .104E+01

  TRUSLD    1        RF5

 LOAD TYPE =        I
  CHANGE IN LENGTH =    .500E-02

  TRUSEL  A4       K4
 DX=   .150E+01      DY=  -.200E+01      L =   .250E+01
 E =   .300E+01      A =   .104E+01

  TRUSEL  A8       K8
 DX=   .450E+01      DY=   .200E+01      L =   .492E+01
 E =   .300E+01      A =   .100E+01

  TRUSEL  A9       K9
 DX=   .450E+01      DY=  -.200E+01      L =   .492E+01
 E =   .300E+01      A =   .100E+01

  TRUSEL  A10      K10
 DX=   .000E+00      DY=   .100E+01      L =   .100E+01
 E =   .300E+01      A =   .100E+11

  TRUSLD    1        RF10

 LOAD TYPE =        I
  CHANGE IN LENGTH =  -.100E-01

  ADDSTF  K          10
 MEMBER K   MEMBER CODE        DEGREES OF FREEDOM
      K1           C1           1      2      3      4
      K1           C2           3      4      5      6
      K3           C3           1      2      7      8
      K4           C4           7      8      3      4
      K3           C5           3      4      9     10
      K4           C6           9     10      5      6
      K1           C7           7      8      9     10
      K8           C8           1      2      9     10
      K9           C9           7      8      5      6
      K10          C10          0      0      5      6
```

****** *Computer Output (continued):*

```
 ADDFEF  R          2
MEMBER FEF    MEMBER CODE
     RF5            C5
     RF10           C10


 PRINT   R          1

THIS IS THE STRUCTURE EQUIVALENT NODAL LOAD VECTOR,R

            1
 1    .0000E+00
 2    .0000E+00
 3  -.3750E-02
 4  -.5000E-02
 5    .0000E+00
 6  -.3000E+09
 7    .0000E+00
 8    .0000E+00
 9    .3750E-02
10    .5000E-02


 BOUNDC  K          R          2
NO. OF B.C. =       2

 NEQ =       1,  DISPL =  .000E+00
 NEQ =       2,  DISPL =  .000E+00

 SOLVE   K          R          D

 PRINT   D          1

THIS IS THE VECTOR OF NODAL DISPLACEMENTS,D.

            1
 1    .0000E+00
 2    .0000E+00
 3  -.5081E-03
 4  -.8125E-02
 5  -.3725E-08
 6  -.1000E-01
 7    .4316E-02
 8  -.3660E-02
 9    .4316E-02
10  -.6340E-02


  REMARK    1

FORCES HAVE EA(0)/3 FACTORED OUT


  TRUSFC  D         10

 MEMBER LOCAL FORCE VECTORS
```

****** *Computer Output (continued):*

```
MEMBER CODE  =   C1

1    .50809E-03
2    .00000E+00
3  -.50809E-03
4    .00000E+00


MEMBER CODE  =   C2

1  -.50809E-03
2    .00000E+00
3    .50809E-03
4    .00000E+00


MEMBER CODE  =   C3

1    .42341E-03
2  -.89548E-10
3  -.42341E-03
4    .89548E-10


MEMBER CODE  =   C4

1  -.84682E-03
2  -.12236E-09
3    .84682E-03
4    .12236E-09


MEMBER CODE  =   C5

1    .84682E-03
2    .25017E-09
3  -.84682E-03
4  -.25017E-09


MEMBER CODE  =   C6

1  -.42341E-03
2  -.89548E-10
3    .42341E-03
4    .89548E-10


MEMBER CODE  =   C7

1  -.46566E-09
2    .00000E+00
3    .46566E-09
4    .00000E+00
```

****** *Computer Output (continued):*

```
MEMBER CODE  =    C8

1  -.83402E-03
2  -.64310E-10
3   .83402E-03
4   .64310E-10


MEMBER CODE  =    C9

1   .83402E-03
2  -.64311E-10
3  -.83402E-03
4   .64311E-10


MEMBER CODE  =    C10

1   .00000E+00
2   .00000E+00
3   .00000E+00
4   .00000E+00
STOP
```

5.3c. Considering Case I, it is seen that

$$d\Delta_0 = \beta\, dx\, t_{\text{c.a.}} = \beta\, dx \left[t_1 + \frac{c_1}{c_1 + c_2}(t_2 - t_1) \right] \tag{5.12}$$

where β is the coefficient of thermal expansion. Integrating over the length gives

$$\Delta_0 = \beta L \left[t_1 + \frac{c_1}{c_1 + c_2}(t_2 - t_1) \right] \tag{5.13}$$

Now, considering Case II,

$$d\phi_0 = \frac{\beta\, dx\,(t_2 - t_{\text{c.a.}}) - \beta\, dx\,(t_1 - t_{\text{c.a.}})}{c_1 + c_2} = \frac{\beta\, dx}{h}(t_2 - t_1) \tag{5.14}$$

Integrating gives

$$\phi_{0j} - \phi_{0i} = \frac{\beta L}{h}(t_2 - t_1) \tag{5.15}$$

Since this thermal loading is symmetric with respect to the member centerline,

$$\phi_{0j} = -\phi_{0i} = \frac{\beta L}{2h}(t_2 - t_1) \tag{5.16}$$

For the special case where $c_1 = c_2$, Eq. 5.13 reduces to

$$\Delta_0 = \frac{\beta L(t_1 + t_2)}{2} \tag{5.13a}$$

For simplicity, only the special case where $c_1 = c_2$ will be considered in subsequent work. Rewriting Eqs. 5.13a and 5.16 together in matrix form gives

$$\begin{pmatrix} \phi_{0i} \\ \phi_{0j} \\ \Delta_0 \end{pmatrix} = \frac{\beta L}{2} \begin{bmatrix} 1/h & -1/h \\ -1/h & 1/h \\ 1 & 1 \end{bmatrix} \begin{pmatrix} t_1 \\ t_2 \end{pmatrix} \tag{5.17}$$

or

$$\mathbf{\Delta}_0 = \boldsymbol{\beta}\mathbf{t} \tag{5.17a}$$

Substituting from Eqs. 2.37, 3.19, and 5.17 into Eq. 5.7 yields

$$\begin{pmatrix} r_{F01} \\ r_{F02} \\ r_{F03} \\ r_{F04} \\ r_{F05} \\ r_{F06} \end{pmatrix} = -\begin{bmatrix} -\sin\alpha/L & -\sin\alpha/L & -\cos\alpha \\ \cos\alpha/L & \cos\alpha/L & -\sin\alpha \\ 1 & 0 & 0 \\ \sin\alpha/L & \sin\alpha/L & \cos\alpha \\ -\cos\alpha/L & -\cos\alpha/L & \sin\alpha \\ 0 & 1 & 0 \end{bmatrix} \frac{2EI}{L} \begin{bmatrix} 2 & 1 & 0 \\ 1 & 2 & 0 \\ 0 & 0 & A/2I \end{bmatrix}$$

$$\times \frac{\beta L}{2} \begin{bmatrix} 1/h & -1/h \\ -1/h & 1/h \\ 1 & 1 \end{bmatrix} \begin{pmatrix} t_1 \\ t_2 \end{pmatrix} \tag{5.18}$$

Carrying out these matrix products, we get

$$\mathbf{r}_{F0} = \Big\langle \beta AE \frac{(t_1 + t_2)}{2}\cos\alpha \quad \beta AE \frac{(t_1 + t_2)}{2}\sin\alpha \quad -\frac{\beta EI}{h}(t_1 - t_2)$$
$$-\beta AE \frac{(t_1 + t_2)}{2}\cos\alpha \quad -\beta AE \frac{(t_1 + t_2)}{2}\sin\alpha \quad \frac{\beta EI}{h}(t_1 - t_2)\Big\rangle \tag{5.18a}$$

or, for the special case where $\alpha = 0$,

$$\mathbf{r}_{F0} = \mathbf{r}'_{F0} = \Big\langle \beta AE \frac{(t_1 + t_2)}{2} \quad 0 \quad -\frac{\beta EI}{h}(t_1 - t_2)$$
$$-\beta AE \frac{(t_1 + t_2)}{2} \quad 0 \quad \frac{\beta EI}{h}(t_1 - t_2)\Big\rangle \tag{5.18b}$$

Equation 5.18a may be specialized to yield the member thermal fixed-end force vector for both truss members and beam members. The vector for truss members is

$$\mathbf{r}_{FO} = \left\langle \beta AE \frac{(t_1 + t_2)}{2} \cos \alpha \quad \beta AE \frac{(t_1 + t_2)}{2} \sin \alpha, \quad -\beta AE \frac{(t_1 + t_2)}{2} \cos \alpha \quad -\beta AE \frac{(t_1 + t_2)}{2} \sin \alpha \right\rangle \tag{5.19}$$

and the vector for beam members is

$$\mathbf{r}_{FO} = \left\langle 0 \quad -\frac{\beta EI}{h}(t_1 - t_2) \quad 0 \quad \frac{\beta EI}{h}(t_1 - t_2) \right\rangle \tag{5.20}$$

Finally, for future work, Eq. 5.1b will be restored to the form of Eq. 4.8a, or

$$\mathbf{r} = \mathbf{kd} + \mathbf{r}_F \tag{5.21}$$

where it is to be understood that the term $\mathbf{r}_F$ has been generalized to include the effects of initial strains as well as distributed forces on the member.

Consider the steel rigid-jointed frame of Fig. 4.7 to be exposed to external and internal temperature variations such that the external surface temperature is 15°C while the internal temperature is 26°C. The zero thermal load temperature is taken as 21°C. This frame is shown again in Fig. 5.4a. Since this is a symmetric structure under a symmetric loading, we may consider only half of the structure, as shown in Fig. 5.4b, which also gives the member, nodal, and coordinate numbering and member directions. Using Eq. 5.18a, we can now determine the member fixed-end forces.

Member 1: $t_1 = -6°\text{C} \qquad t_2 = +5°\text{C} \qquad \alpha = 90°$

$$\mathbf{r}_F^{(1)} = EI_0\beta \left\langle 0 \quad \frac{10(-1) \times 1}{2} \quad \frac{-(-11) \times 1}{.75} \quad 0 \quad \frac{-2(-1) \times 1}{2} \quad \frac{(-11) \times 1}{.75} \right\rangle$$

or

$$\mathbf{r}_F^{(1)} = EI_0\beta \langle 0 \quad -5 \quad 14.67 \quad 0 \quad 5 \quad -14.67 \rangle \tag{A}$$

Member 2: $t_1 = -6°\text{C} \qquad t_2 = +5°\text{C} \qquad \alpha = 0°$

$$\mathbf{r}_F^{(2)} = EI_0\beta \langle -10 \quad 0 \quad 29.33 \quad 10 \quad 0 \quad -29.33 \rangle \tag{B}$$

By coding, it is seen that

$$\mathbf{R}_F = EI_0\beta \langle -10 \quad 5 \quad 14.67 \quad 0 \rangle \tag{C}$$

where $EI_0\beta = 4.68 \text{ kN} \cdot \text{m}^2/°\text{C}$.

Using SMIS3, we can obtain the structure nodal displacements and member-end forces, as shown in the next printout. In Table 5.1 some of these results are compared with corresponding results obtained for the external applied force loading of Fig. 4.7. It is seen that the largest nodal displacement due to the

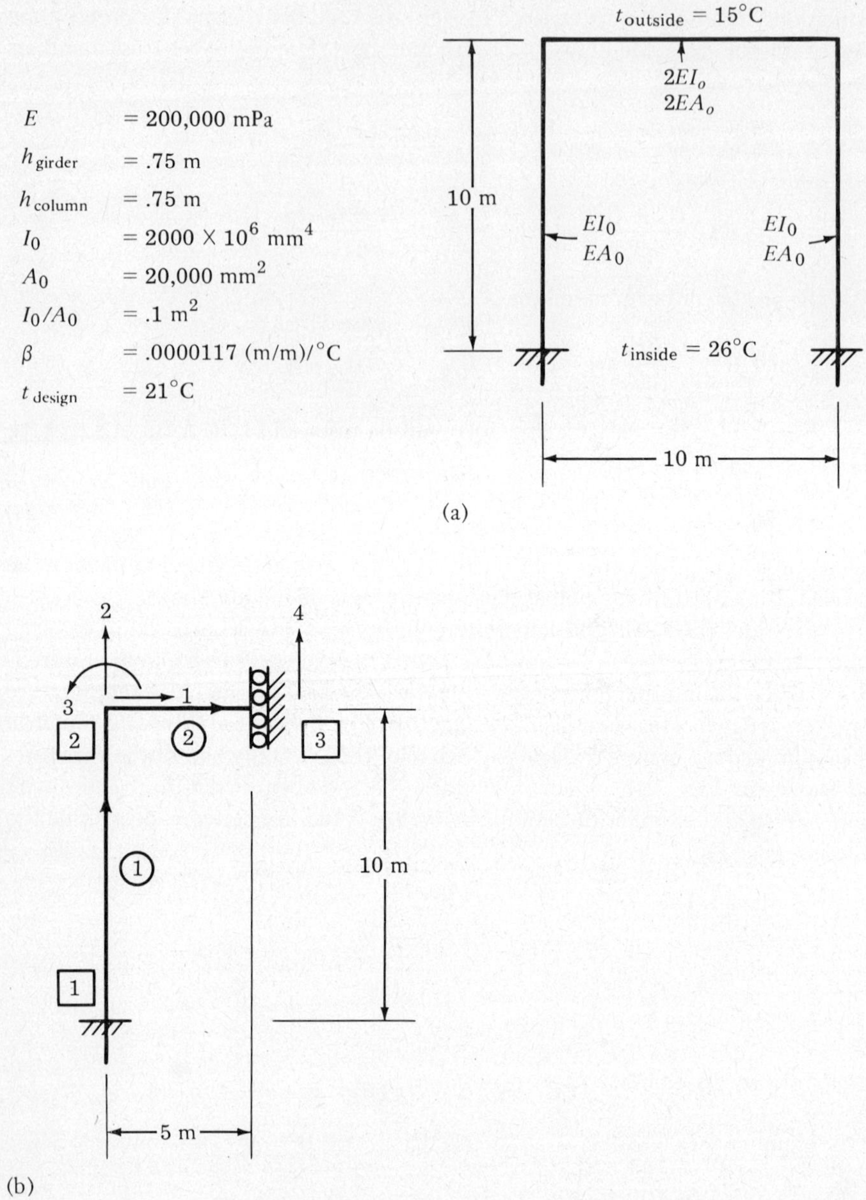

Fig. 5.4 Thermally loaded frame: (a) dimensions and loading; (b) model, left half.

external force loading is, in this case, several orders of magnitude larger than the largest displacement due to the thermal load. However, no such disparity exists between the largest forces in the two load cases. This situation may at first seem contradictory; however, upon reflection it is realized that the highest internal forces will be induced in a member under a given thermal load when the member

****** *Computer Input:*

```
START
REMARK 5
SOLUTION OF SYMMETRIC FRAME UNDER SYMMETRIC THERMAL LOAD.
E = 2E08 KN/M**2, I(0) = .002 M**4, A(0) = .02 M**2,
EI(0)/5 = 8E04 KN/M**2

**** UNITS ARE KILONEWTONS, METERS AND DEGREES CELCIUS ****
INITLZ KUU RU DO 3 3 5 0 2
1 1 0
1 2 0
1 3 0
3 1 0
3 3 0
C1 1 2
C2 2 3
FRAMEL A1 K1 1 RF1
0 10 2E08 .002 .02
T -6 5 .75 .0000117
PRINT RF1 2
THIS IS THE VECTOR OF FIXED-END FORCES,RF1,DUE TO
THERMAL LOADING ON MEMBER 1.
FRAMEL A2 K2 1 RF2
5 0 2E08 .004 .04
T -6 5 .75 .0000117
PRINT RF2 2
THIS IS THE VECTOR OF FIXED-END FORCES,RF2,DUE TO
THERMAL LOADING ON MEMBER 2.
BLDSTF KUU RU DO 2
K1 RF1 C1
K2 RF2 C2
PRINT RU 1
THIS IS THE VECTOR OF EQUIVALENT NODAL LOADS,RU,FOR THE FRAME.
SOLVE KUU RU DU
PRINT DU 1
THIS IS THE VECTOR OF NODAL DISPLACEMENTS,DU,FOR THE FRAME.
FRAMFC DU 2
A1 K1 C1 RF1
A2 K2 C2 RF2
STOP
```

****** *Computer Output:*

```
START

REMARK      5


SOLUTION OF SYMMETRIC FRAME UNDER SYMMETRIC THERMAL LOAD.
E = 2E08 KN/M**2, I(0) = .002 M**4, A(0) = .02 M**2,
EI(0)/5 = 8E04 KN/M**2

**** UNITS ARE KILONEWTONS, METERS AND DEGREES CELCIUS ****


INITLZ  KUU     RU      DO

 NUMBER OF NODES         =     3
 NUMBER OF DOF PER NODE  =     3
 NUMBER OF CONSTRAINTS   =     5
 NUMBER OF NODAL LOADS   =     0
 NUMBER OF ELEMENTS      =     2

 NULL    4 BY     4 STIFFNESS ARRAY, KUU    ,FORMED

 INITIAL DISPL. VECTOR = DO       REARRANGED COORDINATE CODE
```

****** *Computer Output (continued):*

```
        1            0                          -5
        2            0                          -6
        3            0                          -7
        4            0                           1
        5            0                           2
        6            0                           3
        7            0                          -8
        8            0                           4
        9            0                          -9

 NODAL LOAD VECTOR,   RU
          1
 1        0
 2        0
 3        0
 4        0


 MEMBER CODE        DEGREES OF FREEDOM
    C1          -5     -6     -7      1      2      3
    C2           1      2      3     -8      4     -9

FRAMEL  A1      K1          1      RF1
 DX=          0     DY=  1.000E+01      L =  1.000E+01
 E =  2.000E+08     I =  2.000E-03      A =  2.000E-02

 LOAD TYPE = T
  TOP TEMPERATURE = -6.000E+00      BOTTOM TEMPERATURE =  5.000E+00
  MEMBER HEIGHT   =  7.500E-01      THERMAL EXP. COEF. =  1.170E-05

PRINT   RF1         2

THIS IS THE VECTOR OF FIXED-END FORCES,RF1,DUE TO
THERMAL LOADING ON MEMBER 1.

          1
 1        0
 2 -2.3400E+01
 3  6.8640E+01
 4        0
 5  2.3400E+01
 6 -6.8640E+01


FRAMEL  A2      K2          1      RF2
 DX=  5.000E+00     DY=          0      L =  5.000E+00
 E =  2.000E+08     I =  4.000E-03      A =  4.000E-02

 LOAD TYPE = T
  TOP TEMPERATURE = -6.000E+00      BOTTOM TEMPERATURE =  5.000E+00
  MEMBER HEIGHT   =  7.500E-01      THERMAL EXP. COEF. =  1.170E-05

PRINT   RF2         2

THIS IS THE VECTOR OF FIXED-END FORCES,RF2,DUE TO
THERMAL LOADING ON MEMBER 2.

          1
 1 -4.6800E+01
 2        0
 3  1.3728E+02
 4  4.6800E+01
 5        0
 6 -1.3728E+02
```

****** *Computer Output (continued):*

```
BLDSTF  KUU     RU      DO          2

 MEMBER K   MEMBER RF   MEMBER CODE
    K1         RF1           C1
    K2         RF2           C2

 STRUCTURE EQUIVALENT NODAL LOAD VECTOR,  RU
              1
 1   4.6800E+01
 2  -2.3400E+01
 3  -6.8640E+01
 4            0


PRINT   RU            1

THIS IS THE VECTOR OF EQUIVALENT NODAL LOADS,RU,FOR THE FRAME.


              1
 1   4.6800E+01
 2  -2.3400E+01
 3  -6.8640E+01
 4            0


SOLVE   KUU     RU      DU

PRINT   DU            1

THIS IS THE VECTOR OF NODAL DISPLACEMENTS,DU,FOR THE FRAME.


              1
 1   3.2407E-05
 2  -5.8500E-05
 3  -2.1693E-04
 4  -6.0083E-04


FRAMFC  DU            2

 MEMBER LOCAL FORCE VECTORS

 MEMBER CODE  = C1

 1             0
 2  -5.05078E+00
 3   5.20633E+01
 4             0
 5   5.05078E+00
 6  -1.02571E+02

 MEMBER CODE  = C2

 1   5.05078E+00
 2   2.27374E-13
 3   1.02571E+02
 4  -5.05078E+00
 5  -2.27374E-13
 6  -1.02571E+02

STOP
```

Table 5.1 COMPARISON OF RESPONSES OF FRAME TO EXTERNAL FORCE LOADING AND THERMAL LOADING

RESPONSE	FORCE LOADING	THERMAL LOADING
D_1	.02022 m	.00003241 m
D_2	.000161 m	−.0000585 m
D_3	−.0005501 rad	−.0002169 rad
$r_1'^{(1)}$	−143.9 kN	0 kN
$r_2'^{(1)}$	−64.4 kN	−5.051 kN
$r_3'^{(1)}$	541.4 kN · m	52.06 kN · m
$r_4'^{(1)}$	23.88 kN	0 kN
$r_5'^{(1)}$	64.4 kN	5.051 kN
$r_6'^{(1)}$	297.4 kN · m	−102.6 kN · m

ends are *totally* restrained. (Conversely, when the member is free to distort, no end forces will be induced due to thermal loading.) So it is reasonable to expect large internal forces in a thermally loaded structure, especially when nodal displacements are small. The obvious answer to the problem of high thermal stresses is therefore to minimize the constraints on the structure—perhaps even build in releases (expansion joints). In the case of the frame example, it would be best to insert hinges, say, at each support and at the center of the frame. The resulting "three-hinged arch" is a statically determinate structure that can distort freely under this type of thermal loading, resulting in zero thermally induced member-end forces. Clearly, the thermal loading is of considerable significance in this particular case. These results may not be typical for structures and loadings normally encountered by the civil engineer. Nevertheless, they should be a signal as to the possible importance of initial strains whether thermally induced or resulting from fabrication error.

5.3 SPECIAL MEMBERS

One or more of a variety of circumstances may occasionally exist in a structure, making it either necessary or convenient to employ special member force-displacement relationships to account for that circumstance in the analysis of the structure. We will briefly discuss here the most common examples of such circumstances.

5.3.1 Variable Section

The cross section of a structural member may vary along the length of the member in a stepwise fashion, in a piecewise linear fashion, or in some more complex continuous fashion. A frame member is shown in Fig. 5.5 in which the cross section is taken as a general function of x', the local coordinate collinear with the member. Anticipating the use of an energy method for determining the member flexibility and stiffness coefficients, we may rewrite Eq. 3.28 in

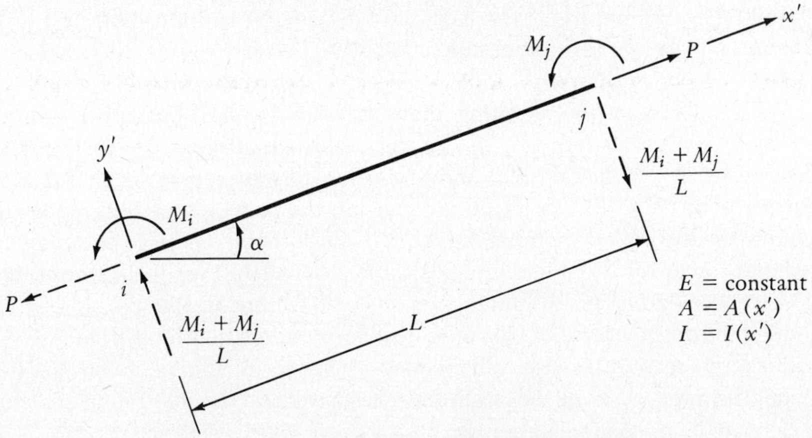

Fig. 5.5 Frame member, variable cross section.

generalized form as

$$U = \int_0^L \frac{M^2\,dx'}{2EI(x')} + \int_0^L \frac{S^2\,dx'}{2A(x')E} \tag{5.21}$$

where $A \rightarrow A(x')$ and $I \rightarrow I(x')$ to emphasize their possible variation over the length of the member. The internal moment and axial force expressions can be shown to be the same as those listed in Eq. 3.29 for the member of constant cross section.

$$\begin{aligned} M &= M_i - (M_i + M_j)\frac{x'}{L} \\ S &= P \end{aligned} \tag{3.29}$$

Substituting from Eq. 3.29 into 5.21 yields a slightly generalized form of Eq. 3.30.

$$U = \int_0^L \left(M_i - (M_i + M_j)\frac{x'}{L}\right)^2 \frac{dx'}{2EI(x')} + \int_0^L \frac{P^2\,dx'}{2A(x')E} \tag{5.22}$$

From previous work (see p. 82), it is known that flexibility coefficients may be obtained by appropriately differentiating the strain energy expression of Eq. 5.22. For example,

$$f_{12} = \frac{\partial^2 U}{\partial M_i\,\partial M_j} \tag{5.23}$$

Therefore

$$f_{12} = \int_0^L \frac{(-x'/L)(1 - x'/L)\,dx'}{EI(x')} \tag{5.24}$$

Depending on the nature of the function $I(x')$, the integration indicated in Eq. 5.24 may now be carried out either formally or by some appropriate numerical technique such as Simpson's rule. Once this has been done for each basic flexibility coefficient, the resulting member basic force-displacement relationships may be inverted to obtain the basic stiffness array. The basic stiffness force-displacement relationships may then be transformed to local and global coordinates as before. Once again we can use the principle of superposition to account for distributed member forces. It is important to notice, however, that fixed-end forces for the member of variable cross section will be different from those for a member of constant cross section. It should also be noticed that although the development of the force-displacement relationships for members of variable cross section is more difficult than it was for members of constant cross section, the application of the resulting relationships to the solution of structural problems will be the same as before.

5.3.2 Shear Distortions

If a frame member is relatively deep, shearing distortions may be nonnegligible relative to flexural distortions and axial distortions. Once again turning to an energy method, we may readily take these shearing distortions into account by including shear strain energy in the energy expression. Thus Eq. 3.28 becomes

$$U = \int_0^L \frac{M^2\,dx'}{2EI} + n\int_0^L \frac{V^2\,dx'}{2AG} + \int_0^L \frac{S^2\,dx'}{2AE} \tag{5.25}$$

where n is a factor that takes into account the nonuniform distribution of shearing stress over the cross section (Ref. 3.3). Augmenting the internal bending moment and axial force expressions of Eq. 3.29 with a corresponding expression for internal shear force in a frame member gives

$$\begin{aligned} M &= M_i - (M_i + M_j)\frac{x'}{L} \\ S &= P \\ V &= \frac{M_i + M_j}{L} \end{aligned} \tag{5.26}$$

Substituting these expressions into Eq. 5.25 yields

$$\begin{aligned} U = &\int_0^L \left(M_i - (M_i + M_j)\frac{x'}{L}\right)^2 \frac{dx'}{2EI} \\ &+ n\int_0^L \left(\frac{M_i + M_j}{L}\right)^2 \frac{dx'}{2AG} + \int_0^L \frac{P^2\,dx'}{2AE} \end{aligned} \tag{5.27}$$

Once again operating appropriately on the strain energy expression, we may obtain the member basic flexibility coefficients. For example,

$$f_{12} = \frac{\partial^2 U}{\partial M_i\,\partial M_j} = -\frac{L}{6EI} + \frac{n}{LAG} \tag{5.28}$$

It is seen that the second term in this equation represents the influence of shear distortions on this flexibility coefficient. Continuing in this fashion, we could obtain the remaining member basic flexibility coefficients. Following the usual steps, we could also obtain the member basic, local, and global stiffness relations. Once again note that the fixed-end forces may also be modified by the shear distortions. Note also that once the modified force-displacement relationships are obtained, application of these relationships to the solution of structural problems will be substantially the same as before. The beam member BEAMEL of the SMIS3 program has shear distortion capability.

5.3.3 Inextensional Members

In classical structural analysis, it is often assumed that members do not distort axially. Such members are referred to as being inextensional. The consideration of true inextensional member behavior has the advantage of reducing the number of degrees of freedom in a problem, since each inextensional member constitutes a geometrical constraint on the behavior of the structure. Unfortunately, inextensional member behavior is accompanied by a significant disadvantage as well: The systematic nature of the statement of the problem is destroyed. Although this difficulty can be readily circumvented for rectangular frames by special use of unconstrained coordinate coding, as will be shown shortly, it remains a problem for general frames. A second disadvantage that may accompany the assumption of inextensional member behavior is simply the approximation inherent in that assumption. For some frames, members must be considered to be extensional to obtain a reasonably accurate solution. To gain an insight as to the possible seriousness of this last disadvantage, it is worthwhile to compare solutions to example frame problems in regard to both extensional and inextensional member behavior.

To this end, consider the frame problem of Fig. 5.6a, which we previously analyzed allowing extensional behavior of the members. This example is reworked here using SMIS3 and taking into account inextensional behavior in two different ways. The solution displayed in the next printout (pp. 235–237) has been carried out by using the model of Fig. 5.6b and assuming very large values for the member cross-sectional areas. This approximate consideration of inextensional member behavior lacks the single advantage associated with the exact consideration of inextensional behavior, since the number of degrees of freedom remain undiminished. Furthermore, this technique possesses the added disadvantage of possible numerical difficulties stemming from large off-diagonal stiffness terms that may appear in the structure/stiffness matrix. Nevertheless, the technique can still be used as an expedient method of obtaining comparative solutions.

In a second solution, displayed in the printout on pages 237–239, the method of accounting for inextensional behavior is limited in application mainly to rectangular frames. This method makes use of the unconstrained coordinate code discussed first in Sec. 4.5.2. It is possible to use the unconstrained coordinate code to model inextensional rectangular frames because the inextensional constraints

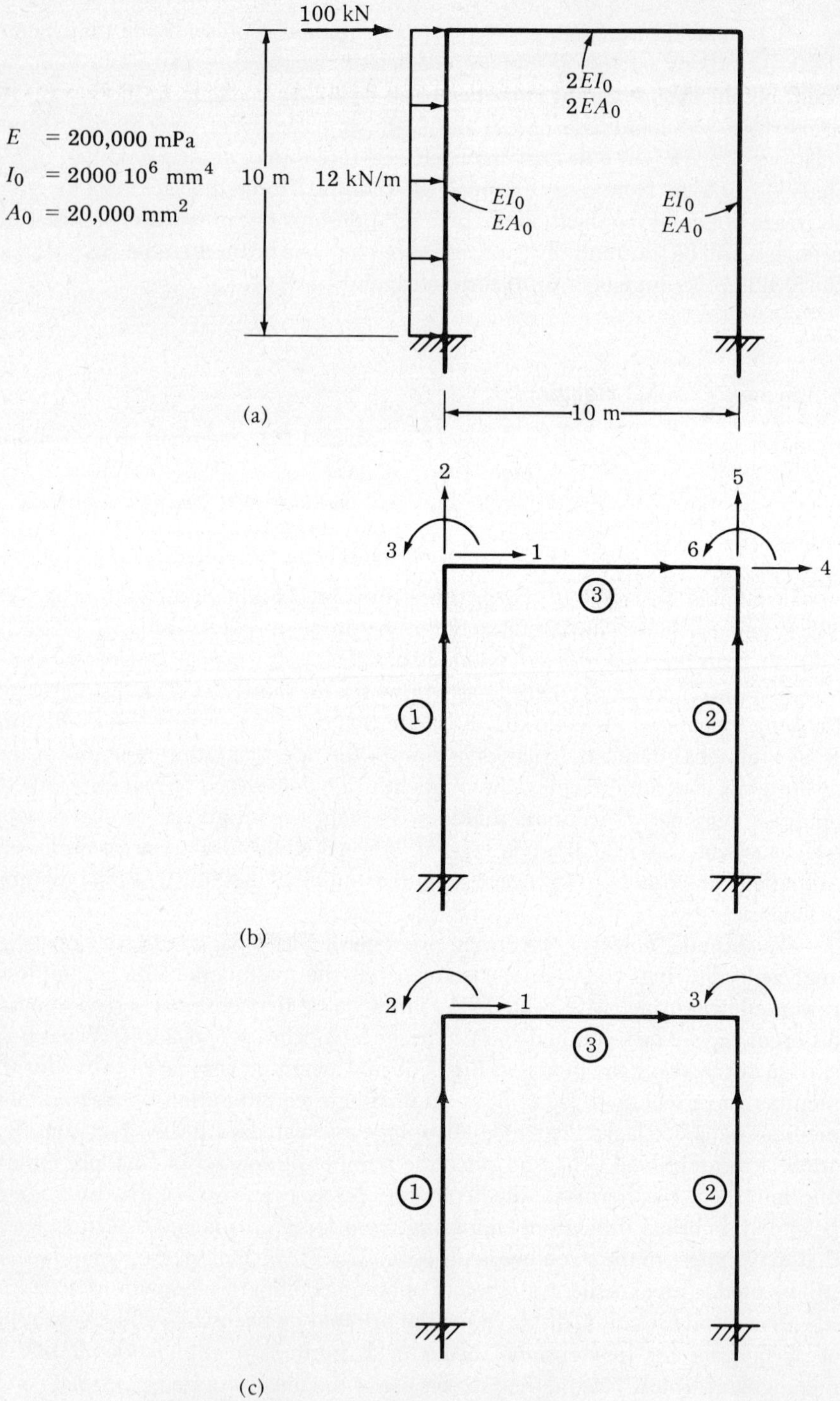

Fig. 5.6 Frame example, inextensional models: (a) frame of Fig. 4.7; (b) approximate inextensional model; (c) exact inextensional model.

****** *Computer Input:*

```
START
REMARK 4
FRAME PROBLEM WITH A(0) = 1E05, APPROXIMATING AN
INEXTENSIONAL SOLUTION.

**** UNITS ARE KILONEWTONS AND METERS ****
FORM K 6 6
LOAD R 6 1
100 0 0 0 0 0
FRAMEL A1 K1 1 RF1
0 10 2E08 .002 1E05
F 0 10 -12 -12 0
FRAMEL A3 K3 0
10 0 2E08 .004 2E05
ADDSTF K 3
K1 C1 0 0 0 1 2 3
K1 C2 0 0 0 4 5 6
K3 C3 1 2 3 4 5 6
ADDFEF R 1
RF1 C1
PRINT K 1
THIS IS THE COMPLETELY ASSEMBLED STRUCTURAL STIFFNESS MATRIX,K.
PRINT R 1
THIS IS THE VECTOR,R,OF EQUIVALENT AND CONCENTRATED NODAL LOADS.
SOLVE K R D
PRINT D 1
THIS IS THE VECTOR,D,OF STRUCTURE NODAL DISPLACEMENTS.
FRAMFC D 3
A1 K1 C1 RF1
A1 K1 C2 0
A3 K3 C3 0
STOP
```

****** *Computer Output:*

```
START

REMARK      4


FRAME PROBLEM WITH A(0) = 1E05, APPROXIMATING AN
INEXTENSIONAL SOLUTION.

**** UNITS ARE KILONEWTONS AND METERS ****



FORM    K              6    6
    6 ROWS,      6 COLUMNS

LOAD    R              6    1
    6 ROWS,      1 COLUMNS

FRAMEL  A1       K1          1      RF1
 DX=            0      DY=  1.000E+01      L =  1.000E+01
 E =  2.000E+08      I =  2.000E-03      A =  1.000E+05

  LOAD TYPE = F
   DIMENSION C =              0      LOAD INTENSITY AT C = -1.200E+01
   DIMENSION D =   1.000E+01      LOAD INTENSITY AT D = -1.200E+01
   LOAD ANGLE  =              0

FRAMEL  A3       K3          0
 DX=  1.000E+01      DY=            0      L =  1.000E+01
 E =  2.000E+08      I =  4.000E-03      A =  2.000E+05
```

****** *Computer Output (continued):*

```
ADDSTF  K           3
 MEMBER K    MEMBER CODE        DEGREES OF FREEDOM
    K1          C1              0     0     0     1     2     3
    K1          C2              0     0     0     4     5     6
    K3          C3              1     2     3     4     5     6

ADDFEF  R           1
 MEMBER FEF    MEMBER CODE
   RF1             C1

PRINT   K           1


THIS IS THE COMPLETELY ASSEMBLED STRUCTURAL STIFFNESS MATRIX,K.


               1          2          3          4          5          6
 1  4.0000E+12          0 2.4000E+04-4.0000E+12          0          0
 2           0 2.0000E+12 4.8000E+04          0-9.6000E+03 4.8000E+04
 3  2.4000E+04 4.8000E+04 4.8000E+05          0-4.8000E+04 1.6000E+05
 4 -4.0000E+12          0          0 4.0000E+12          0 2.4000E+04
 5           0-9.6000E+03-4.8000E+04          0 2.0000E+12-4.8000E+04
 6           0 4.8000E+04 1.6000E+05 2.4000E+04-4.8000E+04 4.8000E+05


PRINT   R           1


THIS IS THE VECTOR,R,OF EQUIVALENT AND CONCENTRATED NODAL LOADS.


               1
 1  1.6000E+02
 2           0
 3  1.0000E+02
 4           0
 5           0
 6           0


SOLVE   K       R       D

PRINT   D           1


THIS IS THE VECTOR,D,OF STRUCTURE NODAL DISPLACEMENTS.


               1
 1  2.0032E-02
 2  3.2308E-11
 3 -5.1683E-04
 4  2.0032E-02
 5 -3.2308E-11
 6 -8.2933E-04


FRAMFC  D           3

 MEMBER LOCAL FORCE VECTORS


 MEMBER CODE   = C1

 1 -6.46157E+01
 2  1.43750E+02
```

****** *Computer Output (continued):*

```
 3  5.39425E+02
 4  6.46157E+01
 5 -2.37503E+01
 6  2.98078E+02

 MEMBER CODE  = C2

 1  6.46157E+01
 2  7.62503E+01
 3  4.14425E+02
 4 -6.46157E+01
 5 -7.62503E+01
 6  3.48078E+02

 MEMBER CODE  = C3

 1  7.62500E+01
 2 -6.46157E+01
 3 -2.98078E+02
 4 -7.62500E+01
 5  6.46157E+01
 6 -3.48078E+02

STOP
```

****** *Computer Input:*

```
START
REMARK 4
FRAME PROBLEM WITH INEXTENSIONAL BEHAVIOR OF MEMBERS
BEING IMPOSED BY USE OF UNCONSTRAINED COORDINATE CODE.

**** UNITS ARE KILONEWTONS AND METERS ****
FORM K 3 3
LOAD R 3 1
100 0 0
FRAMEL A1 K1 1 RF1
0 10 2E08 .002 .02
F 0 10 -12 -12 0
FRAMEL A3 K3 0
10 0 2E08 .004 .04
ADDSTF K 3
K1 C1 0 0 0 1 0 2
K1 C2 0 0 0 1 0 3
K3 C3 0 0 2 0 0 3
ADDFEF R 1
RF1 C1
PRINT K 1
THIS IS THE COMPLETELY ASSEMBLED STRUCTURAL STIFFNESS MATRIX,K.
PRINT R 1
THIS IS THE VECTOR,R,OF EQUIVALENT AND CONCENTRATED NODAL LOADS.
SOLVE K R D
PRINT D 1
THIS IS THE VECTOR,D,OF STRUCTURE NODAL DISPLACEMENTS.
FRAMFC D 3
A1 K1 C1 RF1
A1 K1 C2 0
A3 K3 C3 0
STOP
```

****** *Computer Output:*

```
START

REMARK      4


FRAME PROBLEM WITH INEXTENSIONAL BEHAVIOR OF MEMBERS
BEING IMPOSED BY USE OF UNCONSTRAINED COORDINATE CODE.

**** UNITS ARE KILONEWTONS AND METERS ****



FORM    K               3   3
   3 ROWS,     3 COLUMNS

LOAD    R               3   1
   3 ROWS,     1 COLUMNS

FRAMEL  A1      K1          1     RF1
 DX=          0     DY=  1.000E+01     L =  1.000E+01
 E =  2.000E+08     I =  2.000E-03     A =  2.000E-02

 LOAD TYPE = F
  DIMENSION C =              0     LOAD INTENSITY AT C = -1.200E+01
  DIMENSION D =   1.000E+01     LOAD INTENSITY AT D = -1.200E+01
  LOAD ANGLE  =              0

FRAMEL. A3      K3            0
 DX=  1.000E+01     DY=          0     L =  1.000E+01
 E =  2.000E+08     I =  4.000E-03     A =  4.000E-02

ADDSTF  K           3
 MEMBER K    MEMBER CODE       DEGREES OF FREEDOM
    K1           C1            0     0     0     1     0     2
    K1           C2            0     0     0     1     0     3
    K3           C3            0     0     2     0     0     3

ADDFEF- R            1
 MEMBER FEF    MEMBER CODE
    RF1           C1

PRINT   K            1


THIS IS THE COMPLETELY ASSEMBLED STRUCTURAL STIFFNESS MATRIX,K.


              1          2          3
 1  9.6000E+03 2.4000E+04 2.4000E+04
 2  2.4000E+04 4.8000E+05 1.6000E+05
 3  2.4000E+04 1.6000E+05 4.8000E+05



PRINT   R            1


THIS IS THE VECTOR,R,OF EQUIVALENT AND CONCENTRATED NODAL LOADS.


              1
 1  1.6000E+02
 2  1.0000E+02
 3           0
```

****** *Computer Output (continued):*

```
SOLVE   K        R        D

PRINT   D                 1

THIS IS THE VECTOR,D,OF STRUCTURE NODAL DISPLACEMENTS.

           1
 1  2.0032E-02
 2 -5.1683E-04
 3 -8.2933E-04


FRAMFC  D                 3
 MEMBER LOCAL FORCE VECTORS

 MEMBER CODE   = C1

 1              0
 2  1.43750E+02
 3  5.39423E+02
 4              0
 5 -2.37500E+01
 6  2.98077E+02

 MEMBER CODE   = C2

 1              0
 2  7.62500E+01
 3  4.14423E+02
 4              0
 5 -7.62500E+01
 6  3.48077E+02

 MEMBER CODE   = C3

 1              0
 2 -6.46154E+01
 3 -2.98077E+02
 4              0
 5  6.46154E+01
 6 -3.48077E+02

 STOP
```

on such structures are always in global coordinate directions. For example, the model of Fig. 5.6c shows that the inextensional behaviors of members 1 and 2 cause the vertical displacement at nodes 1 and 2 to be zero, while the inextensional behavior of member 3 causes the horizontal displacements at nodes 1 and 2 to be identical; thus only one coordinate is required to represent both displacements. These facts are reflected in the unconstrained coordinate code for member 3, which is 1 0 2 1 0 3. This simple device yields an "exact" inextensional solution

Table 5.2 COMPARISON OF EXTENSIONAL AND INEXTENSIONAL SOLUTIONS: RECTANGULAR FRAME

RESPONSE	EXTENSIONAL	APPROXIMATE INEXTENSIONAL	"EXACT" INEXTENSIONAL	% ERROR
D_1	.02022 m	.02003 m	.02003 m	.94
D_2	.000161 m	3.231×10^{-11} m	0 m	100
D_3	−.0005501 rad	−.0005168 rad	−.0005168 rad	6.05
D_4	.02014 m	.02003 m	.02003m	.54
D_5	−.000161 m	-3.231×10^{-11} m	0 m	100
D_6	−.0008555 rad	−.0008293 rad	−.0008293 rad	3.06
$r_1'^{(1)}$	−64.4 kN	−64.6 kN	—	—
$r_2'^{(1)}$	−143.88 kN	143.7 kN	143.7 kN	.12
$r_3'^{(1)}$	541.4 kN · m	539.4 kN · m	539.4 kN · m	.36

with greater efficiency than the extensional solution. This approach would not work in general for the case of nonrectangular frames, however, since in these cases the inextensional constraints result in more complex linear dependencies among structure coordinates. Such problems are beyond the scope of this book but are discussed elsewhere (Ref. 5.1) in the literature.

Table 5.2 compares the displacement and force results of extensional and inextensional theory for the current frame problem. These results, which are typical for this frame, indicate that although displacements are slightly in error in the inextensional case, force results are very good. Such results are not, however, necessarily typical for all frames.

In general, inextensional analyses of frames in which high axial forces are developed in the members will exhibit large errors in both nodal displacements and internal forces. Frames that are susceptible to the development of high axial forces in their members include braced frames, multistory frames, and some nonrectangular frames. In this connection it is interesting to consider the frame of Fig. 5.7, which is a braced version of the frame of Fig. 5.6. A comparison of typical results for an inextensional and an extensional analysis of this frame is shown in Table 5.3. These results verify that inextensional analyses of frames must be used with care, since such analyses can in some cases give results that are greatly in error. There is no implication intended here that a competent structural

Table 5.3 COMPARISON OF EXTENSIONAL AND INEXTENSIONAL SOLUTIONS: BRACED FRAME

RESPONSE	EXTENSIONAL	APPROXIMATE INEXTENSIONAL	% ERROR
D_1	.00750 m	0 m	100
D_2	.0000460 m	0 m	100
D_3	−.00007526 rad	.0002344 rad	411
D_4	.00734 m	0 m	100
D_5	−.000295 m	0 m	100
D_6	−.000376 rad	−.00007812 rad	79
$r_1'^{(1)}$	−18.38 kN	7.5 kN	141
$r_2'^{(1)}$	94.18 kN	65.62 kN	30
$r_3'^{(1)}$	273.9 kN · m	118.8 kN · m	57

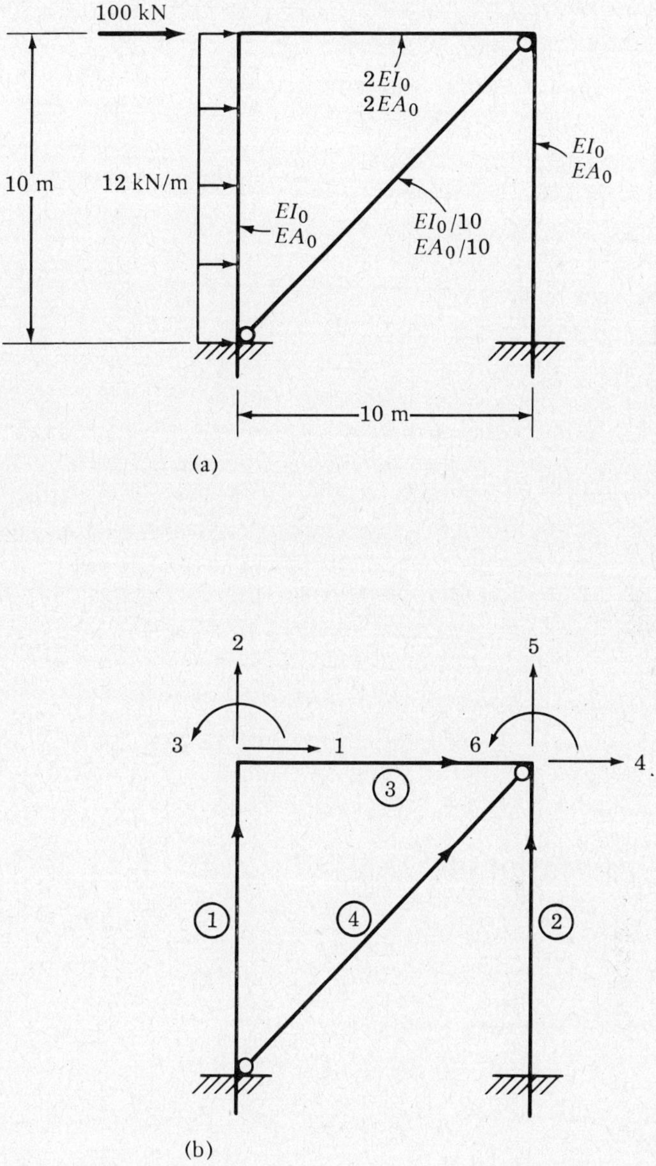

Fig. 5.7 Braced frame: (a) dimensions and loading; (b) approximate inextensional model.

engineer would consider using inextensional analysis on the braced frame of Fig. 5.7. The example is offered only to illustrate the point that inextensional analyses must not be used indiscriminately. In fact, the competent structural engineer would undoubtedly recognize that only member 4 of the example of Fig. 5.7 would likely need to be modeled extensionally. This modeling would make it possible to obtain the accuracy of a totally extensional analysis with the efficiency

****** *Computer Input:*

```
START
REMARK 3
BRACED FRAME,EXTENSIONAL SOLUTION.

**** UNITS ARE KILONEWTONS AND METERS ****
FORM K 6 6
LOAD R 6 1
100 0 0 0 0 0
FRAMEL A1 K1 1 RF1
0 10 2E08 .002 .02
F 0 10 -12 -12 0
FRAMEL A3 K3 0
10 0 2E08 .004 .04
TRUSEL A4 K4
10 10 2E08 .002
ADDSTF K 4
K1 C1 0 0 0 1 2 3
K1 C2 0 0 0 4 5 6
K3 C3 1 2 3 4 5 6
K4 C4 0 0 4 5
ADDFEF R 1
RF1 C1
PRINT K 1
THIS IS THE COMPLETELY ASSEMBLED STRUCTURAL STIFFNESS MATRIX,K.
PRINT R 1
THIS IS THE VECTOR,R,OF EQUIVALENT AND CONCENTRATED NODAL LOADS.
SOLVE K R D
PRINT D 1
THIS IS THE VECTOR,D,OF STRUCTURE NODAL DISPLACEMENTS.
FRAMFC D 3
A1 K1 C1 RF1
A1 K1 C2 0
A3 K3 C3 0
TRUSFC D 1
A4 K4 C4 0
STOP
```

******* *Computer Output:*

```
START

REMARK      3


BRACED FRAME,EXTENSIONAL SOLUTION.

**** UNITS ARE KILONEWTONS AND METERS ****



FORM    K              6    6
    6 ROWS,       6 COLUMNS

LOAD    R              6    1
    6 ROWS,       1 COLUMNS

FRAMEL  A1       K1          1      RF1
 DX=            0      DY=  1.000E+01       L =  1.000E+01
 E =  2.000E+08        I =  2.000E-03       A =  2.000E-02

 LOAD TYPE = F
   DIMENSION C =              0      LOAD INTENSITY AT C = -1.200E+01
   DIMENSION D =  1.000E+01          LOAD INTENSITY AT D = -1.200E+01
   LOAD ANGLE  =              0
```

****** *Computer Output (continued):*

```
FRAMEL  A3       K3          0
 DX=  1.000E+01     DY=          0      L =  1.000E+01
 E =  2.000E+08     I =  4.000E-03      A =  4.000E-02

TRUSEL  A4       K4
 DX=  1.000E+01     DY=  1.000E+01      L =  1.414E+01
 E =  2.000E+08     A =  2.000E-03

ADDSTF  K               4
 MEMBER K   MEMBER CODE       DEGREES OF FREEDOM
    K1           C1           0     0     0     1     2     3
    K1           C2           0     0     0     4     5     6
    K3           C3           1     2     3     4     5     6
    K4           C4           0     0     4     5

ADDFEF  R               1
 MEMBER FEF   MEMBER CODE
    RF1            C1

PRINT   K               1


THIS IS THE COMPLETELY ASSEMBLED STRUCTURAL STIFFNESS MATRIX,K.


            1          2          3          4          5          6
 1  8.0480E+05          0 2.4000E+04-8.0000E+05          0          0
 2           0 4.0960E+05 4.8000E+04          0-9.6000E+03 4.8000E+04
 3  2.4000E+04 4.8000E+04 4.8000E+05          0-4.8000E+04 1.6000E+05
 4 -8.0000E+05          0          0 8.1894E+05 1.4142E+04 2.4000E+04
 5           0-9.6000E+03-4.8000E+04 1.4142E+04 4.2374E+05-4.8000E+04
 6           0 4.8000E+04 1.6000E+05 2.4000E+04-4.8000E+04 4.8000E+05


PRINT   R               1


THIS IS THE VECTOR,R,OF EQUIVALENT AND CONCENTRATED NODAL LOADS.


            1
 1  1.6000E+02
 2           0
 3  1.0000E+02
 4           0
 5           0
 6           0


SOLVE   K       R       D

PRINT   D               1


THIS IS THE VECTOR,D,OF STRUCTURE NODAL DISPLACEMENTS.


            1
 1  7.4964E-03
 2  4.5964E-05
 3 -7.5263E-05
 4  7.3391E-03
 5 -2.9501E-04
 6 -3.7597E-04


FRAMFC  D               3
```

****** *Computer Output (continued):*

```
 MEMBER LOCAL FORCE VECTORS


 MEMBER CODE   = C1

 1 -1.83856E+01
 2  9.41765E+01
 3  2.73893E+02
 4  1.83856E+01
 5  2.58235E+01
 6  6.78718E+01


 MEMBER CODE   = C2

 1  1.18005E+02
 2  2.62046E+01
 3  1.46062E+02
 4 -1.18005E+02
 5 -2.62046E+01
 6  1.15984E+02


 MEMBER CODE   = C3

 1  1.25824E+02
 2 -1.83856E+01
 3 -6.78718E+01
 4 -1.25824E+02
 5  1.83856E+01
 6 -1.15984E+02
TRUSFC  D               1

 MEMBER LOCAL FORCE VECTORS


 MEMBER CODE   = C4

 1 -1.40882E+02
 2            0
 3  1.40882E+02
 4            0

STOP
```

of a totally inextensional analysis. Prob. 5.11 offers you the opportunity to verify this statement.

An interesting sidelight to the example of Fig. 5.7 is the use of truss and frame members together in modeling a structure. The SMIS3 solution for the extensional case of this example is displayed in the next printout to demonstrate how this is accomplished.

5.3.4 Frame Member with Hinge

Although we have, in the interest of simplicity, avoided frame members with internal hinges thus far, it is not that uncommon to encounter them in practice. It is therefore appropriate to develop the hinged-member equilibrium, kinematic, and stiffness relationships at this time. From Fig. 5.8 it is seen that the presence of a hinge reduces the number of member basic forces from three to two. M_i and P have, somewhat arbitrarily, been chosen as the basic forces in this case. Enforcing member overall equilibrium (taking into account the hinge), we can determine the full set of member-end forces shown in Fig. 5.8a.

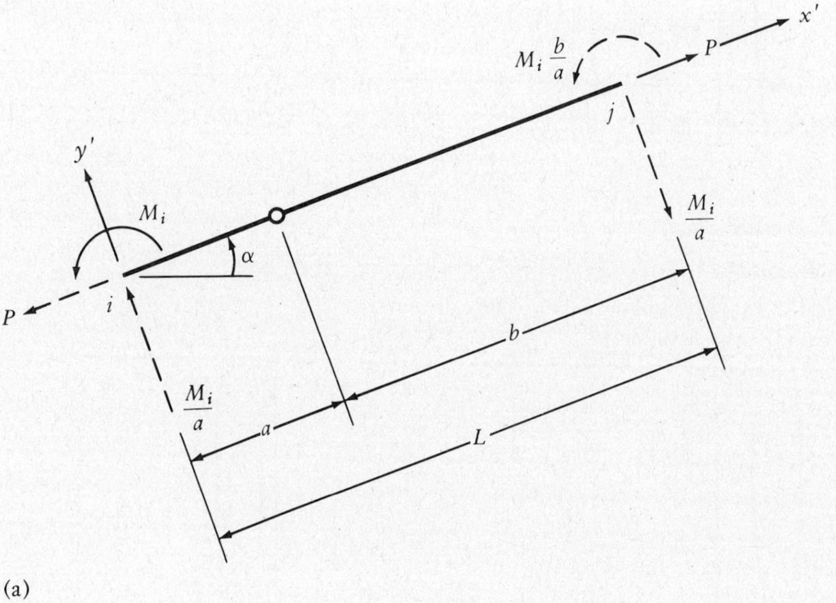

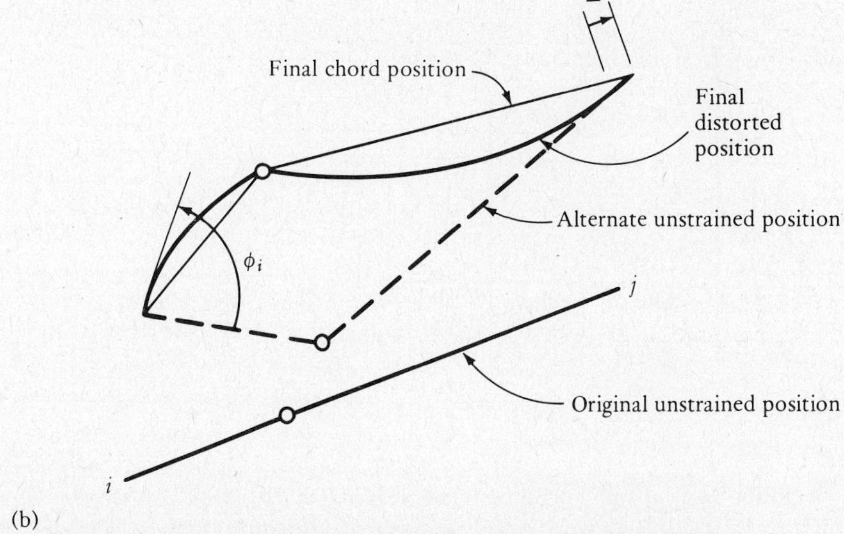

Fig. 5.8 Hinged frame member: (a) basic forces; (b) basic displacements.

Using the usual symbolic representation, we may now write the vector of member-end forces in local coordinates (by comparing Figs. 2.10c and 5.8a) as

$$\begin{pmatrix} r'_1 \\ r'_2 \\ r'_3 \\ r'_4 \\ r'_5 \\ r'_6 \end{pmatrix} = \begin{bmatrix} 0 & -1 \\ 1/a & 0 \\ 1 & 0 \\ 0 & 1 \\ -1/a & 0 \\ b/a & 0 \end{bmatrix} \begin{pmatrix} M_i \\ P \end{pmatrix} \tag{5.29}$$

or

$$\mathbf{r}' = \mathbf{LP} \tag{5.29a}$$

The relationship between member-end forces in global and local coordinates was given in Eq. 2.35 to be

$$\begin{pmatrix} r_1 \\ r_2 \\ r_3 \\ r_4 \\ r_5 \\ r_6 \end{pmatrix} = \begin{bmatrix} \cos\alpha & -\sin\alpha & 0 & 0 & 0 & 0 \\ \sin\alpha & \cos\alpha & 0 & 0 & 0 & 0 \\ 0 & 0 & 1 & 0 & 0 & 0 \\ 0 & 0 & 0 & \cos\alpha & -\sin\alpha & 0 \\ 0 & 0 & 0 & \sin\alpha & \cos\alpha & 0 \\ 0 & 0 & 0 & 0 & 0 & 1 \end{bmatrix} \begin{pmatrix} r'_1 \\ r'_2 \\ r'_3 \\ r'_4 \\ r'_5 \\ r'_6 \end{pmatrix} \tag{2.35}$$

or

$$\mathbf{r} = \boldsymbol{\alpha}\mathbf{r}' \tag{2.35a}$$

Substituting from Eq. 5.29 into 2.35 gives

$$\begin{pmatrix} r_1 \\ r_2 \\ r_3 \\ r_4 \\ r_5 \\ r_6 \end{pmatrix} = \begin{bmatrix} -\sin\alpha/a & -\cos\alpha \\ \cos\alpha/a & -\sin\alpha \\ 1 & 0 \\ \sin\alpha/a & \cos\alpha \\ -\cos\alpha/a & \sin\alpha \\ b/a & 0 \end{bmatrix} \begin{pmatrix} M_i \\ P \end{pmatrix} \tag{5.30}$$

or

$$\mathbf{r} = \mathbf{TP} \tag{5.30a}$$

On the basis of previous experience, it is logical to expect that there will be only two basic displacements, ϕ_i and Δ, corresponding to the two basic forces, M_i

and P_i for the hinged frame member. Although it is clear that the basic displacement Δ is necessary and sufficient to completely describe the axial distortions of this frame member, it is not so obvious that a single basic displacement ϕ_i will be necessary and sufficient to define the flexural distortions. Figure 5.8b shows a typical hinged frame member displaced in an arbitrary and general fashion from its initial unstrained position. This figure also shows that by selecting an alternative unstrained position as a starting point, it is possible to arrive at the same distorted configuration by varying the single flexural displacement parameter ϕ_i along with Δ. The relationship of these basic member displacements to the vector of local member-end displacements shown in Fig. 2.10d may now be shown to be contragredient in nature with respect to the basic-to-local force relationship of Eq. 5.29. Thus

$$\begin{pmatrix} \phi_i \\ \Delta \end{pmatrix} = \begin{bmatrix} 0 & 1/a & 1 & 0 & -1/a & b/a \\ -1 & 0 & 0 & 1 & 0 & 0 \end{bmatrix} \begin{pmatrix} d'_1 \\ d'_2 \\ d'_3 \\ d'_4 \\ d'_5 \\ d'_6 \end{pmatrix} \tag{5.31}$$

or

$$\Delta = \mathbf{L}^T \mathbf{d}' \tag{5.31a}$$

We can demonstrate that this equation must necessarily be true by applying the principle of virtual work as follows.

Imagine that the member of Fig. 5.8 undergoes a virtual displacement $\delta\mathbf{d}'$ such that

$$\delta\mathbf{d}' = \langle \delta d'_1 \; \delta d'_2 \; \delta d'_3 \; \delta d'_4 \; \delta d'_5 \; \delta d'_6 \rangle \tag{5.32}$$

If Eq. 5.31 is correct, it follows that the corresponding variation in the vector of basic displacements will be

$$\delta\boldsymbol{\Delta} = \mathbf{L}^T \, \delta\mathbf{d}' \tag{5.33}$$

or

$$\langle \delta\phi_i \; \delta\Delta \rangle = \mathbf{L}^T \langle \delta d'_1 \; \delta d'_2 \; \delta d'_3 \; \delta d'_4 \; \delta d'_5 \; \delta d'_6 \rangle \tag{5.33a}$$

The virtual work done by the local member-end forces as they move through their respective virtual displacements may be written as

$$\delta W = r'_1 \, \delta d'_1 + r'_2 \, \delta d'_2 + r'_3 \, \delta d'_3 + r'_4 \, \delta d'_4 + r'_5 \, \delta d'_5 + r'_6 \, \delta d'_6 \tag{5.34}$$

or, in matrix form,

$$\delta W = (\delta d_1' \ \delta d_2' \ \delta d_3' \ \delta d_4' \ \delta d_5' \ \delta d_6') \begin{pmatrix} r_1' \\ r_2' \\ r_3' \\ r_4' \\ r_5' \\ r_6' \end{pmatrix} \tag{5.34a}$$

or

$$\delta W = \delta \mathbf{d}'^T \, \mathbf{r}' \tag{5.34b}$$

Since the same distorted position of the member can be produced by the corresponding virtual change in the basic displacements, it follows that the same work must be done by the basic forces. Thus

$$\delta W = M_i \, \delta\phi_i + P\delta\Delta \tag{5.35}$$

or, in matrix form,

$$\delta W = (\delta\phi_i \ \delta\Delta) \begin{pmatrix} M_i \\ P \end{pmatrix} \tag{5.35a}$$

or

$$\delta W = \delta \mathbf{\Delta}^T \, \mathbf{P} \tag{5.35b}$$

Comparing Eqs. 5.34b and 5.35b, we see that

$$\delta \mathbf{d}'^T \, \mathbf{r}' = \delta \mathbf{\Delta}^T \, \mathbf{P} \tag{5.36}$$

Substituting from Eq. 5.33 into 5.36 gives

$$\delta \mathbf{d}'^T \, \mathbf{r}' = \delta \mathbf{d}'^T \, \mathbf{LP} \tag{5.37}$$

or

$$\delta \mathbf{d}'^T \, (\mathbf{r}' - \mathbf{LP}) = 0 \tag{5.37a}$$

But since $\delta \mathbf{d}'^T$ is arbitrary,

$$\mathbf{r}' - \mathbf{LP} = 0 \tag{5.38}$$

and

$$\mathbf{r}' = \mathbf{LP} \tag{5.38a}$$

which is seen to check Eq. 5.29a, thus verifying the contragrediency of this set of force and displacement transformations. Equation 5.31 could, of course, be obtained independently by a comparison of member kinematics in the basic and local systems. You may wish to do this to gain further confidence in these relationships.

The basic member-end displacements may now be related to the global member-end displacements by using Eq. 2.36, which is repeated here:

$$\begin{Bmatrix} d'_1 \\ d'_2 \\ d'_3 \\ d'_4 \\ d'_5 \\ d'_6 \end{Bmatrix} = \begin{bmatrix} \cos\alpha & \sin\alpha & 0 & 0 & 0 & 0 \\ -\sin\alpha & \cos\alpha & 0 & 0 & 0 & 0 \\ 0 & 0 & 1 & 0 & 0 & 0 \\ 0 & 0 & 0 & \cos\alpha & \sin\alpha & 0 \\ 0 & 0 & 0 & -\sin\alpha & \cos\alpha & 0 \\ 0 & 0 & 0 & 0 & 0 & 1 \end{bmatrix} \begin{Bmatrix} d_1 \\ d_2 \\ d_3 \\ d_4 \\ d_5 \\ d_6 \end{Bmatrix} \tag{2.36}$$

or

$$\mathbf{d}' = \boldsymbol{\alpha}^T \mathbf{d} \tag{2.36a}$$

Substituting from Eq. 2.36 into 5.31 gives

$$\begin{Bmatrix} \phi_i \\ \Delta \end{Bmatrix} = \begin{bmatrix} -\sin\alpha/a & \cos\alpha/a & 1 & \sin\alpha/a & -\cos\alpha/a & b/a \\ -\cos\alpha & -\sin\alpha & 0 & \cos\alpha & \cos\alpha & 0 \end{bmatrix} \begin{Bmatrix} d_1 \\ d_2 \\ d_3 \\ d_4 \\ d_5 \\ d_6 \end{Bmatrix} \tag{5.39}$$

or

$$\Delta = \mathbf{T}^T \mathbf{d} \tag{5.39a}$$

Equations 5.30 and 5.39 could now be used to consider hinged members in the development of structure equilibrium and compatibility relationships, respectively. It is left to you, if you are interested in these areas, to generalize the equilibrium and compatibility concepts of Chap. 2 to include hinged members. It is of greater current interest to develop the capability of considering hinged members in building the structure force-displacement relationships.

The hinged-member basic force-displacement relationships may be conveniently developed with the use of energy methods. If we neglect shearing distortions, we can write the internal strain energy in the member as

$$U = \int_0^L \frac{M^2\,dx'}{2EI} + \int_0^L \frac{S^2\,dx'}{2AE} \tag{3.28}$$

where it is seen from Fig. 5.8a that

$$\begin{aligned} M &= M_i - \frac{M_i x'}{a} \\ S &= P \end{aligned} \tag{5.40}$$

Substituting from Eq. 5.40 into 3.28 gives

$$U = \int_0^L \frac{M_i^2 (1 - x'/a)^2\, dx'}{2EI} + \int_0^L \frac{P^2\, dx'}{2AE} \tag{5.41}$$

Integrating gives

$$U = \frac{M_i^2}{2EI}\left(x' - \frac{x'^2}{a} + \frac{x'^3}{3a^2}\right)\Bigg|_0^L + \frac{P^2 x'}{2AE}\Bigg|_0^L \tag{5.41a}$$

or

$$U = \frac{M_i^2}{2EI}\left(L - \frac{L^2}{a} + \frac{L^3}{3a^2}\right) + \frac{P^2 L}{2AE} \tag{5.41b}$$

or

$$U = \frac{M_i^2}{2EI}\left(\frac{a^3 + b^3}{3a^2}\right) + \frac{PL}{2AE} \tag{5.41c}$$

Differentiating Eq. 5.41c appropriately yields

$$\begin{aligned} f_{11} &= \frac{\partial^2 U}{\partial M_i^2} = \frac{a^3 + b^3}{3EIa^2} \\ f_{12} &= f_{21} = \frac{\partial^2 U}{\partial M_i\, \partial P} = 0 \\ f_{22} &= \frac{\partial^2 U}{\partial P^2} = \frac{L}{AE} \end{aligned} \tag{5.42}$$

Using the flexibility coefficients of Eq. 5.42, we may finally write the member basic flexibility relationships as

$$\begin{pmatrix} \phi_i \\ \Delta \end{pmatrix} = \frac{a^3 + b^3}{3EI} \begin{bmatrix} \frac{1}{a^2} & 0 \\ 0 & \frac{3LI}{A(a^3 + b^3)} \end{bmatrix} \begin{pmatrix} M_i \\ P \end{pmatrix} \tag{5.43}$$

or

$$\mathbf{\Delta} = \mathbf{f}\mathbf{P} \tag{5.43a}$$

It follows that the basic stiffness relationships are

$$\begin{pmatrix} M_i \\ P \end{pmatrix} = \frac{3EI}{a^3 + b^3} \begin{bmatrix} a^2 & 0 \\ 0 & \frac{A(a^3 + b^3)}{3LI} \end{bmatrix} \begin{pmatrix} \phi_i \\ \Delta \end{pmatrix} \tag{5.44}$$

or

$$\mathbf{P} = \mathscr{k}\boldsymbol{\Delta} \tag{5.44a}$$

Using Eqs. 5.29, 5.31, and 5.44, we may write the member local stiffness relationships as

$$\begin{pmatrix} r'_1 \\ r'_2 \\ r'_3 \\ r'_4 \\ r'_5 \\ r'_6 \end{pmatrix} = \begin{bmatrix} 0 & -1 \\ \frac{1}{a} & 0 \\ 1 & 0 \\ 0 & 1 \\ \frac{-1}{a} & 0 \\ \frac{b}{a} & 0 \end{bmatrix} \frac{3EI}{a^3+b^3} \begin{bmatrix} a^2 & 0 \\ 0 & \frac{A(a^3+b^3)}{3LI} \end{bmatrix} \begin{bmatrix} 0 & \frac{1}{a} & 1 & 0 & \frac{-1}{a} & \frac{b}{a} \\ -1 & 0 & 0 & 1 & 0 & 0 \end{bmatrix} \begin{pmatrix} d'_1 \\ d'_2 \\ d'_3 \\ d'_4 \\ d'_5 \\ d'_6 \end{pmatrix} \tag{5.45}$$

or

$$\begin{pmatrix} r'_1 \\ r'_2 \\ r'_3 \\ r'_4 \\ r'_5 \\ r'_6 \end{pmatrix} = \frac{3EI}{a^3+b^3} \begin{bmatrix} \frac{A(a^3+b^3)}{3LI} & 0 & 0 & -\frac{A(a^3+b^3)}{3LI} & 0 & 0 \\ 0 & 1 & a & 0 & -1 & b \\ 0 & a & a^2 & 0 & -a & ab \\ -\frac{A(a^3+b^3)}{3LI} & 0 & 0 & \frac{A(a^3+b^3)}{3LI} & 0 & 0 \\ 0 & -1 & -a & 0 & 1 & -b \\ 0 & b & ab & 0 & -b & b^2 \end{bmatrix} \begin{pmatrix} d'_1 \\ d'_2 \\ d'_3 \\ d'_4 \\ d'_5 \\ d'_6 \end{pmatrix} \tag{5.45a}$$

or

$$\mathbf{r}' = \mathbf{k}'\mathbf{d}' \tag{5.45b}$$

Superposing the effect of a distributed loading gives

$$\begin{pmatrix} r'_1 \\ r'_2 \\ r'_3 \\ r'_4 \\ r'_5 \\ r'_6 \end{pmatrix} = \frac{3EI}{a^3+b^3} \begin{bmatrix} \frac{A(a^3+b^3)}{3LI} & 0 & 0 & -\frac{A(a^3+b^3)}{3LI} & 0 & 0 \\ 0 & 1 & a & 0 & -1 & b \\ 0 & a & a^2 & 0 & -a & ab \\ -\frac{A(a^3+b^3)}{3LI} & 0 & 0 & \frac{A(a^3+b^3)}{3LI} & 0 & 0 \\ 0 & -1 & -a & 0 & 1 & -b \\ 0 & b & ab & 0 & -b & b^2 \end{bmatrix} \begin{pmatrix} d'_1 \\ d'_2 \\ d'_3 \\ d'_4 \\ d'_5 \\ d'_6 \end{pmatrix} + \begin{pmatrix} r'_{F1} \\ r'_{F2} \\ r'_{F3} \\ r'_{F4} \\ r'_{F5} \\ r'_{F6} \end{pmatrix} \tag{5.46}$$

or

$$\mathbf{r}' = \mathbf{k}'\mathbf{d}' + \mathbf{r}'_F \tag{5.46a}$$

The local member stiffness relationships of Eq. 5.46a may, of course, be transformed to global coordinates by using Eq. 2.35a and 2.36a, which gives

$$\mathbf{r} = \boldsymbol{\alpha}\mathbf{k}'\boldsymbol{\alpha}^T\mathbf{d} + \boldsymbol{\alpha}\mathbf{r}'_F \tag{5.47}$$

or

$$\mathbf{r} = \mathbf{k}\mathbf{d} + \mathbf{r}_F \tag{5.47a}$$

Carrying the stiffness matrix transformation of Eq. 5.47 out in detail, we get Eq. 5.48 displayed on page 253.

The stiffness relationship for a beam with an internal hinge may be obtained as a special case of Eq. 5.48, yielding

$$\begin{pmatrix} r_1 \\ r_2 \\ r_3 \\ r_4 \end{pmatrix} = \frac{3EI}{a^3 + b^3}\begin{bmatrix} 1 & a & -1 & b \\ a & a^2 & -a & ab \\ -1 & -a & 1 & -b \\ b & ab & -b & b^2 \end{bmatrix}\begin{pmatrix} d_1 \\ d_2 \\ d_3 \\ d_4 \end{pmatrix} + \begin{pmatrix} r_{F1} \\ r_{F2} \\ r_{F3} \\ r_{F4} \end{pmatrix} \tag{5.49}$$

Consider the beam structure example of Fig. 5.9. For this structure, a single member with an internal hinge, the stiffness relationships may be written as

$$\begin{pmatrix} r_1 \\ r_2 \\ r_3 \\ r_4 \end{pmatrix} = \frac{3EI}{2\ell^3}\begin{bmatrix} 1 & \ell & -1 & \ell \\ \ell & \ell^2 & -\ell & \ell^2 \\ -1 & -\ell & 1 & -\ell \\ \ell & \ell^2 & -\ell & \ell^2 \end{bmatrix}\begin{pmatrix} d_1 \\ d_2 \\ d_3 \\ d_4 \end{pmatrix} + \begin{pmatrix} q\ell \\ q\ell^2/2 \\ q\ell \\ -q\ell^2/2 \end{pmatrix} \tag{A}$$

or, in structure notation, taking supports into account,

$$\begin{pmatrix} R_1 \\ R_2 \\ R_3 \\ \hline 0 \end{pmatrix} = \frac{3EI}{2\ell^3}\left[\begin{array}{ccc|c} 1 & \ell & -1 & \ell \\ \ell & \ell^2 & -\ell & \ell^2 \\ -1 & -\ell & 1 & -\ell \\ \hline \ell & \ell^2 & -\ell & \ell^2 \end{array}\right]\begin{pmatrix} 0 \\ 0 \\ 0 \\ \hline D_4 \end{pmatrix} + \begin{pmatrix} q\ell \\ q\ell^2/2 \\ q\ell \\ \hline -q\ell^2/2 \end{pmatrix} \tag{B}$$

The last of these four equations is

$$0 = \frac{3EI}{2\ell^3}(\ell^2 D_4) - q\ell^2/2 \tag{C}$$

or, by solving,

$$D_4 = \frac{2\ell}{3EI} \cdot \frac{q\ell^2}{2} = \frac{q\ell^3}{3EI} \tag{C$'$}$$

$$
\mathbf{k} = \frac{3EI}{a^3 + b^3}
\begin{bmatrix}
\frac{c^2A(a^3 + b^3)}{3LI} + s^2 & \frac{scA(a^3 + b^3)}{3LI} - sc & -sa & \frac{-c^2A(a^3 + b^3)}{3LI} - s^2 & \frac{-scA(a^3 + b^3)}{3LI} - sc & -sb \\
 & \frac{s^2A(a^3 + b^3)}{3LI} + c^2 & ca & \frac{-scA(a^3 + b^3)}{3LI} + sc & \frac{-s^2A(a^3 + b^3)}{3LI} - c^2 & cb \\
 & & a^2 & sa & -ca & ab \\
 & \text{Symmetric} & & \frac{c^2A(a^3 + b^3)}{3LI} + s^2 & \frac{scA(a^3 + b^3)}{3LI} - sc & sb \\
 & & & & \frac{s^2A(a^3 + b^3)}{3LI} + c^2 & -cb \\
 & & & & & b^2
\end{bmatrix}
\qquad (5.48)
$$

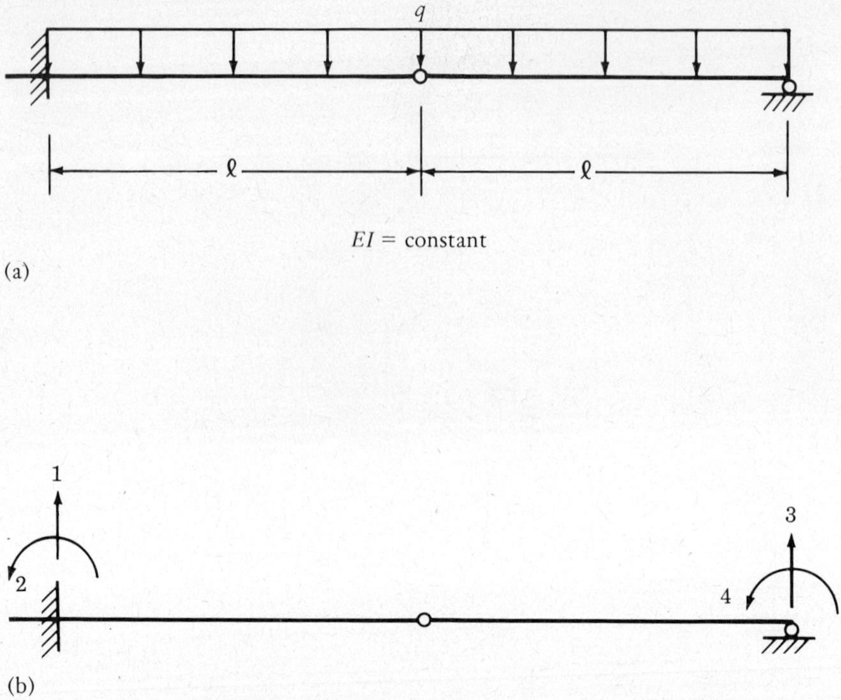

Fig. 5.9 Beam with internal hinge: (a) beam; (b) beam model.

Backsubstituting into the first three scalar equations of matrix Eq. B gives

$$\begin{pmatrix} R_1 \\ R_2 \\ R_3 \end{pmatrix} = \frac{3EI}{2\ell^3}\begin{bmatrix} \ell \\ \ell^2 \\ -\ell \end{bmatrix}\frac{q\ell^3}{3EI} + \begin{pmatrix} q\ell \\ q\ell^2/2 \\ q\ell \end{pmatrix}$$
$$= \begin{pmatrix} q\ell/2 \\ q\ell^2/2 \\ -q\ell/2 \end{pmatrix} + \begin{pmatrix} q\ell \\ q\ell^2/2 \\ q\ell \end{pmatrix} = \begin{pmatrix} 3q\ell/2 \\ q\ell^2 \\ q\ell/2 \end{pmatrix} \quad \text{(D)}$$

which can be verified by statics.

5.4 SKEWED SUPPORTS

A skewed support is a support that provides a constraint to the structure in a nonglobal direction. The effect that such a skewed support has on the structure stiffness relationships can perhaps best be demonstrated by considering a specific example. Fig. 5.10 shows the statically indeterminate truss of Fig. 4.6 with one change: the right-hand support is now skewed.

Modeling this skewed support with a truss member and specializing the

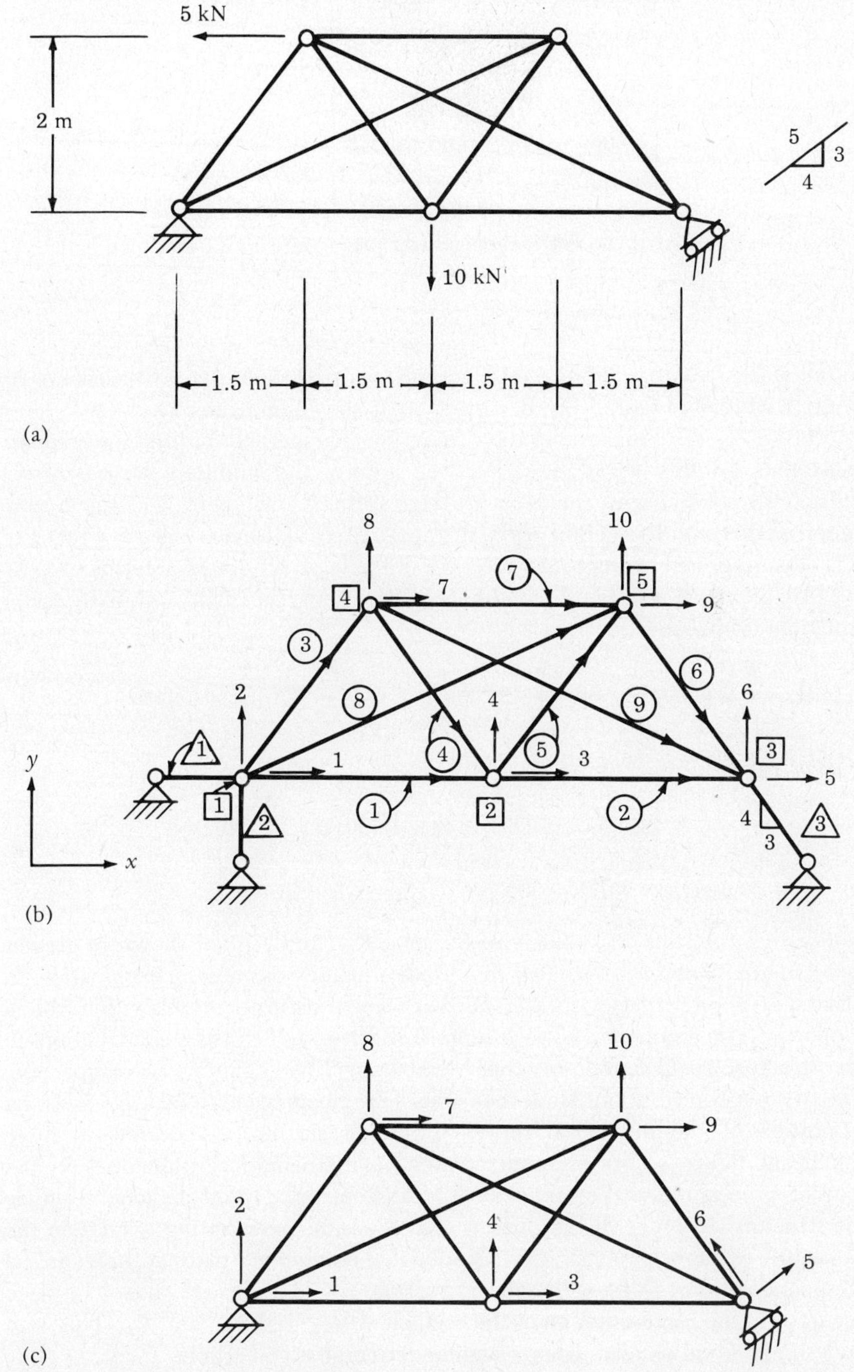

Fig. 5.10 Truss with skewed support: (a) dimensions and loading; (b) global coordinate model; (c) mixed coordinate model.

stiffness array of Eq. 4.1a with $EA/L = (EA_0/3) \cdot \infty$, we get

$$\mathbf{k}_c^{(3)} = \frac{EA_0}{3} \cdot \infty \begin{bmatrix} .36 & -.48 & -.36 & .48 \\ -.48 & .64 & .48 & -.64 \\ -.36 & .48 & .36 & -.48 \\ .48 & -.64 & -.48 & .64 \end{bmatrix} \tag{A}$$

whereas in the case of the vertical constraint previously considered, we had

$$k_c^{(3)} = \frac{EA_0}{3} \cdot \infty \tag{B}$$

Building the structure stiffness relationships by using the support stiffness matrix of Eq. A instead of that of Eq. B, we get Eq. C shown on page 257.

It is now useful to consider the various possible methods of imposing support constraints on this set of equations. As before, the infinitely large support stiffnesses may be approximated with large stiffness coefficients and Eq. C may be solved directly. Inspection of the stiffness matrix, however, will show that Eq. C is ill-conditioned owing to the presence of the large off-diagonal terms.

To demonstrate this ill-conditioning, consider the fifth and sixth scalar equations of matrix Eq. C:

$$R_5 = \frac{EA_0}{3}[-D_3 + (.36\infty + 1.96)D_5 - (.48\infty + .826)D_6 - .509D_7 + .226D_8 - .45D_9 + .6D_{10}] \tag{D}$$

$$R_6 = \frac{EA_0}{3}[-(.48\infty + .826)D_5 + (.64\infty + .901)D_6 + .226D_7 - .101D_8 + .6D_9 - .8D_{10}] \tag{E}$$

Notice that the stiffness coefficients K_{55}, K_{56}, K_{65}, and K_{66} are shown in greater detail in Eqs. D and E than in Eq. C. This discrepancy requires an explanation. In general, structure stiffness coefficients that receive a stiffness contribution from a support will be dominated by that support stiffness, and stiffness contributions to that structure coefficient from other members will be negligible by comparison. The stiffness contributions from the "other" members have, in fact, been lost by truncation of the numerical representation of stiffness coefficients to three significant figures. This situation creates no mathematical difficulties in the solution process unless the support stiffnesses contribute to off-diagonal terms in the structure array, as in the present case. Consider now letting ∞ take on the value 10^{10}, as has been our practice. Notice that obtaining 3-place accuracy in the elimination of, say, D_5 from these two equations would require the use of 14-place accuracy in the representation of the coefficients K_{55}, K_{56}, K_{65}, and K_{66}. This is an undesirable situation even when using modern digital computers.

A straightforward application of the modification method to this set of equations also presents difficulties. Dividing, say, the sixth scalar equation of

$$
\begin{Bmatrix} R_1 \\ R_2 \\ R_3 \\ R_4 \\ R_5 \\ R_6 \\ R_7 \\ R_8 \\ R_9 \\ R_{10} \end{Bmatrix} = \frac{EA_0}{3} \begin{bmatrix}
\infty & .826 & -1 & 0 & \cdot & \cdot & -.45 & -.6 & -.509 & -.226 \\
.826 & \infty & 0 & 0 & \cdot & \cdot & -.6 & -.8 & -.226 & -.101 \\
-1 & 0 & 2.9 & 0 & -1 & 0 & -.45 & .6 & -.45 & -.6 \\
0 & 0 & 0 & 1.6 & 0 & 0 & .6 & -.8 & -.6 & -.8 \\
\cdot & \cdot & -1 & 0 & .36\infty & -.48\infty & -.509 & .226 & -.45 & .6 \\
\cdot & \cdot & 0 & 0 & -.48\infty & .64\infty & .226 & -.101 & .6 & -.8 \\
-.45 & -.6 & -.45 & .6 & -.509 & .226 & 2.41 & -.226 & -1 & 0 \\
-.6 & -.8 & .6 & -.8 & .226 & -.101 & -.226 & 1.7 & 0 & 0 \\
-.509 & -.226 & -.45 & -.6 & -.45 & .6 & -1 & 0 & 2.41 & .226 \\
-.226 & -.101 & -.6 & -.8 & .6 & -.8 & 0 & 0 & .226 & 1.7
\end{bmatrix} \begin{Bmatrix} D_1 \\ D_2 \\ D_3 \\ D_4 \\ D_5 \\ D_6 \\ D_7 \\ D_8 \\ D_9 \\ D_{10} \end{Bmatrix} \quad \text{(C)}
$$

matrix Eq. C by the very large term represented by ∞ yields the alternative equation

$$D_6 = \tfrac{3}{4} D_5 \tag{F}$$

Substituting this result into the fifth scalar equation of Eq. C gives

$$\begin{aligned} R_5 = &-D_3 + (.36\infty + 1.96)D_5 + (-.48\infty - .826)\tfrac{3}{4} D_5 - .509D_7 \\ &+ .226D_8 - .45D_9 + .6D_{10} \end{aligned} \tag{G}$$

or

$$R_5 = -D_3 + 1.36D_5 - .509D_7 + .226D_8 - .45D_9 + .6D_{10} \tag{G$'$}$$

The fifth and sixth scalar equations of matrix Eq. C could now be replaced with alternative Eqs. F and G′; and, as before, the first and second scalar equations of Eq. C, which involve constraints in global coordinate directions, may be replaced by

$$\begin{aligned} 0 &= D_1 \\ 0 &= D_2 \end{aligned} \tag{H}$$

Once again it is seen that the resulting stiffness matrix will be free of support stiffnesses and therefore could be obtained by an appropriate though, in this case, rather awkward modification of the unconstrained structure-stiffness matrix. More important, however, is the fact that the modified stiffness matrix will be unsymmetric about the main diagonal owing to alternative Eqs. F and G′. This lack of symmetry is unacceptable because efficient solution techniques for the simultaneous equations generated by the direct stiffness method are dependent on the stiffness matrix being symmetric, as will be shown in Chap. 7.

The rearrangement method is based on the unconstrained structure stiffness relationships, which are shown in Eq. I on page 259. Eq. I may be represented symbolically as

$$\mathbf{R} = \mathbf{KD} \tag{I$'$}$$

It may be recalled that the rearrangement method involves permutating rows and columns of the arrays in Eq. I so that unknown displacements may be isolated from known displacements by matrix partitioning. This permutation is seen to be impossible in the case of Eq. I because displacements D_5 and D_6 are neither completely known nor completely unknown.

Thus none of the three solution techniques described may be successfully applied to the standard set of global structure stiffness equations arising from a problem involving a constraint in a nonglobal direction. But, as we will now see, transforming the stiffness equations from the global system of coordinates will make it possible to successfully apply any of the three solution methods in the usual fashion.

The relationships between the nodal displacements in the global coordinate directions of Fig. 5.10b and those in the mixed coordinate directions of Fig. 5.10c

$$\begin{pmatrix} R_1 \\ R_2 \\ R_3 \\ R_4 \\ R_5 \\ R_6 \\ R_7 \\ R_8 \\ R_9 \\ R_{10} \end{pmatrix} = \frac{EA_0}{3} \begin{bmatrix} 1.96 & .826 & -1 & 0 & \cdot & \cdot & -.45 & -.6 & -.509 & -.226 \\ .826 & .901 & 0 & 0 & \cdot & \cdot & -.6 & -.8 & -.226 & -.101 \\ -1 & 0 & 2.9 & 0 & -1 & 0 & -.45 & .6 & -.45 & -.6 \\ 0 & 0 & 0 & 1.6 & 0 & 0 & .6 & -.8 & -.6 & -.8 \\ \cdot & \cdot & -1 & 0 & 1.96 & -.826 & -.509 & .226 & -.45 & .6 \\ \cdot & \cdot & 0 & 0 & -.826 & .901 & .226 & -.101 & .6 & -.8 \\ -.45 & -.6 & -.45 & .6 & -.509 & .226 & 2.41 & -.226 & -1 & 0 \\ -.6 & -.8 & .6 & -.8 & .226 & -.101 & -.226 & 1.7 & 0 & 0 \\ -.509 & -.226 & -.45 & -.6 & -.45 & .6 & -1 & 0 & 2.41 & .226 \\ -.226 & -.101 & -.6 & -.8 & .6 & -.8 & 0 & 0 & .226 & 1.7 \end{bmatrix} \begin{pmatrix} D_1 \\ D_2 \\ D_3 \\ D_4 \\ D_5 \\ D_6 \\ D_7 \\ D_8 \\ D_9 \\ D_{10} \end{pmatrix} \quad \text{(l)}$$

are

$$\begin{pmatrix} D_1 \\ D_2 \\ D_3 \\ D_4 \\ D_5 \\ D_6 \\ D_7 \\ D_8 \\ D_9 \\ D_{10} \end{pmatrix} = \begin{bmatrix} 1 & \cdot & \cdot & \cdot & \cdot & \cdot & \cdot & \cdot & \cdot & \cdot \\ \cdot & 1 & \cdot & \cdot & \cdot & \cdot & \cdot & \cdot & \cdot & \cdot \\ \cdot & \cdot & 1 & \cdot & \cdot & \cdot & \cdot & \cdot & \cdot & \cdot \\ \cdot & \cdot & \cdot & 1 & \cdot & \cdot & \cdot & \cdot & \cdot & \cdot \\ \cdot & \cdot & \cdot & \cdot & \frac{4}{5} & -\frac{3}{5} & \cdot & \cdot & \cdot & \cdot \\ \cdot & \cdot & \cdot & \cdot & \frac{3}{5} & \frac{4}{5} & \cdot & \cdot & \cdot & \cdot \\ \cdot & \cdot & \cdot & \cdot & \cdot & \cdot & 1 & \cdot & \cdot & \cdot \\ \cdot & \cdot & \cdot & \cdot & \cdot & \cdot & \cdot & 1 & \cdot & \cdot \\ \cdot & \cdot & \cdot & \cdot & \cdot & \cdot & \cdot & \cdot & 1 & \cdot \\ \cdot & \cdot & \cdot & \cdot & \cdot & \cdot & \cdot & \cdot & \cdot & 1 \end{bmatrix} \begin{pmatrix} \hat{D}_1 \\ \hat{D}_2 \\ \hat{D}_3 \\ \hat{D}_4 \\ \hat{D}_5 \\ \hat{D}_6 \\ \hat{D}_7 \\ \hat{D}_8 \\ \hat{D}_9 \\ \hat{D}_{10} \end{pmatrix} \qquad \text{(J)}$$

The corresponding relationships among forces are

$$\begin{pmatrix} \hat{R}_1 \\ \hat{R}_2 \\ \hat{R}_3 \\ \hat{R}_4 \\ \hat{R}_5 \\ \hat{R}_6 \\ \hat{R}_7 \\ \hat{R}_8 \\ \hat{R}_9 \\ \hat{R}_{10} \end{pmatrix} = \begin{bmatrix} 1 & \cdot & \cdot & \cdot & \cdot & \cdot & \cdot & \cdot & \cdot & \cdot \\ \cdot & 1 & \cdot & \cdot & \cdot & \cdot & \cdot & \cdot & \cdot & \cdot \\ \cdot & \cdot & 1 & \cdot & \cdot & \cdot & \cdot & \cdot & \cdot & \cdot \\ \cdot & \cdot & \cdot & 1 & \cdot & \cdot & \cdot & \cdot & \cdot & \cdot \\ \cdot & \cdot & \cdot & \cdot & \frac{4}{5} & \frac{3}{5} & \cdot & \cdot & \cdot & \cdot \\ \cdot & \cdot & \cdot & \cdot & -\frac{3}{5} & \frac{4}{5} & \cdot & \cdot & \cdot & \cdot \\ \cdot & \cdot & \cdot & \cdot & \cdot & \cdot & 1 & \cdot & \cdot & \cdot \\ \cdot & \cdot & \cdot & \cdot & \cdot & \cdot & \cdot & 1 & \cdot & \cdot \\ \cdot & \cdot & \cdot & \cdot & \cdot & \cdot & \cdot & \cdot & 1 & \cdot \\ \cdot & \cdot & \cdot & \cdot & \cdot & \cdot & \cdot & \cdot & \cdot & 1 \end{bmatrix} \begin{pmatrix} R_1 \\ R_2 \\ R_3 \\ R_4 \\ R_5 \\ R_6 \\ R_7 \\ R_8 \\ R_9 \\ R_{10} \end{pmatrix} \qquad \text{(K)}$$

where in the present case

$$\begin{aligned} \hat{\mathbf{R}} &= \langle \hat{R}_1 \quad \hat{R}_2 \quad 0 \quad -10 \quad 0 \quad \hat{R}_6 \quad -5 \quad 0 \quad 0 \quad 0 \rangle \\ \hat{\mathbf{D}} &= \langle 0 \quad 0 \quad \hat{D}_3 \quad \hat{D}_4 \quad \hat{D}_5 \quad 0 \quad \hat{D}_7 \quad \hat{D}_8 \quad \hat{D}_9 \quad \hat{D}_{10} \rangle \end{aligned} \qquad \text{(L)}$$

Combining Eqs. I, J, K, and L gives Eq. M on page 260. Eq. M may be represented symbolically as

$$\hat{\mathbf{R}} = \hat{\mathbf{K}}\hat{\mathbf{D}} \qquad \text{(M')}$$

As expected, the symmetric transformation of the symmetric $\mathbf{K}$ matrix yielded a symmetric $\hat{\mathbf{K}}$ matrix. Moreover, only rows 5 and 6 were affected by the transformation, which involved only a rotation of coordinates at node 3. Finally (and most important), we see that Eq. M may be rearranged in the usual way.

$$
\begin{pmatrix} \hat{R}_1 \\ \hat{R}_2 \\ 0 \\ -10 \\ 0 \\ \hat{R}_6 \\ -5 \\ 0 \\ 0 \\ 0 \end{pmatrix} = \frac{EA_0}{3} \begin{bmatrix} 1.96 & .826 & -1 & 0 & \cdot & \cdot & -.45 & -.6 & -.509 & -.226 \\ .826 & .901 & 0 & 0 & \cdot & \cdot & -.6 & -.8 & -.226 & -.101 \\ -1 & 0 & 2.9 & 0 & -.8 & .6 & -.45 & .6 & -.45 & -.6 \\ 0 & 0 & 0 & 1.6 & 0 & 0 & .6 & -.8 & -.6 & -.8 \\ \cdot & \cdot & -.8 & 0 & .786 & -.739 & -.272 & .12 & 0 & 0 \\ \cdot & \cdot & .6 & 0 & -.739 & 2.08 & .486 & -.216 & .75 & -1 \\ -.45 & -.6 & -.45 & .6 & -.272 & .486 & 2.41 & -.226 & -1 & 0 \\ -.6 & -.8 & .6 & -.8 & .120 & -.216 & -.226 & 1.7 & 0 & 0 \\ -.509 & -.226 & -.45 & -.6 & 0 & .75 & -1 & 0 & 2.41 & .226 \\ -.226 & -.101 & -.6 & -.8 & 0 & -1 & 0 & 0 & .226 & 1.7 \end{bmatrix} \begin{pmatrix} 0 \\ 0 \\ \hat{D}_3 \\ \hat{D}_4 \\ \hat{D}_5 \\ 0 \\ \hat{D}_7 \\ \hat{D}_8 \\ \hat{D}_9 \\ \hat{D}_{10} \end{pmatrix} \qquad \text{(M)}
$$

Since this procedure would be a straightforward application of the rearrangement method, it will not be carried out here. It should also be noted that the modification method could also be applied to Eq. M in a standard manner. Even the original method of adding large support stiffnesses to Eq. M will now work without ill-conditioning, because there will be no large off-diagonal terms in the stiffness matrix based on this coordinate system.

A somewhat shorter transformation of Eq. I can be used as an alternative to that of Eqs. J and K. The skewed constraint imposes the following conditions on the forces and displacements in the original coordinate system of Fig. 5.10b:

$$\begin{aligned} &D_6 = \tfrac{3}{4} D_5 \\ &\tfrac{4}{5} R_5 + \tfrac{3}{5} R_6 = 0 \longrightarrow R_5 + \tfrac{3}{4} R_6 = 0 \end{aligned} \tag{N}$$

which may be stated in matrix form as

$$\begin{pmatrix} D_1 \\ D_2 \\ D_3 \\ D_4 \\ D_5 \\ D_6 \\ D_7 \\ D_8 \\ D_9 \\ D_{10} \end{pmatrix} = \begin{bmatrix} 1 & \cdot & \cdot & \cdot & \cdot & \cdot & \cdot & \cdot & \cdot \\ \cdot & 1 & \cdot & \cdot & \cdot & \cdot & \cdot & \cdot & \cdot \\ \cdot & \cdot & 1 & \cdot & \cdot & \cdot & \cdot & \cdot & \cdot \\ \cdot & \cdot & \cdot & 1 & \cdot & \cdot & \cdot & \cdot & \cdot \\ \cdot & \cdot & \cdot & \cdot & 1 & \cdot & \cdot & \cdot & \cdot \\ \cdot & \cdot & \cdot & \cdot & \frac{3}{4} & \cdot & \cdot & \cdot & \cdot \\ \cdot & \cdot & \cdot & \cdot & \cdot & 1 & \cdot & \cdot & \cdot \\ \cdot & \cdot & \cdot & \cdot & \cdot & \cdot & 1 & \cdot & \cdot \\ \cdot & \cdot & \cdot & \cdot & \cdot & \cdot & \cdot & 1 & \cdot \\ \cdot & \cdot & \cdot & \cdot & \cdot & \cdot & \cdot & \cdot & 1 \end{bmatrix} \begin{pmatrix} D_1 \\ D_2 \\ D_3 \\ D_4 \\ D_5 \\ D_7 \\ D_8 \\ D_9 \\ D_{10} \end{pmatrix} \tag{O}$$

and

$$\begin{pmatrix} R_1 \\ R_2 \\ R_3 \\ R_4 \\ 0 \\ R_7 \\ R_8 \\ R_9 \\ R_{10} \end{pmatrix} = \begin{bmatrix} 1 & \cdot & \cdot & \cdot & \cdot & \cdot & \cdot & \cdot & \cdot & \cdot \\ \cdot & 1 & \cdot & \cdot & \cdot & \cdot & \cdot & \cdot & \cdot & \cdot \\ \cdot & \cdot & 1 & \cdot & \cdot & \cdot & \cdot & \cdot & \cdot & \cdot \\ \cdot & \cdot & \cdot & 1 & \cdot & \cdot & \cdot & \cdot & \cdot & \cdot \\ \cdot & \cdot & \cdot & \cdot & 1 & \frac{3}{4} & \cdot & \cdot & \cdot & \cdot \\ \cdot & \cdot & \cdot & \cdot & \cdot & \cdot & 1 & \cdot & \cdot & \cdot \\ \cdot & \cdot & \cdot & \cdot & \cdot & \cdot & \cdot & 1 & \cdot & \cdot \\ \cdot & \cdot & \cdot & \cdot & \cdot & \cdot & \cdot & \cdot & 1 & \cdot \\ \cdot & \cdot & \cdot & \cdot & \cdot & \cdot & \cdot & \cdot & \cdot & 1 \end{bmatrix} \begin{pmatrix} R_1 \\ R_2 \\ R_3 \\ R_4 \\ R_5 \\ R_6 \\ R_7 \\ R_8 \\ R_9 \\ R_{10} \end{pmatrix} \tag{P}$$

Using Eqs. I, O, and P and known **R** and **D** values, we obtain Eq. Q as shown on

$$\begin{pmatrix} R_1 \\ R_2 \\ 0 \\ -10 \\ 0 \\ -5 \\ 0 \\ 0 \\ 0 \end{pmatrix} = \frac{EA_0}{3} \begin{bmatrix} 1.96 & .826 & -1 & 0 & \cdot & -.45 & -.6 & -.509 & -.226 \\ .826 & .901 & 0 & 0 & \cdot & -.6 & -.8 & -.226 & -.101 \\ -1 & 0 & 2.9 & 0 & -1 & -.45 & .6 & -.45 & -.6 \\ 0 & 0 & 0 & 1.6 & 0 & .6 & -.8 & -.6 & -.8 \\ \cdot & \cdot & -1 & 0 & 1.23 & -.340 & .150 & 0 & 0 \\ -.45 & -.6 & -.45 & .6 & -.340 & 2.41 & -.226 & -1 & 0 \\ -.6 & -.8 & .6 & -.8 & .150 & -.226 & 1.7 & 0 & 0 \\ -.509 & -.226 & -.45 & -.6 & 0 & -1 & 0 & 2.41 & .226 \\ -.226 & -.101 & -.6 & -.8 & 0 & 0 & 0 & .226 & 1.7 \end{bmatrix} \begin{pmatrix} 0 \\ 0 \\ D_3 \\ D_4 \\ D_5 \\ D_7 \\ D_8 \\ D_9 \\ D_{10} \end{pmatrix} \qquad \text{(Q)}$$

page 263. Thus the imposition of the constraint conditions of Eq. N reduces the original system of 10 degrees of freedom to a system of 9 degrees of freedom. Once again, Eq. Q can be solved by any one of the three standard methods discussed previously.

It is perhaps obvious that in applying either of the transformations described, algorithms could be developed that would make it unnecessary to carry out the matrix products indicated. For example, the last transformation involves replacement of the fifth row (and column) of the stiffness matrix with the sum of the fifth row (column) and $\frac{3}{4}$ of the sixth row (column). This replacement could obviously be accomplished without the need to build, store, or manipulate the large transformation relationships of Eqs. O and P.

It is also possible to build the mixed coordinate stiffness relationships of Eq. M directly, making the transformations at the structure level unnecessary. The global stiffness relationships for the plane truss member of Fig. 5.11a have already been shown to be

$$\begin{pmatrix} r_1 \\ r_2 \\ r_3 \\ r_4 \end{pmatrix}^{(n)} = \frac{EA}{L} \begin{bmatrix} c^2 & cs & -c^2 & -cs \\ cs & s^2 & -cs & -s^2 \\ -c^2 & -cs & c^2 & cs \\ -cs & s^2 & cs & s^2 \end{bmatrix}^{(n)} \begin{pmatrix} d_1 \\ d_2 \\ d_3 \\ d_4 \end{pmatrix}^{(n)} \tag{4.1a}$$

where $c = \cos \alpha_n$ and $s = \sin \alpha_n$. Generalizing the approach of page 90 will yield the stiffness relationships for the truss member in the skewed coordinates of Fig. 5.11b as

$$\begin{pmatrix} \hat{r}_1 \\ \hat{r}_2 \\ \hat{r}_3 \\ \hat{r}_4 \end{pmatrix}^{(n)} = \frac{EA}{L} \begin{bmatrix} c_i^2 & c_i s_i & -c_i c_j & -c_i s_j \\ c_i s_i & s_i^2 & -s_i c_j & -s_i s_j \\ -c_i c_j & -s_i c_j & c_j^2 & c_j s_j \\ -c_i s_j & -s_i s_j & c_j s_j & s_j^2 \end{bmatrix}^{(n)} \begin{pmatrix} \hat{d}_1 \\ \hat{d}_2 \\ \hat{d}_3 \\ \hat{d}_4 \end{pmatrix}^{(n)} \tag{5.50}$$

where

$$c_i = \cos \hat{\alpha}_{in}$$

$$s_i = \sin \hat{\alpha}_{in}$$

$$c_j = \cos \hat{\alpha}_{jn}$$

$$s_j = \sin \hat{\alpha}_{jn}$$

or

$$\hat{\mathbf{r}}^{(n)} = \hat{\mathbf{k}}^{(n)} \hat{\mathbf{d}}^{(n)} \tag{5.51a}$$

The stiffness relationships for the truss of Fig. 5.10 may now be built directly from the individual member stiffness relationships in skewed coordinates. As shown previously on page 100,

$$\mathbf{k}^{(1)} = \mathbf{k}^{(7)} = \frac{EA_0}{3}\begin{bmatrix} 1 & 0 & -1 & 0 \\ 0 & 0 & 0 & 0 \\ -1 & 0 & 1 & 0 \\ 0 & 0 & 0 & 0 \end{bmatrix}$$

$$\mathbf{k}^{(3)} = \mathbf{k}^{(5)} = \frac{EA_0}{3}\begin{bmatrix} .45 & .6 & -.45 & -.6 \\ .6 & .8 & -.6 & -.8 \\ -.45 & -.6 & .45 & .6 \\ -.6 & -.8 & .6 & .8 \end{bmatrix}$$

$$\mathbf{k}^{(4)} = \frac{EA_0}{3}\begin{bmatrix} .45 & -.6 & -.45 & .6 \\ -.6 & .8 & .6 & -.8 \\ -.45 & .6 & .45 & -.6 \\ .6 & -.8 & -.6 & .8 \end{bmatrix}$$

$$\mathbf{k}^{(8)} = \frac{EA_0}{3}\begin{bmatrix} .509 & .226 & -.509 & -.226 \\ .226 & .101 & -.226 & -.101 \\ -.509 & -.226 & .509 & .226 \\ -.226 & -.101 & .226 & .101 \end{bmatrix}$$

Using Eq. 5.50, we may now write

$$\hat{\mathbf{k}}^{(2)} = \frac{EA_0}{3}\begin{bmatrix} 1 & 0 & -.8 & .6 \\ 0 & 0 & 0 & 0 \\ -.8 & 0 & .64 & -.48 \\ .6 & 0 & -.48 & .36 \end{bmatrix}$$

$$\hat{\mathbf{k}}^{(6)} = \frac{EA_0}{3}\begin{bmatrix} .45 & -.6 & 0 & .75 \\ -.6 & .8 & 0 & -1 \\ 0 & .0 & 0 & 0 \\ .75 & -1 & 0 & 1.25 \end{bmatrix}$$

$$\hat{\mathbf{k}}^{(9)} = \frac{EA_0}{3}\begin{bmatrix} .509 & -.226 & -.272 & .486 \\ -.226 & .101 & .120 & -.216 \\ -.272 & .120 & .146 & -.259 \\ .486 & -.216 & -.259 & .464 \end{bmatrix}$$

A few random coefficients of the stiffness matrix of Eq. M may now be checked

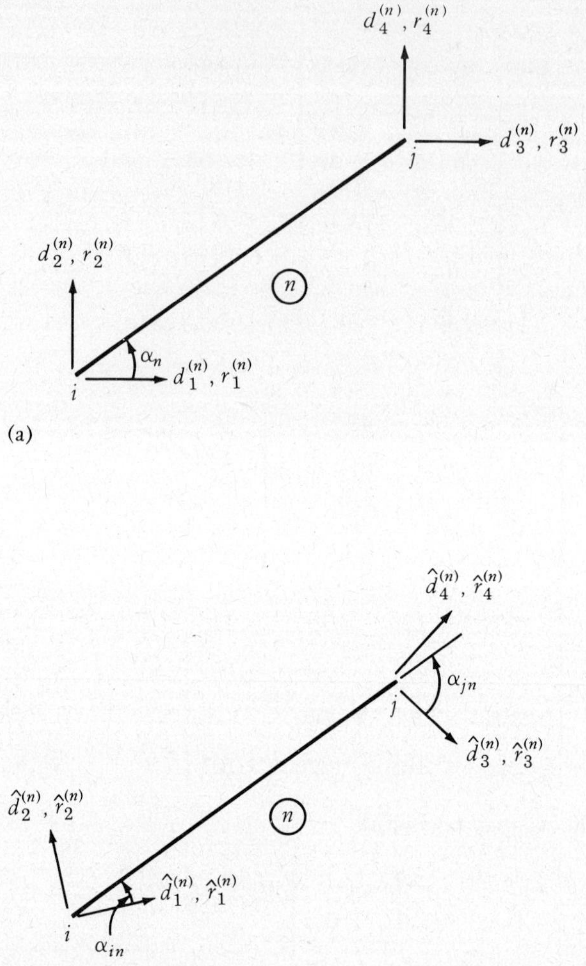

Fig. 5.11 Plane truss member-end forces and displacements: (a) global coordinates; (b) skewed coordinates.

by coordinate coding:

$$\hat{K}_{55} = \hat{k}_{33}^{(2)} + \hat{k}_{33}^{(6)} + \hat{k}_{33}^{(9)} = \frac{EA_0}{3}(.64 + 0 + .146) = \frac{EA_0}{3}(.786)$$

$$\hat{K}_{56} = \hat{k}_{34}^{(2)} + \hat{k}_{34}^{(6)} + \hat{k}_{34}^{(9)} = \frac{EA_0}{3}(-.48 + 0 - .259) = \frac{EA_0}{3}(-.739)$$

$$\hat{K}_{69} = \hat{k}_{41}^{(6)} = \frac{EA_0}{3}(.75)$$

$$\hat{K}_{94} = \hat{k}_{32}^{(5)} = k_{32}^{(5)} = \frac{EA_0}{3}(-.6)$$

These results are seen to check.

5.5 SUMMARY

In this chapter we have considered the so-called special topics of initially strained members, members of variable cross section, members with shear deformation, members with internal hinges, inextensional members, and skewed supports. Although this set is not exhaustive, it does include many subjects encountered in practice. We have studied the effect of each of these considerations on the standard stiffness method presented in Chap. 4. Perhaps the most significant single point to be gleaned from this study is the versatility of the direct stiffness method. Although the development and programming of a stiffness method general enough to take into account all these special considerations is formidable, more important perhaps is the fact that the application of the method remains little changed.

PROBLEMS

5.1 The truss of Fig. P5.1 is identical with the truss of Fig. P4.4; thus $EA = EA_0 =$ constant. The loading is the same as that considered in Prob. 2.14: member 6 has been fabricated .01 m too long, and a support settlement of .02 m downward has occurred in direction 7. Taking $EA_0/3$ as a factor, solve for the nodal displacements and member-end forces by each of the following direct stiffness approaches. Use the appropriate SMIS3 structural analysis commands.

(a) Consider the fabrication error to be an initial strain and the settlement to be an initial displacement. Check the fixed-end force vector for member 6 by hand.

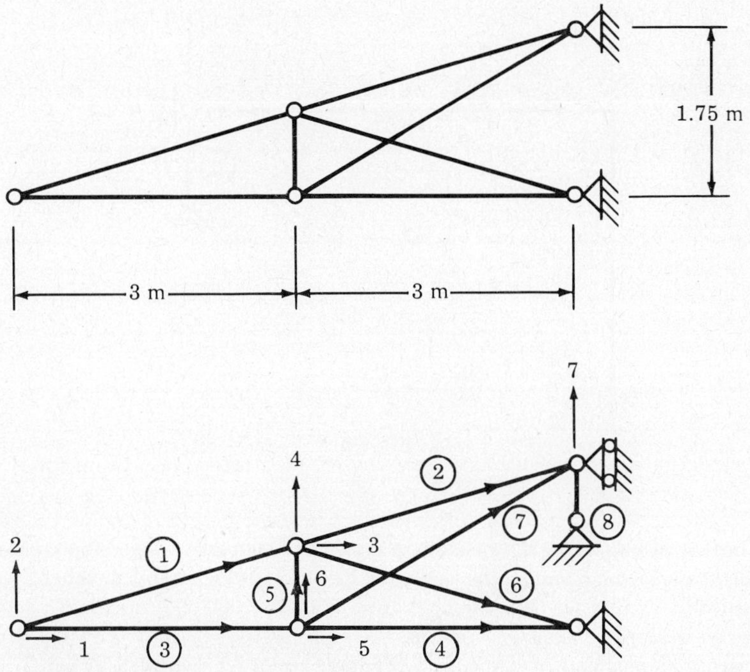

Fig. P5.1

(b) Consider both the fabrication error and the settlement to be initial strains. Take EA/L for the support to be $10^5 \times EA_0/3$.

(c) Consider the fabrication error to be an initial strain, and consider the settlement by preloading the support. Use the same support stiffness as for part b.

5.2 The frame of Fig. P5.2 is the same as the frame of Fig. P4.5; therefore again take $EI_0/2$ as a factor, and let $I_0/A_0 = .005\ \text{m}^2$. The loading in this example consists of a fabrication error that causes member 2 to be .015 m too long and also causes a settlement of .01 m downward at the right-hand support. Obtain the solution to this example by the direct stiffness approaches outlined as parts a, b, and c in Prob. 5.1. Check the fixed-end force vector for member 2 by hand.

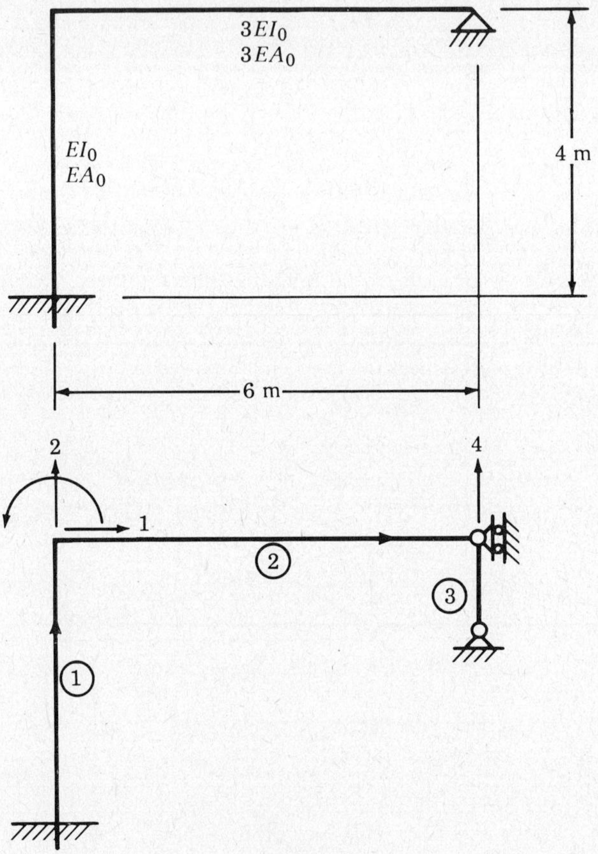

Fig. P5.2

5.3 The truss of Fig. P4.3 is shown again in Fig. P5.3. Again take $EA = EA_0 =$ constant and $EA_0/4$ as a factor. Consider the loading of Prob. 2.13, which is that member 7 has been heated to 60°C above design temperature. With $\beta = .0000117$ (m/m)/°C, determine the nodal displacements and internal forces, using the appropriate SMIS3 structural analysis commands. Check the fixed-end force vector for member 7 by hand.

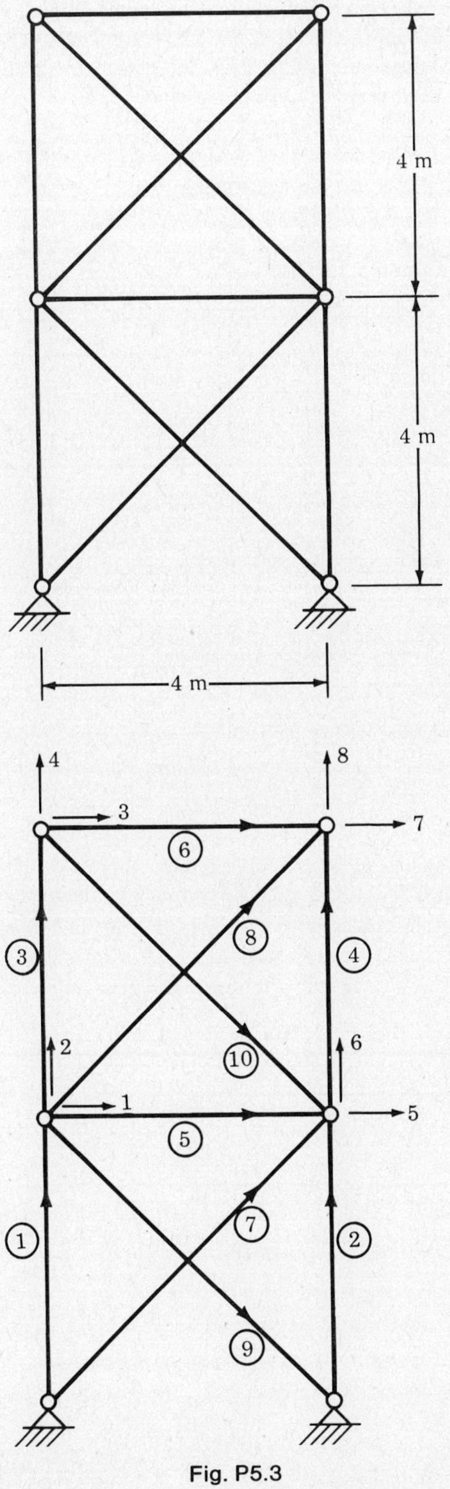

Fig. P5.3

5.4 Using symmetry, solve for the member-end forces due to a thermal load of +40°C on the roof members of the frame of Fig. P5.4. Use SMIS3 structural analysis commands. Check the thermal fixed-end force vector by hand.

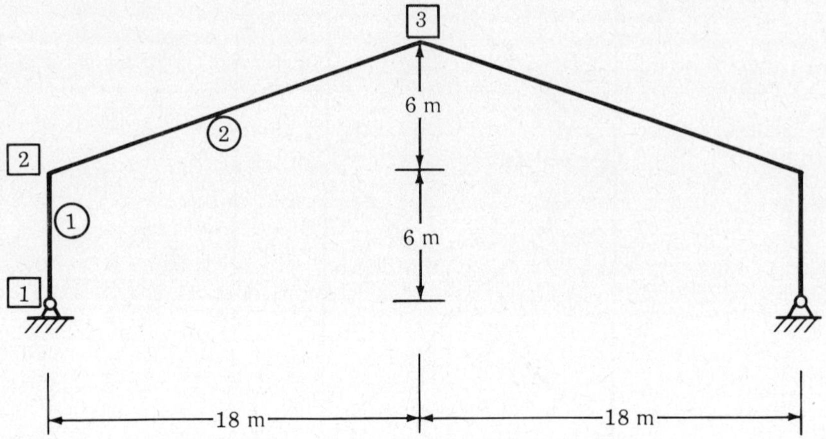

$E = 2 \times 10^8$ kN/m^2

$I = .0024$ m^4

$A = .025$ m^2

$\beta = .0000117$ (m/m)/°C

Fig. P5.4

5.5 The box girder bridge of Fig. P5.5 is subjected to a thermal loading as indicated. Using symmetry and SMIS3 structural analysis commands, determine the nodal displacements and member-end forces. Check the thermal fixed-end force vector by hand.

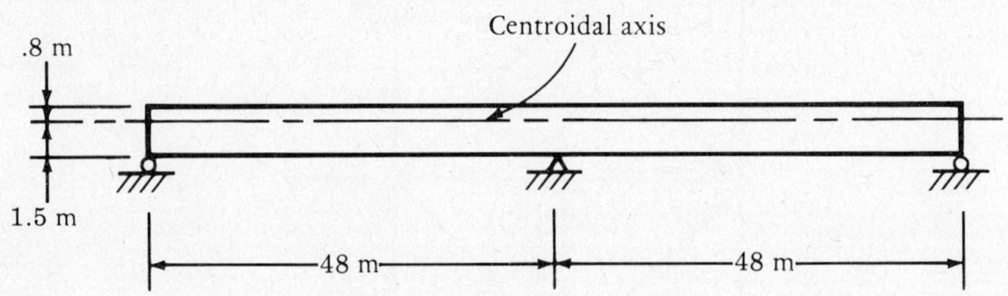

$E = 286 \times 10^6$ kN/m^2

$I_0 = 4$ m^4

$A_0 = 4$ m^2

$\beta = .00099$ (m/m)/°C

$t_t = 45.5$°C

$t_b = 36.6$°C

$t_{\text{zero load}} = 32.2$°C

Fig. P5.5

5.6 The beam of Fig. P5.6 has a rectangular cross section of variable height. Obtain the basic flexibility array $\mathscr{f}$ for the member by using Castigliano's second theorem. Check $\mathscr{f}$ for the special case of c = 0. Evaluate $\mathscr{f}$ for the case of c = h. Obtain the global stiffness array **k** by properly transforming $\mathscr{f}$. Approximate **k** for the case where c = h by using 2, 4, and 8 element meshes of finite elements (uniform beam members).

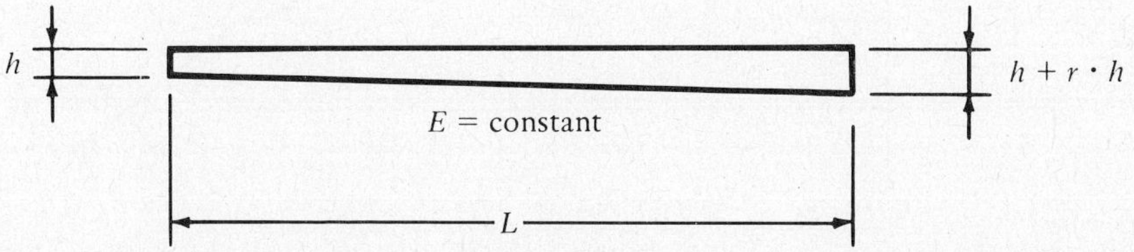

Fig. P5.6

5.7 Determine the fixed-end force vector $\mathbf{r}_F$ for the beam of Fig. P5.6 subjected to a uniformly distributed load of intensity q.

5.8 Rework the beam example of Fig. 4.8 considering the effect of shear deformations. Assume the beam to be of I section with $A_{0,\text{web}}$ = .005 m^2 and $G = 80 \times 10^6$ kN/m^2. Use SMIS3 structural analysis commands. Do not consider support settlements.

5.9 The frame of Fig. P5.9 was previously shown in Fig. P4.5. Again taking $EI_0/2$ as a factor, determine the nodal displacements and member-end forces. Consider the members to behave inextensionally. Compare results with those obtained using extensional theory. Take inextensional behavior into account, using SMIS3 structural analysis commands and an appropriate unconstrained coordinate code.

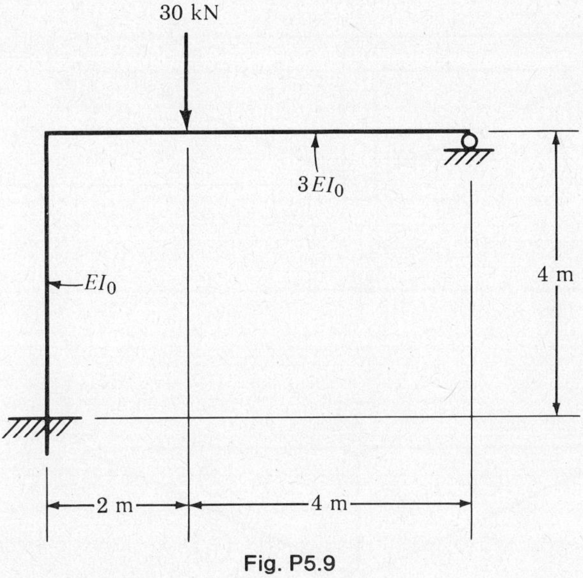

Fig. P5.9

5.10 Follow the instructions of Prob. 5.9 for the frame of Fig. P5.10, which was previously considered in Prob. 4.6.

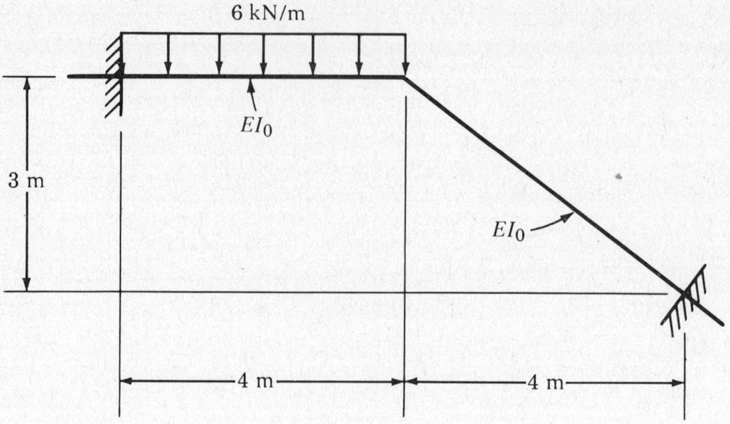

Fig. P5.10

5.11 Rework the example frame of Fig. 5.7 with only member 4 considered extensional. Account for inextensional behavior, using an appropriate unconstrained coordinate code. Use SMIS3 structural analysis commands.

5.12 The queen's post truss of Fig. P5.12 is seen to be a variation of the statically determinate truss of Fig. P2.3. However, the queen's post truss has a continuous upper chord and is therefore statically indeterminate. The queen's post truss also has its load distributed over the upper chord instead of concentrated at the joints. Using for members 1, 2, and 4 $E = 12 \times 10^6\ \text{kN/m}^2$, $I = 2.25 \times 10^{-4}\ \text{m}^4$ and $A = 3 \times 10^{-2}\ \text{m}^2$

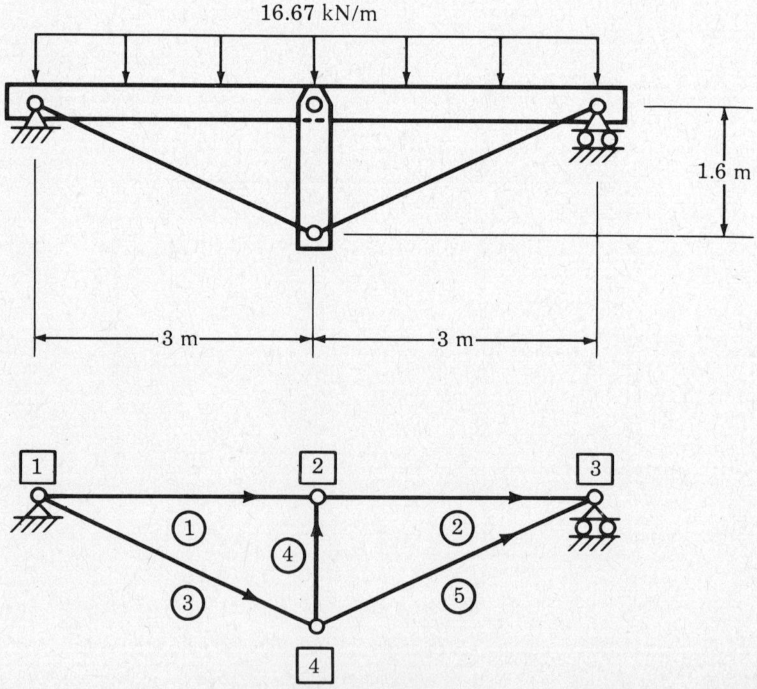

Fig. P5.12

and for members 3 and 5 $E = 200 \times 10^6$ kN/m^2 and $A = 5 \times 10^{-4}$ m^2, analyze the truss using the SMIS3 structural analysis commands. (*Hint:* Members 1 and 2 must be represented by frame members, while members 3, 4 and 5 may be represented by truss members.)

5.13 Solve Prob. 5.12 with a concentrated downward load of 50 kN at node 2 instead of the distributed load. Compare results with those of Prob. 2.3.

5.14 Solve Prob. 5.12 using symmetry conditions and modeling only the right half. (*Hint:* To get correct nodal displacements, a horizontal rigid-body displacement will have to be superposed on these results.)

5.15 The truss of Fig. P5.15 is a variation of the truss of Fig. P2.5, with the current truss having all rigid joints and a uniformly distributed load. Obtain a direct stiffness solution for nodal displacements and member-end forces by using SMIS3 structural analysis commands. Model only the right half of the truss by using symmetry conditions. (See hint for Prob. 5.14.)

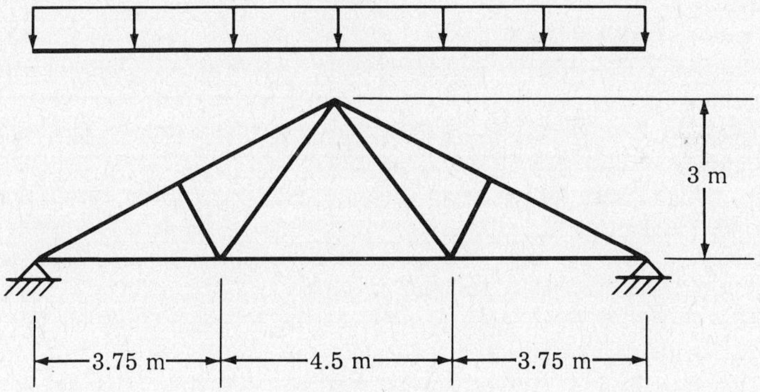

Fig. P5.15

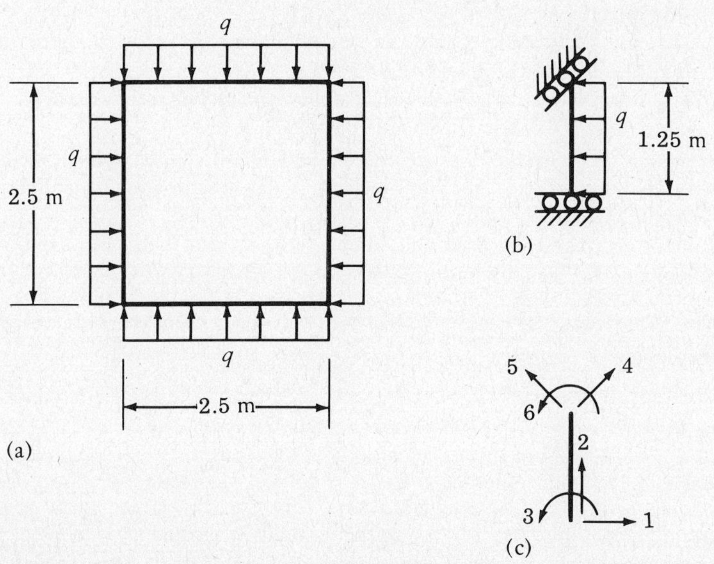

Fig. P5.16

5.16 The frame of Fig. P5.16a may be modeled as shown in Fig. P5.16b if symmetry conditions are imposed. Analyze the single-member structure of Fig. P5.16b by transforming the local frame-member stiffness relationships of Eq. 4.6 to the skewed coordinate system of Fig. P5.16c. Use $I/A = .01\ m^2$ and carry out the analysis by hand calculations.

5.17 Work Prob. 5.15 assuming continuous upper and lower chords except that the leftmost, rightmost, and uppermost joints are pinned. Web members are pinned to chord members. (*Hint:* Use of a hinged frame member will be required.) Use SMIS3 structural analysis commands and consider symmetry. (See hint for Prob. 5.14.)

The SMIS3 program has been written to allow certain subroutines to be conveniently replaced by user-generated subroutines. Problems 5.18 to 5.21 make use of this feature. To successfully accomplish these replacements, it will be necessary to study the appropriate sections of the documentation presented in Sec. A2.6 of App. 2.

5.18 Write a FORTRAN subroutine to generate the vector $\mathbf{r}'_F$ of local fixed-end forces for a truss member subjected to an initial axial strain. Replace subroutine TRFEFI described in part C8 of Sec. A2.6. Check the functioning of this replacement subroutine by working Prob. 5.1b.

5.19 Write a FORTRAN subroutine to generate the vector $\mathbf{r}'_F$ of local fixed-end forces for a frame member subjected to a thermal loading. Replace subroutine FRFEFT described in Part D8 of Sec. A2.6. Check the functioning of this replacement subroutine by working the example of Fig. 5.4.

5.20 Write a FORTRAN subroutine to generate the basic stiffness array $\mathscr{k}$ for a beam member considering the effect of shear distortions. Replace subroutine BMBKF described in part B2 of Sec. A2.6. Check the functioning of this replacement subroutine by working an appropriate example.

5.21 Write FORTRAN subroutines to perform the following functions for a frame member with one internal hinge:

(a) Generate the basic stiffness array $\mathscr{k}$. Replace subroutine FRBKFH described in part D2a of Sec. A2.6.

(b) Generate the basic-local transformation array $\mathbf{L}$. Replace subroutine FRLH described in part D4a of Sec. A2.6.

(c) Generate the local-global transformation array $\boldsymbol{\alpha}$. Replace subroutine FRA described in part D3 of Sec. A2.6.

(d) Transform the basic stiffness array $\mathscr{k}$ into the global stiffness array $\mathbf{k}$. Replace subroutine FORMKT described in part D5 of Sec. A2.6.

(e) Generate the vector $\mathbf{r}'_F$ of local fixed-end forces due to a concentrated force applied tranversely at the hinge location. Replace subroutine FRFEFH described in part D7a of Sec. A2.6.

(f) Transform the local fixed-end force vector $\mathbf{r}'_F$ into the global fixed-end force vector $\mathbf{r}_F$. Replace subroutine FORMRF described in part D9 of Sec. A2.6.

Check the functioning of these replacement subroutines by working an appropriate example.

REFERENCE

5.1 Curiskis, J. I., and Valliappan, S. "A Solution Algorithm for linear Constraint Equations in Finite Element Analysis." In *UNICIV Report No. R160*. Kensington, Australia: University of New South Wales, September 1976.

Chapter 6
Force Method

6.1 INTRODUCTION

In Chap. 3 we developed two mutually inverse relationships between basic member-end forces and corresponding displacements for structural members. One of these relationships, the stiffness relationship, provided the basis for the development of the displacement method of structural analysis presented and discussed at great length in Chaps. 4 and 5. The other relationship, the flexibility relationship, will now be used to develop a second method of structural analysis called the force method. Whereas nodal displacements were the primary unknowns in the displacement method, with internal forces being determined by backsubstitution, internal basic structure forces will be found to be the primary unknowns in the force method, with nodal displacements being determined by backsubstitution.

6.2 STATICALLY DETERMINATE STRUCTURES

As in the case of the displacement method, it is convenient to consider statically determinate structures first in the development of the force method of analysis. In fact, you will see from the following discussion that the solution for internal forces in statically determinate structures as carried out in Chap. 2 was a somewhat incomplete special case application of the force method.

6.2.1 Statically Determinate Truss

The application of the flexibility relationships of Chap. 3 to the solution of statically determinate structures will first be demonstrated for the truss of Fig. 2.3, which is displayed again in Fig. 6.1 for convenience. Since the basic forces in the structure are the primary unknowns in the force method, it is not surprising

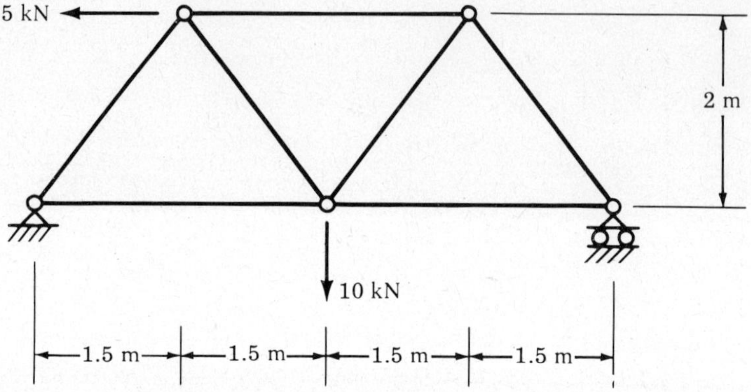

$$A^{(1)} = A^{(2)} = A^{(7)} = A_0$$

$$A^{(3)} = A^{(4)} = A^{(5)} = A^{(6)} = \frac{2.5}{2.4} A_0$$

$$E = \text{constant}$$

(a)

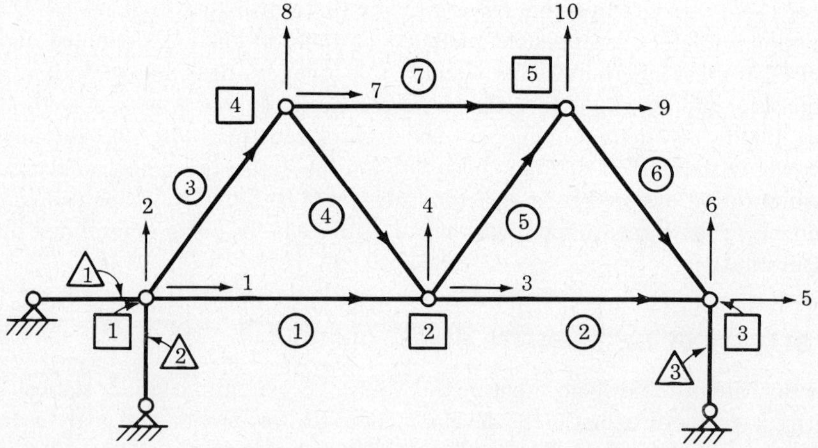

(b)

Fig. 6.1 Truss example (previously shown in Figs. 2.3 and 4.3): (a) dimensions and loading; (b) structural model.

that the first step in this method is the writing of the nodal equilibrium equations. In Chap. 2 we built the structure nodal equilibrium relationships for the example truss from the individual member equilibrium relationships. These relationships were displayed in Eq. K in Sec. 2.5.2 and are repeated here as Eq. A below for completeness.

$$
\begin{array}{c}
\\ \\ \\ \\ \\ \\ \\ \\ \\ \\ \\
\end{array}
\begin{pmatrix} R_1 \\ R_2 \\ R_3 \\ R_4 \\ R_5 \\ R_6 \\ R_7 \\ R_8 \\ R_9 \\ R_{10} \end{pmatrix}
\begin{array}{c} \boxed{1} \\ \\ \boxed{2} \\ \\ \boxed{3} \\ \\ \boxed{4} \\ \\ \boxed{5} \\ \\ \end{array}
=
\begin{array}{cccccccccc}
\textcircled{1} & \textcircled{2} & \textcircled{3} & \textcircled{4} & \textcircled{5} & \textcircled{6} & \textcircled{7} & \triangle_1 & \triangle_2 & \triangle_3 \\
\end{array}
\begin{bmatrix}
-1 & \cdot & -.6 & \cdot & \cdot & \cdot & \cdot & 1 & \cdot & \cdot \\
0 & \cdot & -.8 & \cdot & \cdot & \cdot & \cdot & \cdot & 1 & \cdot \\
1 & -1 & \cdot & .6 & -.6 & \cdot & \cdot & \cdot & \cdot & \cdot \\
0 & 0 & \cdot & -.8 & -.8 & \cdot & \cdot & \cdot & \cdot & \cdot \\
\cdot & 1 & \cdot & \cdot & \cdot & .6 & \cdot & \cdot & \cdot & \cdot \\
\cdot & 0 & \cdot & \cdot & \cdot & -.8 & \cdot & \cdot & \cdot & 1 \\
\cdot & \cdot & .6 & -.6 & \cdot & \cdot & -1 & \cdot & \cdot & \cdot \\
\cdot & \cdot & .8 & .8 & \cdot & \cdot & 0 & \cdot & \cdot & \cdot \\
\cdot & \cdot & \cdot & \cdot & .6 & -.6 & 1 & \cdot & \cdot & \cdot \\
\cdot & \cdot & \cdot & \cdot & .8 & .8 & 0 & \cdot & \cdot & \cdot
\end{bmatrix}
\begin{pmatrix} P^{(1)} \\ P^{(2)} \\ P^{(3)} \\ P^{(4)} \\ P^{(5)} \\ P^{(6)} \\ P^{(7)} \\ P_c^{(1)} \\ P_c^{(2)} \\ P_c^{(3)} \end{pmatrix}
\qquad \text{(A)}
$$

Symbolically, this equation may be represented in general as

$$\mathbf{R} = \mathbf{a}\overline{\mathbf{P}} \qquad (2.18)$$

Since this truss is statically determinate, the basic forces are independent of member flexibilities. Therefore the equilibrium relationships of Eq. A may be solved for the basic forces directly, as was done in Chap. 2, the result being Eq. L in Sec. 2.5.2. This equation is repeated here as Eq. B.

$$
\begin{pmatrix} P^{(1)} \\ P^{(2)} \\ P^{(3)} \\ P^{(4)} \\ P^{(5)} \\ P^{(6)} \\ P^{(7)} \\ P_c^{(1)} \\ P_c^{(2)} \\ P_c^{(3)} \end{pmatrix}
=
\begin{bmatrix}
0 & 0 & 1 & -.375 & 1 & 0 & .75 & -.5625 & .75 & -.1875 \\
0 & 0 & 0 & -.375 & 1 & 0 & .25 & -.1875 & .25 & -.5625 \\
0 & 0 & 0 & .625 & 0 & 0 & .41667 & .9375 & .41667 & .3125 \\
0 & 0 & 0 & -.625 & 0 & 0 & -.41667 & .3125 & -.41667 & -.3125 \\
0 & 0 & 0 & -.625 & 0 & 0 & .41667 & -.3125 & .41667 & .3125 \\
0 & 0 & 0 & .625 & 0 & 0 & -.41667 & .3125 & -.41667 & .9375 \\
0 & 0 & 0 & .75 & 0 & 0 & -.5 & .375 & .5 & .375 \\
1 & 0 & 1 & 0 & 1 & 0 & 1 & 0 & 1 & 0 \\
0 & 1 & 0 & .5 & 0 & 0 & .33333 & .75 & .33333 & .25 \\
0 & 0 & 0 & .5 & 0 & 1 & -.33333 & .25 & -.33333 & .75
\end{bmatrix}
\begin{pmatrix} R_1 \\ R_2 \\ R_3 \\ R_4 \\ R_5 \\ R_6 \\ R_7 \\ R_8 \\ R_9 \\ R_{10} \end{pmatrix}
\qquad \text{(B)}
$$

This result may be represented symbolically as

$$\overline{\mathbf{P}} = \mathbf{a}^{-1}\mathbf{R} \qquad (2.19)$$

To simplify future matrix manipulations, it is now convenient to let $\mathbf{b} = \mathbf{a}^{-1}$, giving

$$\mathbf{P} = \mathbf{bR} \tag{6.1}$$

Substituting the vector of nodal loads,

$$\mathbf{R} = \langle 0 \quad 0 \quad 0 \quad -10 \quad 0 \quad 0 \quad -5 \quad 0 \quad 0 \quad 0 \rangle \tag{C}$$

into Eq. B gives the vector of primary unknown basic forces:

$$\overline{\mathbf{P}} = \langle 0 \quad 2.5 \quad -9.33 \quad 8.33 \quad 4.17 \quad -4.17 \quad -5 \quad -5 \quad -6.67 \quad -3.33 \rangle \tag{D}$$

This is as far as we were able to carry the solution to this truss example in Chap. 2, because we had not yet developed member force-displacement relationships at that point. The solution may now be completed by making use of the flexibility relationships of Chap. 3 to obtain the structure nodal displacements.

In Chap. 3 we expressed the truss member and support flexibility relationships, respectively, as

$$\boldsymbol{\Delta} = \boldsymbol{f}\mathbf{P} \tag{3.5a}$$

and

$$\Delta_c = f_c P_c \tag{3.20}$$

A superscript may now be added to each of these equations to emphasize that they apply to an individual member or constraint of the structure, giving, respectively,

$$\boldsymbol{\Delta}^{(n)} = \boldsymbol{f}^{(n)}\mathbf{P}^{(n)} \quad \text{and} \quad \Delta_c^{(n)} = f_c^{(n)} P_c^{(n)} \tag{6.2}$$

Writing Eq. 6.2 for the collection of all members and constraints of the truss structure of Fig. 6.1 gives

$$\begin{pmatrix} \boldsymbol{\Delta}^{(1)} \\ \boldsymbol{\Delta}^{(2)} \\ \boldsymbol{\Delta}^{(3)} \\ \boldsymbol{\Delta}^{(4)} \\ \boldsymbol{\Delta}^{(5)} \\ \boldsymbol{\Delta}^{(6)} \\ \boldsymbol{\Delta}^{(7)} \\ \Delta_c^{(1)} \\ \Delta_c^{(2)} \\ \Delta_c^{(3)} \end{pmatrix} = \begin{bmatrix} \boldsymbol{f}^{(1)} & & & & & & & & & \\ & \boldsymbol{f}^{(2)} & & & & & & & & \\ & & \boldsymbol{f}^{(3)} & & & & & \boldsymbol{\Theta} & & \\ & & & \boldsymbol{f}^{(4)} & & & & & & \\ & & & & \boldsymbol{f}^{(5)} & & & & & \\ & & & & & \boldsymbol{f}^{(6)} & & & & \\ & & & & & & \boldsymbol{f}^{(7)} & & & \\ & \boldsymbol{\Theta} & & & & & & f_c^{(1)} & & \\ & & & & & & & & f_c^{(2)} & \\ & & & & & & & & & f_c^{(3)} \end{bmatrix} \begin{pmatrix} \mathbf{P}^{(1)} \\ \mathbf{P}^{(2)} \\ \mathbf{P}^{(3)} \\ \mathbf{P}^{(4)} \\ \mathbf{P}^{(5)} \\ \mathbf{P}^{(6)} \\ \mathbf{P}^{(7)} \\ \mathbf{P}_c^{(1)} \\ \mathbf{P}_c^{(2)} \\ \mathbf{P}_c^{(3)} \end{pmatrix} \tag{E}$$

or, more briefly,

$$\overline{\boldsymbol{\Delta}} = \overline{\boldsymbol{f}}\,\overline{\mathbf{P}} \tag{6.3}$$

The flexibility relationship for an individual truss member was given in detail in Eq. 3.5 as

$$\Delta = \frac{L}{EA} P \tag{3.5}$$

The physical properties shown in Fig. 6.1 for the example truss are the same as those used before. Using these properties we find the individual truss member flexibilities (on the basis of Eq. 3.5) to be

$$f^{(1)} = f^{(2)} = f^{(7)} = \frac{3}{EA_0}$$

$$f^{(3)} = f^{(4)} = f^{(5)} = f^{(6)} = \frac{2.5}{(2.5/2.4)EA_0} = \frac{2.4}{EA_0}$$

$$f_c^{(1)} = f_c^{(2)} = f_c^{(3)} = 0 \qquad \text{(Supports are assumed to be rigid.)}$$

Substituting this information into Eq. E gives

$$\begin{pmatrix} \Delta^{(1)} \\ \Delta^{(2)} \\ \Delta^{(3)} \\ \Delta^{(4)} \\ \Delta^{(5)} \\ \Delta^{(6)} \\ \Delta^{(7)} \\ \Delta_c^{(1)} \\ \Delta_c^{(2)} \\ \Delta_c^{(3)} \end{pmatrix} = \frac{3}{EA_0} \begin{bmatrix} 1 & & & & & & & & & \\ & 1 & & & & & & & & \\ & & .8 & & & & \Theta & & & \\ & & & .8 & & & & & & \\ & & & & .8 & & & & & \\ & & & & & .8 & & & & \\ & & & & & & 1 & & & \\ & & \Theta & & & & & 0 & & \\ & & & & & & & & 0 & \\ & & & & & & & & & 0 \end{bmatrix} \begin{pmatrix} P^{(1)} \\ P^{(2)} \\ P^{(3)} \\ P^{(4)} \\ P^{(5)} \\ P^{(6)} \\ P^{(7)} \\ P_c^{(1)} \\ P_c^{(2)} \\ P_c^{(3)} \end{pmatrix} \tag{F}$$

From joint compatibility considerations, in Chap. 2 we developed Eq. 2.24 as

$$\overline{\mathbf{\Delta}} = \mathbf{a}^T \mathbf{D} \tag{2.24}$$

Solving for nodal displacements **D** gives

$$\mathbf{D} = [\mathbf{a}^T]^{-1} \overline{\mathbf{\Delta}} \tag{6.4}$$

or

$$\mathbf{D} = [\mathbf{a}^{-1}]^T \overline{\mathbf{\Delta}} \tag{6.4a}$$

or, since $\mathbf{b} = \mathbf{a}^{-1}$,

$$\mathbf{D} = \mathbf{b}^T \overline{\mathbf{\Delta}} \tag{6.4b}$$

Substituting from Eq. 6.1 into 6.3 and then substituting that result into Eq. 6.4b gives

$$\mathbf{D} = \mathbf{b}^T \bar{\mathbf{f}} \mathbf{b} \mathbf{R} \tag{6.5}$$

or

$$\mathbf{D} = \mathbf{F} \mathbf{R} \tag{6.5a}$$

where $\mathbf{F} = \mathbf{b}^T \bar{\mathbf{f}} \mathbf{b}$ is the structure flexibility matrix and Eq. 6.5a represents the structure force-displacement relationships in terms of flexibility coefficients. The matrix product indicated on the right-hand side of Eq. 6.5 has been carried out by using SMIS3, as shown in the next printout. The solution is

$$\mathbf{D} = \frac{3}{EA_0} \langle 0 \quad 0 \quad 0 \quad -17.19 \quad 2.5 \quad 0 \quad .3472 \quad -8.594 \quad -4.653 \quad -9.531 \rangle \tag{G}$$

which is seen to agree with the stiffness solution shown on page 105.

****** *Computer Input:*

```
START
REMARK 4
SOLUTION FOR INTERNAL FORCES AND NODAL DISPLACEMENTS
OF STATICALLY DETERMINATE TRUSS BY FORCE METHOD

**** UNITS ARE KILONEWTONS AND METERS ****
LOAD A 10 10
-1 0 -.6 4*0 1 4*0 -.8 5*0 1 0 1 -1 0 .6 -.6
8*0 -.8 -.8 6*0 1 3*0 .6 9*0 -.8 3*0 1 0 0 .6 -.6
0 0 -1 5*0 .8 .8 10*0 .6 -.6 1 7*0 .8 .8 4*0
PRINT A 1
THIS IS THE MATRIX OF FORCE INFLUENCE COEFFECIENTS, A
INVERT A
DUPL A B
DELETE A
TRANS B BT
FORM FBAR 10 10
LOAD FDG 1 10
1 1 .8 .8 .8 .8 1 0 0 0
STODG FBAR FDG
PRINT FBAR 2
THIS IS THE FLEXIBILITY SUPERMATRIX, FBAR, FOR THE COLLECTION
OF ALL MEMBERS AND CONSTRAINTS WITH 3/EA(0) FACTORED OUT.
LOAD R 10 1
0 0 0 -10 0 0 -5 0 0 0
PRINT R 1
THIS IS THE LOAD VECTOR, R.
MULT B R PBAR
PRINT PBAR 2
THIS THE VECTOR  OF BASIC FORCES, PBAR, FOR THE
COLLECTION OF ALL MEMBERS AND CONSTRAINTS.
MULT FBAR PBAR FBARBR
MULT BT FBARBR D
PRINT D 2
THIS IS THE VECTOR OF STRUCTURE NODAL DISPLACEMENTS,
D, WITH 3/EA(0) FACTORED OUT.
STOP
```

****** *Computer Output:*

START

REMARK 4

SOLUTION FOR INTERNAL FORCES AND NODAL DISPLACEMENTS
OF STATICALLY DETERMINATE TRUSS BY FORCE METHOD

**** UNITS ARE KILONEWTONS AND METERS ****

LOAD A 10 10
10 ROWS, 10 COLUMNS

PRINT A 1

THIS IS THE MATRIX OF FORCE INFLUENCE COEFFECIENTS, A

	1	2	3	4	5	6
1	-1.0000E+00	0	-6.0000E-01	0	0	0
2	0	0	-8.0000E-01	0	0	0
3	1.0000E+00	-1.0000E+00	0	6.0000E-01	-6.0000E-01	0
4	0	0	0	-8.0000E-01	-8.0000E-01	0
5	0	1.0000E+00	0	0	0	6.0000E-01
6	0	0	0	0	0	-8.0000E-01
7	0	0	6.0000E-01	-6.0000E-01	0	0
8	0	0	8.0000E-01	8.0000E-01	0	0
9	0	0	0	0	6.0000E-01	-6.0000E-01
10	0	0	0	0	8.0000E-01	8.0000E-01

	7	8	9	10
1	0	1.0000E+00	0	0
2	0	0	1.0000E+00	0
3	0	0	0	0
4	0	0	0	0
5	0	0	0	0
6	0	0	0	1.0000E+00
7	-1.0000E+00	0	0	0
8	0	0	0	0
9	1.0000E+00	0	0	0
10	0	0	0	0

INVERT A

DUPL A B

DELETE A

TRANS B BT

FORM FBAR 10 10
10 ROWS, 10 COLUMNS

LOAD FDG 1 10
1 ROWS, 10 COLUMNS

STODG FBAR FDG

PRINT FBAR 2

****** *Computer Output (continued):*

THIS IS THE FLEXIBILITY SUPERMATRIX, FBAR, FOR THE COLLECTION
OF ALL MEMBERS AND CONSTRAINTS WITH 3/EA(0) FACTORED OUT.

	1	2	3	4	5	6
1	1.0000E+00	0	0	0	0	0
2	0	1.0000E+00	0	0	0	0
3	0	0	8.0000E-01	0	0	0
4	0	0	0	8.0000E-01	0	0
5	0	0	0	0	8.0000E-01	0
6	0	0	0	0	0	8.0000E-01
7	0	0	0	0	0	0
8	0	0	0	0	0	0
9	0	0	0	0	0	0
10	0	0	0	0	0	0

	7	8	9	10
1	0	0	0	0
2	0	0	0	0
3	0	0	0	0
4	0	0	0	0
5	0	0	0	0
6	0	0	0	0
7	1.0000E+00	0	0	0
8	0	0	0	0
9	0	0	0	0
10	0	0	0	0

```
LOAD     R            10     1

   10 ROWS,       1 COLUMNS

PRINT    R             1
```

THIS IS THE LOAD VECTOR, R.

	1
1	0
2	0
3	0
4	-1.0000E+01
5	0
6	0
7	-5.0000E+00
8	0
9	0
10	0

```
MULT     B        R         PBAR

PRINT    PBAR          2
```

THIS THE VECTOR OF BASIC FORCES, PBAR, FOR THE
COLLECTION OF ALL MEMBERS AND CONSTRAINTS.

****** *Computer Output (continued)*:

```
             1
 1 -2.8422E-14
 2  2.5000E+00
 3 -8.3333E+00
 4  8.3333E+00
 5  4.1667E+00
 6 -4.1667E+00
 7 -5.0000E+00
 8 -5.0000E+00
 9 -6.6667E+00
10 -3.3333E+00


MULT     FBAR    PBAR    FBARBR

MULT     BT      FBARBR  D

PRINT    D            2


THIS IS THE VECTOR OF STRUCTURE NODAL DISPLACEMENTS,
D, WITH 3/EA(0) FACTORED OUT.


             1
 1           0
 2           0
 3 -2.8422E-14
 4 -1.7187E+01
 5  2.5000E+00
 6           0
 7  3.4722E-01
 8 -8.5937E+00
 9 -4.6528E+00
10 -9.5312E+00

STOP
```

It is true, of course, that the analysis of statically determinate structures by either the force method or the displacement method is of secondary interest compared to the analysis of statically indeterminate structures by either of these methods. Even so, it is possible to gain important additional insight into these structural analysis techniques by comparing their respective applications to the solution of the example statically determinate truss. To this end you may now wish to briefly review the displacement method solution of the example truss as outlined in Sec. 4.3.1. When we compare the force method solutions and the displacement method solutions, certain similarities and differences become apparent. For example, the nodal compatibility and equilibrium relationships are inverted (solved) before being used in the force method, whereas they are used directly in the displacement method. Moreover, the order of enforcing nodal equilibrium and compatibility conditions in the force method is just the reverse of that used in the displacement method. This reversal is not surprising, since the basic force-displacement relationships, which are at the very core of either method, are in inverted order in one method as opposed to the other. It is also seen that the member flexibility relationships are not (and indeed cannot) be transformed to global coordinates before assembly of the structure flexibility relationships.*

Together these changes make the force method somewhat less convenient to use than the displacement method. You will find that, in the main, the coding methods used to enforce nodal equilibrium and compatibility, and thereby to build the structure stiffness relationships from the member stiffness relationships in the direct stiffness method, can no longer be used in the force (or flexibility) method. That is, there is no such thing as a "direct" flexibility method corresponding to the direct stiffness method. You can deduce this fact in a superficial way by simply observing that in the force method the member flexibility relationships are, of necessity, stated in the basic system, whereas the structure flexibility relationships are, again of necessity, in the global system; thus direct coding is impossible. Of course, the somewhat more awkward code of Chap. 2 can be used to build the structure equilibrium (and compatibility) relationships from the individual member equilibrium (and kinematic) relationships. The structure flexibility relationships must, however, be developed by the more inefficient method involving matrix transformations.

Finally, again note that when we use the force method, we first determine internal forces ($\overline{\mathrm{P}} = \mathbf{bR}$) and then backsubstitute to get nodal displacements, whereas, in using the displacement method, we determine nodal displacements first and then backsubstitute to obtain internal forces. The major consequence of this procedure is that when there are more unknown internal forces than there are equations of nodal equilibrium (statically indeterminate structures) the force method will have to be significantly modified. This modification will be shown subsequently; first, however, it is important to broaden the force method to be able to consider distributed loadings.

6.2.2 Statically Determinate Frame—Distributed Loading

As in the case of the stiffness method, responses to distributed loadings are handled by superposition. However, in the case of the force method, it is better to apply superposition at the structure level rather than at the member level. This approach was shown in Sec. 2.11.2 in connection with the statically determinate frame example of Fig. 2.13, which is shown again in Fig. 6.2. Notice that in Fig. 6.2 the general problem has been broken into two cases: Case I, a fixed-end case subjected to distributed loading; and Case II, the original structure subjected only to nodal loading, which, when superposed on the Case I loading, yields the general

*The reason for this situation is that there exist no inverses of the relations $\mathbf{r} = \mathbf{TP}$ or $\mathbf{\Delta} = \mathbf{T}^T\mathbf{d}$, which would be required to accomplish the transformation in question. It may be seen that, physically, no inverses of these relationships can exist because (1) an arbitrarily prescribed set of global forces $\mathbf{r}$ could not, in general, be consistent with any set of basic forces $\mathbf{P}$, as the arbitrary global set would likely violate member overall equilibrium, and (2) a prescribed set of basic member displacements $\mathbf{\Delta}$ could not guarantee a unique displaced position $\mathbf{d}$ of the member; that is, there would be an infinite number of member positions that would be consistent with a given member distortion.

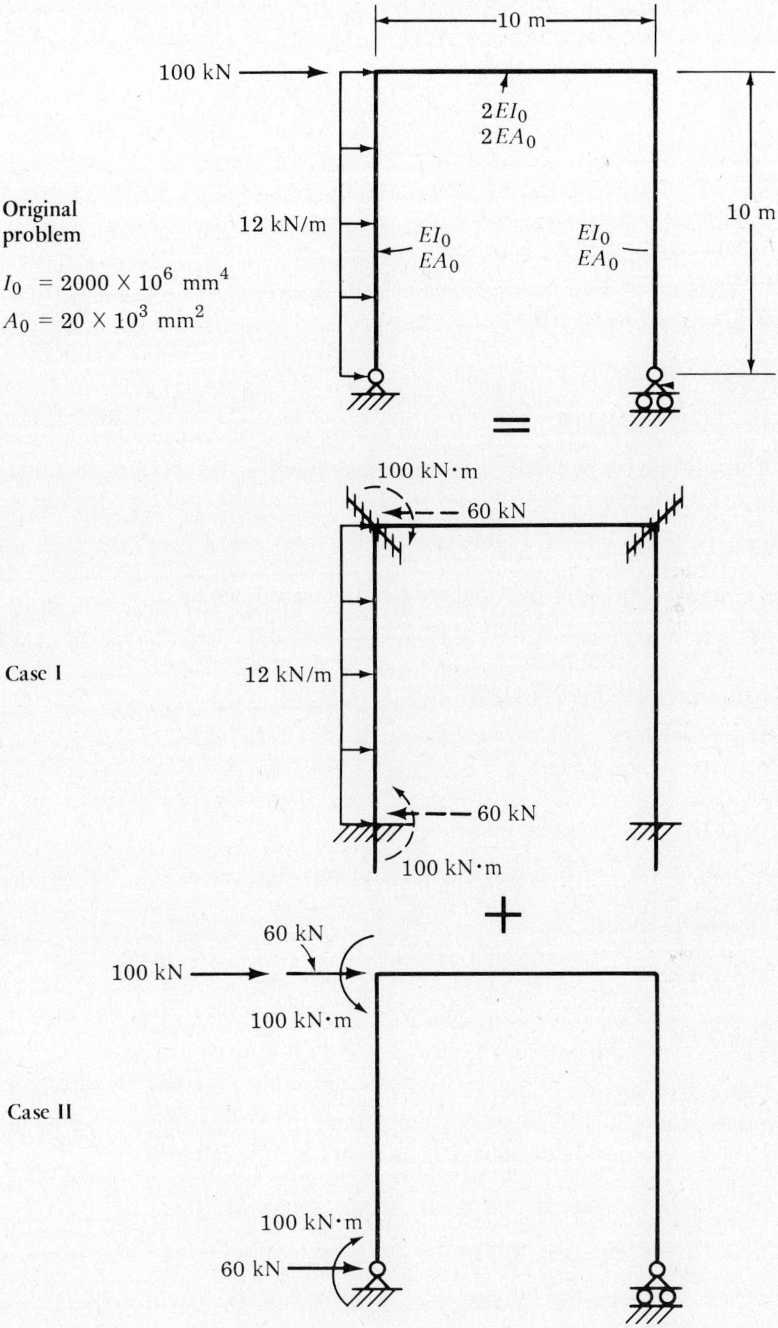

Fig. 6.2 Statically determinate frame with distributed loading (shown previously in Fig. 2.13).

loading. In Sec. 2.11.2 we obtained the following equations:

$$\overline{\mathbf{P}} = \overline{\mathbf{P}}_{\mathrm{I}} + \overline{\mathbf{P}}_{\mathrm{II}} \tag{2.41}$$

where

$$\overline{\mathbf{P}}_{\mathrm{I}} = \overline{\mathbf{P}}_F \tag{2.42}$$

and

$$\overline{\mathbf{P}}_{\mathrm{II}} = \mathbf{a}^{-1}(\mathbf{R} - \mathbf{R}_F) \tag{2.51}$$

or

$$\overline{\mathbf{P}} = \mathbf{a}^{-1}(\mathbf{R} - \mathbf{R}_F) + \overline{\mathbf{P}}_F \tag{2.52}$$

Once again letting $\mathbf{b} = \mathbf{a}^{-1}$, we get

$$\overline{\mathbf{P}} = \mathbf{b}(\mathbf{R} - \mathbf{R}_F) + \overline{\mathbf{P}}_F \tag{6.6}$$

The solution for nodal displacements may now also be obtained by superposition as

$$\mathbf{D} = \mathbf{D}_{\mathrm{I}} + \mathbf{D}_{\mathrm{II}} \tag{6.7}$$

Since all nodal degrees of freedom are fixed in Case I, we have

$$\mathbf{D}_{\mathrm{I}} = \mathbf{\Theta} \tag{6.8}$$

Applying Eq. 6.5 to Case II gives

$$\mathbf{D}_{\mathrm{II}} = \mathbf{b}^T\!\!\not{\mathbf{f}}\,\mathbf{b}\mathbf{R}_{\mathrm{II}} \tag{6.9}$$

where

$$\mathbf{R}_{\mathrm{II}} = \mathbf{R} - \mathbf{R}_F \tag{2.48b}$$

Thus it follows from Eqs. 6.7, 6.8, 6.9, and 2.48b that

$$\mathbf{D} = \mathbf{b}^T\!\!\not{\mathbf{f}}\,\mathbf{b}(\mathbf{R} - \mathbf{R}_F) \tag{6.10}$$

or

$$\mathbf{D} = \mathbf{F}\,(\mathbf{R} - \mathbf{R}_F) \tag{6.11}$$

Using the member and coordinate numbering of Fig. 2.12b and the constraint numbering on page 50, we obtained the vectors $\overline{\mathbf{P}}_F$, $\mathbf{R}$, and $\mathbf{R}_F$ in Sec. 2.11.2. These vectors are displayed again here for completeness.

$$\begin{aligned} \overline{\mathbf{P}}_F &= \langle 100 \quad -100 \quad 0 \quad 0 \quad 0 \quad 0 \quad 0 \quad 0 \quad 0 \quad 0 \quad 0 \quad 0\rangle \\ \mathbf{R} &= \langle 0 \quad 0 \quad 0 \quad 100 \quad 0 \quad 0 \quad 0 \quad 0 \quad 0 \quad 0 \quad 0 \quad 0\rangle \\ \mathbf{R}_F &= \langle -60 \quad 0 \quad 100 \quad -60 \quad 0 \quad -100 \quad 0 \quad 0 \quad 0 \quad 0 \quad 0 \quad 0\rangle \end{aligned} \tag{A}$$

Substituting into Eq. 6.6 gives the solution for $\overline{\mathbf{P}}$ as obtained in the printout on pages 59–62 and previously displayed on page 58 as

$$\overline{\mathbf{P}} = \langle 0 \quad 1600 \quad 160 \quad 0 \quad 0 \quad -160 \quad -1600 \quad 0 \quad 0 \quad 220 \quad 160 \quad -160\rangle \tag{B}$$

The frame-member flexibility relationship was given in Eq. 3.18 as

$$\begin{pmatrix}\phi_i\\ \phi_j\\ \Delta\end{pmatrix} = \frac{L}{6EI}\begin{bmatrix}2 & -1 & 0\\ -1 & 2 & 0\\ 0 & 0 & 6I/A\end{bmatrix}\begin{pmatrix}M_i\\ M_j\\ P\end{pmatrix} \tag{3.18}$$

Using the member physical properties shown in Fig. 6.2, we may write the individual member basic flexibility matrices (on the basis of Eq. 3.18) as

$$\begin{aligned} f^{(1)} = f^{(2)} &= \frac{L}{6EI_0}\begin{bmatrix}2 & -1 & 0\\ -1 & 2 & 0\\ 0 & 0 & .6\end{bmatrix}\\ f^{(3)} &= \frac{L}{12EI_0}\begin{bmatrix}2 & -1 & 0\\ -1 & 2 & 0\\ 0 & 0 & .6\end{bmatrix} = \frac{L}{6EI_0}\begin{bmatrix}1 & -.5 & 0\\ -.5 & 1 & 0\\ 0 & 0 & .3\end{bmatrix}\end{aligned} \tag{C}$$

Once again all support flexibilities are taken as zero. It follows that

$$\bar{f} = \frac{L}{6EI_0}\begin{bmatrix}
2 & -1 & 0 & \cdot & \cdot & \cdot & \cdot & \cdot & \cdot & \cdot & \cdot & \cdot\\
-1 & 2 & 0 & \cdot & \cdot & \cdot & \cdot & \cdot & \cdot & \cdot & \cdot & \cdot\\
0 & 0 & .6 & \cdot & \cdot & \cdot & \cdot & \cdot & \cdot & \cdot & \cdot & \cdot\\
\cdot & \cdot & \cdot & 2 & -1 & 0 & \cdot & \cdot & \cdot & \cdot & \cdot & \cdot\\
\cdot & \cdot & \cdot & -1 & 2 & 0 & \cdot & \cdot & \cdot & \cdot & \cdot & \cdot\\
\cdot & \cdot & \cdot & 0 & 0 & .6 & \cdot & \cdot & \cdot & \cdot & \cdot & \cdot\\
\cdot & \cdot & \cdot & \cdot & \cdot & \cdot & 1 & -.5 & 0 & \cdot & \cdot & \cdot\\
\cdot & \cdot & \cdot & \cdot & \cdot & \cdot & -.5 & 1 & 0 & \cdot & \cdot & \cdot\\
\cdot & \cdot & \cdot & \cdot & \cdot & \cdot & 0 & 0 & .3 & \cdot & \cdot & \cdot\\
\cdot & \cdot & \cdot & \cdot & \cdot & \cdot & \cdot & \cdot & \cdot & 0 & \cdot & \cdot\\
\cdot & \cdot & \cdot & \cdot & \cdot & \cdot & \cdot & \cdot & \cdot & \cdot & 0 & \cdot\\
\cdot & \cdot & \cdot & \cdot & \cdot & \cdot & \cdot & \cdot & \cdot & \cdot & \cdot & 0
\end{bmatrix} \tag{D}$$

Substituting from Eqs. A and D and from page 54 (for **b**) into Eq. 6.10 and using SMIS3 as shown in the next printout, we obtain a solution for the nodal displacements:

$$\mathbf{D} = \frac{L}{6EI_0}\langle 0 \quad 0 \quad -7019 \quad 51192 \quad 96 \quad -1619 \quad 51192 \quad -96 \quad 780.8 \quad 59000 \quad 0 \quad 780.8\rangle \tag{E}$$

This result is verified by the SMIS3 stiffness solution displayed for comparison in the printout shown on pages 293–296.

****** *Computer Input:*

```
START
REMARK 5
SOLUTION FOR THE INTERNAL FORCES AND NODAL
DISPLACEMENTS OF THE STATICALLY DETERMINATE FRAME
SUBJECTED TO DISTRIBUTED LOADING, FORCE METHOD.

**** UNITS ARE KILONEWTONS AND METERS ****
REMARK 1
CASE I SOLUTION
LOAD PBARI 12 1
100 -100 10*0
PRINT PBARI 2
THIS IS THE VECTOR OF BASIC FORCES, PBARI, FOR THE COLLECTION
OF ALL MEMBERS AND CONSTRAINTS, OBTAINED BY INSPECTION, CASE I.
REMARK 1
CASE II SOLUTION
FORM A 12 12
LOAD T1 6 3
-.1 -.1 3*0 -1 1 2*0 .1 .1 3*0 1 0 1 0
LOAD T2 6 3
.1 .1 3*0 1 1 2*0 -.1 -.1 3*0 -1 0 1 0
LOAD T3 6 3
2*0 -1 .1 .1 0 1 4*0 1 -.1 -.1 2*0 1 0
LOAD CONSTR 1 1
1
STOSM A T1 1 1
STOSM A T2 7 4
STOSM A T3 4 7
STOSM A CONSTR 1 10
STOSM A CONSTR 2 11
STOSM A CONSTR 11 12
PRINT A 2
THIS IS THE MATRIX, A, OF STRUCTURE FORCE
INFLUENCE COEFFECIENTS.
INVERT A
DUPL A B
DELETE A
TRANS B BT
FORM FBAR 12 12
LOAD FN 3 3
2 -1 0 -1 2 0 0 0 .6
STOSM FBAR FN 1 1
STOSM FBAR FN 4 4
LOAD FN 3 3
1 -.5 0 -.5 1 0 0 0 .3
STOSM FBAR FN 7 7
PRINT FBAR 2
THIS THE FLEXIBILITY SUPERMATRIX, FBAR, FOR THE COLLECTION
OF ALL MEMBERS AND CONSTRAINTS WITH L/6EI(0) FACTORED OUT.
LOAD RII 12 1
60 0 -100 160 0 100 6*0
PRINT RII 1
THIS IS THE CASE II LOAD VECTOR, RII
MULT  B RII PBARII
PRINT PBARII 2
THIS IS THE VECTOR OF BASIC FORCES, PBARII, FOR THE COLLECTION
OF ALL MEMBERS AND CONSTRAINTS, CASE II.
MULT FBAR PBARII TEM
MULT BT TEM DII
PRINT DII 2
THIS IS THE VECTOR OF STRUCTURE NODAL DISPLACEMENTS,
WITH L/6EI(0) FACTORED OUT, CASE II.
REMARK 1
THE TOTAL SOLUTION IS NOW OBTAINED BY SUPERPOSITION.
ADD PBARI PBARII
DUPL PBARI PBAR
```

****** *Computer Input (continued):*

```
DELETE PBARI
PRINT PBAR 2
THIS THE VECTOR OF BASIC FORCES, PBAR, FOR
THE COLLECTION OF ALL MEMBERS AND CONSTRAINTS.
REMARK 3
SINCE DI IS NULL, THE TOTAL NODAL
DISPLACEMENTS ARE EQUAL TO THE CASE II
NODAL DISPLACEMENTS PRINTED ABOVE.
STOP
```

****** *Computer Output:*

```
START

REMARK      5


SOLUTION FOR THE INTERNAL FORCES AND NODAL
DISPLACEMENTS OF THE STATICALLY DETERMINATE FRAME
SUBJECTED TO DISTRIBUTED LOADING, FORCE METHOD.

**** UNITS ARE KILONEWTONS AND METERS ****


REMARK      1


CASE I SOLUTION


LOAD     PBARI        12     1
   12 ROWS,      1 COLUMNS

PRINT    PBARI        2


THIS IS THE VECTOR OF BASIC FORCES, PBARI, FOR THE COLLECTION
OF ALL MEMBERS AND CONSTRAINTS, OBTAINED BY INSPECTION, CASE I.


             1
  1  1.0000E+02
  2 -1.0000E+02
  3           0
  4           0
  5           0
  6           0
  7           0
  8           0
  9           0
 10           0
 11           0
 12           0


REMARK      1


CASE II SOLUTION
```

****** *Computer Output (continued):*

```
FORM     A           12   12
   12 ROWS,     12 COLUMNS

LOAD     T1           6    3
    6 ROWS,      3 COLUMNS

LOAD     T2           6    3
    6 ROWS,      3 COLUMNS

LOAD     T3           6    3
    6 ROWS,      3 COLUMNS

LOAD     CONSTR        1    1
    1 ROWS,      1 COLUMNS

STOSM    A        T1           1    1
 ROW NUMBER    1,  COLUMN NUMBER    1

STOSM    A        T2           7    4
 ROW NUMBER    7,  COLUMN NUMBER    4

STOSM    A        T3           4    7
 ROW NUMBER    4,  COLUMN NUMBER    7

STOSM    A        CONSTR       1   10
 ROW NUMBER    1,  COLUMN NUMBER   10

STOSM    A        CONSTR       2   11
 ROW NUMBER    2,  COLUMN NUMBER   11

STOSM    A        CONSTR      11   12
 ROW NUMBER   11,  COLUMN NUMBER   12

PRINT    A            2


THIS IS THE MATRIX, A, OF STRUCTURE FORCE
INFLUENCE COEFFECIENTS.
```

	1	2	3	4	5	6
1	-1.0000E-01	-1.0000E-01	0	0	0	0
2	0	0	-1.0000E+00	0	0	0
3	1.0000E+00	0	0	0	0	0
4	1.0000E-01	1.0000E-01	0	0	0	0
5	0	0	1.0000E+00	0	0	0
6	0	1.0000E+00	0	0	0	0
7	0	0	0	1.0000E-01	1.0000E-01	0
8	0	0	0	0	0	1.0000E+00
9	0	0	0	1.0000E+00	0	0
10	0	0	0	-1.0000E-01	-1.0000E-01	0
11	0	0	0	0	0	-1.0000E+00
12	0	0	0	0	1.0000E+00	0

	7	8	9	10	11	12
1	0	0	0	1.0000E+00	0	0
2	0	0	0	0	1.0000E+00	0
3	0	0	0	0	0	0
4	0	0	-1.0000E+00	0	0	0
5	1.0000E-01	1.0000E-01	0	0	0	0
6	1.0000E+00	0	0	0	0	0
7	0	0	1.0000E+00	0	0	0
8	-1.0000E-01	-1.0000E-01	0	0	0	0
9	0	1.0000E+00	0	0	0	0
10	0	0	0	0	0	0
11	0	0	0	0	0	1.0000E+00
12	0	0	0	0	0	0

****** *Computer Output (continued):*

```
INVERT  A

DUPL    A       B

DELETE  A

TRANS   B       BT

FORM    FBAR          12   12
   12 ROWS,     12 COLUMNS

LOAD    FN             3    3
    3 ROWS,      3 COLUMNS

STOSM   FBAR    FN          1    1
 ROW NUMBER   1,  COLUMN NUMBER   1

STOSM   FBAR    FN          4    4
 ROW NUMBER   4,  COLUMN NUMBER   4

LOAD    FN             3    3
    3 ROWS,      3 COLUMNS

STOSM   FBAR    FN          7    7
 ROW NUMBER   7,  COLUMN NUMBER   7

PRINT   FBAR           2
```

THIS THE FLEXIBILITY SUPERMATRIX, FBAR, FOR THE COLLECTION
OF ALL MEMBERS AND CONSTRAINTS WITH L/6EI(0) FACTORED OUT.

	1	2	3	4	5	6
1	2.0000E+00	-1.0000E+00	0	0	0	0
2	-1.0000E+00	2.0000E+00	0	0	0	0
3	0	0	6.0000E-01	0	0	0
4	0	0	0	2.0000E+00	-1.0000E+00	0
5	0	0	0	-1.0000E+00	2.0000E+00	0
6	0	0	0	0	0	6.0000E-01
7	0	0	0	0	0	0
8	0	0	0	0	0	0
9	0	0	0	0	0	0
10	0	0	0	0	0	0
11	0	0	0	0	0	0
12	0	0	0	0	0	0

	7	8	9	10	11	12
1	0	0	0	0	0	0
2	0	0	0	0	0	0
3	0	0	0	0	0	0
4	0	0	0	0	0	0
5	0	0	0	0	0	0
6	0	0	0	0	0	0
7	1.0000E+00	-5.0000E-01	0	0	0	0
8	-5.0000E-01	1.0000E+00	0	0	0	0
9	0	0	3.0000E-01	0	0	0
10	0	0	0	0	0	0
11	0	0	0	0	0	0
12	0	0	0	0	0	0

```
LOAD    RII           12    1
   12 ROWS,      1 COLUMNS

PRINT   RII            1
```

****** *Computer Output (continued):*

THIS IS THE CASE II LOAD VECTOR, RII

```
           1
 1  6.0000E+01
 2           0
 3 -1.0000E+02
 4  1.6000E+02
 5           0
 6  1.0000E+02
 7           0
 8           0
 9           0
10           0
11           0
12           0
```

```
MULT    B       RII     PBARII

PRINT   PBARII          2
```

THIS IS THE VECTOR OF BASIC FORCES, PBARII, FOR THE COLLECTION
OF ALL MEMBERS AND CONSTRAINTS, CASE II.

```
           1
 1 -1.0000E+02
 2  1.7000E+03
 3  1.6000E+02
 4           0
 5           0
 6 -1.6000E+02
 7 -1.6000E+03
 8           0
 9           0
10  2.2000E+02
11  1.6000E+02
12 -1.6000E+02
```

```
MULT    FBAR    PBARII  TEM

MULT    BT      TEM     DII

PRINT   DII             2
```

THIS IS THE VECTOR OF STRUCTURE NODAL DISPLACEMENTS,
WITH L/6EI(0) FACTORED OUT, CASE II.

```
           1
 1           0
 2           0
 3 -7.0192E+03
 4  5.1192E+04
 5  9.6000E+01
 6 -1.6192E+03
 7  5.1192E+04
 8 -9.6000E+01
 9  7.8080E+02
10  5.9000E+04
11           0
12  7.8080E+02
```

****** *Computer Output (continued):*

```
REMARK      1

THE TOTAL SOLUTION IS NOW OBTAINED BY SUPERPOSITION.

ADD     PBARI   PBARII

DUPL    PBARI   PBAR

DELETE  PBARI

PRINT   PBAR         2

THIS THE VECTOR OF BASIC FORCES, PBAR, FOR
THE COLLECTION OF ALL MEMBERS AND CONSTRAINTS.

              1
 1            0
 2  1.6000E+03
 3  1.6000E+02
 4            0
 5            0
 6 -1.6000E+02
 7 -1.6000E+03
 8            0
 9            0
10  2.2000E+02
11  1.6000E+02
12 -1.6000E+02


REMARK      3

SINCE DI IS NULL, THE TOTAL NODAL
DISPLACEMENTS ARE EQUAL TO THE CASE II
NODAL DISPLACEMENTS PRINTED ABOVE.

STOP
```

****** *Computer Input:*

```
START
REMARK 6
CHECK SOLUTION OF STATICALLY DETERMINATE FRAME,
DIRECT STIFFNESS METHOD ( VALUES OF  E, I(0) AND A(0) ARE
TAKEN SUCH THAT L/6EI(0) = 1 AND MAY BE CONSIDERED
AS A FACTOR OF THE SOLUTION. )

**** UNITS ARE KILONEWTONS AND METERS ****
INITLZ KUU RU D0 4 3 3 1 3
1 1 0
1 2 0
4 2 0
2 1 100
C1 1 2
C2 4 3
C3 2 3
```

****** *Computer Input (continued):*

```
FRAMEL A1 K1 1 RF1
0 10 1.666667 1 10
F 0 10 -12 -12 0
FRAMEL A3 K3 0
10 0 1.666667 2 20
BLDSTF KUU RU D0 3
K1 RF1 C1
K1   0 C2
K3   0 C3
SOLVE KUU RU DU
PRINT DU 2
THIS IS THE VECTOR, DU, OF STRUCTURE NODAL DISPLACEMENTS
WITH L/6EI(0) FACTORED OUT
FRAMFC DU 3
A1 K1 C1 RF1
A1 K1 C2 0
A3 K3 C3 0
STOP
```

****** *Computer Output:*

```
START

REMARK      6

CHECK SOLUTION OF STATICALLY DETERMINATE FRAME,
DIRECT STIFFNESS METHOD ( VALUES OF  E, I(0) AND A(0) ARE
TAKEN SUCH THAT L/6EI(0) = 1 AND MAY BE CONSIDERED
AS A FACTOR OF THE SOLUTION. )

**** UNITS ARE KILONEWTONS AND METERS ****


INITLZ  KUU     RU      D0

 NUMBER OF NODES          =     4
 NUMBER OF DOF PER NODE =     3
 NUMBER OF CONSTRAINTS  =     3
 NUMBER OF NODAL LOADS  =     1
 NUMBER OF ELEMENTS      =     3

 NULL    9 BY      9 STIFFNESS ARRAY, KUU    ,FORMED

 INITIAL DISPL. VECTOR = D0        REARRANGED COORDINATE CODE
       1              0                          -10
       2              0                          -11
       3              0                            1
       4              0                            2
       5              0                            3
       6              0                            4
       7              0                            5
       8              0                            6
       9              0                            7
      10              0                            8
      11              0                          -12
      12              0                            9
```

****** *Computer Output (continued):*

```
 NODAL LOAD VECTOR,  RU
               1
 1             0
 2  1.0000E+02
 3             0
 4             0
 5             0
 6             0
 7             0
 8             0
 9             0


 MEMBER CODE        DEGREES OF FREEDOM
    C1           -10   -11     1     2     3     4
    C2             8   -12     9     5     6     7
    C3             2     3     4     5     6     7

FRAMEL  A1      K1           1     RF1
 DX=          0     DY=  1.000E+01     L =  1.000E+01
 E =  1.667E+00     I =  1.000E+00     A =  1.000E+01

 LOAD TYPE = F
  DIMENSION C =             0    LOAD INTENSITY AT C = -1.200E+01
  DIMENSION D =  1.000E+01    LOAD INTENSITY AT D = -1.200E+01
  LOAD ANGLE  =             0

FRAMEL  A3      K3           0
 DX=  1.000E+01     DY=          0     L =  1.000E+01
 E =  1.667E+00     I =  2.000E+00     A =  2.000E+01

BLDSTF  KUU     RU      D0          3

 MEMBER K   MEMBER RF   MEMBER CODE
    K1         RF1          C1
    K1         0            C2
    K3         0            C3

 STRUCTURE EQUIVALENT NODAL LOAD VECTOR,  RU
               1
 1 -1.0000E+02
 2  1.6000E+02
 3             0
 4  1.0000E+02
 5             0
 6             0
 7             0
 8             0
 9             0


SOLVE   KUU     RU      DU

PRINT   DU           2

THIS IS THE VECTOR, DU, OF STRUCTURE NODAL DISPLACEMENTS
WITH L/6EI(0) FACTORED OUT
```

****** *Computer Output (continued):*

```
                1
 1 -7.0192E+03
 2  5.1192E+04
 3  9.6000E+01
 4 -1.6192E+03
 5  5.1192E+04
 6 -9.6000E+01
 7  7.8080E+02
 8  5.9000E+04
 9  7.8080E+02


FRAMFC  DU          3

 MEMBER LOCAL FORCE VECTORS


 MEMBER CODE   = C1

 1 -1.60000E+02
 2  2.20000E+02
 3  1.09139E-11
 4  1.60000E+02
 5 -1.00000E+02
 6  1.60000E+03

 MEMBER CODE   = C2

 1  1.60000E+02
 2  5.45697E-12
 3  9.09495E-12
 4 -1.60000E+02
 5 -5.45697E-12
 6  1.81899E-11

 MEMBER CODE   = C3

 1            0
 2 -1.60000E+02
 3 -1.60000E+03
 4            0
 5  1.60000E+02
 6  1.45519E-11
STOP
```

6.3 STATICALLY INDETERMINATE STRUCTURES

As a first example of the analysis of statically indeterminate structures by the force method, it is interesting and useful to consider a statically indeterminate variation of the statically determinate truss just analyzed in Sec. 6.2.1.

6.3.1 Statically Indeterminate Truss

The truss of Fig. 4.6, which is shown again in Fig. 6.3 for convenience, was previously analyzed by the displacement method in Sec. 4.4.1. The structure equilibrium equations for this truss can be built from the individual member

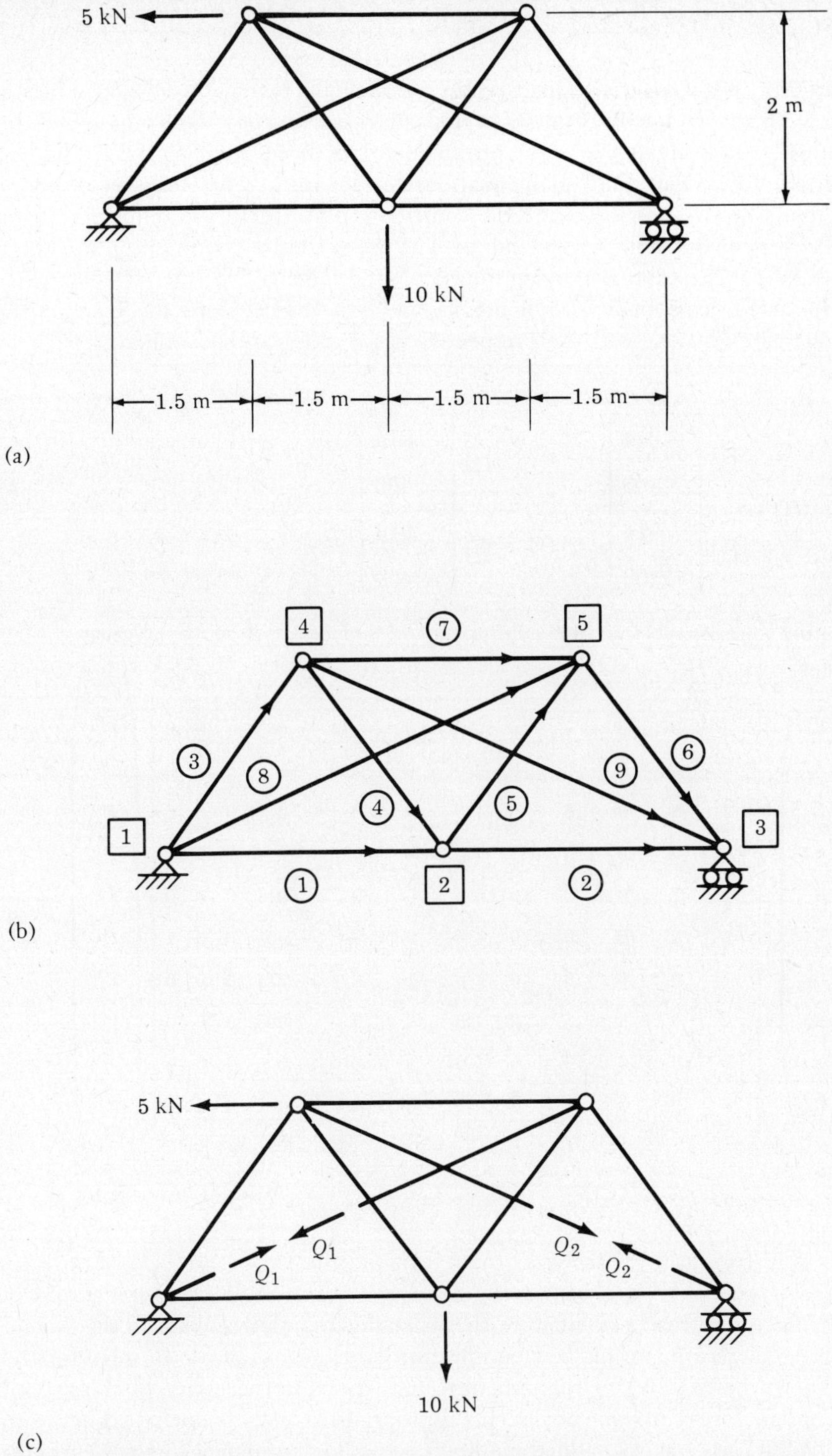

Fig. 6.3 Statically indeterminate truss (previously shown in Fig. 4.6): (a) dimensions and loading; (b) structural model; (c) cut-back structure.

equilibrium equations by using, say, the nodal coding technique developed in Sec. 2.5.2. This procedure has already been partially carried out in the building of the statically determinate truss equilibrium equations in Eq. A of Sec. 6.2.1. Thus we can now obtain the equilibrium equations for the statically indeterminate truss by augmenting that equation with the contribution to structure equilibrium of the added members numbers 8 and 9. The equilibrium relationships for these members may be obtained by appropriately specializing Eq. 2.10a, which applies to a general truss member. For brevity and convenience, only the **T** arrays for each member, along with their respective nodal codes, are shown here.

$$\mathbf{T}^{(8)} = \begin{array}{c} \boxed{1} \\ \\ \boxed{5} \end{array} \overset{\textcircled{8}}{\begin{bmatrix} -.914 \\ -.406 \\ \hdashline .914 \\ .406 \end{bmatrix}} \qquad \mathbf{T}^{(9)} = \begin{array}{c} \boxed{4} \\ \\ \boxed{3} \end{array} \overset{\textcircled{9}}{\begin{bmatrix} -.914 \\ .406 \\ \hdashline .914 \\ -.406 \end{bmatrix}} \tag{A}$$

Coding these results into new columns 8 and 9 of Eq. A of Sec. 6.2.1 gives

$$\begin{Bmatrix} R_1 \\ R_2 \\ R_3 \\ R_4 \\ R_5 \\ R_6 \\ R_7 \\ R_8 \\ R_9 \\ R_{10} \end{Bmatrix} = \begin{bmatrix} -1 & 0 & -.6 & 0 & 0 & 0 & 0 & -.914 & 0 & 1 & 0 & 0 \\ 0 & 0 & -.8 & 0 & 0 & 0 & 0 & -.406 & 0 & 0 & 1 & 0 \\ 1 & -1 & 0 & .6 & -.6 & 0 & 0 & 0 & 0 & 0 & 0 & 0 \\ 0 & 0 & 0 & -.8 & -.8 & 0 & 0 & 0 & 0 & 0 & 0 & 0 \\ 0 & 1 & 0 & 0 & 0 & .6 & 0 & 0 & .914 & 0 & 0 & 0 \\ 0 & 0 & 0 & 0 & 0 & -.8 & 0 & 0 & -.406 & 0 & 0 & 1 \\ 0 & 0 & .6 & -.6 & 0 & 0 & -1 & 0 & -.914 & 0 & 0 & 0 \\ 0 & 0 & .8 & .8 & 0 & 0 & 0 & 0 & .406 & 0 & 0 & 0 \\ 0 & 0 & 0 & 0 & .6 & -.6 & 1 & .914 & 0 & 0 & 0 & 0 \\ 0 & 0 & 0 & 0 & .8 & .8 & 0 & .406 & 0 & 0 & 0 & 0 \end{bmatrix} \begin{Bmatrix} P^{(1)} \\ P^{(2)} \\ P^{(3)} \\ P^{(4)} \\ P^{(5)} \\ P^{(6)} \\ P^{(7)} \\ P^{(8)} \\ P^{(9)} \\ P_c^{(1)} \\ P_c^{(2)} \\ P_c^{(3)} \end{Bmatrix} \tag{B}$$

or

$$\mathbf{R} = \mathbf{a}\overline{\mathbf{P}} \tag{2.18}$$

Since the truss is a structure that is statically indeterminate to the second degree, there are two too few equilibrium equations available to solve for the unknown basic forces of the structure. It is therefore necessary to choose two redundant forces. In this case it seems appropriate, though somewhat arbitrary, to select the basic forces of added members 8 and 9 as redundant forces. Rearranging Eq. B so that the vector of unknown basic forces may be conveniently partitioned into redundant and nonredundant sets, we get

$$\begin{pmatrix} R_1 \\ R_2 \\ R_3 \\ R_4 \\ R_5 \\ R_6 \\ R_7 \\ R_8 \\ R_9 \\ R_{10} \end{pmatrix} = \left[\begin{array}{cccccccccc|cc} -1 & 0 & -.6 & 0 & 0 & 0 & 0 & 1 & 0 & 0 & -.914 & 0 \\ 0 & 0 & -.8 & 0 & 0 & 0 & 0 & 0 & 1 & 0 & -.406 & 0 \\ 1 & -1 & 0 & .6 & -.6 & 0 & 0 & 0 & 0 & 0 & 0 & 0 \\ 0 & 0 & 0 & -.8 & -.8 & 0 & 0 & 0 & 0 & 0 & 0 & 0 \\ 0 & 1 & 0 & 0 & 0 & .6 & 0 & 0 & 0 & 0 & 0 & .914 \\ 0 & 0 & 0 & 0 & 0 & -.8 & 0 & 0 & 0 & 1 & 0 & -.406 \\ 0 & 0 & .6 & -.6 & 0 & 0 & -1 & 0 & 0 & 0 & 0 & -.914 \\ 0 & 0 & .8 & .8 & 0 & 0 & 0 & 0 & 0 & 0 & 0 & .406 \\ 0 & 0 & 0 & 0 & .6 & -.6 & 1 & 0 & 0 & 0 & .914 & 0 \\ 0 & 0 & 0 & 0 & .8 & .8 & 0 & 0 & 0 & 0 & .406 & 0 \end{array}\right] \left(\begin{array}{c} P^{(1)} \\ P^{(2)} \\ P^{(3)} \\ P^{(4)} \\ P^{(5)} \\ P^{(6)} \\ P^{(7)} \\ P_c^{(1)} \\ P_c^{(2)} \\ \hline P_c^{(3)} \\ P^{(8)} \\ P^{(9)} \end{array}\right) \qquad \text{(B}'\text{)}$$

Partitioned Eq. B′ may be represented symbolically as

$$\mathbf{R} = [\mathbf{a}_R \,|\, \mathbf{a}_Q] \left(\begin{array}{c} \overline{\mathbf{P}}_R \\ \hline \overline{\mathbf{P}}_Q \end{array}\right) \qquad (6.12)$$

Using Eq. 6.12, we can now obtain the basic unknowns represented by the vector $\overline{\mathbf{P}}_R$ in terms of the applied loads $\mathbf{R}$ and the redundant forces $\overline{\mathbf{P}}_Q$

$$\mathbf{R} = \mathbf{a}_R \overline{\mathbf{P}}_R + \mathbf{a}_Q \overline{\mathbf{P}}_Q \qquad (6.12a)$$

and

$$\overline{\mathbf{P}}_R = \mathbf{a}_R^{-1}\mathbf{R} - \mathbf{a}_R^{-1}\mathbf{a}_Q \overline{\mathbf{P}}_Q \qquad (6.13)$$

Carrying out in detail the operations indicated in Eq. 6.13, we have the solution of Eq. B′ shown as Eq. C on page 300. (Notice that, owing to the choice of redundants, $\mathbf{a}_R^{-1}$ in Eq. C is the same as $\mathbf{a}^{-1}$ for the statically determinate truss.) Augmenting Eq. C with the equations

$$P^{(8)} = Q_1$$
$$P^{(9)} = Q_2 \qquad \text{(D)}$$

and rearranging the resulting equation so that the left-hand side vector is $\overline{\mathbf{P}}$, we obtain. Eq. E shown on page 301. Partitioned Eq. E may be represented symbolically as

$$\overline{\mathbf{P}} = [\mathbf{b}_R \,|\, \mathbf{b}_Q] \left(\begin{array}{c} \mathbf{R} \\ \hline \mathbf{Q} \end{array}\right) \qquad (6.14)$$

Notice that Eq. E may be considered to represent a solution for the internal forces in a statically determinate "cutback" form of the original structure. Figure 6.3c

$$\begin{pmatrix} P^{(1)} \\ P^{(2)} \\ P^{(3)} \\ P^{(4)} \\ P^{(5)} \\ P^{(6)} \\ P^{(7)} \\ P_c^{(1)} \\ P_c^{(2)} \\ P_c^{(3)} \end{pmatrix} = \begin{bmatrix} 0 & 0 & 1 & -.375 & 1 & 0 & .75 & -.5625 & .75 & -.1875 \\ 0 & 0 & 0 & -.375 & 1 & 0 & .25 & -.1875 & .25 & -.5625 \\ 0 & 0 & 0 & .625 & 0 & 0 & .4167 & .9375 & .4167 & .3125 \\ 0 & 0 & 0 & -.625 & 0 & 0 & -.4167 & .3125 & -.4167 & -.3125 \\ 0 & 0 & 0 & -.625 & 0 & 0 & .4167 & -.3125 & .4167 & .3125 \\ 0 & 0 & 0 & .625 & 0 & 0 & -.4167 & .3125 & -.4167 & .9375 \\ 0 & 0 & 0 & .75 & 0 & 0 & -.5 & .375 & .5 & .375 \\ 1 & 0 & 1 & 0 & 1 & 0 & 1 & 0 & 1 & 0 \\ 0 & 1 & 0 & .5 & 0 & 0 & .3333 & .75 & .3333 & .25 \\ 0 & 0 & 0 & .5 & 0 & 1 & -.3333 & .25 & -.3333 & .75 \end{bmatrix} \begin{pmatrix} R_1 \\ R_2 \\ R_3 \\ R_4 \\ R_5 \\ R_6 \\ R_7 \\ R_8 \\ R_9 \\ R_{10} \end{pmatrix} - \begin{bmatrix} .6093 & 0 \\ 0 & .6093 \\ .5077 & 0 \\ -.5077 & .5077 \\ .5077 & -.5077 \\ 0 & .5077 \\ .6093 & .6093 \\ 0 & 0 \\ 0 & 0 \\ 0 & 0 \end{bmatrix} \begin{pmatrix} P^{(8)} \\ P^{(9)} \end{pmatrix} \quad \text{(C)}$$

$$
\begin{pmatrix} P^{(1)} \\ P^{(2)} \\ P^{(3)} \\ P^{(4)} \\ P^{(5)} \\ P^{(6)} \\ P^{(7)} \\ P^{(8)} \\ P^{(9)} \\ P_c^{(1)} \\ P_c^{(2)} \\ P_c^{(3)} \end{pmatrix}
=
\left[\begin{array}{cccccccccc:cc}
0 & 0 & 1 & -.375 & 1 & 0 & .75 & -.5625 & .75 & -.1875 & -.6093 & 0 \\
0 & 0 & 0 & -.375 & 1 & 0 & .25 & -.1875 & .25 & -.5625 & 0 & -.6093 \\
0 & 0 & 0 & .625 & 0 & 0 & .4167 & .9375 & .4167 & .3125 & -.5077 & 0 \\
0 & 0 & 0 & -.625 & 0 & 0 & -.4167 & .3125 & -.4167 & -.3125 & .5077 & -.5077 \\
0 & 0 & 0 & -.625 & 0 & 0 & .4167 & -.3125 & .4167 & .3125 & -.5077 & .5077 \\
0 & 0 & 0 & .625 & 0 & 0 & -.4167 & .3125 & -.4167 & .9375 & 0 & -.5077 \\
0 & 0 & 0 & .75 & 0 & 0 & -.5 & .375 & .5 & .375 & -.6093 & -.6093 \\
0 & 0 & 0 & 0 & 0 & 0 & 0 & 0 & 0 & 0 & 1 & 0 \\
0 & 0 & 0 & 0 & 0 & 0 & 0 & 0 & 0 & 0 & 0 & 1 \\
1 & 0 & 1 & 0 & 1 & 0 & 1 & 0 & 1 & 0 & 0 & 0 \\
0 & 1 & 0 & .5 & 0 & 0 & .3333 & .75 & .3333 & .25 & 0 & 0 \\
0 & 0 & 0 & .5 & 0 & 1 & -.3333 & .25 & -.3333 & .75 & 0 & 0
\end{array}\right]
\begin{pmatrix} R_1 \\ R_2 \\ R_3 \\ R_4 \\ R_5 \\ R_6 \\ R_7 \\ R_8 \\ R_9 \\ R_{10} \\ \hdashline Q_1 \\ Q_2 \end{pmatrix}
\quad \text{(E)}
$$

shows this cutback structure loaded with the external applied loads plus the redundant forces that are exposed in the cutting-back process. By the contragredient law it follows that

$$\begin{pmatrix} \mathbf{D} \\ \hline \mathbf{D}_Q \end{pmatrix} = \begin{bmatrix} \mathbf{b}_R^T \\ \hline \mathbf{b}_Q^T \end{bmatrix} \bar{\boldsymbol{\Delta}} \tag{6.15}$$

where **D** is, of course, the vector of nodal displacements corresponding to **R**, the vector of nodal forces; and $\mathbf{D}_Q$ is the vector of relative displacements at the cuts corresponding to **Q**, the vector of redundant forces exposed at the cuts. Using Eqs. 6.14 and 6.15 along with Eq. 6.3, we may now write

$$\begin{pmatrix} \mathbf{D} \\ \hline \mathbf{D}_Q \end{pmatrix} = \begin{bmatrix} \mathbf{b}_R^T \\ \hline \mathbf{b}_Q^T \end{bmatrix} \bar{\mathbf{f}} [\mathbf{b}_R \mid \mathbf{b}_Q] \begin{pmatrix} \mathbf{R} \\ \hline \mathbf{Q} \end{pmatrix} \tag{6.16}$$

or

$$\begin{pmatrix} \mathbf{D} \\ \hline \mathbf{D}_Q \end{pmatrix} = \left[\begin{array}{c|c} \mathbf{b}_R^T \bar{\mathbf{f}} \mathbf{b}_R & \mathbf{b}_R^T \bar{\mathbf{f}} \mathbf{b}_Q \\ \hline \mathbf{b}_Q^T \bar{\mathbf{f}} \mathbf{b}_R & \mathbf{b}_Q^T \bar{\mathbf{f}} \mathbf{b}_Q \end{array}\right] \begin{pmatrix} \mathbf{R} \\ \hline \mathbf{Q} \end{pmatrix} \tag{6.16a}$$

or

$$\begin{pmatrix} \mathbf{D} \\ \hline \mathbf{D}_Q \end{pmatrix} = \left[\begin{array}{c|c} \mathbf{F}_{RR} & \mathbf{F}_{RQ} \\ \hline \mathbf{F}_{QR} & \mathbf{F}_{QQ} \end{array}\right] \begin{pmatrix} \mathbf{R} \\ \hline \mathbf{Q} \end{pmatrix} \tag{6.16b}$$

where

$$\begin{aligned} \mathbf{F}_{RR} &= \mathbf{b}_R^T \bar{\mathbf{f}} \mathbf{b}_R \\ \mathbf{F}_{RQ} &= \mathbf{F}_{QR}^T = \mathbf{b}_R^T \bar{\mathbf{f}} \mathbf{b}_Q \\ \mathbf{F}_{QQ} &= \mathbf{b}_Q^T \bar{\mathbf{f}} \mathbf{b}_Q \end{aligned} \tag{6.17}$$

In order that the cutback structure behave identically to the statically indeterminate parent structure, the redundant forces at the cuts must take on values such that no relative displacements actually occur at the cuts. Thus

$$\mathbf{D}_Q = \boldsymbol{\Theta} \tag{6.18}$$

Substituting this information into Eq. 6.16b and expanding the resulting equation gives

$$\begin{aligned} \mathbf{D} &= \mathbf{F}_{RR}\mathbf{R} + \mathbf{F}_{RQ}\mathbf{Q} \\ \boldsymbol{\Theta} &= \mathbf{F}_{QR}\mathbf{R} + \mathbf{F}_{QQ}\mathbf{Q} \end{aligned} \tag{6.19}$$

The second equation of Eq. 6.19 may now be solved for the redundant forces that would restore structural continuity at the release points:

$$\mathbf{Q} = -\mathbf{F}_{QQ}^{-1}\mathbf{F}_{QR}\mathbf{R} \tag{6.20}$$

Substituting these results into Eq. 6.14 yields a solution for the complete set of unknown basic forces for the structure:

$$\overline{\mathbf{P}} = [\mathbf{b}_R \mid \mathbf{b}_Q] \begin{pmatrix} \mathbf{R} \\ \hline -\mathbf{F}_{QQ}^{-1}\mathbf{F}_{QR}\mathbf{R} \end{pmatrix} \tag{6.21}$$

or

$$\overline{\mathbf{P}} = \mathbf{b}_R\mathbf{R} - \mathbf{b}_Q\mathbf{F}_{QQ}^{-1}\mathbf{F}_{QR}\mathbf{R} \tag{6.21a}$$

or

$$\overline{\mathbf{P}} = [\mathbf{b}_R - \mathbf{b}_Q\mathbf{F}_{QQ}^{-1}\mathbf{F}_{QR}]\mathbf{R} \tag{6.21b}$$

Now, backsubstituting from Eq. 6.20 into the first equation of Eq. 6.19, we obtain a solution for the nodal displacements:

$$\mathbf{D} = \mathbf{F}_{RR}\mathbf{R} - \mathbf{F}_{RQ}\mathbf{F}_{QQ}^{-1}\mathbf{F}_{QR}\mathbf{R} \tag{6.22}$$

or

$$\mathbf{D} = [\mathbf{F}_{RR} - \mathbf{F}_{RQ}\mathbf{F}_{QQ}^{-1}\mathbf{F}_{QR}]\mathbf{R} \tag{6.22a}$$

Before obtaining a detailed solution for the truss as outlined symbolically in Eq. 6.20, 6.21b, and 6.22a, we must first build the basic flexibility matrix for the collection of all members of the truss. The physical properties of the truss members are the same as those listed for the statically determinate truss in Fig. 6.1, with the addition that $A^{(8)} = A^{(9)} = A_0$ for members 8 and 9. The flexibilities for members 8 and 9 (on the basis of Eq. 3.5) are therefore

$$f^{(8)} = f^{(9)} = \frac{L^{(8)}}{EA_0} = \frac{(4.5^2 + 2^2)^{1/2}}{EA_0} = \frac{4.924}{EA_0}$$

Since the remaining members and supports are unchanged from the statically determinate truss, we can obtain the flexibilities for the collection of all members of the statically indeterminate truss by appropriately augmenting Eq. F of Sec. 6.2.1, using the coefficient just given. Thus we have

$$\begin{pmatrix} \Delta^{(1)} \\ \Delta^{(2)} \\ \Delta^{(3)} \\ \Delta^{(4)} \\ \Delta^{(5)} \\ \Delta^{(6)} \\ \Delta^{(7)} \\ \Delta^{(8)} \\ \Delta^{(9)} \\ \Delta_c^{(1)} \\ \Delta_c^{(2)} \\ \Delta_c^{(3)} \end{pmatrix} = \frac{3}{EA_0} \begin{bmatrix} 1 &&&&&&&&&&& \\ & 1 &&&&&&&&&& \\ && .8 &&&&&&&&& \\ &&& .8 &&&& \Theta &&&& \\ &&&& .8 &&&&&&& \\ &&&&& .8 &&&&&& \\ &&&&&& 1 &&&&& \\ &&&&&&& 1.641 &&&& \\ && \Theta &&&&&& 1.641 &&& \\ &&&&&&&&& 0 && \\ &&&&&&&&&& 0 & \\ &&&&&&&&&&& 0 \end{bmatrix} \begin{pmatrix} P^{(1)} \\ P^{(2)} \\ P^{(3)} \\ P^{(4)} \\ P^{(5)} \\ P^{(6)} \\ P^{(7)} \\ P^{(8)} \\ P^{(9)} \\ P_c^{(1)} \\ P_c^{(2)} \\ P_c^{(3)} \end{pmatrix} \tag{F}$$

or

$$\bar{\Delta} = \bar{f}\bar{\mathbf{P}} \tag{6.3}$$

The structure flexibility submatrices required for the detailed solution of the example truss may now be obtained by substituting appropriately into Eq. 6.17:

$$\mathbf{F}_{RR} = \frac{3}{EA_0}\begin{bmatrix} 0 & 0 & 0 & 0 & 0 & 0 & 0 & 0 & 0 & 0 \\ 0 & 0 & 0 & 0 & 0 & 0 & 0 & 0 & 0 & 0 \\ 0 & 0 & 1 & -.375 & 1 & 0 & .75 & -.5625 & .75 & -.1875 \\ 0 & 0 & -.375 & 2.094 & -.75 & 0 & -.75 & 1.188 & 0 & 1.188 \\ 0 & 0 & 1 & -.75 & 2 & 0 & 1 & -.75 & 1 & -.75 \\ 0 & 0 & 0 & 0 & 0 & 0 & 0 & 0 & 0 & 0 \\ 0 & 0 & .75 & -.75 & 1 & 0 & 1.431 & -.6563 & .9306 & -.4688 \\ 0 & 0 & -.5625 & 1.188 & -.75 & 0 & -.6563 & 1.430 & -.2813 & .6641 \\ 0 & 0 & .75 & 0 & 1 & 0 & .9306 & -.2813 & 1.431 & -.09375 \\ 0 & 0 & -.1875 & 1.188 & -.75 & 0 & -.4688 & .6641 & -.09375 & 1.430 \end{bmatrix} \tag{G}$$

$$\mathbf{F}_{QR} = \frac{3}{EA_0}\begin{bmatrix} 0 & 0 & -.6092 & -.4822 & -.6092 & 0 & -.6600 & -.01267 & -1.269 & -.4950 \\ 0 & 0 & 0 & -.4822 & -.6092 & 0 & .6600 & -.4949 & .05076 & -.01267 \end{bmatrix}$$

$$\mathbf{F}_{QQ} = \frac{3}{EA_0}\begin{bmatrix} 3.002 & -.04124 \\ -.04124 & 3.002 \end{bmatrix}$$

Finally, by inspecting Fig. 6.3, we can directly obtain the load vector:

$$\mathbf{R} = \langle 0 \quad 0 \quad 0 \quad -10 \quad 0 \quad 0 \quad -5 \quad 0 \quad 0 \quad 0 \rangle \tag{H}$$

Substituting appropriately from Eqs. G and H into Eqs. 6.20, 6.21b, and 6.22a and carrying out the indicated operations gives the redundant forces,

$$\mathbf{Q} = \langle -2.713 \quad -.5445 \rangle \tag{I}$$

the basic unknown structure forces,

$$\bar{\mathbf{P}} = \langle 1.653 \quad 2.832 \quad -6.956 \quad 7.232 \quad 5.268 \quad -3.890 \quad -3.015 \quad -2.713 \quad -.5446 \quad -5 \quad -6.667 \quad -3.333 \rangle \tag{J}$$

and the structure nodal displacements,

$$\mathbf{D} = \frac{3}{EA_0}\langle 0 \quad 0 \quad 1.653 \quad -15.62 \quad 4.485 \quad 0 \quad 1.779 \quad -8.290 \quad -1.237 \quad -8.181 \rangle \tag{K}$$

respectively. The detailed solutions for the vectors of Eqs. I, J, and K were carried out by using SMIS3, as shown in the next printout.

****** *Computer Input:*

```
START
REMARK 4
SOLUTION OF STATICALLY INDETERMINATE TRUSS, FORCE
METHOD, ARBITRARILY SELECTED REDUNDANT FORCES

**** UNITS ARE KILONEWTONS AND METERS ****
LOAD AT 12 12
-1 0 1 11*0 -1 0 1 7*0 -.6 -.8 4*0 .6 .8
6*0 .6 -.8 0 0 -.6 .8 6*0 -.6 -.8 4*0
.6 .8 6*0 .6 -.8 0 0 -.6 .8 8*0 -1 0 1
0 0 0 -.9138 -.4061 6*0 .9138 .4061 1 5*0 .9138 -.4061
-.9138 .4061 0 0 0 1 1 12*0 1 15*0 1 6*0
TRANS AT A
DELETE AT
PRINT A 3
THIS THE EQUILIBRIUM MATRIX, A, AUGMENTED WITH ROWS
WHICH RELATE REDUNDANT LOADS , Q, TO REDUNDANT
FORCES, PQ.
INVERT A
RMVSM A BR 1 1 12 10
RMVSM A BQ 1 11 12 2
PRINT BR 1
THIS IS THE MATRIX, BR
PRINT BQ 1
THIS IS THE MATRIX, BQ
FORM FBAR 12 12
LOAD FDG 1 12
1 1 .8 .8 .8 .8 1 1.641 1.641 0 0 0
STODG FBAR FDG
PRINT FBAR 2
THIS IS THE FLEXIBILITY SUPERMATRIX, FBAR, FOR THE COLLECTION
OF ALL MEMBERS AND CONSTRAINTS WITH 3/EA(0) FACTORED OUT.
MULT FBAR BR TEM1
TRANS TEM1 TEM2
DELETE TEM1
MULT TEM2 BR FRR
MULT TEM2 BQ FQR
TRANS FQR FRQ
DELETE TEM2
MULT FBAR BQ TEM1
TRANS TEM1 TEM2
DELETE TEM1
MULT TEM2 BQ FQQ
DELETE TEM2
PRINT FRR 2
THIS IS THE FLEXIBILITY SUBMATRIX, FRR,
WITH 3/EA(0) FACTORED OUT.
PRINT FRQ 2
THIS IS THE FLEXIBILITY SUBMATRIX, FRQ,
WITH 3/EA(0) FACTORED OUT.
PRINT FQQ 2
THIS IS THE FLEXIBILITY SUBMATRIX, FQQ,
WITH 3/EA(0) FACTORED OUT.
SYMINV FQQ
LOAD R 10 1
0 0 0 -10 0 0 -5 0 0 0
MULT FRQ R TEM1
MULT FQQ TEM1 Q
DELETE TEM1
SCALE Q -1
PRINT Q 1
THIS IS THE VECTOR OF REDUNDANT FORCES, Q.
MULT BR R PBAR
MULT BQ Q TEM1
ADD PBAR TEM1
DELETE TEM1
PRINT PBAR 2
THIS IS THE VECTOR OF INTERNAL BASIC FORCES, PBAR,
FOR THE COLLECTION OF ALL MEMBERS AND CONSTRAINTS.
```

****** *Computer Input (continued):*

```
MULT FRR R D
MULT FQR Q TEM1
ADD D TEM1
DELETE TEM1
PRINT D 2
THIS IS THE VECTOR OF STRUCTURE NODAL DISPLACEMENTS,
WITH 3/EA(0) FACTORED OUT.
STOP
```

****** *Computer Output:*

```
START

REMARK      4


SOLUTION OF STATICALLY INDETERMINATE TRUSS, FORCE
METHOD, ARBITRARILY SELECTED REDUNDANT FORCES

**** UNITS ARE KILONEWTONS AND METERS ****



LOAD    AT          12   12
   12 ROWS,    12 COLUMNS

TRANS   AT      A

DELETE  AT

PRINT   A           3


THIS THE EQUILIBRIUM MATRIX, A, AUGMENTED WITH ROWS
WHICH RELATE REDUNDANT LOADS , Q, TO REDUNDANT
FORCES, PQ.


             1          2          3          4          5          6
 1 -1.0000E+00          0-6.0000E-01          0          0          0
 2           0          0-8.0000E-01          0          0          0
 3  1.0000E+00-1.0000E+00          0 6.0000E-01-6.0000E-01          0
 4           0          0          0-8.0000E-01-8.0000E-01          0
 5           0 1.0000E+00          0          0          0 6.0000E-01
 6           0          0          0          0          0-8.0000E-01
 7           0          0 6.0000E-01-6.0000E-01          0          0
 8           0          0 8.0000E-01 8.0000E-01          0          0
 9           0          0          0          0 6.0000E-01-6.0000E-01
10           0          0          0          0 8.0000E-01 8.0000E-01
11           0          0          0          0          0          0
12           0          0          0          0          0          0
             7          8          9         10         11         12
 1           0-9.1380E-01          0 1.0000E+00          0          0
 2           0-4.0610E-01          0          0 1.0000E+00          0
 3           0          0          0          0          0          0
 4           0          0          0          0          0          0
 5           0          0 9.1380E-01          0          0          0
 6           0          0-4.0610E-01          0          0 1.0000E+00
 7 -1.0000E+00          0-9.1380E-01          0          0          0
 8           0          0 4.0610E-01          0          0          0
 9  1.0000E+00 9.1380E-01          0          0          0          0
10           0 4.0610E-01          0          0          0          0
11           0 1.0000E+00          0          0          0          0
12           0          0 1.0000E+00          0          0          0
```

****** *Computer Output (continued):*

```
INVERT  A

RMVSM   A       BR            1    1   12   10
 ROW NUMBER    1,  COLUMN NUMBER   1
   12 ROWS,     10 COLUMNS

RMVSM   A       BQ            1   11   12    2
 ROW NUMBER    1,  COLUMN NUMBER  11
   12 ROWS,      2 COLUMNS

PRINT   BR            1

THIS IS THE MATRIX, BR

             1          2          3          4          5          6
  1          0          0 1.0000E+00-3.7500E-01 1.0000E+00          0
  2          0          0          0-3.7500E-01 1.0000E+00          0
  3          0          0          0 6.2500E-01          0          0
  4          0          0          0-6.2500E-01          0          0
  5          0          0          0-6.2500E-01          0          0
  6          0          0          0 6.2500E-01          0          0
  7          0          0          0 7.5000E-01          0          0
  8          0          0          0          0          0          0
  9          0          0          0          0          0          0
 10 1.0000E+00          0 1.0000E+00-4.4409E-15 1.0000E+00          0
 11          0 1.0000E+00          0 5.0000E-01          0          0
 12          0          0          0 5.0000E-01          0 1.0000E+00
             7          8          9         10
  1 7.5000E-01-5.6250E-01 7.5000E-01-1.8750E-01
  2 2.5000E-01-1.8750E-01 2.5000E-01-5.6250E-01
  3 4.1667E-01 9.3750E-01 4.1667E-01 3.1250E-01
  4-4.1667E-01 3.1250E-01-4.1667E-01-3.1250E-01
  5 4.1667E-01-3.1250E-01 4.1667E-01 3.1250E-01
  6-4.1667E-01 3.1250E-01-4.1667E-01 9.3750E-01
  7-5.0000E-01 3.7500E-01 5.0000E-01 3.7500E-01
  8          0          0          0          0
  9          0          0          0          0
 10 1.0000E+00 1.3323E-15 1.0000E+00-2.2204E-15
 11 3.3333E-01 7.5000E-01 3.3333E-01 2.5000E-01
 12-3.3333E-01 2.5000E-01-3.3333E-01 7.5000E-01


PRINT   BQ            1

THIS IS THE MATRIX, BQ

             1          2
  1-6.0921E-01-1.8750E-05
  2-1.8750E-05-6.0921E-01
  3-5.0766E-01 3.1250E-05
  4 5.0766E-01-5.0766E-01
  5-5.0766E-01 5.0766E-01
  6 3.1250E-05-5.0766E-01
  7-6.0919E-01-6.0919E-01
  8 1.0000E+00          0
  9          0 1.0000E+00
 10 5.3018E-15-1.0604E-14
 11-2.5000E-05 2.5000E-05
 12 2.5000E-05-2.5000E-05
```

****** *Computer Output (continued):*

```
LOAD     FDG           1    12
    1 ROWS,     12 COLUMNS

STODG    FBAR     FDG

PRINT    FBAR           2


THIS IS THE FLEXIBILITY SUPERMATRIX, FBAR, FOR THE COLLECTION
OF ALL MEMBERS AND CONSTRAINTS WITH 3/EA(0) FACTORED OUT.


               1           2           3           4           5           6
  1   1.0000E+00           0           0           0           0           0
  2            0  1.0000E+00           0           0           0           0
  3            0           0  8.0000E-01           0           0           0
  4            0           0           0  8.0000E-01           0           0
  5            0           0           0           0  8.0000E-01           0
  6            0           0           0           0           0  8.0000E-01
  7            0           0           0           0           0           0
  8            0           0           0           0           0           0
  9            0           0           0           0           0           0
 10            0           0           0           0           0           0
 11            0           0           0           0           0           0
 12            0           0           0           0           0           0
               7           8           9          10          11          12
  1            0           0           0           0           0           0
  2            0           0           0           0           0           0
  3            0           0           0           0           0           0
  4            0           0           0           0           0           0
  5            0           0           0           0           0           0
  6            0           0           0           0           0           0
  7   1.0000E+00           0           0           0           0           0
  8            0  1.6410E+00           0           0           0           0
  9            0           0  1.6410E+00           0           0           0
 10            0           0           0           0           0           0
 11            0           0           0           0           0           0
 12            0           0           0           0           0           0


MULT     FBAR     BR       TEM1

TRANS    TEM1     TEM2

DELETE   TEM1

MULT     TEM2     BR       FRR

MULT     TEM2     BQ       FQR

TRANS    FQR      FRQ

DELETE   TEM2

MULT     FBAR     BQ       TEM1

TRANS    TEM1     TEM2

DELETE   TEM1

MULT     TEM2     BQ       FQQ

DELETE   TEM2

PRINT    FRR            2
```

****** *Computer Output (continued):*

```
THIS IS THE FLEXIBILITY SUBMATRIX, FRR,
WITH 3/EA(0) FACTORED OUT.

               1            2            3            4            5            6
  1            0            0            0            0            0            0
  2            0            0            0            0            0            0
  3            0            0 1.0000E+00-3.7500E-01 1.0000E+00            0
  4            0            0-3.7500E-01 2.0937E+00-7.5000E-01            0
  5            0            0 1.0000E+00-7.5000E-01 2.0000E+00            0
  6            0            0            0            0            0            0
  7            0            0 7.5000E-01-7.5000E-01 1.0000E+00            0
  8            0            0-5.6250E-01 1.1875E+00-7.5000E-01            0
  9            0            0 7.5000E-01-5.3291E-15 1.0000E+00            0
 10            0            0-1.8750E-01 1.1875E+00-7.5000E-01            0
               7            8            9           10
  1            0            0            0            0
  2            0            0            0            0
  3  7.5000E-01-5.6250E-01 7.5000E-01-1.8750E-01
  4 -7.5000E-01 1.1875E+00-3.5527E-15 1.1875E+00
  5  1.0000E+00-7.5000E-01 1.0000E+00-7.5000E-01
  6            0            0            0            0
  7  1.4306E+00-6.5625E-01 9.3056E-01-4.6875E-01
  8 -6.5625E-01 1.4297E+00-2.8125E-01 6.6406E-01
  9  9.3056E-01-2.8125E-01 1.4306E+00-9.3750E-02
 10 -4.6875E-01 6.6406E-01-9.3750E-02 1.4297E+00


PRINT   FRQ          2

THIS IS THE FLEXIBILITY SUBMATRIX, FRQ,
WITH 3/EA(0) FACTORED OUT.

               1            2            3            4            5            6
  1            0            0-6.0921E-01-4.8224E-01-6.0922E-01            0
  2            0            0-1.8750E-05-4.8224E-01-6.0923E-01            0
               7            8            9           10
  1 -6.5998E-01-1.2670E-02-1.2692E+00-4.9493E-01
  2  6.5994E-01-4.9493E-01 5.0757E-02-1.2670E-02


PRINT   FQQ          2

THIS IS THE FLEXIBILITY SUBMATRIX, FQQ,
WITH 3/EA(0) FACTORED OUT.

               1            2
  1  3.0018E+00-4.1237E-02
  2 -4.1237E-02 3.0018E+00


SYMINV  FQQ

LOAD    R            10    1
   10 ROWS,      1 COLUMNS

MULT    FRQ     R          TEM1

MULT    FQQ     TEM1       Q

DELETE  TEM1

SCALE   Q            -1.00000
 SCALAR= -1.0000000E+00
PRINT   Q            1
```

****** *Computer Output (continued):*

```
THIS IS THE VECTOR OF REDUNDANT FORCES, Q.


            1
 1 -2.7133E+00
 2 -5.4455E-01


MULT    BR      R       PBAR

MULT    BQ      Q       TEM1

ADD     PBAR    TEM1

DELETE  TEM1

PRINT   PBAR        2


THIS IS THE VECTOR OF INTERNAL BASIC FORCES, PBAR,
FOR THE COLLECTION OF ALL MEMBERS AND CONSTRAINTS.


            1
 1  1.6530E+00
 2  2.8318E+00
 3 -6.9559E+00
 4  7.2323E+00
 5  5.2677E+00
 6 -3.8903E+00
 7 -3.0153E+00
 8 -2.7133E+00
 9 -5.4455E-01
10 -5.0000E+00
11 -6.6666E+00
12 -3.3334E+00


MULT    FRR     R       D

MULT    FQR     Q       TEM1

ADD     D       TEM1

DELETE  TEM1

PRINT   D           2


THIS IS THE VECTOR OF STRUCTURE NODAL DISPLACEMENTS,
WITH 3/EA(0) FACTORED OUT.


            1
 1           0
 2           0
 3  1.6530E+00
 4 -1.5616E+01
 5  4.4848E+00
 6           0
 7  1.7786E+00
 8 -8.2899E+00
 9 -1.2367E+00
10 -8.1814E+00

STOP
```

Notice that, as expected, this solution checks the displacement method solution of Sec. 4.4.1 satisfactorily. It is interesting at this point to further compare the force and displacement methods of analyzing statically indeterminate structures. A brief review of Secs. 4.3.2 and 4.4.1 will remind you that the displacement method was applied in exactly the same way to statically indeterminate structures as it was to statically determinate structures. The principal reason for this identical application is that the number of structure nodal displacements, the primary unknowns in the displacement method, is always equal to the number of structure stiffness equations available to solve for them. On the other hand, the force method of solution of statically indeterminate structures as outlined in this section differs in one important way from the force method of solution of statically determinate structures as described in Sec. 6.2.1. That is, the equilibrium equations of Eq. 6.12 had to be augmented with an appropriate number of elastic continuity equations—from the second equation of Eq. 6.19—before a solution for the primary unknowns could be completed. This augmentation was required because the primary unknowns for the force method are the set of basic forces in the structure that will exceed the number of nodal equilibrium equations available to solve for them by the number of redundant basic forces present in the structure.

Thus, in addition to the disadvantages already listed in Sec. 6.2.1, the further complications just mentioned seem to make the force method even less attractive than the displacement method. Nevertheless, there are structures that may be solved more efficiently (if, perhaps, not more conveniently) by the force method than by the displacement method.

For statically indeterminate structures there will, as already stated, always be more equations to solve in the force method than in the displacement method. In the force method, the equations requiring solution are in two groups—equilibrium equations and continuity equations—whereas in the displacement method, only stiffness equations are generated. If the number of redundant forces is small compared to the number of nodal degrees of freedom, the number of continuity equations required in the force method will be relatively small and the total number of equations in each method will be close to the same. Furthermore, although the equilibrium equations in the force method possess an unsymmetric coefficient matrix $\mathbf{a}_R$, their $\mathbf{a}_R$ array will be more sparsely populated than will the $\mathbf{K}$ array of the displacement method. The $\mathbf{a}_R$ array may also be numerically better conditioned for solution than the $\mathbf{K}$ array. Thus in some cases the equilibrium equations may be more easily solved in the force method than the stiffness equations in the displacement method. This savings may offset the penalty of having to solve continuity equations in the force method if the number of redundants is small, and the force method solution may therefore sometimes be obtained more efficiently than the displacement method solution. Which solution approach will be most efficient in a given case is best determined by experience.

Convenience is another matter. The choice of the most convenient method will depend mainly on the availability of computer programs and on the personal preference of the analyst.

It is now appropriate to consider the application of the force method to the solution of a statically indeterminate structure having distributed loadings.

6.3.2 Statically Indeterminate Beam

Equations 6.6 and 6.10 of Sec. 6.2.2, which were developed for the solution of statically determinate structures under distributed loadings, could now be extended to form the basis for the solution of statically indeterminate structures under distributed loadings. In actual application this may be the desired approach; however, for purposes of presenting the solution here, it is both conceptually and notationally clearer to revert to the superposition approach (Case I plus Case II) that led to the equations in Sec. 6.2.2.

Consider the continuous beam example of Fig. 4.8, which is shown again in Fig. 6.4.

Case I

The vector $\overline{\mathbf{P}}_F$ of basic fixed-end forces for the collection of members and constraints of the beam may be determined by inspection with the use of familiar equations.

Thus since Eq. 2.42 gives

$$\overline{\mathbf{P}}_{\mathrm{I}} = \overline{\mathbf{P}}_F \tag{2.42}$$

it follows that

$$\overline{\mathbf{P}}_{\mathrm{I}} = \langle 288 \quad -288 \quad 160 \quad -80 \quad 0 \quad 0 \quad 0 \quad 0 \quad 0 \quad 0 \quad 0 \quad 0 \rangle \tag{A}$$

The structure nodal fixing forces $\mathbf{R}_F$ may now be obtained from member and nodal equilibrium considerations. The result is

$$\mathbf{R}_F = \langle 72 \quad 288 \quad 98.67 \quad -128 \quad 9.33 \quad -80 \quad 0 \quad 0 \rangle \tag{B}$$

The solution to Case I is completed by noting that the nodal displacements are, by definition, all zero. Thus

$$\mathbf{D}_{\mathrm{I}} = \langle 0 \quad 0 \quad 0 \quad 0 \quad 0 \quad 0 \quad 0 \quad 0 \rangle \tag{C}$$

Case II

The vector $\mathbf{R}$ of nodal loads on the original structure may be determined by inspection as

$$\mathbf{R} = \langle 0 \quad 0 \quad 0 \quad 0 \quad 0 \quad -20 \quad 0 \quad 0 \rangle \tag{D}$$

And since

$$\mathbf{R}_{\mathrm{II}} = \mathbf{R} - \mathbf{R}_F \tag{2.48b}$$

it follows that

$$\mathbf{R}_{\mathrm{II}} = \langle -72 \quad -288 \quad -98.67 \quad 128 \quad -9.33 \quad 60 \quad 0 \quad 0 \rangle \tag{E}$$

Now that we have the Case II load vector, it is appropriate to develop the structure equilibrium equations for this case. We can do this by using the member equilibrium relationships and the support constraint conditions. Beam-member

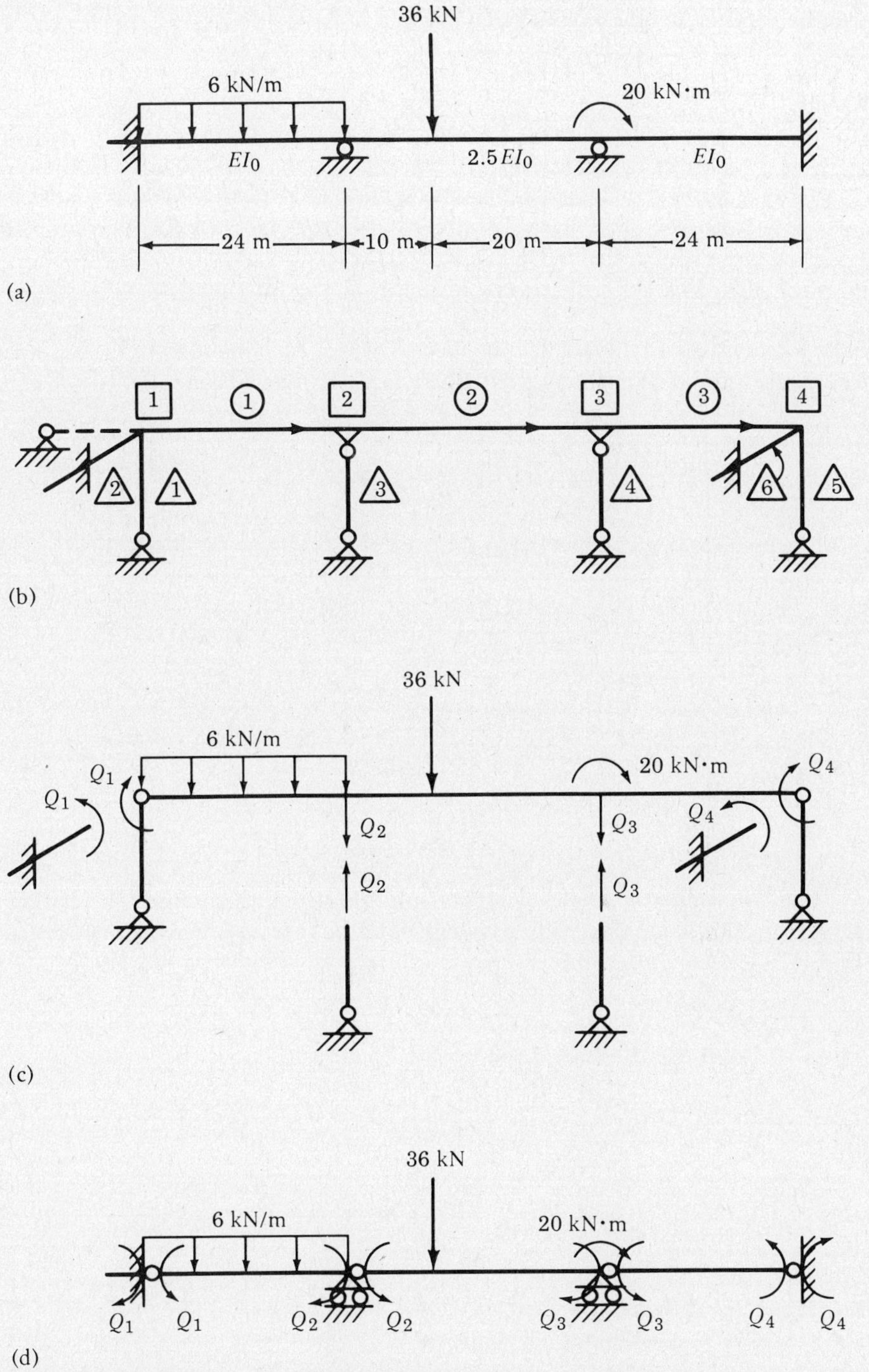

Fig. 6.4 Continuous beam: (a) dimensions and loading; (b) structural model; (c) cut-back structure; (d) automatically generated cut-back structure.

equilibrium relationships were given in Eq. 2.27a as

$$\begin{pmatrix} r_1 \\ r_2 \\ r_3 \\ r_4 \end{pmatrix} = \begin{bmatrix} 1/L & 1/L \\ 1 & 0 \\ -1/L & -1/L \\ 0 & 1 \end{bmatrix} \begin{pmatrix} M_i \\ M_j \end{pmatrix} \tag{2.27a}$$

or

$$\mathbf{r} = \mathbf{TP} \tag{2.27b}$$

The **T** arrays based on Eq. 2.27a are specialized for each member of the example beam and are shown here along with their respective nodal codes:

$$\mathbf{T}^{(1)} = \begin{matrix} \boxed{1} \\ \\ \boxed{2} \\ \\ \end{matrix} \overset{①}{\begin{bmatrix} .04167 & .04167 \\ 1 & 0 \\ -.04167 & -.04167 \\ 0 & 1 \end{bmatrix}} \qquad \mathbf{T}^{(2)} = \begin{matrix} \boxed{2} \\ \\ \boxed{3} \\ \\ \end{matrix} \overset{②}{\begin{bmatrix} .03333 & .03333 \\ 1 & 0 \\ -.03333 & -.03333 \\ 0 & 1 \end{bmatrix}}$$

$$\mathbf{T}^{(3)} = \begin{matrix} \boxed{3} \\ \\ \boxed{4} \\ \\ \end{matrix} \overset{③}{\begin{bmatrix} .04167 & .04167 \\ 1 & 0 \\ -.04167 & -.04167 \\ 0 & 1 \end{bmatrix}}$$

We can determine the effect of supports on the structure equilibrium relationships by inspection of Fig. 6.4b, which gives us the following constraint information.

Constraint List

Constraint	Node	Coordinate
1	1	1
2	1	2
3	2	1
4	3	1
5	4	1
6	4	2

On the basis of the member and support information that we now have, we can build the Case II structure equilibrium relationships shown in Eq. F on page 315. Eq. F may be represented symbolically as

$$\mathbf{R}_{II} = \mathbf{a}\overline{\mathbf{P}}_{II} \tag{6.23}$$

Choosing, again somewhat arbitrarily, the redundant basic forces as shown on the cutback structure of Fig. 6.4c and rearranging Eq. F so that the redundants may

$$\begin{pmatrix} -72 \\ -288 \\ -98.67 \\ 128 \\ -9.33 \\ 60 \\ 0 \\ 0 \end{pmatrix}_{\text{II}} = \begin{bmatrix} .04167 & .04167 & \cdot & \cdot & \cdot & \cdot & 1 & \cdot & \cdot & \cdot & \cdot & \cdot \\ 1 & 0 & \cdot & \cdot & \cdot & \cdot & \cdot & 1 & \cdot & \cdot & \cdot & \cdot \\ -.04167 & -.04167 & .03333 & .03333 & \cdot & \cdot & \cdot & \cdot & 1 & \cdot & \cdot & \cdot \\ 0 & 1 & 1 & 0 & \cdot & \cdot & \cdot & \cdot & \cdot & \cdot & \cdot & \cdot \\ \cdot & \cdot & -.03333 & -.03333 & .04167 & .04167 & \cdot & \cdot & \cdot & 1 & \cdot & \cdot \\ \cdot & \cdot & 0 & 1 & 1 & 0 & \cdot & \cdot & \cdot & \cdot & \cdot & \cdot \\ \cdot & \cdot & \cdot & \cdot & -.04167 & -.04167 & \cdot & \cdot & \cdot & \cdot & 1 & \cdot \\ \cdot & \cdot & \cdot & \cdot & 0 & 1 & \cdot & \cdot & \cdot & \cdot & \cdot & 1 \end{bmatrix} \begin{pmatrix} M_i^{(1)} \\ M_j^{(1)} \\ M_i^{(2)} \\ M_j^{(2)} \\ M_i^{(3)} \\ M_j^{(3)} \\ P_c^{(1)} \\ M_c^{(2)} \\ P_c^{(3)} \\ P_c^{(4)} \\ P_c^{(5)} \\ M_c^{(6)} \end{pmatrix}_{\text{II}} \qquad \text{(F)}$$

be partitioned from the remaining forces in the $\overline{\mathbf{P}}_{\text{II}}$ vector, we get Eq. F′ on page 317. Partitioned Eq. F′ may be symbolically represented as

$$\mathbf{R}_{\text{II}} = [\mathbf{a}_R \,|\, \mathbf{a}_Q] \begin{pmatrix} \overline{\mathbf{P}}_R \\ \hline \overline{\mathbf{P}}_Q \end{pmatrix}_{\text{II}} \tag{6.24}$$

Solving for $\overline{\mathbf{P}}_R$ in terms of the applied load $\mathbf{R}_{\text{II}}$ and the redundant forces $\overline{\mathbf{P}}_Q$ and rearranging gives

$$\overline{\mathbf{P}}_{R\text{II}} = \mathbf{a}_R^{-1}\mathbf{R}_{\text{II}} - \mathbf{a}_R^{-1}\mathbf{a}_Q\overline{\mathbf{P}}_{Q\text{II}} \tag{6.25}$$

Carrying out the solution represented by Eq. 6.25 in detail yields

$$\begin{pmatrix} M_i^{(1)} \\ M_j^{(1)} \\ M_i^{(2)} \\ M_j^{(2)} \\ M_i^{(3)} \\ M_j^{(3)} \\ P_c^{(1)} \\ P_c^{(5)} \end{pmatrix}_{\text{II}} = \begin{bmatrix} 0 & 1 & 0 & 0 & 0 & 0 & 0 & 0 \\ 0 & -.6923 & -16.62 & .3077 & -7.385 & .3077 & 0 & .3077 \\ 0 & .6923 & 16.62 & .6923 & 7.385 & -.3077 & 0 & -.3077 \\ 0 & -.3077 & -7.385 & -.3077 & -16.62 & .6923 & 0 & .6923 \\ 0 & .3077 & 7.385 & .3077 & 16.62 & .3077 & 0 & -.6923 \\ 0 & 0 & 0 & 0 & 0 & 0 & 0 & 1 \\ 1 & -.0128 & .6923 & -.0128 & .3077 & -.0128 & 0 & -.0128 \\ 0 & .0128 & .3077 & .0128 & .6923 & .0128 & 1 & .0128 \end{bmatrix}$$

$$\times \begin{pmatrix} -72 \\ -288 \\ -98.67 \\ 128 \\ -9.33 \\ 60 \\ 0 \\ 0 \end{pmatrix}_{\text{II}} - \begin{bmatrix} 1 & 0 & 0 & 0 \\ -.6923 & -16.62 & -7.385 & .3077 \\ .6923 & 16.62 & 7.385 & -.3077 \\ -.3077 & -7.385 & -16.62 & -.6923 \\ .3077 & 7.385 & 16.62 & -.6923 \\ 0 & 0 & 0 & 1 \\ -.0128 & .6923 & .3077 & -.0128 \\ .0128 & .3077 & .6923 & .0128 \end{bmatrix} \begin{pmatrix} M_c^{(2)} \\ P_c^{(3)} \\ P_c^{(4)} \\ M_c^{(6)} \end{pmatrix}_{\text{II}} \tag{G}$$

Augmenting this set of equations with the equations

$$M_c^{(1)} = Q_1 \quad P_c^{(2)} = Q_2 \quad P_c^{(3)} = Q_3 \quad M_c^{(4)} = Q_4 \tag{H}$$

and rearranging the vector on the left-hand side to obtain $\overline{\mathbf{P}}_{\text{II}}$, we get Eq. I on page 118. Partitioning Eq. I as shown,

$$\overline{\mathbf{P}}_{\text{II}} = [\mathbf{b}_R \,|\, \mathbf{b}_Q] \begin{pmatrix} \mathbf{R} \\ \hline \mathbf{Q} \end{pmatrix}_{\text{II}} \tag{6.26}$$

Having the $\mathbf{b}_b$ and $\mathbf{b}_Q$ arrays, we can now appropriately consider applying Eqs. 6.20, 6.21b, and 6.22a to the solution of Case II. In this context these equations

$$
\begin{pmatrix} -72 \\ -288 \\ -98.67 \\ 128 \\ -9.33 \\ 60 \\ 0 \\ 0 \end{pmatrix}_{\text{II}} = \left[\begin{array}{cccccccc|cccc}
.04167 & .04167 & \cdot & \cdot & \cdot & \cdot & 1 & \cdot & \cdot & \cdot & \cdot & \cdot \\
1 & 0 & \cdot & \cdot & \cdot & \cdot & \cdot & \cdot & 1 & \cdot & \cdot & \cdot \\
-.04167 & -.04167 & .03333 & .03333 & \cdot & \cdot & \cdot & \cdot & \cdot & 1 & \cdot & \cdot \\
0 & 1 & 1 & 0 & \cdot & \cdot & \cdot & \cdot & \cdot & \cdot & \cdot & \cdot \\
\cdot & \cdot & -.03333 & -.03333 & .04167 & .04167 & \cdot & \cdot & \cdot & \cdot & 1 & \cdot \\
\cdot & \cdot & 0 & 1 & 1 & 0 & \cdot & \cdot & \cdot & \cdot & \cdot & \cdot \\
\cdot & \cdot & \cdot & \cdot & -.04167 & -.04167 & \cdot & 1 & \cdot & \cdot & \cdot & \cdot \\
\cdot & \cdot & \cdot & \cdot & 0 & 1 & \cdot & \cdot & \cdot & \cdot & \cdot & 1
\end{array}\right] \begin{pmatrix} M_i^{(1)} \\ M_j^{(1)} \\ M_i^{(2)} \\ M_j^{(2)} \\ M_i^{(3)} \\ M_j^{(3)} \\ P_c^{(1)} \\ P_c^{(5)} \\ \hdashline M_c^{(2)} \\ P_c^{(3)} \\ P_c^{(4)} \\ M_c^{(6)} \end{pmatrix}_{\text{II}} \qquad \text{(F}'\text{)}
$$

$$
\begin{pmatrix} M_i^{(1)} \\ M_j^{(1)} \\ M_i^{(2)} \\ M_j^{(2)} \\ M_i^{(3)} \\ M_j^{(3)} \\ P_c^{(1)} \\ M_c^{(2)} \\ P_c^{(3)} \\ P_c^{(4)} \\ P_c^{(5)} \\ M_c^{(6)} \end{pmatrix}_{\text{II}} = \left[\begin{array}{cccccccc|cccc}
0 & 1 & 0 & 0 & 0 & 0 & 0 & 0 & -1 & 0 & 0 & 0 \\
0 & -.6923 & -16.62 & .3077 & -7.385 & .3077 & 0 & .3077 & .6923 & 16.62 & 7.385 & -.3077 \\
0 & .6923 & 16.62 & .6923 & 7.385 & -.3077 & 0 & -.3077 & -.6923 & -16.62 & -7.385 & .3077 \\
0 & -.3077 & -7.385 & -.3077 & -16.62 & .6923 & 0 & .6923 & .3077 & 7.385 & 16.62 & -.6923 \\
0 & .3077 & 7.385 & .3077 & 16.62 & .3077 & 0 & -.6923 & -.3077 & -7.385 & -16.62 & .6923 \\
0 & 0 & 0 & 0 & 0 & 0 & 0 & 1 & 0 & 0 & 0 & -1 \\
1 & -.0128 & .6923 & -.0128 & .3077 & -.0128 & 0 & -.0128 & .0128 & -.6923 & -.3077 & .0128 \\
0 & 0 & 0 & 0 & 0 & 0 & 0 & 0 & 1 & 0 & 0 & 0 \\
0 & 0 & 0 & 0 & 0 & 0 & 0 & 0 & 0 & 1 & 0 & 0 \\
0 & 0 & 0 & 0 & 0 & 0 & 0 & 0 & 0 & 0 & 1 & 0 \\
0 & -.0128 & .3077 & .0128 & .6923 & .0128 & 1 & .0128 & -.0128 & -.3077 & -.6923 & -.0128 \\
0 & 0 & 0 & 0 & 0 & 0 & 0 & 0 & 0 & 0 & 0 & 1
\end{array}\right] \begin{pmatrix} -72 \\ -288 \\ -98.67 \\ 128 \\ -9.33 \\ 60 \\ 0 \\ 0 \\ \hline Q_1 \\ Q_2 \\ Q_3 \\ Q_4 \end{pmatrix}_{\text{II}} \qquad \text{(I)}
$$

may be written, respectively, as

$$\mathbf{Q}_{II} = -\mathbf{F}_{QQ}^{-1}\mathbf{F}_{QR}\mathbf{R}_{II} \tag{6.27}$$

$$\overline{\mathbf{P}}_{II} = [\mathbf{b}_R - \mathbf{b}_Q\mathbf{F}_{QQ}^{-1}\mathbf{F}_{QR}]\mathbf{R}_{II} \tag{6.28}$$

$$\mathbf{D}_{II} = [\mathbf{F}_{RR} - \mathbf{F}_{RQ}\mathbf{F}_{QQ}^{-1}\mathbf{F}_{QR}]\mathbf{R}_{II} \tag{6.29}$$

where

$$\begin{aligned} \mathbf{F}_{RR} &= \mathbf{b}_R^T \, f \, \mathbf{b}_R \\ \mathbf{F}_{QR} &= \mathbf{F}_{RQ}^T = \mathbf{b}_Q^T \, f \, \mathbf{b}_R \\ \mathbf{F}_{QQ} &= \mathbf{b}_Q^T \, f \, \mathbf{b}_Q \end{aligned} \tag{6.17}$$

To complete this Case II solution, we must first obtain the basic flexibility matrix for the collection of members and constraints of the beam. Equation 3.16 states the basic flexibility force-displacement relationships for a beam member as

$$\begin{pmatrix} \phi_i \\ \phi_j \end{pmatrix} = \frac{L}{6EI}\begin{bmatrix} 2 & -1 \\ -1 & 2 \end{bmatrix}\begin{pmatrix} M_i \\ M_j \end{pmatrix} \tag{3.16}$$

or

$$\Delta = f\mathbf{P} \tag{3.16a}$$

Specializing this relationship for the individual members of the example structure gives for the flexibility arrays

$$f^{(1)} = f^{(3)} = \frac{4}{EI_0}\begin{bmatrix} 2 & -1 \\ -1 & 2 \end{bmatrix} \qquad f^{(2)} = \frac{5}{2.5EI_0}\begin{bmatrix} 2 & -1 \\ -1 & 2 \end{bmatrix} = \frac{4}{EI_0}\begin{bmatrix} 1 & -.5 \\ -.5 & .1 \end{bmatrix}$$

All support flexibility coefficients are, of course, zero. Collecting these results for all members and supports gives

$$\begin{pmatrix} \phi_i^{(1)} \\ \phi_j^{(1)} \\ \phi_i^{(2)} \\ \phi_j^{(2)} \\ \phi_i^{(3)} \\ \phi_j^{(3)} \\ \Delta_c^{(1)} \\ \phi_c^{(2)} \\ \Delta_c^{(3)} \\ \Delta_c^{(4)} \\ \Delta_c^{(5)} \\ \phi_c^{(6)} \end{pmatrix}_{II} = \frac{4}{EI_0}\begin{bmatrix} 2 & -1 & \cdot & \cdot & \cdot & \cdot & \cdot & \cdot & \cdot & \cdot & \cdot & \cdot \\ -1 & 2 & \cdot & \cdot & \cdot & \cdot & \cdot & \cdot & \cdot & \cdot & \cdot & \cdot \\ \cdot & \cdot & 1 & -.5 & \cdot & \cdot & \cdot & \cdot & \cdot & \cdot & \cdot & \cdot \\ \cdot & \cdot & -.5 & 1 & \cdot & \cdot & \cdot & \cdot & \cdot & \cdot & \cdot & \cdot \\ \cdot & \cdot & \cdot & \cdot & 2 & -1 & \cdot & \cdot & \cdot & \cdot & \cdot & \cdot \\ \cdot & \cdot & \cdot & \cdot & -1 & 2 & \cdot & \cdot & \cdot & \cdot & \cdot & \cdot \\ \cdot & \cdot & \cdot & \cdot & \cdot & \cdot & 0 & \cdot & \cdot & \cdot & \cdot & \cdot \\ \cdot & \cdot & \cdot & \cdot & \cdot & \cdot & \cdot & 0 & \cdot & \cdot & \cdot & \cdot \\ \cdot & \cdot & \cdot & \cdot & \cdot & \cdot & \cdot & \cdot & 0 & \cdot & \cdot & \cdot \\ \cdot & \cdot & \cdot & \cdot & \cdot & \cdot & \cdot & \cdot & \cdot & 0 & \cdot & \cdot \\ \cdot & \cdot & \cdot & \cdot & \cdot & \cdot & \cdot & \cdot & \cdot & \cdot & 0 & \cdot \\ \cdot & \cdot & \cdot & \cdot & \cdot & \cdot & \cdot & \cdot & \cdot & \cdot & \cdot & 0 \end{bmatrix}\begin{pmatrix} M_i^{(1)} \\ M_j^{(1)} \\ M_i^{(2)} \\ M_j^{(2)} \\ M_i^{(3)} \\ M_j^{(3)} \\ P_c^{(1)} \\ M_c^{(2)} \\ P_c^{(3)} \\ P_c^{(4)} \\ P_c^{(5)} \\ M_c^{(6)} \end{pmatrix}_{II} \tag{J}$$

which, of course, may be represented symbolically as

$$\overline{\Delta}_{II} = \overline{\mathscr{f}}\,\overline{\mathbf{P}}_{II} \tag{6.30}$$

The structure flexibility submatrices required to obtain a detailed Case II solution for the example may now be obtained by substituting appropriately from Eqs. I and J into Eq. 6.17, giving

$$\mathbf{F}_{RR} = \frac{4}{EI_0}\begin{bmatrix} 0 & 0 & 0 & 0 & 0 & 0 & 0 & 0 \\ 0 & 5.32 & 63.0 & .243 & 44.9 & -1.26 & 0 & -2.18 \\ 0 & 63.0 & 1114. & 13.2 & 901. & -22.7 & 0 & -44.9 \\ 0 & .243 & 13.2 & 1.16 & 22.8 & .334 & 0 & -1.26 \\ 0 & 44.9 & 901. & 22.8 & 1114. & -13.2 & 0 & -63.0 \\ 0 & -1.26 & -22.7 & .334 & -13.2 & 1.16 & 0 & .243 \\ 0 & 0 & 0 & 0 & 0 & 0 & 0 & 0 \\ 0 & -2.18 & -44.9 & -1.26 & -63.0 & .243 & 0 & 5.32 \end{bmatrix}$$

$$\mathbf{F}_{QR} = \frac{4}{EI_0}\begin{bmatrix} 0 & 5.32 & 63.0 & .243 & 44.9 & -1.26 & 0 & -2.18 \\ 0 & 63.0 & 1114. & 13.2 & 901. & -22.8 & 0 & -44.9 \\ 0 & 44.9 & 901. & 22.8 & 1114. & -13.2 & 0 & -63.0 \\ 0 & -2.18 & -44.9 & -1.26 & -63.0 & .243 & 0 & 5.32 \end{bmatrix} \tag{K}$$

$$\mathbf{F}_{QQ} = \frac{4}{EI_0}\begin{bmatrix} 5.32 & 63.0 & 44.9 & -2.18 \\ 63.0 & 1114. & 901. & -44.9 \\ 44.9 & 901. & 1114. & -63.0 \\ -2.18 & -44.9 & -63.0 & 5.32 \end{bmatrix}$$

Substituting from Eqs. E and K into Eqs. 6.27, 6.28, and 6.29 gives the Case II solution for redundant forces, basic forces, and nodal displacements, respectively. The detailed calculations required to accomplish these results have been carried out and are shown in the next printout, giving

$$\mathbf{Q}_{II} = \langle -308.2 \quad -100.8 \quad -5.037 \quad -3.250 \rangle \tag{L}$$

$$\overline{\mathbf{P}}_{II} = \langle 20.25 \quad 40.5 \quad 87.5 \quad 53.5 \quad 6.5 \quad 3.25 \quad -74.53 \quad -308.2 \quad -100.8 \quad -5.037 \quad .4063 \quad -3.25 \rangle \tag{M}$$

and

$$\mathbf{D}_{II} = \frac{4}{EI_0}\langle 0 \quad 0 \quad 0 \quad 60.75 \quad 0 \quad 9.75 \quad 0 \quad 0 \rangle \tag{N}$$

****** *Computer Input:*

```
START
REMARK 6
SOLUTION FOR THE INTERNAL FORCES AND NODAL DISPLACEMENTS OF THE
STATICALLY INDETERMINATE BEAM, FORCE METHOD, ARBITRARY REDUNDANTS

**** UNITS ARE KILONEWTONS AND METERS ****

CASE I SOLUTION
LOAD PBARI 12 1
288 -288 160 -80 0 0 0 0 0 0 0 0
PRINT PBARI 1
THIS IS THE VECTOR OF BASIC FIXED-END FORCES, PBARI, CASE I
REMARK 1
CASE II SOLUTION
LOAD RII 8 1
-72 -288 -98.67 128 -9.33 60 0 0
PRINT RII 1
THIS IS THE CASE II LOAD VECTOR, RII
LOAD AR 8 8
.04167 .04167 4*0 1 0 1 7*0 -.04167 -.04167 .03333 .03333 5*0 1 1 7*0
-.03333 -.03333  .04167 .04167 5*0 1 1 7*0 -.04167 -.04167 0 1 5*0 1 0 0
PRINT AR 1
THIS IS THE EQUILIBRIUM SUBMATRIX, AR
LOAD AQ 8 4
4*0 1 4*0 1 8*0 1 12*0 1
PRINT AQ 1
THIS IS THE EQUILIBRIUM SUBMATRIX, AQ
INVERT AR
MULT AR AQ TEM1
SCALE TEM1 -1
FORM BR 12 8
FORM BQ 12 4
RMVSM AR TEM2 1 1 7 8
RMVSM AR TEM3 8 1 1 8
STOSM BR TEM2 1 1
STOSM BR TEM3 11 1
PRINT BR 1
THIS IS THE SUBMATRIX, BR
DELETE TEM2
DELETE TEM3
RMVSM TEM1 TEM2 1 1 7 4
RMVSM TEM1 TEM3 8 1 1 4
DELETE TEM1
LOAD TEM4 5 4
1 4*0 1 4*0 1 8*0 1
STOSM BQ TEM4 8 1
STOSM BQ TEM3 11 1
STOSM BQ TEM2 1 1
PRINT BQ 1
THIS IS THE SUBMATRIX, BQ
DELETE TEM2
DELETE TEM3
DELETE TEM4
LOAD FBAR 12 12
2 -1 10*0 -1 2 12*0 1 -.5 10*0 -.5 1 12*0 2 -1 10*0 -1 2 78*0
PRINT FBAR 1
THIS IS THE FLEXIBILITY SUPERMATRIX, FBAR, WITH 4/EI(0) FACTORED OUT
MULT FBAR BR TEM
TRANS BR BRT
MULT BRT TEM FRR
PRINT FRR 1
THIS IS THE FLEXIBILITY SUBMATRIX, FRR, WITH 4/EI(0) FACTORED OUT
TRANS BQ BQT
MULT BQT TEM FQR
PRINT FQR 1
THIS IS THE FLEXIBILITY SUBMATRIX, FQR, WITH 4/EI(0) FACTORED OUT
DELETE TEM
```

****** *Computer Input (continued):*

```
MULT FBAR BQ TEM
MULT BQT TEM FQQ
PRINT FQQ 1
THIS IS THE FLEXIBILITY SUBMATRIX, FQQ, WITH 4/EI(0) FACTORED OUT
SYMINV FQQ
SCALE FQQ -1
MULT FQR RII TEM
MULT FQQ TEM QII
DELETE TEM
PRINT QII 1
THIS IS THE VECTOR OF REDUNDANT FORCES, QII, CASE II
MULT BQ QII TEM
MULT BR RII PBARII
ADD PBARII TEM
DELETE TEM
PRINT PBARII 1
THIS IS THE VECTOR OF INTERNAL BASIC FORCES, PBARII, CASE II
TRANS FQR FRQ
MULT FRQ QII TEM
MULT FRR RII DII
ADD DII TEM
PRINT DII 1
THIS IS THE VECTOR OF STRUCTURE NODAL DISPLACEMENTS, D, TOTAL SOLUTION
ADD PBARI PBARII
PRINT PBARI 1
THIS IS THE VECTOR OF INTERNAL BASIC FORCES, PBAR, TOTAL SOLUTION
STOP
```

****** *Computer Output:*

```
START

REMARK      6


SOLUTION FOR THE INTERNAL FORCES AND NODAL DISPLACEMENTS OF THE
STATICALLY INDETERMINATE BEAM, FORCE METHOD, ARBITRARY REDUNDANTS

**** UNITS ARE KILONEWTONS AND METERS ****

CASE I SOLUTION


LOAD    PBARI       12    1
   12 ROWS,      1 COLUMNS

PRINT   PBARI        1


THIS IS THE VECTOR OF BASIC FIXED-END FORCES, PBARI, CASE I


             1
 1  2.8800E+02
 2 -2.8800E+02
 3  1.6000E+02
 4 -8.0000E+01
 5           0
 6           0
 7           0
 8           0
 9           0
10           0
11           0
12           0
```

****** *Computer Output (continued):*

```
REMARK       1

CASE II SOLUTION


LOAD    RII          8    1
    8 ROWS,      1 COLUMNS

PRINT   RII          1

THIS IS THE CASE II LOAD VECTOR, RII

            1
 1 -7.2000E+01
 2 -2.8800E+02
 3 -9.8670E+01
 4  1.2800E+02
 5 -9.3300E+00
 6  6.0000E+01
 7           0
 8           0


LOAD    AR           8    8
    8 ROWS,      8 COLUMNS

PRINT   AR           1

THIS IS THE EQUILIBRIUM SUBMATRIX, AR

            1           2           3           4           5           6
 1  4.1670E-02 4.1670E-02           0           0           0           0
 2  1.0000E+00           0           0           0           0           0
 3 -4.1670E-02-4.1670E-02 3.3330E-02 3.3330E-02           0           0
 4           0 1.0000E+00 1.0000E+00           0           0           0
 5           0           0-3.3330E-02-3.3330E-02 4.1670E-02 4.1670E-02
 6           0           0           0 1.0000E+00 1.0000E+00           0
 7           0           0           0           0-4.1670E-02-4.1670E-02
 8           0           0           0           0           0 1.0000E+00
            7           8
 1  1.0000E+00           0
 2           0           0
 3           0           0
 4           0           0
 5           0           0
 6           0           0
 7           0 1.0000E+00
 8           0           0


LOAD    AQ           8    4
    8 ROWS,      4 COLUMNS

PRINT   AQ           1

THIS IS THE EQUILIBRIUM SUBMATRIX, AQ
```

****** *Computer Output (continued):*

```
             1            2            3            4
 1           0            0            0            0
 2  1.0000E+00            0            0            0
 3           0   1.0000E+00            0            0
 4           0            0            0            0
 5           0            0   1.0000E+00            0
 6           0            0            0            0
 7           0            0            0            0
 8           0            0            0   1.0000E+00

INVERT   AR

MULT     AR       AQ       TEM1

SCALE    TEM1        -1.00000
 SCALAR= -1.0000000E+00

FORM     BR           12    8
   12 ROWS,      8 COLUMNS

FORM     BQ           12    4
   12 ROWS,      4 COLUMNS

RMVSM    AR       TEM2         1     1     7     8
 ROW NUMBER    1,  COLUMN NUMBER    1
    7 ROWS,      8 COLUMNS

RMVSM    AR       TEM3         8     1     1     8
 ROW NUMBER    8,  COLUMN NUMBER    1
    1 ROWS,      8 COLUMNS

STOSM    BR       TEM2         1     1
 ROW NUMBER    1,  COLUMN NUMBER    1

STOSM    BR       TEM3        11     1
 ROW NUMBER   11,  COLUMN NUMBER    1

PRINT    BR           1


THIS IS THE SUBMATRIX, BR


             1           2           3           4           5           6
 1           0 1.0000E+00           0           0           0           0
 2           0-6.9233E-01-1.6615E+01 3.0767E-01-7.3835E+00 3.0767E-01
 3           0 6.9233E-01 1.6615E+01 6.9233E-01 7.3835E+00-3.0767E-01
 4           0-3.0767E-01-7.3835E+00-3.0767E-01-1.6615E+01 6.9233E-01
 5           0 3.0767E-01 7.3835E+00 3.0767E-01 1.6615E+01 3.0767E-01
 6           0           0           0           0           0           0
 7  1.0000E+00-1.2821E-02 6.9233E-01-1.2821E-02 3.0767E-01-1.2821E-02
 8           0           0           0           0           0           0
 9           0           0           0           0           0           0
10           0           0           0           0           0           0
11           0 1.2821E-02 3.0767E-01 1.2821E-02 6.9233E-01 1.2821E-02
12           0           0           0           0           0           0
             7           8
 1           0           0
 2           0 3.0767E-01
 3           0-3.0767E-01
 4           0 6.9233E-01
 5           0-6.9233E-01
 6           0 1.0000E+00
 7           0-1.2821E-02
 8           0           0
 9           0           0
10           0           0
11  1.0000E+00 1.2821E-02
12           0           0
```

****** *Computer Output (continued):*

```
DELETE   TEM2

DELETE   TEM3

RMVSM    TEM1     TEM2         1    1    7    4
 ROW NUMBER   1,  COLUMN NUMBER   1
    7 ROWS,     4 COLUMNS

RMVSM    TEM1     TEM3         8    1    1    4
 ROW NUMBER   8,  COLUMN NUMBER   1
    1 ROWS,     4 COLUMNS

DELETE   TEM1

LOAD     TEM4          5    4
    5 ROWS,     4 COLUMNS

STOSM    BQ       TEM4         8    1
 ROW NUMBER   8,  COLUMN NUMBER   1

STOSM    BQ       TEM3        11    1
 ROW NUMBER  11,  COLUMN NUMBER   1

STOSM    BQ       TEM2         1    1
 ROW NUMBER   1,  COLUMN NUMBER   1

PRINT    BQ            1


THIS IS THE SUBMATRIX, BQ


             1           2           3           4
 1 -1.0000E+00           0           0           0
 2  6.9233E-01 1.6615E+01 7.3835E+00-3.0767E-01
 3 -6.9233E-01-1.6615E+01-7.3835E+00 3.0767E-01
 4  3.0767E-01 7.3835E+00 1.6615E+01-6.9233E-01
 5 -3.0767E-01-7.3835E+00-1.6615E+01 6.9233E-01
 6           0           0           0-1.0000E+00
 7  1.2821E-02-6.9233E-01-3.0767E-01 1.2821E-02
 8  1.0000E+00           0           0           0
 9           0 1.0000E+00           0           0
10           0           0 1.0000E+00           0
11 -1.2821E-02-3.0767E-01-6.9233E-01-1.2821E-02
12           0           0           0 1.0000E+00


DELETE   TEM2

DELETE   TEM3

DELETE   TEM4

LOAD     FBAR         12   12
   12 ROWS,    12 COLUMNS

PRINT    FBAR          1


THIS IS THE FLEXIBILITY SUPERMATRIX, FBAR, WITH 4/EI(0) FACTORED OUT
```

****** *Computer Output (continued):*

```
             1           2           3           4           5           6
  1  2.0000E+00-1.0000E+00           0           0           0           0
  2 -1.0000E+00 2.0000E+00           0           0           0           0
  3           0           0 1.0000E+00-5.0000E-01           0           0
  4           0           0-5.0000E-01 1.0000E+00           0           0
  5           0           0           0           0 2.0000E+00-1.0000E+00
  6           0           0           0           0-1.0000E+00 2.0000E+00
  7           0           0           0           0           0           0
  8           0           0           0           0           0           0
  9           0           0           0           0           0           0
 10           0           0           0           0           0           0
 11           0           0           0           0           0           0
 12           0           0           0           0           0           0
             7           8           9          10          11          12
  1           0           0           0           0           0           0
  2           0           0           0           0           0           0
  3           0           0           0           0           0           0
  4           0           0           0           0           0           0
  5           0           0           0           0           0           0
  6           0           0           0           0           0           0
  7           0           0           0           0           0           0
  8           0           0           0           0           0           0
  9           0           0           0           0           0           0
 10           0           0           0           0           0           0
 11           0           0           0           0           0           0
 12           0           0           0           0           0           0


MULT     FBAR     BR       TEM

TRANS    BR       BRT

MULT     BRT      TEM      FRR

PRINT    FRR          1

THIS IS THE FLEXIBILITY SUBMATRIX, FRR, WITH 4/EI(0) FACTORED OUT


             1           2           3           4           5           6
  1           0           0           0           0           0           0
  2           0 5.3196E+00 6.3050E+01 2.4262E-01 4.4942E+01-1.2574E+00
  3           0 6.3050E+01 1.1144E+03 1.3206E+01 9.0132E+02-2.2791E+01
  4           0 2.4262E-01 1.3206E+01 1.1656E+00 2.2791E+01-3.3436E-01
  5           0 4.4942E+01 9.0132E+02 2.2791E+01 1.1144E+03-1.3206E+01
  6           0-1.2574E+00-2.2791E+01-3.3436E-01-1.3206E+01 1.1656E+00
  7           0           0           0           0           0           0
  8           0-2.1804E+00-4.4942E+01-1.2574E+00-6.3050E+01 2.4262E-01
             7           8
  1           0           0
  2           0-2.1804E+00
  3           0-4.4942E+01
  4           0-1.2574E+00
  5           0-6.3050E+01
  6           0 2.4262E-01
  7           0           0
  8           0 5.3196E+00


TRANS    BQ       BQT

MULT     BQT      TEM      FQR

PRINT    FQR          1
```

****** *Computer Output (continued):*

```
THIS IS THE FLEXIBILITY SUBMATRIX, FQR, WITH 4/EI(0) FACTORED OUT

            1          2          3          4          5          6
 1           0-5.3196E+00-6.3050E+01-2.4262E-01-4.4942E+01 1.2574E+00
 2           0-6.3050E+01-1.1144E+03-1.3206E+01-9.0132E+02 2.2791E+01
 3           0-4.4942E+01-9.0132E+02-2.2791E+01-1.1144E+03 1.3206E+01
 4           0 2.1804E+00 4.4942E+01 1.2574E+00 6.3050E+01-2.4262E-01
            7          8
 1           0 2.1804E+00
 2           0 4.4942E+01
 3           0 6.3050E+01
 4           0-5.3196E+00


DELETE   TEM

MULT     FBAR     BQ       TEM

MULT     BQT      TEM      FQQ

PRINT    FQQ            1

THIS IS THE FLEXIBILITY SUBMATRIX, FQQ, WITH 4/EI(0) FACTORED OUT

            1          2          3          4
 1  5.3196E+00 6.3050E+01 4.4942E+01-2.1804E+00
 2  6.3050E+01 1.1144E+03 9.0132E+02-4.4942E+01
 3  4.4942E+01 9.0132E+02 1.1144E+03-6.3050E+01
 4 -2.1804E+00-4.4942E+01-6.3050E+01 5.3196E+00


SYMINV   FQQ

SCALE    FQQ          -1.00000
 SCALAR= -1.0000000E+00

MULT     FQR      RII      TEM

MULT     FQQ      TEM      QII

DELETE   TEM

PRINT    QII            1

THIS IS THE VECTOR OF REDUNDANT FORCES, QII, CASE II

            1
 1 -3.0825E+02
 2 -1.0084E+02
 3 -5.0368E+00
 4 -3.2500E+00


MULT     BQ       QII      TEM

MULT     BR       RII      PBARII

ADD      PBARII   TEM

DELETE   TEM

PRINT    PBARII         1
```

****** *Computer Output (continued):*

```
THIS IS THE VECTOR OF INTERNAL BASIC FORCES, PBARII, CASE II

             1
 1   2.0250E+01
 2   4.0500E+01
 3   8.7500E+01
 4   5.3500E+01
 5   6.5000E+00
 6   3.2500E+00
 7  -7.4531E+01
 8  -3.0825E+02
 9  -1.0084E+02
10  -5.0368E+00
11   4.0628E-01
12  -3.2500E+00

TRANS    FQR      FRQ

MULT     FRQ      QII      TEM

MULT     FRR      RII      DII

ADD      DII      TEM

PRINT    DII         1

THIS IS THE VECTOR OF STRUCTURE NODAL DISPLACEMENTS, D, TOTAL SOLUTION

             1
 1            0
 2  -3.4925E-10
 3  -8.3819E-09
 4   6.0750E+01
 5  -6.5193E-09
 6   9.7500E+00
 7            0
 8   4.0745E-10

ADD      PBARI    PBARII

PRINT    PBARI       1

THIS IS THE VECTOR OF INTERNAL BASIC FORCES, PBAR, TOTAL SOLUTION

             1
 1   3.0825E+02
 2  -2.4750E+02
 3   2.4750E+02
 4  -2.6500E+01
 5   6.5000E+00
 6   3.2500E+00
 7  -7.4531E+01
 8  -3.0825E+02
 9  -1.0084E+02
10  -5.0368E+00
11   4.0628E-01
12  -3.2500E+00

STOP
```

TOTAL SOLUTION

Finally, by superposing the results of the Case I and Case II solutions, the total solution to the original example is obtained as

$$\overline{\mathbf{P}} = \overline{\mathbf{P}}_{\mathrm{I}} + \overline{\mathbf{P}}_{\mathrm{II}} = [\mathbf{b}_R - \mathbf{b}_Q\mathbf{F}_{QQ}^{-1}\mathbf{F}_{QR}](\mathbf{R} - \mathbf{R}_F) + \overline{\mathbf{P}}_F \tag{6.31}$$

or

$$\overline{\mathbf{P}} = \langle 308.2 \quad -247.5 \quad 247.5 \quad -26.5 \quad 6.5 \quad 3.25 \quad -74.53 \quad -308.2 \quad -100.8 \quad -5.037 \quad .4063 \quad -3.25 \rangle \tag{O}$$

and

$$\mathbf{D} = \mathbf{D}_{\mathrm{I}} + \mathbf{D}_{\mathrm{II}} = [\mathbf{F}_{RR} - \mathbf{F}_{RQ}\mathbf{F}_{QQ}^{-1}\mathbf{F}_{QR}](\mathbf{R} - \mathbf{R}_F) \tag{6.32}$$

or

$$\mathbf{D} = \frac{4}{EI_0}\langle 0 \quad 0 \quad 0 \quad 60.75 \quad 0 \quad 9.75 \quad 0 \quad 0 \rangle \tag{P}$$

These results are seen to check the results of the displacement method solution shown on pages 138–140.

In carrying out the solutions to both the statically indeterminate truss of Sec. 6.3.1 and the continuous beam of this section, we chose the redundants somewhat arbitrarily (see Figs. 6.3c and 6.4c). It is apparent that many alternative sets of forces could have been selected as redundants. It is, however, perhaps somewhat less obvious that one of these sets would be computationally superior to all of the others for each example. The reason for this may be seen by recalling that the flexibility solution for statically indeterminate structures requires the solution of two sets of simultaneous equations, the coefficients of which are represented by two matrices: $\mathbf{a}_R$ of Eq. 6.13 or 6.25 and $\mathbf{F}_{QQ}$ of Eq. 6.20 or 6.27. It is easily seen that the character of each of these arrays will be strongly influenced by the choice of redundant forces. For computational accuracy, it is of course desirable that each of these arrays be numerically well-conditioned. One attribute of a well-conditioned array is a strong main diagonal; that is, diagonal terms should be significantly larger than off-diagonal terms, or it should be possible to interchange rows and columns of the array to make them so. Inspection of both $\mathbf{a}_R$ and $\mathbf{F}_{QQ}$ for the example truss shows that each diagonal is, or can be made, quite strong. In the case of the beam, however, each diagonal is in fact very weak. Thus our choice of redundants was computationally good in the case of the truss and poor in the case of the beam.

The preceding discussion gives rise to two critically important questions: First, is there some way of choosing redundant forces so as to guarantee that both the $\mathbf{a}_R$ matrix and the $\mathbf{F}_{QQ}$ matrix will be as well-conditioned as possible for the given structure? Second, can this be accomplished in a manner suitable for automatic computation? Fortunately, the answer to both these questions is yes. For purposes of clarity, we will consider first obtaining a well-conditioned $\mathbf{a}_R$ matrix without regard to the $\mathbf{F}_{QQ}$ matrix. We will then consider a modified approach that will ensure a well-conditioned $\mathbf{F}_{QQ}$ matrix as well. Both methods will be developed for automatic computation.

6.3.3 Structure-Cutter Method

In actual practice, the matrix $\mathbf{a}_R$ is never inverted; instead, the scalar equations of matrix Eq. 6.13 or 6.25 are solved simultaneously for $\overline{\mathbf{P}}_R$ in terms of **R** and **Q** by using, say, a gaussian elimination technique.* Well-conditioned equations are obtained by employing either complete or partial pivoting during elimination. Complete pivoting is accomplished by searching the entire **a** matrix before each elimination step to determine the largest remaining coefficient, which may then be moved to the desired diagonal position by appropriately interchanging rows and/or columns of **a** (and **R** and $\overline{\mathbf{P}}$). Partial pivoting involves a search of only the row currently being used in the elimination process to find the largest coefficient in that row, which is then moved to the diagonal position by a column interchange in **a** (and a row interchange in $\overline{\mathbf{P}}$). Complete pivoting will yield a better-conditioned $\mathbf{a}_R$ matrix but will be relatively more difficult to program and more time-consuming to perform. Partial pivoting will usually yield a reasonably well-conditioned $\mathbf{a}_R$ matrix, and it has the relative advantages of simplicity of programming and efficiency of execution.

Using either pivoting technique, we see that the redundants will be automatically determined as those basic forces with the smaller coefficients in the **a** matrix. This method of selecting redundants has been called the structure-cutter method (Ref. 6.1). To demonstrate this method, consider again the Case II solution for the continuous beam example, this time using gaussian elimination with partial pivoting.

Equation F of Sec. 6.3.2 is rewritten as Eq. A on page 331 for convenience. Column numbers for array **a** have been added to this equation to aid in keeping track of interchanges. (The Case II subscripting is assumed to be understood here and is dropped.)

Columns of the **a** array are now interchanged to produce the strongest possible diagonal for the $\mathbf{a}_R$ array. For reasons that will be explained shortly, whenever there is an arbitrary choice of columns, one of which is associated with a support force, the interchange will always be prejudiced in favor of moving the column associated with the support force into the $\mathbf{a}_R$ array. This gives Eq. A′ on page 332. The $\mathbf{a}_R$, $\mathbf{a}_Q$, $\overline{\mathbf{P}}_R$, and $\overline{\mathbf{P}}_Q$ arrays have now been systematically determined for this example, and it is seen that the vector of redundant forces is

$$\mathbf{Q} \equiv \overline{\mathbf{P}}_Q = \langle M_i^{(1)} \quad M_i^{(2)} \quad M_i^{(3)} \quad M_j^{(3)} \rangle \tag{B}$$

The cutback structure generated by this automatic choice of redundants is shown in Fig. 6.4d.

It should be mentioned here that it is not typical to be able to obtain $\mathbf{a}_R$ in upper triangular form by interchanging columns (or even both columns and rows). Normally, an elimination or decomposition step would be required along with the pivoting process to produce an upper triangular form of $\mathbf{a}_R$. This procedure was not necessary in this example because of the special geometry of the beam. (A nonrectangular structure such as the truss of Sec. 6.3.1 would require the more complete treatment.) In this case we need only to first rearrange Eq. A′, letting $\mathbf{P}_Q \rightarrow \mathbf{Q}$, as shown in Eq. C, on page 333.

*See App. 1 and Chap. 7.

$$
\begin{bmatrix}
\scriptstyle 1 & \scriptstyle 2 & \scriptstyle 3 & \scriptstyle 4 & \scriptstyle 5 & \scriptstyle 6 & \scriptstyle 7 & \scriptstyle 8 & \scriptstyle 9 & \scriptstyle 10 & \scriptstyle 11 & \scriptstyle 12 \\
.04167 & .04167 & \cdot & \cdot & \cdot & \cdot & 1 & \cdot & \cdot & \cdot & \cdot & \cdot \\
1 & 0 & \cdot & \cdot & \cdot & \cdot & \cdot & 1 & \cdot & \cdot & \cdot & \cdot \\
-.04167 & -.04167 & .03333 & .03333 & \cdot & \cdot & \cdot & \cdot & 1 & \cdot & \cdot & \cdot \\
0 & 1 & 1 & 0 & \cdot & \cdot & \cdot & \cdot & \cdot & \cdot & \cdot & \cdot \\
\cdot & \cdot & -.03333 & -.03333 & .04167 & .04167 & \cdot & \cdot & \cdot & 1 & \cdot & \cdot \\
\cdot & \cdot & 0 & 1 & 1 & 0 & \cdot & \cdot & \cdot & \cdot & \cdot & \cdot \\
\cdot & \cdot & \cdot & \cdot & -.04167 & -.04167 & \cdot & \cdot & \cdot & \cdot & 1 & \cdot \\
\cdot & \cdot & \cdot & \cdot & 0 & 1 & \cdot & \cdot & \cdot & \cdot & \cdot & 1
\end{bmatrix}
\begin{pmatrix}
M_i^{(1)} \\ M_j^{(1)} \\ M_i^{(2)} \\ M_j^{(2)} \\ M_i^{(3)} \\ M_j^{(3)} \\ P_c^{(1)} \\ M_c^{(2)} \\ P_c^{(3)} \\ P_c^{(4)} \\ P_c^{(5)} \\ M_c^{(6)}
\end{pmatrix}
=
\begin{pmatrix}
-72 \\ -228 \\ -98.67 \\ 128 \\ -9.33 \\ .60 \\ 0 \\ 0
\end{pmatrix}
\qquad \text{(A)}
$$

a (arrow pointing to the 0 in row 8, column 5)

$$
\begin{array}{c}
\begin{array}{cccccccc|cccc}
{\scriptstyle 7} & {\scriptstyle 8} & {\scriptstyle 9} & {\scriptstyle 2} & {\scriptstyle 10} & {\scriptstyle 4} & {\scriptstyle 11} & {\scriptstyle 12} & {\scriptstyle 1} & {\scriptstyle 3} & {\scriptstyle 5} & {\scriptstyle 6} \\
1 & \cdot & \cdot & .04167 & \cdot & \cdot & \cdot & \cdot & .04167 & \cdot & \cdot & \cdot \\
\cdot & 1 & \cdot & 0 & \cdot & \cdot & \cdot & \cdot & 1 & \cdot & \cdot & \cdot \\
\cdot & \cdot & 1 & -.04167 & \cdot & .03333 & \cdot & \cdot & -.04167 & .03333 & \cdot & \cdot \\
\cdot & \cdot & \cdot & 1 & \cdot & 0 & \cdot & \cdot & 0 & 1 & \cdot & \cdot \\
\cdot & \cdot & \cdot & \cdot & 1 & -.03333 & \cdot & \cdot & \cdot & -.03333 & .04167 & .04167 \\
\cdot & \cdot & \cdot & \cdot & \cdot & 1 & \cdot & \cdot & \cdot & 0 & 1 & 0 \\
\cdot & \cdot & \cdot & \cdot & \cdot & \cdot & 1 & \cdot & \cdot & \cdot & -.04167 & -.04167 \\
\cdot & \cdot & \cdot & \cdot & \cdot & \underset{\uparrow \, \mathbf{a}_R}{\cdot} & \cdot & 1 & \cdot & \cdot & \underset{\uparrow \, \mathbf{a}_Q}{0} & 1
\end{array}
\end{array}
\begin{pmatrix} P_c^{(1)} \\ M_c^{(2)} \\ P_c^{(3)} \\ M_j^{(1)} \\ P_c^{(4)} \\ M_j^{(2)} \\ P_c^{(5)} \\ M_c^{(6)} \\ \hline M_i^{(1)} \\ M_i^{(2)} \\ M_i^{(3)} \\ M_j^{(3)} \end{pmatrix}
=
\begin{pmatrix} -72 \\ -228 \\ -98.67 \\ 128 \\ -9.33 \\ 60 \\ 0 \\ 0 \end{pmatrix}
\qquad \text{(A}'\text{)}
$$

(Arrows: $\overline{\mathbf{P}}_R$ points to the upper part of the vector, from $P_c^{(1)}$; $\overline{\mathbf{P}}_Q$ points to the lower part, ending at $M_j^{(3)}$.)

$$\begin{bmatrix} 1 & \cdot & \cdot & .04167 & \cdot & \cdot & \cdot & \cdot \\ \cdot & 1 & \cdot & 0 & \cdot & \cdot & \cdot & \cdot \\ \cdot & \cdot & 1 & -.04167 & \cdot & .03333 & \cdot & \cdot \\ \cdot & \cdot & \cdot & 1 & \cdot & 0 & \cdot & \cdot \\ \cdot & \cdot & \cdot & \cdot & 1 & -.03333 & \cdot & \cdot \\ \cdot & \cdot & \cdot & \cdot & \cdot & 1 & \cdot & \cdot \\ \cdot & \cdot & \cdot & \cdot & \cdot & \cdot & 1 & \cdot \\ \cdot & \cdot & \cdot & \cdot & \cdot & \cdot & \cdot & 1 \end{bmatrix} \begin{pmatrix} P_c^{(1)} \\ M_c^{(2)} \\ P_c^{(3)} \\ M_j^{(1)} \\ P_c^{(4)} \\ M_j^{(2)} \\ P_c^{(5)} \\ M_c^{(6)} \end{pmatrix} = \begin{pmatrix} -72 \\ -228 \\ -98.67 \\ 128 \\ -9.33 \\ 60 \\ 0 \\ 0 \end{pmatrix} = \begin{bmatrix} .04167 & \cdot & \cdot & \cdot \\ 1 & \cdot & \cdot & \cdot \\ -.04167 & .03333 & \cdot & \cdot \\ 0 & 1 & \cdot & \cdot \\ \cdot & -.03333 & .04167 & .04167 \\ \cdot & 0 & 1 & 0 \\ \cdot & \cdot & -.04167 & -.04167 \\ \cdot & \cdot & 0 & 1 \end{bmatrix} \begin{pmatrix} Q_1 \\ Q_2 \\ Q_3 \\ Q_4 \end{pmatrix} \qquad \text{(C)}$$

and then to carry out a backward substitution process, starting from the bottom row and working upward a row at a time. The result is

$$\begin{pmatrix} P_c^{(1)} \\ M_c^{(2)} \\ P_c^{(3)} \\ M_j^{(1)} \\ P_c^{(4)} \\ M_j^{(2)} \\ P_c^{(5)} \\ M_c^{(6)} \end{pmatrix} = \begin{pmatrix} -77.33 \\ -228 \\ -95.34 \\ 128 \\ -7.33 \\ 60 \\ 0 \\ 0 \end{pmatrix} + \begin{bmatrix} -.04167 & .04167 & \cdot & \cdot \\ -1 & \cdot & \cdot & \cdot \\ .04167 & -.075 & .03333 & \cdot \\ 0 & -1 & \cdot & \cdot \\ \cdot & .03333 & -.075 & -.04167 \\ \cdot & 0 & -1 & 0 \\ \cdot & \cdot & .04167 & .04167 \\ \cdot & \cdot & 0 & 1 \end{bmatrix} \begin{pmatrix} Q_1 \\ Q_2 \\ Q_3 \\ Q_4 \end{pmatrix} \qquad \text{(D)}$$

Rearranging the vector of basic forces on the left-hand side while augmenting it appropriately to yield $\overline{\mathbf{P}}$ gives

$$\begin{pmatrix} M_i^{(1)} \\ M_j^{(1)} \\ M_i^{(2)} \\ M_j^{(2)} \\ M_i^{(3)} \\ M_j^{(3)} \\ P_c^{(1)} \\ M_c^{(2)} \\ P_c^{(3)} \\ P_c^{(4)} \\ P_c^{(5)} \\ M_c^{(6)} \end{pmatrix} = \begin{pmatrix} 0 \\ 128 \\ 0 \\ 60 \\ 0 \\ 0 \\ -77.33 \\ -228 \\ -95.34 \\ -7.33 \\ 0 \\ 0 \end{pmatrix} + \begin{bmatrix} 1 & \cdot & \cdot & \cdot \\ 0 & -1 & \cdot & \cdot \\ \cdot & 1 & \cdot & \cdot \\ \cdot & 0 & -1 & 0 \\ \cdot & \cdot & 1 & \cdot \\ \cdot & \cdot & \cdot & 1 \\ -.04167 & .04167 & \cdot & \cdot \\ -1 & \cdot & \cdot & \cdot \\ .04167 & -.075 & .03333 & \cdot \\ \cdot & .03333 & -.075 & -.04167 \\ \cdot & \cdot & .04167 & .04167 \\ \cdot & \cdot & 0 & -1 \end{bmatrix} \begin{pmatrix} Q_1 \\ Q_2 \\ Q_3 \\ Q_4 \end{pmatrix} \qquad \text{(E)}$$

or

$$\overline{\mathbf{P}} = \mathbf{b}_R\mathbf{R} + \mathbf{b}_Q\mathbf{Q} \qquad (6.26)$$

where $\mathbf{b}_Q$ and the product $\mathbf{b}_R\mathbf{R}$ are shown in detail in Eq. E. (If you want to determine $\mathbf{b}_R$ separately, you can follow the same procedure, letting $\mathbf{R} \rightarrow \mathbf{IR}$.) Recalling, $\mathbf{\not{f}}$ from Eq. J on page 319 gives

$$\mathbf{b}_Q^T\overline{\mathbf{f}} = \frac{4}{EI_0}\begin{bmatrix} 2 & -1 & \cdot & \cdot & \cdot & \cdot & \cdot & \cdot & \cdot & \cdot & \cdot & \cdot \\ 1 & -2 & 1 & -.5 & \cdot & \cdot & \cdot & \cdot & \cdot & \cdot & \cdot & \cdot \\ \cdot & \cdot & .5 & -1 & 2 & -1 & \cdot & \cdot & \cdot & \cdot & \cdot & \cdot \\ \cdot & \cdot & \cdot & \cdot & -1 & 2 & \cdot & \cdot & \cdot & \cdot & \cdot & \cdot \end{bmatrix}$$

$$\mathbf{F}_{QQ} = \mathbf{b}_Q^T\overline{\mathbf{f}}\,\mathbf{b}_Q = \frac{4}{EI_0}\begin{bmatrix} 2 & 1 & \cdot & \cdot \\ 1 & 3 & .5 & \cdot \\ \cdot & .5 & 3 & -1 \\ \cdot & \cdot & -1 & 2 \end{bmatrix} \tag{F}$$

$$\mathbf{F}_{QR}\mathbf{R} = \mathbf{b}_Q^T\overline{\mathbf{f}}\,\mathbf{b}_R\mathbf{R} = \frac{4}{EI_0}\langle -128 \quad -286 \quad -60 \quad 0\rangle$$

Substituting from Eq. F into Eq. 6.27 gives

$$\mathbf{Q} = \begin{pmatrix} M_i^{(1)} \\ M_i^{(2)} \\ M_i^{(3)} \\ M_j^{(3)} \end{pmatrix} = -\begin{bmatrix} .60417 & -.20833 & .041666 & .020834 \\ -.20833 & .41666 & -.083333 & -.041666 \\ .041667 & -.083333 & .41666 & .20834 \\ .020834 & -.041666 & .20834 & .60417 \end{bmatrix}\begin{pmatrix} -128 \\ -286 \\ -60 \\ 0 \end{pmatrix} = \begin{pmatrix} 20.25 \\ 87.5 \\ 6.5 \\ 3.25 \end{pmatrix} \tag{G}$$

which is seen to agree with the previous solution for redundant forces shown in Eq. L of Sec. 6.3.2. It is clear that further backsubstitution from Eqs. E and F into Eq. 6.28 would produce results identical with those obtained previously and displayed in Eq. M of Sec. 6.3.2.

Notice that although a conscious effort was made to obtain only an $\mathbf{a}_R$ matrix with a strong main diagonal, the structure-cutter method also yielded an $\mathbf{F}_{QQ}$ matrix with a strong main diagonal. This result was not entirely accidental. As you may have noticed from the last expression of Eq. K in Sec. 6.3.2, an ill-conditioned $\mathbf{F}_{QQ}$ matrix will result from choosing redundant forces associated with very flexible released degrees of freedom; and in this case the second and third released degrees of freedom are extremely flexible. Therefore, a unit force even in the direction of either of the stiffer first and fourth released degrees of freedom will cause a relatively large response in these other two more flexible ones. Thus ill-conditioning of $\mathbf{F}_{QQ}$ is due to a choice of redundants that require releasing too much structural stiffness resulting in a relatively flexible cut-back structure.

How may the release of structural stiffness be minimized? To answer this question, it is first necessary to answer another more fundamental question: What properties of a structure contribute to its stiffness? The answer to this last question is structure geometry and individual member stiffness. The influence of structure geometry on structure stiffness is indicated by the relative sizes of the force influence coefficients in the a matrix; thus, if we retain larger terms in $\mathbf{a}_R$ and relegate smaller terms to $\mathbf{a}_Q$ (and choose $\mathbf{Q}$ on this basis), relatively little geometric stiffness will be released and a cutback structure possessing relatively high geometric stiffness will be developed. On the other hand, individual member stiffness is not indicated in the $\mathbf{a}$ matrix but rather in the f matrices. To ensure a well-conditioned $\mathbf{F}_{QQ}$ matrix, it is therefore necessary to modify the pivoting technique to take into account relative member stiffness. This modification may be accomplished by weighting the $\mathbf{a}$ coefficients (for pivoting purposes only and not for elimination purposes) by a factor equal to the reciprocal of the ratio of their corresponding flexibilities (direct ratio of their corresponding stiffnesses). The effect is to make the $\mathbf{a}$ coefficients associated with stiff members appear to be larger (for pivoting purposes), which biases the pivoting process toward retaining these coefficients in $\mathbf{a}_R$. The releases associated with $\mathbf{Q}$ will therefore occur in more flexible members, thus, once again, making for the retention of more stiffness in the cutback structure. Obviously, forces associated with supports will never be chosen as redundants following this approach, which explains our previous bias in this respect. It may be seen that this approach will result in the development of a well-conditioned $\mathbf{F}_{QQ}$ matrix at the expense of achieving an $\mathbf{a}_R$ matrix that is somewhat less than optimally conditioned. The best-conditioned $\mathbf{a}_R$ matrix would result if no weighting were used. Experience shows, however, that this compromise is beneficial.

How would this weighted approach have affected the structure-cutter solution of the continuous beam? Recall that in this example the pivoting process was biased in favor of excluding supports as redundants without specifying why. Other relative member stiffnesses were, however, not considered in carrying out the pivoting process. If this had been done, $M_j^{(2)}$ would have been chosen as a redundant force instead of $M_i^{(3)}$. But this choice would not have altered $\mathbf{a}_R$ significantly or $\mathbf{F}_{QQ}$ at all, since the release associated with $M_i^{(3)}$ is, from a structure stiffness point of view, precisely the same as the release associated with $M_j^{(2)}$.

You may wish to verify that the structure-cutter method applied to the truss of Sec. 6.3.1 will lead to exactly the same choice of redundants as were selected arbitrarily in that solution. This situation explains why $\mathbf{a}_R$ and $\mathbf{F}_{QQ}$ were, as a matter of good fortune, so well-conditioned in that example.

In concluding, it is well to notice that the structure-cutter method provides a systematic method of selecting the "best" set of redundant forces. The force method is therefore more attractive for automatic computation than would otherwise be the case.

6.4 SUMMARY—FORCE METHOD

Individual Member

Hooke's law, internal member equilibrium and compatibility	$\boldsymbol{\Delta} = f\mathbf{P}$	(3.5a, etc.)
Member overall equilibrium	$\mathbf{r} = \mathbf{T}\mathbf{P}$	(2.10b)
Member kinematics	$\boldsymbol{\Delta} = \mathbf{T}^T\mathbf{d}$	(2.11b)

Individual Support

Special case of Eq. 3.5a	$\Delta_c = f_c P_c$	(3.20)
Special case of Eq. 2.10b	$r_c = P_c$	(2.12)
Special case of Eq. 2.11b	$\Delta_c = d_c$	(2.13)

Collection of All Members and Supports

Based on Eqs. 3.5a and 3.20	$\bar{\mathbf{\Delta}} = \bar{\boldsymbol{f}}\,\bar{\mathbf{P}}$	(6.3)
Based on Eqs. 2.10b and 2.12	$\bar{\mathbf{r}} = \bar{\mathbf{T}}\bar{\mathbf{P}}$	(2.15)
Based on Eqs. 2.11b and 2.13	$\bar{\mathbf{\Delta}} = \bar{\mathbf{T}}^T\bar{\mathbf{d}}$	(2.21)

Structure

Nodal equilibrium	$\mathbf{R} = \mathbf{A}\bar{\mathbf{r}}$	(2.16)
Combining Eqs. 2.15 and 2.16	$\mathbf{R} = \mathbf{A}\bar{\mathbf{T}}\bar{\mathbf{P}}$	(2.17)
$\mathbf{a} = \mathbf{A}\bar{\mathbf{T}}$	$\mathbf{R} = \mathbf{a}\bar{\mathbf{P}}$	(2.18)
Rearranging Eq. 2.18, $\bar{\mathbf{P}}_Q$ = redundants	$\mathbf{R} = \mathbf{a}_R\bar{\mathbf{P}}_R + \mathbf{a}_Q\bar{\mathbf{P}}_Q$	(6.12a)
Solution of Eq. 6.12a	$\bar{\mathbf{P}}_R = \mathbf{a}_R^{-1}\mathbf{R} - \mathbf{a}_R^{-1}\,\mathbf{a}_Q\bar{\mathbf{P}}_Q$	(6.13)
Augmenting Eq. 6.13 with $\bar{\mathbf{P}}_Q = \mathbf{Q}$ and rearranging	$\bar{\mathbf{P}} = [\mathbf{b}_R \mid \mathbf{b}_Q]\left(\begin{array}{c}\mathbf{R}\\ \hline \mathbf{Q}\end{array}\right)$	(6.14)
Nodal compatibility	$\bar{\mathbf{d}} = \mathbf{A}^T\mathbf{D}$	(2.22)
Combining Eqs. 2.21 and 2.22	$\bar{\mathbf{\Delta}} = \bar{\mathbf{T}}^T\mathbf{A}^T\mathbf{D}$	(2.23)
$\mathbf{a}^T = \bar{\mathbf{T}}^T\mathbf{A}^T$	$\bar{\mathbf{\Delta}} = \mathbf{a}^T\mathbf{D}$	(2.24)
Manipulation of Eq. 2.24; analagous to obtaining Eq. 6.14 from Eq. 2.18	$\left(\begin{array}{c}\mathbf{D}\\ \hline \mathbf{D}_Q\end{array}\right) = \left[\begin{array}{c}\mathbf{b}_R^T\\ \hline \mathbf{b}_Q^T\end{array}\right]\bar{\mathbf{\Delta}}$	(6.15)
Combining Eqs. 6.3, 6.14, and 6.15	$\left(\begin{array}{c}\mathbf{D}\\ \hline \mathbf{D}_Q\end{array}\right) = \left[\begin{array}{c}\mathbf{b}_R^T\\ \hline \mathbf{b}_Q^T\end{array}\right]\bar{\boldsymbol{f}}\,[\mathbf{b}_R \mid \mathbf{b}_Q]\left(\begin{array}{c}\mathbf{R}\\ \hline \mathbf{Q}\end{array}\right)$	(6.16)
$\mathbf{F}_{RR} = \mathbf{b}_R^T\,\bar{\boldsymbol{f}}\mathbf{b}_R, \ldots$ (6.17)	$\left(\begin{array}{c}\mathbf{D}\\ \hline \mathbf{D}_Q\end{array}\right) = \left[\begin{array}{c\|c}\mathbf{F}_{RR} & \mathbf{F}_{RQ}\\ \hline \mathbf{F}_{QR} & \mathbf{F}_{QQ}\end{array}\right]\left(\begin{array}{c}\mathbf{R}\\ \hline \mathbf{Q}\end{array}\right)$	(6.16b)
Expansion of Eq. 6.16b	$\mathbf{D} = \mathbf{F}_{RR}\mathbf{R} + \mathbf{F}_{RQ}\mathbf{Q}$	(6.19)
Noting $\mathbf{D}_Q \equiv \mathbf{\Theta}$	$\mathbf{\Theta} = \mathbf{F}_{QR}\mathbf{R} + \mathbf{F}_{QQ}\mathbf{Q}$	
Solution of second equation of Eq. 6.19	$\mathbf{Q} = -\,\mathbf{F}_{QQ}^{-1}\,\mathbf{F}_{QR}\mathbf{R}$	(6.20)
Backsubstituting Eq. 6.20 into 6.14	$\bar{\mathbf{P}} = [\mathbf{b}_R - \mathbf{b}_Q\mathbf{F}_{QQ}^{-1}\mathbf{F}_{QR}]\mathbf{R}$	(6.21b)
Backsubstituting Eq. 6.20 into second equation of Eq. 6.19	$\mathbf{D} = [\mathbf{F}_{RR} - \mathbf{F}_{RQ}\mathbf{F}_{QQ}^{-1}\mathbf{F}_{QR}]\mathbf{R}$	(6.22a)

It should be noted that the solution summary just presented is for a structure subjected to nodal loading only. If a structure is subjected also to distributed loading, then the summary represents a Case II solution onto which the Case I fixed-end solution must be superposed as follows:

$$\overline{\mathbf{P}} = [\mathbf{b}_R - \mathbf{b}_Q\mathbf{F}_{QQ}^{-1}\mathbf{F}_{QR}](\mathbf{R} - \mathbf{R}_F) + \overline{\mathbf{P}}_F \tag{6.31}$$

$$\mathbf{D} = [\mathbf{F}_{RR} - \mathbf{F}_{RQ}\mathbf{F}_{QQ}^{-1}\mathbf{F}_{QR}](\mathbf{R} - \mathbf{R}_F) \tag{6.32}$$

PROBLEMS

6.1 Solve for the basic forces for all members and constraints of the example beam of Fig. P6.1. Also solve for the nodal displacements. Use the force method, taking $1/3EI_0$ as a factor. SMIS3 may be used for matrix manipulation as desired.

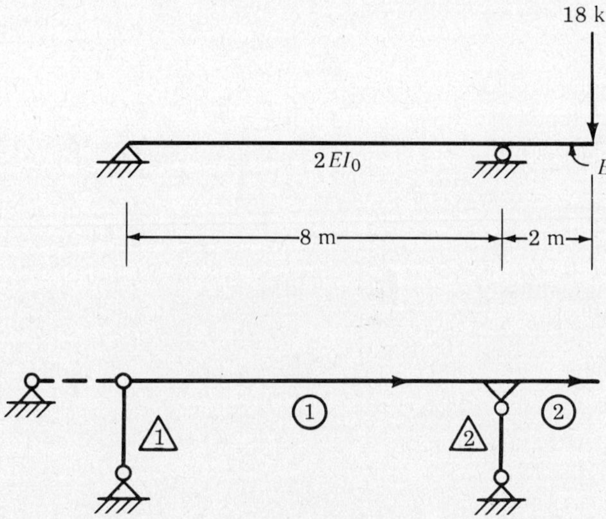

Fig. P6.1

6.2 Following the instructions of Prob. 6.1, consider the beam of Fig. P6.1 subjected to a uniformly distributed load of intensity 9 kN/m. This situation is seen to be identical with that of the structure and loading of Fig. P2.7.

6.3 Solve for the basic forces for all members and constraints of the example beam of Fig. P6.3 Also solve for the nodal displacements. Use the force method, taking constraint force 3 as the redundant force. SMIS3 may be used for matrix manipulation with $1/3EI_0$ taken as a factor.

6.4 Following the instructions of Prob. 6.3, consider the example beam of Fig. P6.3 subjected to a uniformly distributed load of intensity 9 kN/m instead of the concentrated load. This situation is seen to be identical with the structure and loading of Fig. P4.7. Take the basic force in support member 3 as the redundant force.

6.5 Work Prob. 6.3 using the structure-cutter method with partial pivoting to select the redundant force instead of choosing the redundant force as described in Prob. 6.3

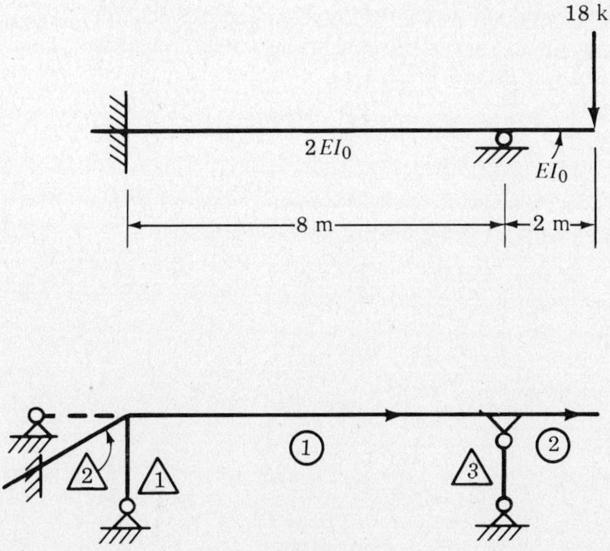

Fig. P6.3

6.6 Work Prob. 6.4 using the structure-cutter method with partial pivoting to select the redundant force instead of choosing the redundant force as described in Prob. 6.4.

6.7 Solve for the member basic forces and constraint forces for the example beam of Fig. P6.7. SMIS3 may be used for matrix manipulation with $1/EI_0$ taken as a factor. Select redundants by the structure-cutter method using partial pivoting and weight functions. Take $W_1 = 1$ and $W_{\text{supports}} = 2 \times$ next highest W.

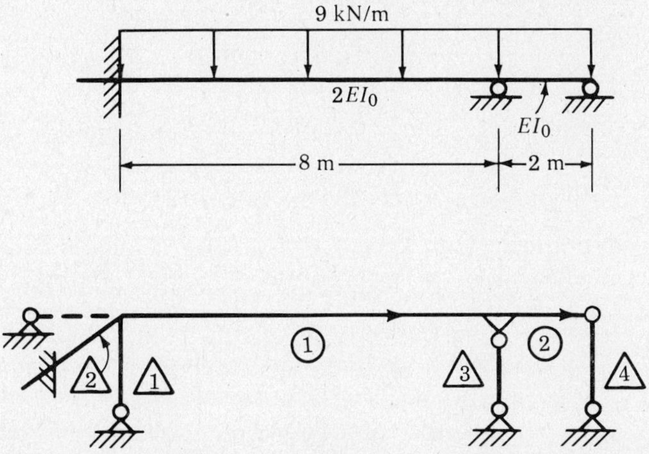

Fig. P6.7

6.8 Work the truss example of Fig. 6.3, selecting the redundant forces by the structure-cutter method with partial pivoting. Use weight functions taking $W_1 = 1$ and $W_{\text{supports}} = 2 \times$ next highest W.

REFERENCE

6.1 Denke, P. H. "A General Digital Computer Analysis of Statically Indeterminate Structures." *NASA Technical Note D-1666*, 1962.

Chapter 7
Solution of Equations

7.1 INTRODUCTION

We have seen that sets of simultaneous algebraic equations are generated in applications of both the displacement and the force methods of structural analysis. Furthermore, since the structures that we considered were linear structures, the equations generated were also linear. Up to now we have carried out solutions of these sets of equations, for the most part, with very little discussion of the detailed steps required. Before embarking on such a discussion now, it is important to put current solution capabilities in some historical perspective. Back in the days (not so long ago) when the mechanical desk calculator was king, the rule of thumb for estimating the time required to solve a general set of simultaneous equations was

$$t = .001\, N^4 \tag{A}$$

where t is time in *hours* and N is the number of equations to be solved (Ref. 7.1). Thus, to solve a set of 10 simultaneous equations, it would have taken 10 hours, whereas today, with the use of a modern digital computer, these same calculations would take a fraction of a second. This astounding increase in computational capability has allowed the structural engineer to routinely solve problems of a magnitude that would have been unthinkable just a few years ago. Thus although each example considered in this text involves only a small number of simultaneous equations, in practice it is not unusual to encounter structures each of whose analyses will result in the generation of hundreds or even thousands of equations. It is still true, therefore, that the solution of the set of equations generated will often be the single most time-consuming, most space-consuming, and therefore most expensive computational step in the analysis. It has been estimated (Refs. 7.2 and 7.3) that the solution of equations consumes 20 to 50 percent of the time required to solve linear static structural analysis problems and as much as 80 percent of the time required for the solution of linear dynamic or nonlinear static problems. It is for this reason that this solution step is singled out for special attention at this point. Because of the overwhelming importance of the displacement method of analysis, we will consider only equations generated by the direct stiffness method in this chapter.

Sets of simultaneous equations may be solved by either direct or iterative methods. Experience has shown that for most structural applications, direct methods of solution are superior to iterative methods (Ref. 7.4); we will therefore discuss only the former. Of the many direct methods available, gaussian elimination is the most widely used because of its broad capabilities and because there is no other method that requires fewer arithmetic operations in its application to the solution of a general set of equations (Ref. 7.4).

7.2 GAUSSIAN ELIMINATION

Gaussian elimination is discussed briefly in App. 1 and was used in the solution of force method equilibrium equations in Sec. 6.3.3. For completeness, this elimination technique will be presented again here in connection with the solution of a set of equations arising from an application of the displacement method. It will be shown that special simplifying rules can be adopted for applying gaussian elimination in this context, the result being significant economies.

For this purpose, consider the frame structure of Fig. 7.1. Specializing Eqs. 4.9a and 4.9b, with $2EI_0/L$ equal to unity, gives for the stiffness matrices of the girders and columns, respectively,

$$\mathbf{k}^{(\text{girder})} = \begin{bmatrix} A/I_0 & 0 & 0 & -A/I_0 & 0 & 0 \\ 0 & 12/L^2 & 6/L & 0 & -12/L^2 & 6/L \\ 0 & 6/L & 4 & 0 & -6/L & 2 \\ -A/I_0 & 0 & 0 & A/I_0 & 0 & 0 \\ 0 & -12/L^2 & -6/L & 0 & 12/L^2 & -6/L \\ 0 & 6/L & 2 & 0 & -6/L & 4 \end{bmatrix}$$

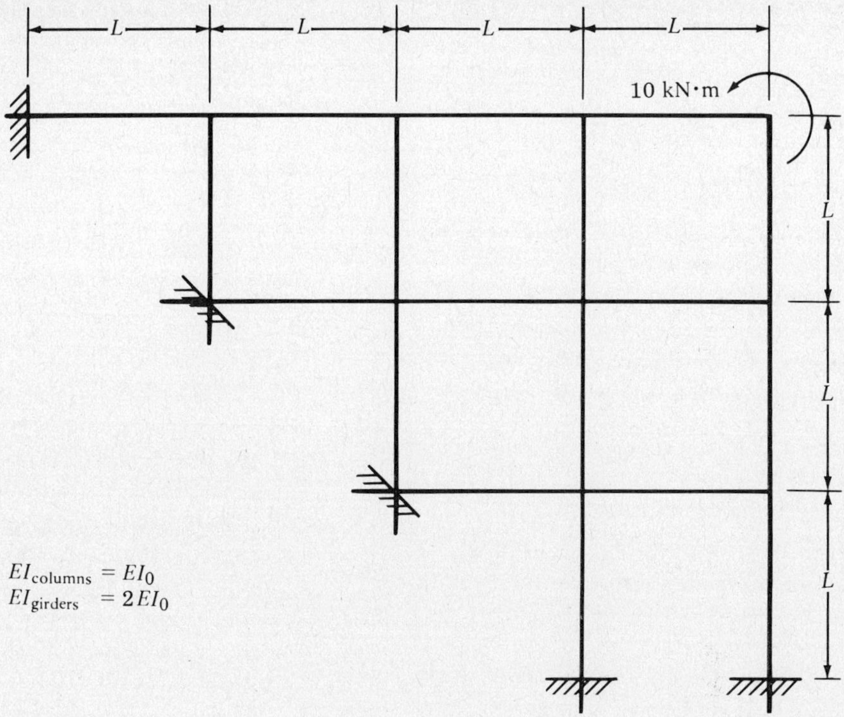

(a)

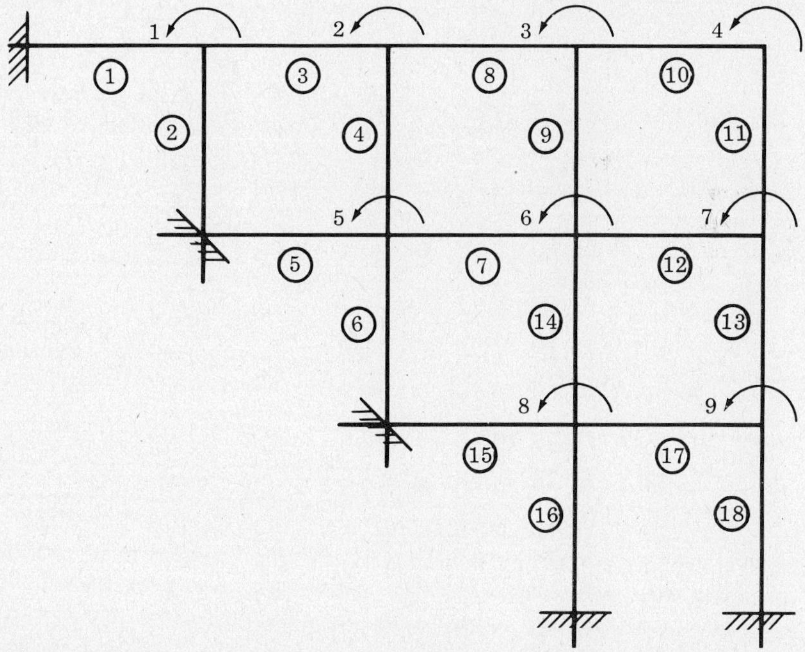

(b)

Fig. 7.1 Frame example: (a) dimensions, physical properties, and loading; (b) structural model.

and

$$\mathbf{k}^{(\text{column})} = \begin{bmatrix} 6/L^2 & 0 & -3/L & -6/L^2 & 0 & -3/L \\ 0 & A/2I_0 & 0 & 0 & -A/2I_0 & 0 \\ -3/L & 0 & 2 & 3/L & 0 & 1 \\ -6/L^2 & 0 & 3/L & 6/L^2 & 0 & 3/L \\ 0 & -A/2I_0 & 0 & 0 & A/2I_0 & 0 \\ -3/L & 0 & 1 & 3/L & 0 & 2 \end{bmatrix}$$

Taking all girder member directions positive to the right and all column member directions positive upward, and assuming inextensional behavior, we find the unconstrained coordinate codes for the members of this structure to be

Member number ↓ \ Member coordinate →	1	2	3	4	5	6
1	0	0	0	0	0	1
2	0	0	0	0	0	1
3	0	0	1	0	0	2
4	0	0	5	0	0	2
5	0	0	0	0	0	5
6	0	0	0	0	0	5
7	0	0	5	0	0	6
8	0	0	2	0	0	3
9	0	0	6	0	0	3
10	0	0	3	0	0	4
11	0	0	7	0	0	4
12	0	0	6	0	0	7
13	0	0	9	0	0	7
14	0	0	8	0	0	6
15	0	0	0	0	0	8
16	0	0	0	0	0	8
17	0	0	8	0	0	9
18	0	0	0	0	0	9

It is worth mention that both structure coordinates and members have been numbered in special sequences for purposes that will be explained subsequently. Using the above codes, we can now build the structure stiffness relationships.

$$\begin{bmatrix} 10 & 2 & \cdot & \cdot & \cdot & \cdot & \cdot & \cdot & \cdot \\ 2 & 10 & 2 & \cdot & 1 & \cdot & \cdot & \cdot & \cdot \\ \cdot & 2 & 10 & 2 & \cdot & 1 & \cdot & \cdot & \cdot \\ \cdot & \cdot & 2 & 6 & \cdot & \cdot & 1 & \cdot & \cdot \\ \cdot & 1 & \cdot & \cdot & 12 & 2 & \cdot & \cdot & \cdot \\ \cdot & \cdot & 1 & \cdot & 2 & 12 & 2 & 1 & \cdot \\ \cdot & \cdot & \cdot & 1 & \cdot & 2 & 8 & \cdot & 1 \\ \cdot & \cdot & \cdot & \cdot & \cdot & 1 & \cdot & 12 & 2 \\ \cdot & \cdot & \cdot & \cdot & \cdot & \cdot & 1 & 2 & 8 \end{bmatrix} \begin{pmatrix} D_1 \\ D_2 \\ D_3 \\ D_4 \\ D_5 \\ D_6 \\ D_7 \\ D_8 \\ D_9 \end{pmatrix} = \begin{pmatrix} 0 \\ 0 \\ 0 \\ 10 \\ 0 \\ 0 \\ 0 \\ 0 \\ 0 \end{pmatrix} \tag{A}$$

The set of nine scalar equations represented by this matrix equation will now be solved by gaussian elimination in two sequences: forward elimination (eight steps) and backward substitution (nine steps).

7.2.1 Forward Elimination

The ith forward elimination step consists of subtracting an appropriate multiple of the ith scalar equation of matrix Eq. A from all succeeding equations to eliminate nonzero coefficients below the ith row in the ith column of the **K** array. These manipulations can be carried out as a series of row operations on **K** and **R**. If this brief general description is unclear, consider the following detailed steps.

Step 1: row 2 – (2/10) × row 1 eliminates K_{21}

$$\begin{bmatrix} 10 & 2 & \cdot & \cdot & \cdot & \cdot & \cdot & \cdot & \cdot \\ \cdot & 9.6 & 2 & \cdot & 1 & \cdot & \cdot & \cdot & \cdot \\ \cdot & 2 & 10 & 2 & \cdot & 1 & \cdot & \cdot & \cdot \\ \cdot & \cdot & 2 & 6 & \cdot & \cdot & 1 & \cdot & \cdot \\ \cdot & 1 & \cdot & \cdot & 12 & 2 & \cdot & \cdot & \cdot \\ \cdot & \cdot & 1 & \cdot & 2 & 12 & 2 & 1 & \cdot \\ \cdot & \cdot & \cdot & 1 & \cdot & 2 & 8 & \cdot & 1 \\ \cdot & \cdot & \cdot & \cdot & \cdot & 1 & \cdot & 12 & 2 \\ \cdot & \cdot & \cdot & \cdot & \cdot & \cdot & 1 & 2 & 8 \end{bmatrix} \begin{pmatrix} D_1 \\ D_2 \\ D_3 \\ D_4 \\ D_5 \\ D_6 \\ D_7 \\ D_8 \\ D_9 \end{pmatrix} = \begin{pmatrix} 0 \\ 0 \\ 0 \\ 10 \\ 0 \\ 0 \\ 0 \\ 0 \\ 0 \end{pmatrix} \tag{A}$$

Step 2: row 3 – (2/9.6) × row 2 eliminates K_{32}
row 5 – (1/9.6) × row 2 eliminates K_{52}

$$
\begin{bmatrix}
10 & 2 & \cdot & \cdot & \cdot & \cdot & \cdot & \cdot & \cdot \\
\cdot & 9.6 & 2 & \cdot & 1 & \cdot & \cdot & \cdot & \cdot \\
\cdot & \cdot & 9.583 & 2 & -.2083 & 1 & \cdot & \cdot & \cdot \\
\cdot & \cdot & 2 & 6 & \cdot & \cdot & 1 & \cdot & \cdot \\
\cdot & \cdot & -.2083 & \cdot & 11.90 & 2 & \cdot & \cdot & \cdot \\
\cdot & \cdot & 1 & \cdot & 2 & 12 & 2 & 1 & \cdot \\
\cdot & \cdot & \cdot & 1 & \cdot & 2 & 8 & \cdot & 1 \\
\cdot & \cdot & \cdot & \cdot & \cdot & 1 & \cdot & 12 & 2 \\
\cdot & \cdot & \cdot & \cdot & \cdot & \cdot & 1 & 2 & 8
\end{bmatrix}
\begin{pmatrix} D_1 \\ D_2 \\ D_3 \\ D_4 \\ D_5 \\ D_6 \\ D_7 \\ D_8 \\ D_9 \end{pmatrix}
=
\begin{pmatrix} 0 \\ 0 \\ 0 \\ 10 \\ 0 \\ 0 \\ 0 \\ 0 \\ 0 \end{pmatrix}
\qquad \text{(B)}
$$

Step 3: row 4 $-$ (2/9.583) $\times$ row 3 eliminates K_{43}
row 5 $-$ ($-$.2083/9.583) $\times$ row 3 eliminates K_{53}
row 6 $-$ (1/9.583) $\times$ row 3 eliminates K_{63}

$$
\begin{bmatrix}
10 & 2 & \cdot & \cdot & \cdot & \cdot & \cdot & \cdot & \cdot \\
\cdot & 9.6 & 2 & \cdot & 1 & \cdot & \cdot & \cdot & \cdot \\
\cdot & \cdot & 9.583 & 2 & -.2083 & 1 & \cdot & \cdot & \cdot \\
\cdot & \cdot & \cdot & 5.583 & .04348 & -.2088 & 1 & \cdot & \cdot \\
\cdot & \cdot & \cdot & .04347 & 11.90 & 2.022 & \cdot & \cdot & \cdot \\
\cdot & \cdot & \cdot & -.2087 & 2.022 & 11.90 & 2 & 1 & \cdot \\
\cdot & \cdot & \cdot & 1 & \cdot & 2 & 8 & \cdot & 1 \\
\cdot & \cdot & \cdot & \cdot & \cdot & 1 & \cdot & 12 & 2 \\
\cdot & \cdot & \cdot & \cdot & \cdot & \cdot & 1 & 2 & 8
\end{bmatrix}
\begin{pmatrix} D_1 \\ D_2 \\ D_3 \\ D_4 \\ D_5 \\ D_6 \\ D_7 \\ D_8 \\ D_9 \end{pmatrix}
=
\begin{pmatrix} 0 \\ 0 \\ 0 \\ 10 \\ 0 \\ 0 \\ 0 \\ 0 \\ 0 \end{pmatrix}
\qquad \text{(C)}
$$

Step 4: row 5 $-$ (.04347/5.583) $\times$ row 4 eliminates K_{54}
row 6 $-$ ($-$.2087/5.583) $\times$ row 4 eliminates K_{64}
row 7 $-$ (1/5.583) $\times$ row 4 eliminates K_{74}

$$
\begin{bmatrix}
10 & 2 & \cdot & \cdot & \cdot & \cdot & \cdot & \cdot & \cdot \\
\cdot & 9.6 & 2 & \cdot & 1 & \cdot & \cdot & \cdot & \cdot \\
\cdot & \cdot & 9.583 & 2 & -.2083 & 1 & \cdot & \cdot & \cdot \\
\cdot & \cdot & \cdot & 5.583 & .04348 & -.2088 & 1 & \cdot & \cdot \\
\cdot & \cdot & \cdot & \cdot & 11.90 & 2.024 & -.007785 & \cdot & \cdot \\
\cdot & \cdot & \cdot & \cdot & 2.024 & 11.89 & 2.037 & 1 & \cdot \\
\cdot & \cdot & \cdot & \cdot & -.007788 & 2.037 & 7.821 & \cdot & 1 \\
\cdot & \cdot & \cdot & \cdot & \cdot & 1 & \cdot & 12 & 2 \\
\cdot & \cdot & \cdot & \cdot & \cdot & \cdot & 1 & 2 & 8
\end{bmatrix}
\begin{pmatrix} D_1 \\ D_2 \\ D_3 \\ D_4 \\ D_5 \\ D_6 \\ D_7 \\ D_8 \\ D_9 \end{pmatrix}
=
\begin{pmatrix} 0 \\ 0 \\ 0 \\ 10 \\ -.07735 \\ .3738 \\ -1.791 \\ 0 \\ 0 \end{pmatrix}
\qquad \text{(D)}
$$

Step 5: row 6 − (2.024/11.90) × row 5 eliminates K_{65}
row 7 − (−.007788/11.90) × row 5 eliminates K_{75}

$$\begin{bmatrix} 10 & 2 & \cdot & \cdot & \cdot & \cdot & \cdot & \cdot & \cdot \\ \cdot & 9.6 & 2 & \cdot & 1 & \cdot & \cdot & \cdot & \cdot \\ \cdot & \cdot & 9.583 & 2 & -.2083 & 1 & \cdot & \cdot & \cdot \\ \cdot & \cdot & \cdot & 5.583 & .04348 & -.2088 & 1 & \cdot & \cdot \\ \cdot & \cdot & \cdot & \cdot & 11.90 & 2.024 & -.007785 & \cdot & \cdot \\ \cdot & \cdot & \cdot & \cdot & \cdot & 11.54 & 2.038 & 1 & \cdot \\ \cdot & \cdot & \cdot & \cdot & \cdot & 2.038 & 7.821 & \cdot & 1 \\ \cdot & \cdot & \cdot & \cdot & \cdot & 1 & \cdot & 12 & 2 \\ \cdot & \cdot & \cdot & \cdot & \cdot & \cdot & 1 & 2 & 8 \end{bmatrix} \begin{pmatrix} D_1 \\ D_2 \\ D_3 \\ D_4 \\ D_5 \\ D_6 \\ D_7 \\ D_8 \\ D_9 \end{pmatrix} = \begin{pmatrix} 0 \\ 0 \\ 0 \\ 10 \\ -.07785 \\ .3870 \\ -1.859 \\ -.03354 \\ 0 \end{pmatrix} \qquad \text{(E)}$$

Step 6: row 7 − (2.038/11.54) × row 6 eliminates K_{76}
row 8 − (1/11.54) × row 6 eliminates K_{86}

$$\begin{bmatrix} 10 & 2 & \cdot & \cdot & \cdot & \cdot & \cdot & \cdot & \cdot \\ \cdot & 9.6 & 2 & \cdot & 1 & \cdot & \cdot & \cdot & \cdot \\ \cdot & \cdot & 9.583 & 2 & -.2083 & 1 & \cdot & \cdot & \cdot \\ \cdot & \cdot & \cdot & 5.583 & .04348 & -.2088 & 1 & \cdot & \cdot \\ \cdot & \cdot & \cdot & \cdot & 11.90 & 2.024 & -.007785 & \cdot & \cdot \\ \cdot & \cdot & \cdot & \cdot & \cdot & 11.54 & 2.038 & 1 & \cdot \\ \cdot & \cdot & \cdot & \cdot & \cdot & \cdot & 7.461 & -.1766 & 1 \\ \cdot & \cdot & \cdot & \cdot & \cdot & \cdot & -.1766 & 11.91 & 2 \\ \cdot & \cdot & \cdot & \cdot & \cdot & \cdot & 1 & 2 & 8 \end{bmatrix} \begin{pmatrix} D_1 \\ D_2 \\ D_3 \\ D_4 \\ D_5 \\ D_6 \\ D_7 \\ D_8 \\ D_9 \end{pmatrix} = \begin{pmatrix} 0 \\ 0 \\ 0 \\ 10 \\ -.07785 \\ .3870 \\ -1.859 \\ -.03354 \\ 0 \end{pmatrix} \qquad \text{(F)}$$

Step 7: row 8 − (−.1766/7.461) × row 7 eliminates K_{87}
row 9 − (1/7.461) × row 7 eliminates K_{97}

$$\begin{bmatrix} 10 & 2 & \cdot & \cdot & \cdot & \cdot & \cdot & \cdot & \cdot \\ \cdot & 9.6 & 2 & \cdot & 1 & \cdot & \cdot & \cdot & \cdot \\ \cdot & \cdot & 9.583 & 2 & -.2083 & 1 & \cdot & \cdot & \cdot \\ \cdot & \cdot & \cdot & 5.583 & .04348 & -.2088 & 1 & \cdot & \cdot \\ \cdot & \cdot & \cdot & \cdot & 11.90 & 2.024 & -.007785 & \cdot & \cdot \\ \cdot & \cdot & \cdot & \cdot & \cdot & 11.54 & 2.038 & 1 & \cdot \\ \cdot & \cdot & \cdot & \cdot & \cdot & \cdot & 7.461 & -.1766 & 1 \\ \cdot & \cdot & \cdot & \cdot & \cdot & \cdot & \cdot & 11.91 & 2.024 \\ \cdot & \cdot & \cdot & \cdot & \cdot & \cdot & \cdot & 2.024 & 7.866 \end{bmatrix} \begin{pmatrix} D_1 \\ D_2 \\ D_3 \\ D_4 \\ D_5 \\ D_6 \\ D_7 \\ D_8 \\ D_9 \end{pmatrix} = \begin{pmatrix} 0 \\ 0 \\ 0 \\ 10 \\ -.07785 \\ .3870 \\ -1.859 \\ -.07755 \\ .2492 \end{pmatrix} \qquad \text{(G)}$$

Step 8: row 9 − (2.024/11.91) × row 8 eliminates K_{98}

$$\begin{bmatrix} 10 & 2 & \cdot & \cdot & \cdot & \cdot & \cdot & \cdot & \cdot \\ & 9.6 & 2 & \cdot & 1 & \cdot & \cdot & \cdot & \cdot \\ & & 9.583 & 2 & -.2083 & 1 & \cdot & \cdot & \cdot \\ & & & 5.583 & .04348 & -.2088 & 1 & \cdot & \cdot \\ & & & & 11.90 & 2.024 & -.007785 & \cdot & \cdot \\ & & & & & 11.54 & 2.038 & 1 & \cdot \\ & & \mathbf{0} & & & & 7.461 & -.1766 & 1 \\ & & & & & & & 11.91 & 2.024 \\ & & & & & & & & 7.522 \end{bmatrix} \begin{pmatrix} D_1 \\ D_2 \\ D_3 \\ D_4 \\ D_5 \\ D_6 \\ D_7 \\ D_8 \\ D_9 \end{pmatrix} = \begin{pmatrix} 0 \\ 0 \\ 0 \\ 10 \\ -.07785 \\ .3870 \\ -1.859 \\ -.07755 \\ .2624 \end{pmatrix} \qquad \text{(H)}$$

This completes the forward elimination process; backward substitution may now be carried out.

7.2.2 Backward Substitution

Backward substitution is simpler than forward elimination in that each row of the coefficient array (and vector on the right-hand side) need be modified only once during backsubstitution. Each step consists of substituting all currently known displacement values into each of the nine equations, in turn, and solving for the remaining unknown in each equation, starting with the last equation and working backward to the first. Following are the results of these steps.

Step 1:

$$D_9 = \frac{1}{7.522}(.2624) = .03488$$

Step 2:

$$D_8 = \frac{1}{11.91}[-.07755 - 2.024(.03488)]$$

$$= -.01244$$

Step 3:

$$D_7 = \frac{1}{7.461}[-1.859 - (-.1766)(-.01244) - 1(.03488)]$$

$$= -.2541$$

Step 4:

$$D_6 = \frac{1}{11.54}[.3870 - 2.038(-.2541) - 1(-.01244)]$$

$$= .07949$$

Step 5:

$$D_5 = \frac{1}{11.90}[-.07785 - 2.024(.07949) - (-.007785)(-.2541)]$$

$$= -.02023$$

Step 6:

$$D_4 = \frac{1}{5.583}[10 - .4348(-.02023) - (-.2088)(.07949) - 1(-.2541)]$$

$$= 1.840$$

Step 7:

$$D_3 = \frac{1}{9.583}[0 - 2(1.840) - (-.2083)(-.02023) - 1(.07949)]$$

$$= -.3927$$

Step 8:

$$D_2 = \frac{1}{9.6}[0 - 2(-.3927) - 1(-.02023)]$$

$$= .08392$$

Step 9:

$$D_1 = \frac{1}{10}[0 - 2(.08392)]$$

$$= -.01678$$

This completes the solution for the unknown displacements. Note that the backsubstitution process described could alternatively have been accomplished in matrix form by performing appropriate row operations on Eq. H. Problem 7.1 gives you the opportunity to prove this statement.

7.2.3 Gaussian Elimination Algorithms

Now that we have gone through the detailed steps of gaussian elimination, it is appropriate to formulate general rules or algorithms governing both the forward elimination and backward substitution processes. You have undoubtedly observed that the steps of both of these sequences are repetitious and lend themselves to formalized statements. The algorithms for the rth forward elimination step are

$$\left.\begin{aligned} K_{ij} &\rightarrow K_{ij} - \frac{K_{ir}}{K_{rr}} K_{rj} \\ R_i &\rightarrow R_i - \frac{K_{ir}}{K_{rr}} R_r \end{aligned}\right\} \quad \begin{aligned} r &= 1, 2, \ldots, N-1 \\ i &= r+1, r+2, \ldots, N \\ j &= r+1, r+2, \ldots, N \end{aligned} \qquad (7.1)$$

and the algorithms for the ith backward substitution step are

$$D_i = \begin{cases} \dfrac{R_N}{K_{NN}} & i = N \\ \dfrac{1}{K_{ii}}\left(R_i - \sum\limits_{j=i+1}^{N} K_{ij}D_j\right) & i = N-1, N-2, \ldots, 1 \end{cases} \tag{7.2}$$

where N is the number of simultaneous equations being solved. You are encouraged to verify that these algorithms function properly by checking one or two steps of the example that was just finished. Further practice with these algorithms is provided in Prob. 7.2.

The effectiveness of a solution algorithm may be judged by its accuracy and its efficiency. The question of accuracy will be considered in Sec. 7.5. Solution efficiency is measured by the extent to which demands on computation resources are minimized. The major computation resources are (1) central processor time, (2) input/output time (including time for transfer of data between main and secondary storage), (3) core space, and (4) disk space. Which of these four is most important in a given computation environment will depend on the manner in which costs are calculated in that environment. It has been found (Ref. 7.5) that solution efficiency is not nearly as strongly influenced by the choice of solution algorithm (assuming the use of a reasonably good algorithm) as it is by the choice of data handling technique. We will therefore turn our attention to the subject of data handling next, using the gaussian elimination algorithm to demonstrate various data handling alternatives. Additional discussion of solution algorithms will be presented in Sec. 7.4.

7.3 DATA HANDLING TECHNIQUES

In our initial discussion of data handling techniques, it is convenient to assume that it is economically and physically feasible to accomplish the entire solution in core. Later these methods will be modified to apply to out-of-core solutions as well. For the time being, therefore, we need be concerned only with minimizing core storage and central processor time.

The most obvious strategy for reducing core storage is to take advantage of the fact that the structure stiffness matrix is symmetric, making it possible to store only the upper "half" of this array (including the diagonal terms). In this connection it is important to notice not only that the stiffness matrix is symmetric before the elimination process starts, but also that the portion of the stiffness matrix not yet reduced to upper triangular form remains symmetric (except for round-off error) during the elimination process. This may be verified by inspecting Eqs. A to H of Sec. 7.2.1

The gaussian elimination algorithms of Eq. 7.1 may now be rewritten by making the substitution $K_{ir} = K_{ri}$:

$$\left.\begin{aligned} K_{ij} &\rightarrow K_{ij} - \frac{K_{ri}}{K_{rr}} K_{rj} \\ R_i &\rightarrow R_i - \frac{K_{ri}}{K_{rr}} R_r \end{aligned}\right\} \qquad \begin{aligned} r &= 1, 2, \ldots, N-1 \\ i &= r+1, r+2, \ldots, N \\ j &= i, i+1, \ldots, N \end{aligned} \tag{7.1a}$$

As $r < i \leq j$, these algorithms require no coefficient from below the diagonal in the stiffness matrix.

A second strategy for reducing core storage may be deduced from a strategy for reducing central processor time. Since terms that are zero in the stiffness matrix contribute nothing to the elimination process, arithmetic operations with, and therefore storage of, zero stiffness coefficients should be avoided. It turns out that there are two distinct categories of zero coefficients in the stiffness array: those that remain zero during the elimination process and those that do not (or are subject to "infill"). It is fortunately often the case that all nonzero terms in a stiffness matrix are clustered in a rather narrow band near the main diagonal. Such a matrix is referred to as a "banded" matrix. The stiffness matrix of Eq. A of Sec. 7.2 is seen to be banded. Once again examining the elimination of steps of Eqs. A to H of Sec. 7.2.1, we see that zeros that are initially outside of the nonzero band of coefficients remain zero during the elimination process, while some of those within the band will become nonzero. These observations can be generalized, and we can adopt the procedure of storing and manipulating not the whole upper half but only the upper half-band of the stiffness matrix.

7.3.1 In-Core Banded Solution

The half- or semi-bandwidth of a stiffness matrix is a measure of the size of the band. The semi-bandwidth of the stiffness matrix of Eq. A of Sec. 7.2 is shown graphically here.

$\longleftarrow$ Semi-bandwidth $\longrightarrow$

$$\begin{bmatrix} 10 & 2 & \cdot & \cdot & \cdot & \cdot & \cdot & \cdot & \cdot \\ & 10 & 2 & \cdot & 1 & \cdot & \cdot & \cdot & \cdot \\ & & 10 & 2 & \cdot & 1 & \cdot & \cdot & \cdot \\ & & & 6 & \cdot & \cdot & 1 & \cdot & \cdot \\ & & & & 12 & 2 & \cdot & \cdot & \cdot \\ & & & & & 12 & 2 & 1 & \cdot \\ & \text{Symmetric} & & & & & 8 & \cdot & 1 \\ & & & & & & & 12 & 2 \\ & & & & & & & & 8 \end{bmatrix} \begin{pmatrix} D_1 \\ D_2 \\ D_3 \\ D_4 \\ D_5 \\ D_6 \\ D_7 \\ D_8 \\ D_9 \end{pmatrix} = \begin{pmatrix} 0 \\ 0 \\ 0 \\ 10 \\ 0 \\ 0 \\ 0 \\ 0 \\ 0 \end{pmatrix} \tag{A}$$

The semi-bandwidth is seen to be equal to four in this case. This value can be calculated as follows:

$$m_B = N_D(1 + \text{MDIF}) \tag{7.3}$$

where m_B = semi-bandwidth
N_D = number of degrees of freedom per node
MDIF = maximum difference in structure nodal numbers associated with any one member of the structure

Examination of Fig. 7.1b gives in this case

$$m_B = 1(1 + 3) = 4$$

as just observed. It follows from Eq. 7.3 that the bandwidth of a stiffness matrix will be very sensitive to the ordering of nodal numbers. The nodes of the example of Fig. 7.1 have been numbered to produce a minimum bandwidth. The subject of bandwidth minimization will be explored in greater depth in Sec. 7.3.3.

The stiffness coefficients of the upper semi-band augmented with the load vector(s)* may now be stored as a rectangular array, as shown in Fig. 7.2. The diagonals of **K** become columns in the rectangle, and the columns of **K** become diagonals in the rectangle. Rows in **K** remain rows in the rectangle. Simply stated, coefficients in the ith row of **K** are shifted $i - 1$ columns to the left before being placed in the rectangular array. Knowing this, we can modify previously developed coding techniques so that only the upper semi-band of the structure stiffness matrix is built—and it is built directly in rectangular form. For example, the member stiffness coefficient that the coordinate code indicates is destined to be added to $K_{\ell m}$ in the full stiffness array would instead be added to $K_{\ell n}$ in the rectangular array where $n = m - \ell + 1$ and where only structure coefficients for which $m > \ell$ and $n \le m_B$ would be considered. The load vector would be coded into column $m_B + 1$ of the rectangular array. Thus, for a narrowly banded stiffness matrix, considerable storage can be saved. Furthermore, it is also possible to modify the gaussian algorithms so that they will operate directly on the rectangular array, thus saving computational time also.

Subtracting the row index minus 1 from all the column indices of all stiffness coefficients and providing an appropriate column index for the load vector terms of Eq. 7.1a gives

$$K_{i,j-i+1} \rightarrow K_{i,j-i+1} - \frac{K_{r,i-r+1}}{K_{r,1}} K_{r,j-r+1}$$

$$R_{i,m_B+1} \rightarrow R_{i,m_B+1} - \frac{K_{r,i-r+1}}{K_{r,1}} R_{r,m_B+1} \tag{7.1b}$$

$$r = 1, 2, \ldots, N - 1$$

$$i = r + 1, r + 2, \ldots, N \qquad i \le r + m_B - 1$$

$$j = i, i + 1, \ldots, N \qquad j \le r + m_B - 1$$

*As a matter of simplicity, we will continue to consider only a single load vector in the following discussion, although multiple load vectors can readily be handled by any of the methods that are considered. Handling of multiple load vectors is demonstrated in Sec. 7.4.2.

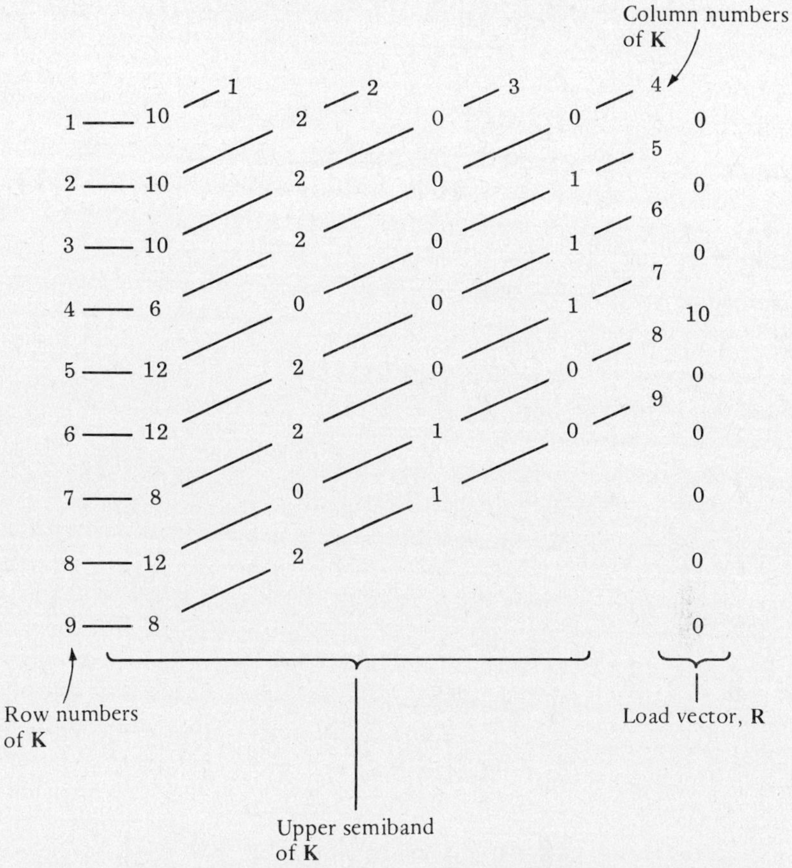

Fig. 7.2 Rectangular form of upper semi-band of **K** augmented with **R**.

For clarity, we have retained the symbols K and R for the stiffness and load portions, respectively, of the rectangular array; in the computer, these would be represented by a single array name. The backsubstitution algorithm of Eq. 7.2 must be similarly modified:

$$D_i = \begin{cases} \dfrac{R_{N,m_B+1}}{K_{N,1}} & i = N \\ \dfrac{1}{K_{i,1}}\left(R_{i,m_B+1} - \displaystyle\sum_{j=i+1}^{N} K_{i,j-i+1}\, D_j\right) & \begin{array}{l} i = N-1, N-2, \ldots, 1 \\ j \leq i + m_B - 1 \end{array} \end{cases} \tag{7.2a}$$

The following is a direct application of the modified gaussian forward elimination algorithm of Eq. 7.1b to the rectangular array of Fig. 7.2.

$r = 1 \qquad i = r + 1 = 2$:

reduction of stiffness array:

$j = i = 2$:

$$K_{21} \rightarrow K_{21} - \frac{K_{12}}{K_{11}} K_{12}$$

$$K_{21} = 10 - \frac{2}{10} 2 = 9.6$$

$j = i + 1 = 3$:

$$K_{22} \rightarrow K_{22} - \frac{K_{12}}{K_{11}} K_{13}$$

$$K_{22} = 2 - \frac{2}{10} 0 = 2$$

$j = i + 2 = 4$:

$$K_{23} \rightarrow K_{23} - \frac{K_{12}}{K_{11}} K_{14}$$

$$K_{23} = 0 - \frac{2}{10} 0 = 0$$

$j = i + 3 = 5 > r + m_B - 1 = 4$

reduction of force vector:

$$R_{25} \rightarrow R_{25} - \frac{K_{12}}{K_{11}} R_{15}$$

$$R_{25} = 0 - \frac{2}{10} 0 = 0$$

$i = r + 2 = 3$:

reduction of stiffness array:

$j = i = 3$:

$$K_{31} \rightarrow K_{31} - \frac{K_{13}}{K_{11}} K_{13}$$

$$K_{31} = 10 - \frac{0}{10} 0 = 10$$

$j = i + 1 = 4$:

$$K_{32} \rightarrow K_{32} - \frac{K_{13}}{K_{11}} K_{14}$$

$$K_{32} = 2 - \frac{0}{10} 0 = 2$$

$j = i + 2 = 5 > r + m_B - 1 = 4$

reduction of force vector:

$$R_{35} \longrightarrow R_{35} - \frac{K_{13}}{K_{11}} R_{15}$$

$$R_{35} = 0 - \frac{0}{10} 0 = 0$$

$i = r + 3 = 4$:

reduction of stiffness array:

$j = i = 4$:

$$K_{41} \longrightarrow K_{41} - \frac{K_{14}}{K_{11}} K_{14}$$

$$K_{41} = 6 - \frac{0}{10} 0 = 6$$

$j = i + 1 = 5 > r + m_B - 1 = 4$

reduction of force vector:

$$R_{45} \longrightarrow R_{45} - \frac{K_{14}}{K_{11}} R_{15}$$

$$R_{45} = 10 - \frac{0}{10} 0 = 10$$

$r = 2$ $i = r + 1 = 3$:

reduction of stiffness array:

$j = i = 3$:

$$K_{31} \longrightarrow K_{31} - \frac{K_{22}}{K_{21}} K_{22}$$

$$K_{31} = 10 - \frac{2}{9.6} 2 = 9.583$$

$j = i + 1 = 4$:

$$K_{32} \longrightarrow K_{32} - \frac{K_{22}}{K_{21}} K_{23}$$

$$K_{32} = 2 - \frac{2}{9.6} 0 = 2$$

$j = i + 2 = 5$:

$$K_{33} \longrightarrow K_{33} - \frac{K_{22}}{K_{21}} K_{24}$$

$$K_{33} = 0 - \frac{2}{9.6} 1 = -.2083$$

$$j = i + 3 = 6 > r + m_B - 1 = 5$$

reduction of force vector:

$$R_{35} \rightarrow R_{35} - \frac{K_{22}}{K_{21}} R_{25}$$

$$R_{35} = 0 - \frac{2}{9.6} 0 = 0$$

and so on.

Continuing in this fashion, we could develop the entire reduced structure stiffness matrix and load vector in rectangular form as shown in Fig. 7.3. Notice that in the computer the reduced matrix of Fig. 7.3 can be overwritten on the original array of Fig. 7.2, because only the most recent values of each coefficient need be saved. We could now demonstrate the application of the backsubstitution algorithm of Eq. 7.2a to the array of Fig. 7.3. However, since that is a relatively straightforward process, which you can carry out as an exercise, we will go on to consider additional improvements to the solution process.

7.3.1.1 Skyline Concept

It has probably not gone unnoticed that even with rectangular storage and appropriately modified gaussian algorithms, a number of zero terms must still be stored and manipulated during the solution process. The manipulation of zero terms can be reduced by placing further limitations on the range over which the gaussian forward elimination algorithms are applied. To impose these limitations, it is convenient to also change the sequencing of the application of the algorithms. As the indexing is stated in Eq. 7.1b, the rth row is used to perform all possible eliminations, row by row, for all rows below the rth row; this process is then repeated for row $r + 1$, and so on. This procedure will now be changed so that all eliminations are performed on the ith row by all rows above the ith row; this process will then be repeated for row $i + 1$, and so on. The gaussian algorithms of Eq. 7.1b are now repeated as Eq. 7.1c with the changed indexing.

10	2	0	0	0
9.6	2	0	1	0
9.583	2	−.2083	1	0
5.583	.04348	−.2088	1	10
11.90	2.024	−.007785	0	−.07785
11.54	2.038	1	0	.3870
7.461	−.1766	1		−1.859
11.91	2.024			−.07755
7.522				.2624
Upper semi-band of reduced **K**				Load vector, **R** corresponding to reduced **K**

Fig. 7.3 Reduced array of Fig. 7.2.

$$K_{i,j-i+1} \rightarrow K_{i,j-i+1} - \frac{K_{r,i-r+1}}{K_{r,1}} K_{r,j-r+1}$$
$$R_{i,m_B+1} \rightarrow R_{i,m_B+1} - \frac{K_{r,i-r+1}}{K_{r,1}} R_{r,m_B+1} \tag{7.1c}$$

$$i = 2, 3, \ldots, N$$
$$j = i, i+1, \ldots, N \qquad j \leq i + m_B - 2$$
$$r = m_m, m_m + 1, \ldots, i - 1$$

The term m_m is the larger of the terms m_i and m_j, where m_i and m_j are the row numbers of the first nonzero coefficients in columns i and j, respectively. The upper limit $i + m_B - 2$ on the index j prevents the algorithm from being applied to modify stiffness coefficients to the right of column $m_B - 1$ of the rectangular array (see Fig. 7.4). This situation is seen to be reasonable because the coefficients in column m_B of the rectangular array are not changed during the elimination process. (Compare the fourth columns in Fig. 7.3 and 7.4.) The lower limit m_m on row index r prevents the algorithm from being applied to cause zero coefficients

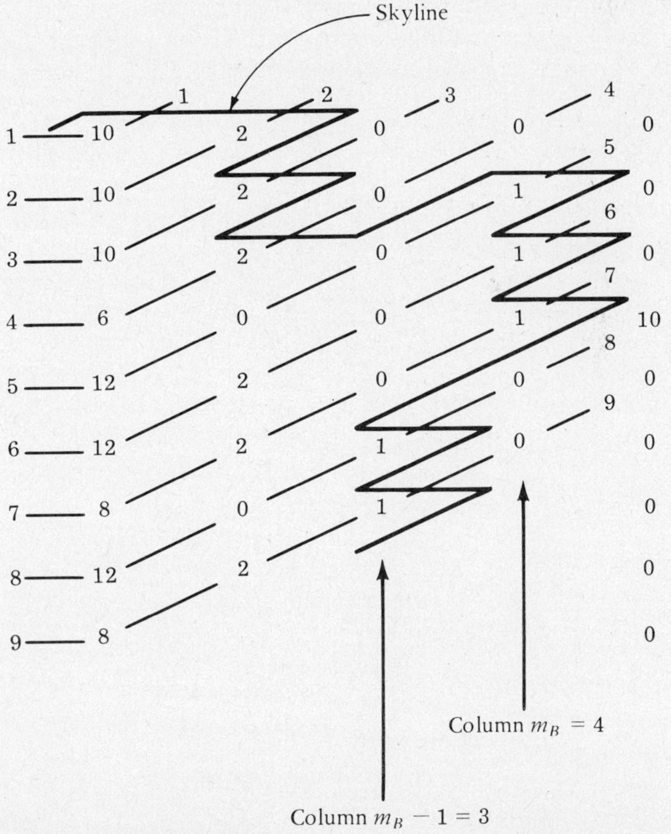

Fig. 7.4 Array of Fig. 7.2 showing skyline.

above the skyline of the stiffness matrix to be involved in the elimination process. The skyline (Ref. 7.6) of the stiffness matrix is shown in the next equation and also in Fig. 7.4.

Skyline

$$\begin{bmatrix} 10 & 2 & \cdot & \cdot & \cdot & \cdot & \cdot & \cdot & \cdot \\ & 10 & 2 & \cdot & 1 & \cdot & \cdot & \cdot & \cdot \\ & & 10 & 2 & \cdot & 1 & \cdot & \cdot & \cdot \\ & & & 6 & \cdot & \cdot & 1 & \cdot & \cdot \\ & & & & 12 & 2 & \cdot & \cdot & \cdot \\ & & & & & 12 & 2 & 1 & \cdot \\ & & & & & & 8 & \cdot & 1 \\ & & & & & & & 12 & 2 \\ & \text{Symmetric} & & & & & & & 8 \end{bmatrix} \begin{pmatrix} D_1 \\ D_2 \\ D_3 \\ D_4 \\ D_5 \\ D_6 \\ D_7 \\ D_8 \\ D_9 \end{pmatrix} = \begin{pmatrix} 0 \\ 0 \\ 0 \\ 10 \\ 0 \\ 0 \\ 0 \\ 0 \\ 0 \end{pmatrix} \quad \text{(B)}$$

This strategy eliminates manipulations with all zeros except those embedded within the semi-band, which will be subject to infill anyway.

It is interesting and useful to now apply the latest version of the gaussian elimination algorithm as shown in the expressions of Eq. 7.1c to the equations represented by Fig. 7.4.

$i = 2$:

reduction of stiffness array:

$j = i = 2 \qquad r = m_m = m_2 = 1$:

$$K_{21} \longrightarrow K_{21} - \frac{K_{12}}{K_{11}} K_{12}$$

$$K_{21} = 10 - \frac{2}{10} 2 = 9.6$$

$r = m_2 + 1 = 2 > i - 1 = 1$:
upper bound of r exceeded

$j = i + 1 = 3 \qquad r = m_m = m_3 = 2 > i - 1 = 1$:
upper bound of r exceeded

$j = i + 2 = 4 \qquad r = m_m = m_4 = 3 > i - 1 = 1$:
upper bound of r exceeded

$j = i + 3 = 5 > i + m_B - 2 = 4$:
upper bound of j exceeded

reduction of force vector:

$$r = m_m = m_{10} = 4 > i - 1 = 1:$$
upper bound of r exceeded

$i = 3$:

reduction of stiffness array:

$j = i = 3$ $\quad r = m_m = m_3 = 2$:

$$K_{31} \longrightarrow K_{31} - \frac{K_{22}}{K_{21}} K_{22}$$

$$K_{31} = 10 - \frac{2}{9.6} 2 = 9.583$$

$$r = m_3 + 1 = 3 > i - 1 = 2:$$
upper bound of r exceeded

$j = i + 1 = 4$ $\quad r = m_m = m_4 = 3 > i - 1 = 2$:
upper bound of r exceeded

$j = i + 2 = 5$ $\quad r = m_m = m_5 = 2$:

$$K_{33} \longrightarrow K_{33} - \frac{K_{22}}{K_{21}} K_{24}$$

$$K_{33} = 0 - \frac{2}{9.6} 1 = -.2083 \qquad \text{(infilled zero)}$$

$$r = m_5 + 1 = 3 > i - 1 = 2:$$
upper bound of r exceeded

$j = i + 3 = 6 > i + m_B - 2 = 5$:
upper bound of j exceeded

reduction of force vector:

$$r = m_m = m_{10} = 4 > i - 1 = 2:$$
upper bound of r exceeded

and so on.

A comparison of these results with those obtained previously (pp. 353–356) indicates a significant increase in data manipulation efficiency of the latest elimination scheme over the former one. The three nontrivial elimination steps accomplished in the latest scheme were accompanied by eight trivial steps in the previous case. It is true that more checks for zeros and checks on limits are required in this latter approach, but these can be made much more rapidly than can the extra calculations of the former method.

It has probably already occurred to you that if we don't need these zeros above the skyline of the stiffness array for calculations, it would be doubly nice if we could also eliminate them from storage. To accomplish this goal, a new data handling scheme (Ref. 7.6) is required, as shown in Fig. 7.5. The stiffness terms

M	
1	10
2	10
3	2
4	10
5	2
6	6
7	2
8	12
9	0
10	0
11	1
12	12
13	2
14	0
15	1
16	8
17	2
18	0
19	1
20	12
21	0
22	1
23	8
24	2
25	1

One-dimensional array of stiffness coefficients below skyline of semiband of **K**

(a)

N		
1	1	$2 - 1 = 1$
2	2	$4 - 2 = 2$
3	4	$6 - 4 = 2$
4	6	$8 - 6 = 2$
5	8	$12 - 8 = 4$
6	12	$16 - 12 = 4$
7	16	$20 - 16 = 4$
8	20	$23 - 20 = 3$
9	23	$26 - 23 = 3$
10	26	

Number of coefficients below skyline in Nth column of **K**

Addresses of diagonal coefficients in one-dimensional stiffness array (M for K_{NN})

(b)

Fig. 7.5 One-dimensional stiffness and address arrays: (a) stiffness array; (b) address array.

are seen to be stored in a one-dimensional array, columnwise, starting with the diagonal term of each column and proceeding to the skyline. For the example frame, this procedure gives the stiffness array shown in Fig. 7.5a. To properly locate each term in the one-dimensional array, a second one-dimensional array is required to list the addresses of each diagonal stiffness term in order. Once again, for the current example this gives the array shown in Fig. 7.5b. Thus, for instance, K_{77}, the seventh diagonal term of **K**, is in the sixteenth position in the one-dimensional stiffness array. Notice that the address array implicitly contains the column heights from the diagonal to the skyline for each column of **K** also. Thus the height of the seventh column of **K** is $20 - 16 = 4$. Finally, with the use of this data storage scheme, the load vector may be stored after the stiffness coefficients in the one-dimensional array.

For the current example, the 2 one-dimensional arrays require a total of 35 storage locations, whereas the rectangular array used previously required 36. Although this savings in storage is modest, the savings can be substantial for larger examples. As was the case for the rectangular storage scheme, the one-dimensional stiffness array and its companion address array can be built directly by an appropriately modified coding technique. In addition, the gaussian algorithms can be modified so that elimination can be carried out directly on the one-dimensional stiffness array. Since the stiffness coefficients are arranged columnwise in the one-dimensional array, the gaussian algorithms will, as a matter of convenience, need to be modified so that elimination proceeds columnwise rather than row-wise. This adjustment is not difficult; but the method does require somewhat more elaborate "bookkeeping" than the method that used the rectangular array. Although the one-dimensional stiffness and address arrays contain sufficient information to allow this bookkeeping to be accomplished, details will not be presented here. (For additional information on this method, see Ref. 7.6.) Columnwise reduction of a stiffness matrix will, however, be discussed in Sec. 7.4.1.

7.3.2 Out-of-Core Banded Solution

Even with all of the efficiencies we have discussed, it may still happen that for very large sets of equations, storage requirements exceed available core. Such problems can be handled by retaining in core at any one time only a portion of the full set of stiffness coefficients and load terms. Various methods are used to retain different "portions" of the stiffnesses and loads.

An out-of-core banded solution makes use of the previously defined rectangular array representation of the semi-band of stiffness coefficients and force vector terms. This solution technique is perhaps best explained in the context of a specific example. Consider the rectangular array of Fig. 7.2, which is shown again in Fig. 7.6. Notice that the array has been divided into three blocks of $m_B - 1$ equations each (where $m_B = 4$ in this case). Major steps in a block-by-block, out-of-core, banded solution of this set of equations are displayed in Fig. 7.7. Inspection of this figure will reveal that only two blocks of coefficients are retained in core at any one time. In addition, it is seen from Fig. 7.7a that only the terms in block 1 are assembled at the beginning of the solution process. Figure 7.7b shows that the first $m_B - 1 = 3$ elimination steps affect only those terms in blocks 1 and 2. At the end of these first three steps, block 1 is completely reduced, and block 2 contains all modifications resulting from the block 1 reduction. (The

10	2	0	0	0	
10	2	0	1	0	Block 1
10	2	0	1	0	
6	0	0	1	10	
12	2	0	0	0	Block 2
12	2	1	0	0	
8	0	1		0	
12	2			0	Block 3
8				0	

Fig. 7.6 Rectangular array of Fig. 7.2, divided into blocks.

10	2	0	0	0	Block 1: assembled in core
10	2	0	1	0	
10	2	0	1	0	
0	0	0	0	0	Block 2: available in core
0	0	0	0	0	
0	0	0	0	0	

(a)

10	2	0	0	0	Block 1: modified by elimination step 1
9.6	2	0	1	0	
10	2	0	1	0	
0	0	0	0	0	Block 2: not affected by elimination step 1
0	0	0	0	0	
0	0	0	0	0	

(b)

10	2	0	0	0	Block 1: completely reduced
9.6	2	0	1	0	
9.583	2	−.2083	1	0	
0	0	0	0	0	Block 2: modified by elimination step 2
−.1	0	0	0	0	
0	0	0	0	0	

(c)

10	2	0	0	0	Block 1: completely reduced
9.6	2	0	1	0	
9.583	2	−.2083	1	0	
−.4174	.04348	−.2088	0	0	Block 2: modified by elimination steps 2 and 3
−.1	.022	0	0	0	
−.1	0	0	0	0	

(d)

10	2	0	0	0	Block 1: moved to secondary storage
9.6	2	0	1	0	
9.583	2	−.2083	1	0	
5.583	.04348	−.2088	1	10	Block 2: assembled and partially reduced; in core
11.90	2.022	0	0	0	
11.90	2	1	0	0	
0	0	0	0	0	Block 3: available in core
0	0	0	0	0	
0	0	0	0	0	

(e)

Fig. 7.7 Block reduction: (a) first two blocks prior to elimination; (b) elimination step 1; (c) elimination step 2; (d) elimination step 3; (e) elimination step 4; (f) housekeeping step.

numbers displayed in the steps of Fig. 7.7 may be deduced easily by inspection of the detailed solution to this set of equations previously shown in Eqs. A to H of Sec. 7.2.1. The gaussian elimination algorithms are therefore not applied again in this case.) At the conclusion of elimination step 3, there is a housekeeping step. This step, displayed in Fig. 7.7e, consists of (1) moving the completely reduced block 1 to secondary storage, (2) moving block 2 into the former position of block 1 and adding the appropriate stiffness coefficients to the modifications already present in block 2, and (3) zeroing former block 2, which becomes block 3. To complete the solution, elimination must continue through block 2 with corresponding modifications to block 3. Another housekeeping step would then take place, followed by a final elimination step. (It is important to recognize that for a larger example, there would be more blocks, and many cycles of elimination and housekeeping would be required.) The backsubstitution process is simpler than the elimination process, as it requires only one block in core at a time. Since the completion of this example from this point is quite straightforward, it will not be shown here.

In the preceding example, the core storage requirement using the out-of-core banded solution was 24 stiffness coefficients, as opposed to 36 coefficients for the previous in-core solution. In general, savings that can be expected to result will depend on the ratio of m_B to N. This ratio is often quite small, leading to significant savings in storage. To provide further efficiency to the scheme for an out-of-core banded solution, a technique of avoiding storage and manipulation of zeros, such as was discussed earlier, could be included.

There are other ways of subdividing the banded array, but the block-by-block method described does have an advantage. The block size we used ($m_B - 1$ equations) is the smallest that will allow complete reduction of the block without requiring out-of-core communication. The scheme would remain virtually unchanged if larger block sizes were used. The best block size to use will depend on whether it is more important to limit core storage or to limit data transfer requirements.

For very large sets of equations, especially those that are poorly banded, it can happen that even a single block $(m_B - 1) \times (m_B + 1)$, will not fit into core. So further subdividing of blocks will be required, with an attendant rise in housekeeping chores. To avoid this, it may be worthwhile to consider data handling techniques that are not dependent on good banding for efficiency of operation. Before taking this step, however, it might be wise to first consider the conditions under which we can expect to have a well-banded stiffness array. In this connection the previously mentioned ratio, m_B/N, will be an important indicator of the quality of banding, since it is equal to the fraction of the total $N \times N$ stiffness matrix for which space must be provided in the $m_B \times N$ stiffness portion of the rectangular array.

7.3.3 Banding Quality

A variety of frame structures are shown in Fig. 7.8 along with representations of their respective stiffness arrays. For comparison purposes, each structure has 20

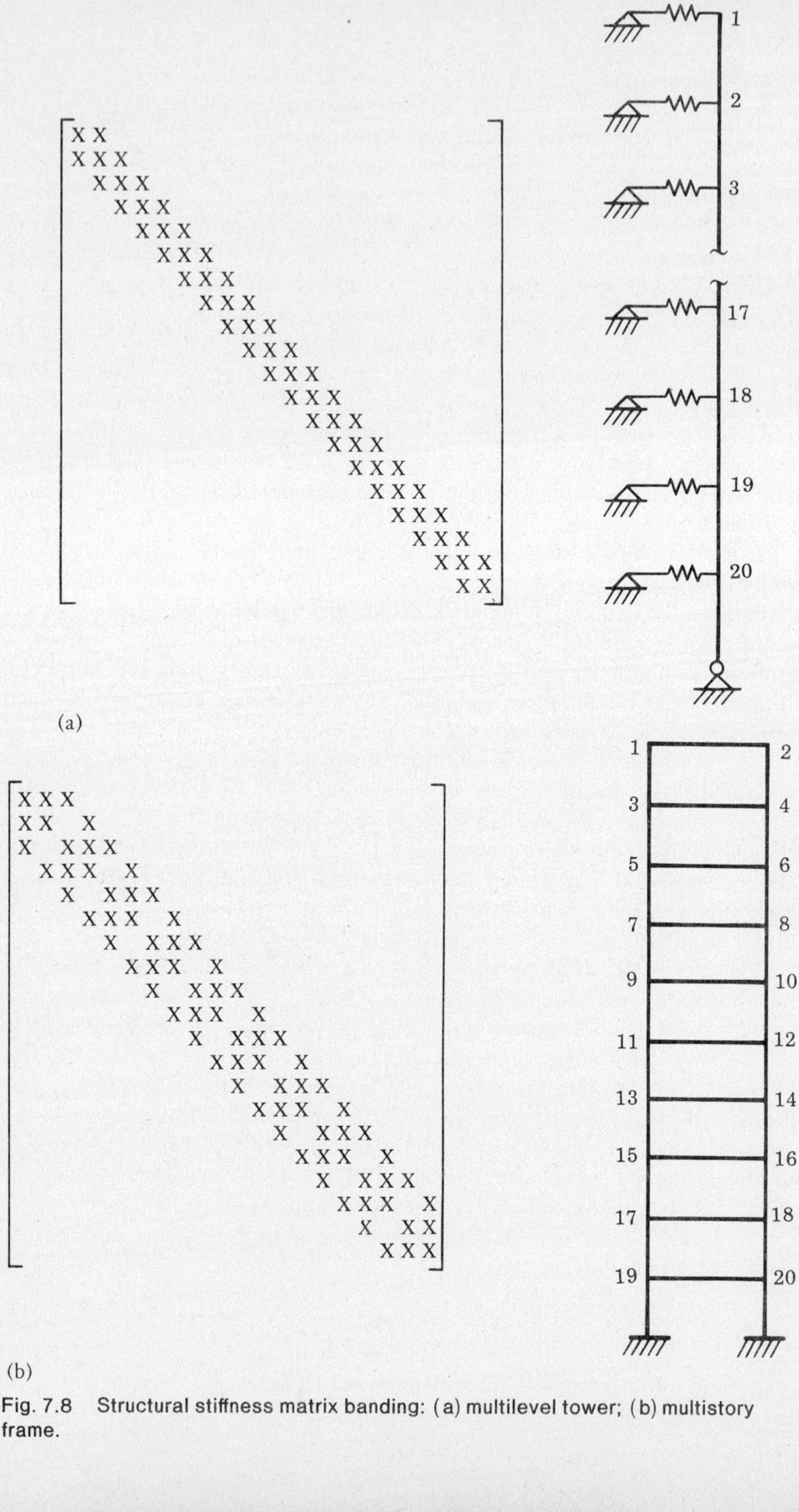

Fig. 7.8 Structural stiffness matrix banding: (a) multilevel tower; (b) multistory frame.

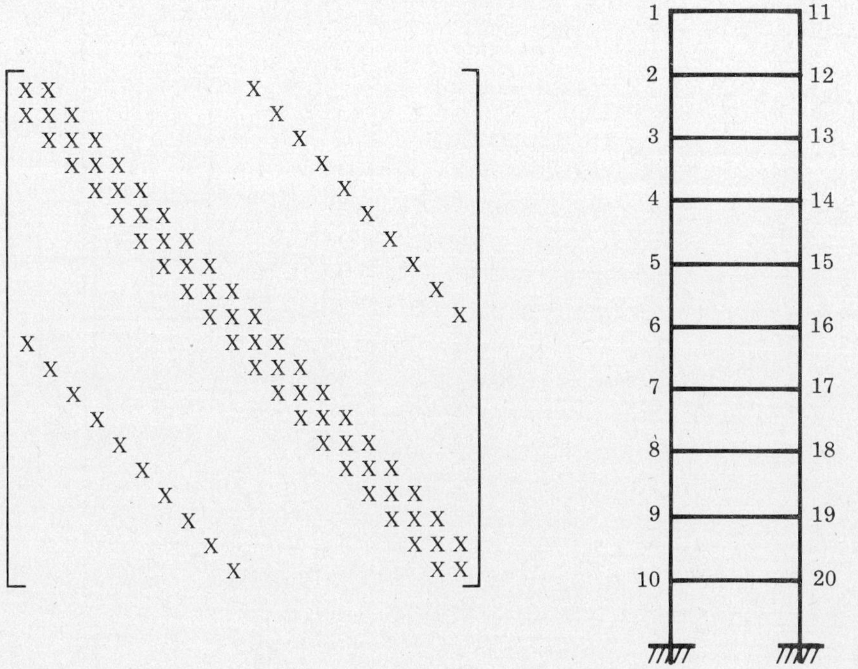

(c)

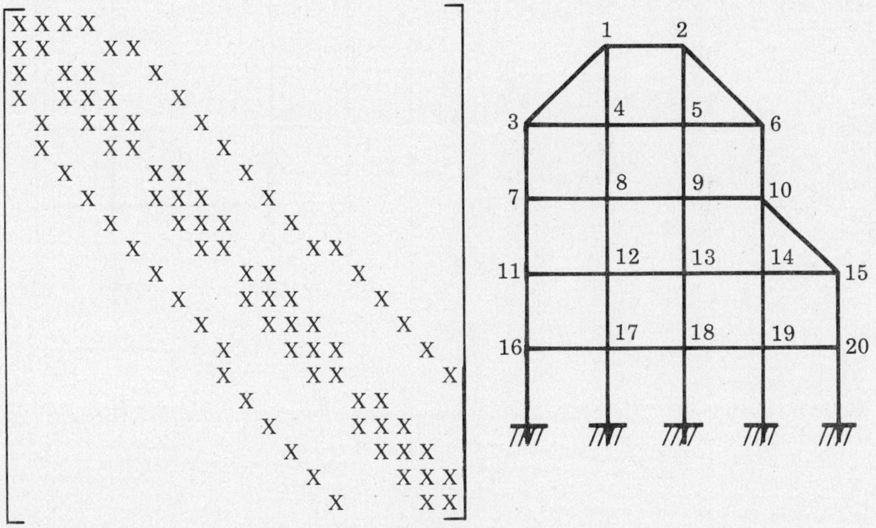

(d)

Fig. 7.8 (continued) Structural stiffness matrix banding: (c) multistory frame, alternate numbering; (d) frame with aspect ratio near unity.

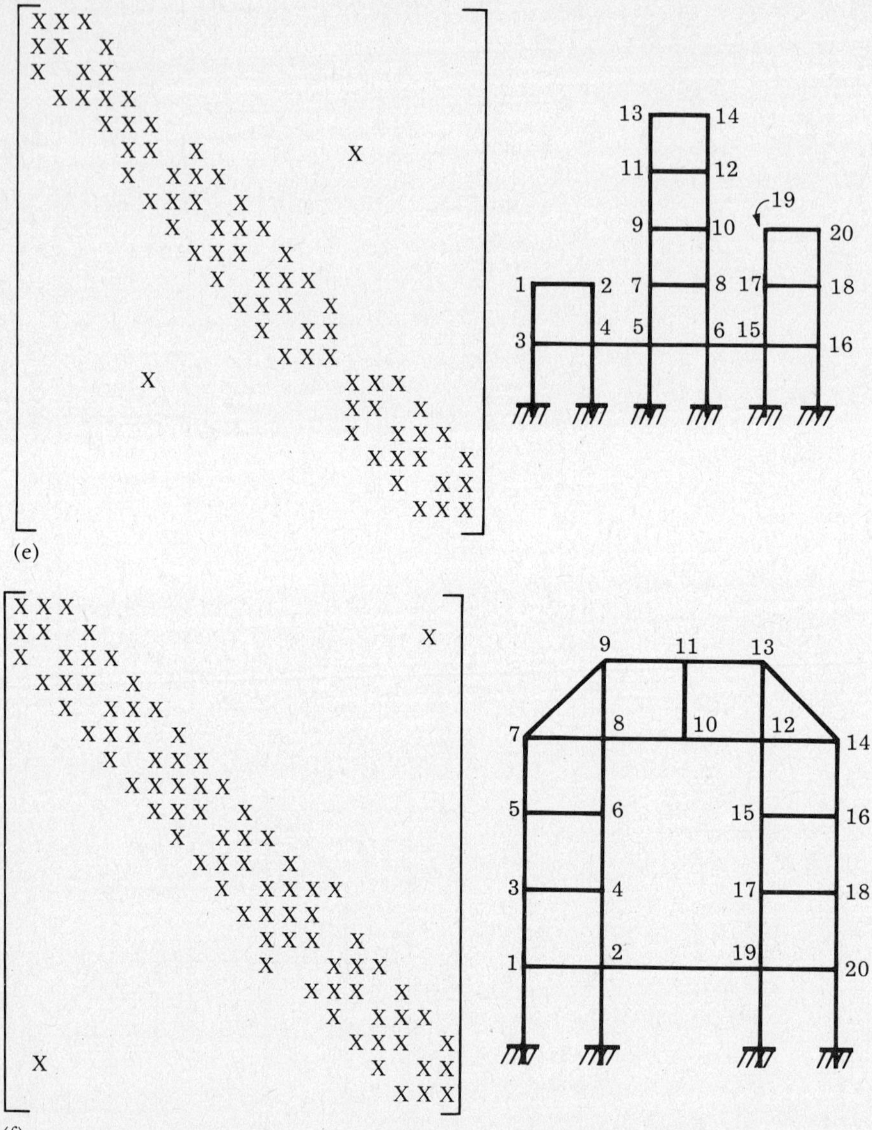

Fig. 7.8 (continued) Structural stiffness matrix banding: (e) branched structure; (f) looped structure.

unconstrained nodes with 3 degrees of freedom each. Thus each stiffness array is 60×60 in size; however, for brevity, the stiffness arrays are represented as consisting of 20 rows and 20 columns of 3×3 subarrays. Only the subarrays that will contain nonzero terms are shown, and they are indicated by the symbol X. Table 7.1 provides a numerical comparison of the degree of banding for the examples of Fig. 7.8. The semi-bandwidth m_B that is displayed in this table was calculated using Eq. 7.3

Table 7.1 COMPARISON OF BANDING IN STIFFNESS MATRICES FOR VARIOUS FRAME STRUCTURES

STRUCTURE	N	N_D	MDIF	m_B	m_B/N
Multilevel tower	60	3	1	6	.10
Multistory frame (b)	60	3	2	9	.15
Multistory frame (c)	60	3	10	33	.55
Frame, aspect ratio $\simeq$ 1	60	3	5	18	.30
Branched frame	60	3	9	30	.50
Looped frame	60	3	17	54	.90

The stiffness matrix for the tower of Fig. 7.8a has a very special form of banding known as block tridiagonal. If the X subarrays in this example were 1×1 instead of 3×3, the matrix would be purely tridiagonal, the most narrowly banded structural stiffness array possible. The block tridiagonal array for this example is also very well banded, as seen from the last column of Table 7.1, which indicates an m_B/N ratio of .10. This degree of banding means that the portion of the rectangular array used to store the stiffness coefficients will have to provide only 10 percent of the space required for the full stiffness array.

The stiffness array for the multistory frame of Fig. 7.8b is also block tridiagonal, but this time the blocks are 6×6 instead of 3×3 as in the previous example. Although this results in a somewhat wider band, this array is still very well banded, as seen from its .15 m_B/N ratio in Table 7.1. The frame of Fig. 7.8c is identical with that of Fig. 7.8b, but the nodal numbering has been changed. The result is a poorly banded stiffness matrix. Thus the ordering of nodes turns out to be crucial in determining the quality of banding of the stiffness matrix for a structure. Numbering should always proceed in the short direction rather than in the long direction.

The maximum number of nodes in the horizontal direction is the same as in the vertical direction for the structure of Fig. 7.8d, and therefore whichever direction nodal numbering proceeds in first is irrelevant to banding. It is seen that the storage requirement for this example is double that of the properly numbered multistory frame. Generalizing, we can say that it follows that the banding of stiffness arrays for a structure of aspect ratio near unity is innately inferior to that for a structure of high aspect ratio.

Branched and looped structures such as the frames of Fig. 7.8e and f, respectively, present banding difficulties also. Numbering these frames in the narrow directions of their loop or branch segments is seen to produce nicely banded arrays, except at isolated spots in the arrays. Thus what might have seemed like a good idea turns out not to be so good after all. It is more effective in these cases to number nodes in the overall narrow direction of the structure. Since these frames have overall aspect ratios near unity, this approach will yield storage requirements similar to those for the frame of Fig. 7.8d.

These few examples do not by any means cover the full range of types of stiffness matrices that might be encountered in practice. They should, however, provide you with some insight as to how various aspects of the structure or its nodal numbering influence the detailed configuration of its stiffness matrix. Note that it may not always be desirable to have to number structure nodes in a sequence that will produce a minimum bandwith. For example, such a numbering scheme may not lend itself to convenient interpretation of results, or it may conflict with a numbering scheme convenient for automatic generation of nodal data. Recognizing these conflicts, some programmers have provided for automatic bandwidth minimization to take place within the computer, with the necessary internal reordering of degrees of freedom hidden from the user.

Nodal numbering to minimize bandwidth does not, however, necessarily result in minimization of computational effort (Ref. 7.7). Because of this situation and because all structures do not possess well-banded stiffness arrays, some programmers have employed data handling techniques that do not rely on narrow banding for their efficiency. One such technique is known as the frontal, or wavefront method (Ref. 7.8).

7.3.4 Frontal Method

The frontal method of data management is shown in Fig. 7.9 making use of gaussian elimination of solve the stiffness equations arising from the example of Fig. 7.1. A careful study of Fig. 7.9 will reveal the following important aspects of this method. Stiffness equations are assembled member by member as usual; but as soon as any one equation is completely assembled, the unknown displacement associated with that degree of freedom is eliminated. Consequently, the order of member numbering determines the sequence of equation solution. It also determines the number of equations partially assembled and unsolved at any one time, which in turn determines the storage required. This situation explains the rather strange member numbering order of Fig. 7.1b. Notice that for this example, the frontal method requires that a maximum of three equations be active at any one time. A banded solution process with elimination done equation by equation (instead of block by block as in the previous section) would have required a maximum of $m_B = 4$ active equations at any one time. Thus even in this very small problem, the potential efficiency of the frontal technique is indicated. A disadvantage of the frontal method is the rather formidable housekeeping tasks associated with keeping track of all the equations in their various stages of development and moving them in and out of secondary storage. For this reason the method is recommended for use only when extremely large sets of equations are encountered, especially if they are poorly banded.

The frontal method gets its name from the fact that the solution sweeps through the structure rather like a wavefront. As a matter of interest, the position of this wavefront at each stage of the elimination process for the preceding example is shown in Fig. 7.10.

The backsubstitution process is, once again, relatively straightforward and will therefore not be shown here.

7.3.5 Substructuring and Static Condensation

Another method of obtaining solutions to very large sets of equations is to physically partition the structure that gives rise to this set of equations into a

Step 1: Effects of members 1, 2 and 3 added; D_1 eliminated:

Equation ↓ / member →	1	2	3	Elim. D_1
1	$4D_1 = 0$	$6D_1 = 0$	$10D_1 + 2D_2 = 0$	
2			$2D_1 + 4D_2 = 0$	$3.6D_2 = 0$

Step 2: Effects of members 4, 5, 6 and 7 added; D_5 eliminated:

Equation ↓ / member →	4	5	6	7	Elim. D_5
2	$5.6D_2 + D_5 = 0$	$5.6D_2 + D_5 = 0$	$5.6D_2 + D_5 = 0$	$5.6D_2 + D_5 = 0$	$5.517D_2 - .1667D_6 = 0$
5	$D_2 + 2D_5 = 0$	$D_2 + 6D_5 = 0$	$D_2 + 8D_5 = 0$	$D_2 + 12D_5 + 2D_6 = 0$	
6				$2D_5 + 4D_6 = 0$	$-.1667D_2 + 3.667D_6 = 0$

Step 3: Effects of member 8 added; D_2 eliminated:

Equation ↓ / member →	8	Elim. D_2
2	$9.517D_2 + 2D_3 - .1667D_6 = 0$	
6	$-.1667D_2 + 3.667D_6 = 0$	$.03504D_3 + 3.664D_6 = 0$
3	$2D_2 + 4D_3 = 0$	$3.580D_3 + .03504D_6 = 0$

Step 4: Effects of members 9 and 10; D_3 eliminated:

Equation ↓ / member →	9	10	Elim. D_3
6	$1.035D_3 + 5.664D_6 = 0$	$1.035D_3 + 5.664D_6 = 0$	$-.2160D_4 + 5.552D_6 = 0$
3	$5.580D_3 + 1.035D_6 = 0$	$9.58D_3 + 2D_4 + 1.035D_6 = 0$	
4		$2D_3 + 4D_4 = 10$	$3.582D_4 - .2160D_6 = 10$

Step 5: Effects of member 11 added; D_4 eliminated:

Equation ↓ / member →	11	Elim. D_4
6	$-.2160D_4 + 5.552D_6 = 0$	$5.546D_6 + .03868D_7 = .3870$
4	$5.582D_4 - .2160D_6 + D_7 = 10$	
7	$D_4 + 2D_7 = 0$	$.03868D_6 + 1.821D_7 = -1.791$

Step 6: Effects of members 12 and 13 added; D_7 eliminated:

Equation ↓ / member →	12	13	Elim. D_7
6	$9.546D_6 + 2.039D_7 = .3870$	$9.546D_6 + 2.039D_7 = .3870$	$9.014D_6 - .2607D_9 = .8542$
7	$2.039D_6 + 5.821D_7 = -1.791$	$2.039D_6 + 7.821D_7 + D_9 = -1.791$	
9		$D_7 + 2D_9 = 0$	$-.2607D_6 + 1.872D_9 = .2291$

Step 7: Effects of member 14 added; D_6 eliminated:

Equation ↓ / member →	14	Elim. D_6
6	$11.01D_6 + D_8 - .2607D_9 = .8542$	
9	$-.2607D_6 + 1.872D_9 = .2291$	$.02367D_8 + 1.868D_9 = .2494$
8	$D_6 + 2D_8 = 0$	$1.909D_8 + .02367D_9 = -.07755$

Step 8: Effects of members 15, 16, 17 and 18 added; D_8 eliminated:

Equation ↓ / member →	15, 16, 17 and 18	Elim. D_8
9	$2.024D_8 + 7.868D_9 = .2494$	$7.523D_9 = .2624$
8	$11.91D_8 + 2.024D_9 = -.07755$	

Fig. 7.9 Frontal method elimination steps.

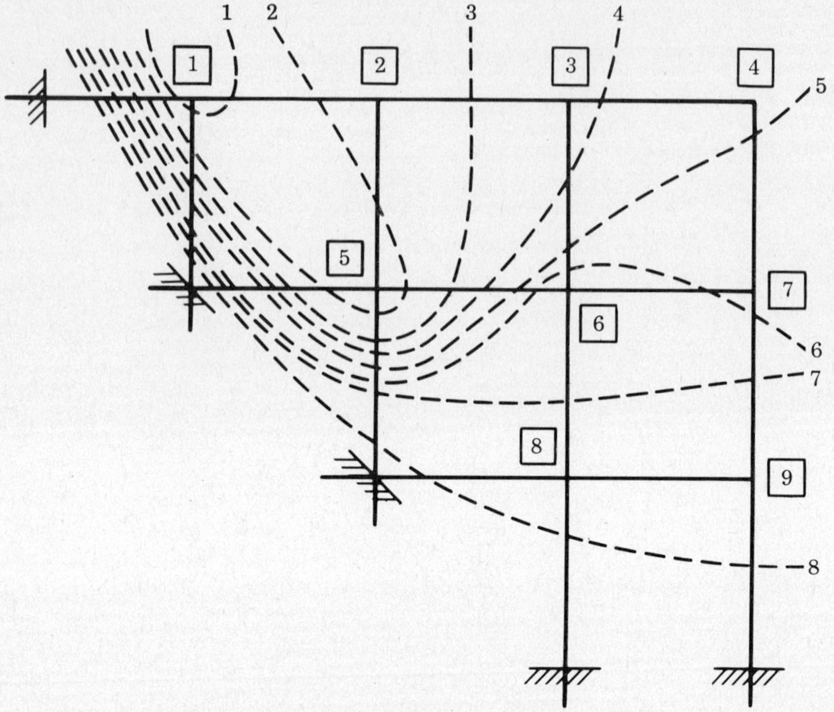

Fig. 7.10 Wavefront steps.

collection of smaller structures called substructures. To demonstrate the substructure concept, we will discuss the special case of substructuring known as static condensation. Consider the variable-section, continuous beam shown in Fig. 7.11a. Instead of representing the continuous beam by the set of six members shown in Fig. 7.11b, we can use the two more complex members shown in Fig. 7.11c. We can then develop the stiffness relationships for these complex members by means of static condensation. Consider the substructure of Fig. 7.11d, where nodal numbering has been chosen so as to make static condensation convenient. (Rearrangement of equations is not required.) As given in Chap. 4, the stiffness relationships for a general uniform beam member are

$$\begin{pmatrix} r_1 \\ r_2 \\ r_3 \\ r_4 \end{pmatrix} = \frac{2EI}{L} \begin{bmatrix} 6/L^2 & 3/L & -6/L^2 & 3/L \\ 3/L & 2 & -3/L & 1 \\ -6/L^2 & -3/L & 6/L^2 & -3/L \\ 3/L & 1 & -3/L & 2 \end{bmatrix} \begin{pmatrix} d_1 \\ d_2 \\ d_3 \\ d_4 \end{pmatrix} + \begin{pmatrix} r_{F1} \\ r_{F2} \\ r_{F3} \\ r_{F4} \end{pmatrix} \qquad (4.4a)$$

Using Eq. 4.4a, we may now write the stiffness relationships for the three sections of the substructure of Fig. 7.11d. For the first and third sections, we get

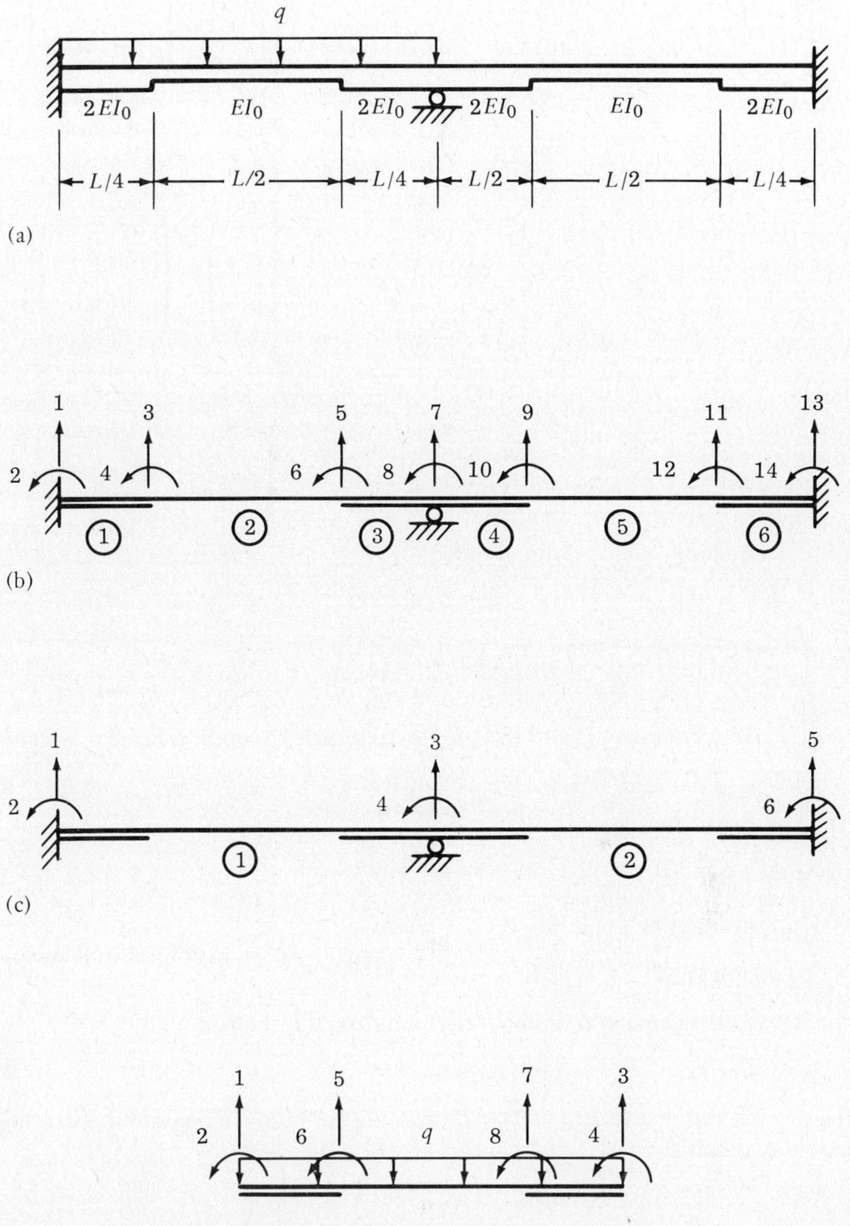

Fig. 7.11 Continuous beam of variable cross section: (a) beam physical properties; (b) conventional member and coordinate representation; (c) substructure representation; (d) typical substructure.

$$\begin{pmatrix} r_1 \\ r_2 \\ r_3 \\ r_4 \end{pmatrix} = \frac{16EI_0}{L}\begin{bmatrix} 96/L^2 & 12/L & -96/L^2 & 12/L \\ 12/L & 2 & -12/L & 1 \\ -96/L^2 & -12/L & 96/L^2 & -12/L \\ 12/L & 1 & -12/L & 2 \end{bmatrix}\begin{pmatrix} d_1 \\ d_2 \\ d_3 \\ d_4 \end{pmatrix} + \begin{pmatrix} qL/8 \\ qL^2/192 \\ qL/8 \\ -qL^2/192 \end{pmatrix} \quad \text{(A)}$$

and for the second section, we get

$$\begin{pmatrix} r_1 \\ r_2 \\ r_3 \\ r_4 \end{pmatrix} = \frac{4EI_0}{L}\begin{bmatrix} 24/L^2 & 6/L & -24/L^2 & 6/L \\ 6/L & 2 & -6/L & 1 \\ -24/L^2 & -6/L & 24/L^2 & -6/L \\ 6/L & 1 & -6/L & 2 \end{bmatrix}\begin{pmatrix} d_1 \\ d_2 \\ d_3 \\ d_4 \end{pmatrix} + \begin{pmatrix} qL/4 \\ qL^2/48 \\ qL/4 \\ -qL^2/48 \end{pmatrix} \quad \text{(B)}$$

Building the substructure stiffness relationships by inspection gives Eq. C on page 373. Since coordinates 5 to 8 are internal to the substructure and therefore not directly associated with any other substructure, the unknowns $d_5, \ldots, d_8$ may be solved for in terms of the applied loads and $d_1, \ldots, d_4$ by using the last four scalar equations matrix of Eq. C. Backsubstituting these results into the remaining four stiffness equations will eliminate unknowns $d_5, \ldots, d_8$ from the problem. Symbolically, partitioned Eq. C may be represented as

$$\begin{pmatrix} \mathbf{r}_E \\ \hline \mathbf{\Theta} \end{pmatrix} = \left[\begin{array}{c|c} \mathbf{k}_{EE} & \mathbf{k}_{EI} \\ \hline \mathbf{k}_{IE} & \mathbf{k}_{II} \end{array}\right]\begin{pmatrix} \mathbf{d}_E \\ \hline \mathbf{d}_I \end{pmatrix} + \begin{pmatrix} \mathbf{r}_{FE} \\ \hline \mathbf{r}_{FI} \end{pmatrix} \quad (7.4)$$

where I stands for internal and E stands for external. Expanding Eq. 7.4, we get

$$\mathbf{r}_E = \mathbf{k}_{EE}\mathbf{d}_E + \mathbf{k}_{EI}\mathbf{d}_I + \mathbf{r}_{FE} \quad \text{and} \quad \mathbf{\Theta} = \mathbf{k}_{IE}\mathbf{d}_E + \mathbf{k}_{II}\mathbf{d}_I + \mathbf{r}_{FI} \quad (7.5), (7.6)$$

Solving Eq. 7.6 gives

$$\mathbf{d}_I = -\mathbf{k}_{II}^{-1}\mathbf{k}_{IE}\mathbf{d}_E - \mathbf{k}_{II}^{-1}\mathbf{r}_{FI} \quad (7.7)$$

Backsubstituting this result into Eq. 7.5 gives

$$\mathbf{r}_E = (-\mathbf{k}_{EI}\mathbf{k}_{II}^{-1}\mathbf{k}_{IE} + \mathbf{k}_{EE})\,\mathbf{d}_E + (\mathbf{r}_{FE} - \mathbf{k}_{EI}\mathbf{k}_{II}^{-1}\mathbf{r}_{FI}) \quad (7.8)$$

which may be represented symbolically in the usual fashion as

$$\mathbf{r} = \mathbf{k}\mathbf{d} + \mathbf{r}_F \quad (7.9)$$

The process outlined in Eqs. 7.4 to 7.9 constitutes static condensation. Carrying this out in detail gives

$$\begin{pmatrix} r_1 \\ r_2 \\ r_3 \\ r_4 \end{pmatrix} = \frac{4EI_0}{3L}\begin{bmatrix} 16/L^2 & 8/L & -16/L^2 & 8/L \\ 8/L & 5 & -8/L & 3 \\ -16/L^2 & -8/L & 16/L^2 & -8/L \\ 8/L & 3 & -8/L & 5 \end{bmatrix}\begin{pmatrix} d_1 \\ d_2 \\ d_3 \\ d_4 \end{pmatrix} + \begin{pmatrix} .5qL \\ .09375qL^2 \\ .5qL \\ -.09375qL^2 \end{pmatrix} \quad \text{(D)}$$

$$
\begin{pmatrix} r_1 \\ r_2 \\ r_3 \\ r_4 \\ \hline 0 \\ 0 \\ 0 \\ 0 \end{pmatrix}
= \frac{4EI_0}{L}
\left[\begin{array}{cccc|cccc}
384/L^2 & 48/L & 0 & 0 & -384/L^2 & 48/L & 0 & 0 \\
48/L & 8 & 0 & 0 & -48/L & 4 & 0 & 0 \\
0 & 0 & 384/L^2 & -48/L & 0 & 0 & -384/L^2 & -48/L \\
0 & 0 & -48/L & 8 & 0 & 0 & 48/L & 4 \\
\hline
-384/L^2 & -48/L & 0 & 0 & 408/L^2 & -42/L & -24/L^2 & 6/L \\
48/L^2 & 4 & 0 & 0 & -42/L & 10 & -6/L & 1 \\
0 & 0 & -384/L^2 & 48/L & -24/L^2 & -6/L & 408/L^2 & 42/L \\
0 & 0 & -48/L & 4 & 6/L & 1 & 42/L & 10
\end{array}\right]
\begin{pmatrix} d_1 \\ d_2 \\ d_3 \\ d_4 \\ \hline d_5 \\ d_6 \\ d_7 \\ d_8 \end{pmatrix}
+
\begin{pmatrix} qL/8 \\ qL^2/192 \\ qL/8 \\ -qL^2/192 \\ \hline 3qL/8 \\ qL^2/64 \\ 3qL/8 \\ -qL^2/64 \end{pmatrix}
\qquad \text{(C)}
$$

which reduces the substructure to four degrees of freedom. In the case of the example continuous beam structure, we are fortunate to have a repeated substructure. When this situation arises in a structure, the substructure method is particularly efficient. We may therefore use the condensed stiffness relationship of Eq. D for both nonuniform members 1 and 2 of Fig. 7.11c.

Building the structure stiffness relationships (considering only the unconstrained degrees of freedom for brevity) gives

$$0 = \frac{4EI_0}{3L}(5 + 5)D_4 - .09375qL^2 \tag{E}$$

Thus the structure stiffness relationships are expressed in a single equation in this case. Solving Eq. E gives

$$D_4 = .00703125qL^3/EI_0 \tag{F}$$

The total solution to this problem by substructuring has involved the solution to one set of four equations for the repeated substructure and one set involving a single equation for the structure—a significant economy over a solution of nine simultaneous equations for this structure without substructuring. For added efficiency in the case of more complex examples, the substructures themselves can be divided into smaller substructures, and so on. This method is known as multilevel substructuring. It must be pointed out, however, that even if the stiffness matrix for the unsubstructured structure is well banded, the stiffness matrix for the substructured structure will likely not be banded. Thus the increase in storage and manipulation efficiency with the use of substructuring may not always be as significant as first expected if a band solver is being used.

Finally, it is important to notice that substructuring simply amounts to first solving certain subsets of the full set of stiffness equations and then solving the remaining altered sets. Substructuring is also a method of modularizing a problem, and as such it may lend itself to efficient use of the next generation of computers (Ref. 7.9). For a more extensive treatment of the substructure method, see Ref. 7.10.

7.3.6 Other Methods

Other very fine methods of data management are in use in addition to those already described. In fact two methods, sparse matrix methods and hypermatrix methods, are employed in the very important NASTRAN and ASKA structural analysis codes, respectively. Both these methods function well whether or not the stiffness array is banded.

Sparse matrix methods (Ref. 7.11) take their name from the fact that they allow arrays that are sparsely populated with nonzero coefficients to be efficiently stored and manipulated. These methods may be considered in a sense to be elaborations of the skyline concept discussed in Sec. 7.3.1.1. Nonzero coefficients are stored in compact or packed form along with their original sparse array addresses. Algorithms for the various matrix operations are developed to work directly with these packed arrays. Thus both storage and manipulation of zero coefficients in arrays are avoided. As we have seen, these are important factors when dealing with arrays that are sparsely populated with nonzero coefficients.

Hypermatrix methods (Ref. 7.12) involve partitioning of the stiffness array into subarrays whose addresses are stored in another array. Zero addresses for subarrays indicate that they are null and not required for manipulation; hence they are also not stored. In what is called a multilevel hypermatrix method, the address array may also be partitioned with addresses of its subarrays stored in yet a third array, and so on. Once again, algorithms are developed to allow manipulation of the coefficients stored in this form. Storage and manipulation of most zero coefficients can thus again be avoided.

As a matter of historical interest, one further data handling technique should be mentioned. The tridiagonalization method (Ref. 7.3) is very effective when applied to examples having tridiagonal stiffness arrays (see Sec. 7.3.3). It will also work well on block tridiagonal arrays if the blocks are not too large. Even before the advent of the computer, this method allowed the solution of very large sets of equations that had such stiffness arrays. However, since its range of efficient application is limited, the tridiagonalization method is no longer widely used.

7.4 OTHER SOLUTION ALGORITHMS

Up to now we have examined various data handling techniques that use a single solution algorithm: gaussian elimination. Although gaussian elimination is the most widely used solution algorithm, it is important to have some familiarity with other commonly used algorithms as well. As a matter of convenience, these additional algorithms will be developed for the reduction of symmetric arrays but will not be modified to apply to array forms encountered in complicated data handling schemes. This modification could be accomplished in a manner similar to that by which the gaussian algorithm was modified previously. As a basis for this discussion, it is useful to first demonstrate that the gaussian algorithm may be developed as a factorization or decomposition scheme as well as an elimination scheme.

7.4.1 Gaussian Decomposition

In Chap. 4 we demonstrated that the stiffness force-displacement relationships for a structure such as the truss of Fig. 7.12 would be of the form

$$\mathbf{KD} = \mathbf{R} \tag{4.14a}$$

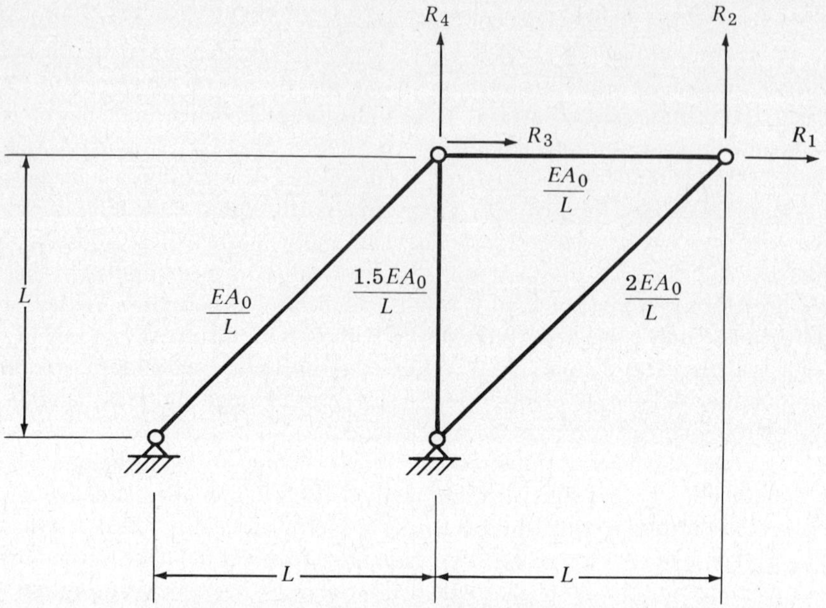

Fig. 7.12 Truss example.

Considering only the unconstrained degrees of freedom and making use of the direct stiffness method, we can obtain these relationships in detail as

$$\begin{bmatrix} 2 & -1 & -1 & .0 \\ -1 & 1 & 0 & 0 \\ -1 & 0 & 1.5 & -.5 \\ 0 & 0 & -.5 & 2 \end{bmatrix} \begin{pmatrix} D_1 \\ D_2 \\ D_3 \\ D_4 \end{pmatrix} = \begin{pmatrix} R_1 \\ R_2 \\ R_3 \\ R_4 \end{pmatrix} \tag{A}$$

Carrying out the first step of gaussian elimination gives

$$\begin{bmatrix} 2 & -1 & -1 & 0 \\ 0 & .5 & -.5 & 0 \\ 0 & -.5 & 1 & -.5 \\ 0 & 0 & -.5 & 2 \end{bmatrix} \begin{pmatrix} D_1 \\ D_2 \\ D_3 \\ D_4 \end{pmatrix} = \begin{pmatrix} R_1 \\ .5R_1 + R_2 \\ .5R_1 + R_3 \\ R_4 \end{pmatrix} \tag{B}$$

Continuing with the second step, we get

$$\begin{bmatrix} 2 & -1 & -1 & 0 \\ 0 & .5 & -.5 & 0 \\ 0 & 0 & .5 & -.5 \\ 0 & 0 & -.5 & 2 \end{bmatrix} \begin{pmatrix} D_1 \\ D_2 \\ D_3 \\ D_4 \end{pmatrix} = \begin{pmatrix} R_1 \\ .5R_1 + R_2 \\ R_1 + R_2 + R_3 \\ R_4 \end{pmatrix} \tag{C}$$

Finally, the third elimination step yields

$$\begin{bmatrix} 2 & -1 & -1 & 0 \\ 0 & .5 & -.5 & 0 \\ 0 & 0 & .5 & -.5 \\ 0 & 0 & 0 & 1.5 \end{bmatrix} \begin{pmatrix} D_1 \\ D_2 \\ D_3 \\ D_4 \end{pmatrix} = \begin{pmatrix} R_1 \\ .5R_1 + R_2 \\ R_1 + R_2 + R_3 \\ R_1 + R_2 + R_3 + R_4 \end{pmatrix} \tag{D}$$

It is now useful to rewrite Eq. D as

$$\begin{bmatrix} 2 & -1 & -1 & 0 \\ 0 & .5 & -.5 & 0 \\ 0 & 0 & .5 & -.5 \\ 0 & 0 & 0 & 1.5 \end{bmatrix} \begin{pmatrix} D_1 \\ D_2 \\ D_3 \\ D_4 \end{pmatrix} = \begin{bmatrix} 1 & 0 & 0 & 0 \\ .5 & 1 & 0 & 0 \\ 1 & 1 & 1 & 0 \\ 1 & 1 & 1 & 1 \end{bmatrix} \begin{pmatrix} R_1 \\ R_2 \\ R_3 \\ R_4 \end{pmatrix} \tag{E}$$

Representing Eq. E symbolically as

$$\mathcal{G}\mathbf{D} = [\mathcal{L}^T]^{-1}\mathbf{R} \tag{7.10}$$

and multiplying Eq. 7.10 by $\mathcal{L}^T$ gives

$$\mathcal{L}^T\mathcal{G}\mathbf{D} = \mathbf{R} \tag{7.11}$$

Carrying this last step out in detail, we get

$$\begin{bmatrix} 1 & 0 & 0 & 0 \\ -.5 & 1 & 0 & 0 \\ -.5 & -1 & 1 & 0 \\ 0 & 0 & -1 & 1 \end{bmatrix} \begin{bmatrix} 2 & -1 & -1 & 0 \\ 0 & .5 & -.5 & 0 \\ 0 & 0 & .5 & -.5 \\ 0 & 0 & 0 & 1.5 \end{bmatrix} \begin{pmatrix} D_1 \\ D_2 \\ D_3 \\ D_4 \end{pmatrix} = \begin{pmatrix} R_1 \\ R_2 \\ R_3 \\ R_4 \end{pmatrix} \tag{F}$$

Comparing Eq. 4.14a and 7.11, we see that the stiffness matrix has indeed been decomposed or factorized into the product of two arrays, or

$$\mathbf{K} = \mathcal{L}^T\mathcal{G} \tag{7.12}$$

where $\mathcal{L}^T$ is a lower triangular matrix with unity on the diagonal and $\mathcal{G}$ is an upper triangular matrix. Although this decomposition was obtained by an elimination process, it can also be accomplished directly. To demonstrate this, it is useful to rewrite Eq. 7.12 in somewhat more detailed symbolic form as

$$\begin{bmatrix} K_{11} & K_{12} & K_{13} & K_{14} \\ K_{21} & K_{22} & K_{23} & K_{24} \\ K_{31} & K_{32} & K_{33} & K_{34} \\ K_{41} & K_{42} & K_{43} & K_{44} \end{bmatrix} = \begin{bmatrix} 1 & & & \\ \ell_{12} & 1 & & \\ \ell_{13} & \ell_{23} & 1 & \\ \ell_{14} & \ell_{24} & \ell_{34} & 1 \end{bmatrix} \begin{bmatrix} g_{11} & g_{12} & g_{13} & g_{14} \\ & g_{22} & g_{23} & g_{24} \\ & & g_{33} & g_{34} \\ & & & g_{44} \end{bmatrix} \tag{7.12a}$$

Carrying out the matrix product indicated on the right-hand side of Eq. 7.12a and equating right- and left-hand sides term by term gives

$$
\begin{aligned}
&K_{11} = \mathcal{G}_{11} \quad && K_{21} = \ell_{12}\mathcal{G}_{11} && K_{31} = \ell_{13}\mathcal{G}_{11}\\
&K_{12} = \mathcal{G}_{12} && K_{22} = \ell_{12}\mathcal{G}_{12} + \mathcal{G}_{22} && K_{32} = \ell_{13}\mathcal{G}_{12} + \ell_{23}\mathcal{G}_{22}\\
&K_{13} = \mathcal{G}_{13} && K_{23} = \ell_{12}\mathcal{G}_{13} + \mathcal{G}_{23} && K_{33} = \ell_{13}\mathcal{G}_{13} + \ell_{23}\mathcal{G}_{23} + \mathcal{G}_{33}\\
&K_{14} = \mathcal{G}_{14} && K_{24} = \ell_{12}\mathcal{G}_{14} + \mathcal{G}_{24} && K_{34} = \ell_{13}\mathcal{G}_{14} + \ell_{23}\mathcal{G}_{24} + \mathcal{G}_{34}
\end{aligned}
\tag{7.13}
$$

$$
\begin{aligned}
K_{41} &= \ell_{14}\mathcal{G}_{11}\\
K_{42} &= \ell_{14}\mathcal{G}_{12} + \ell_{24}\mathcal{G}_{22}\\
K_{43} &= \ell_{14}\mathcal{G}_{13} + \ell_{24}\mathcal{G}_{23} + \ell_{34}\mathcal{G}_{33}\\
K_{44} &= \ell_{14}\mathcal{G}_{14} + \ell_{24}\mathcal{G}_{24} + \ell_{34}\mathcal{G}_{34} + \mathcal{G}_{44}
\end{aligned}
$$

Generalizing these results, we may write

$$
\begin{aligned}
K_{1j} &= \mathcal{G}_{1j}\\
\mathcal{G}_{ij} &= \mathcal{G}_{ii}\ell_{ij}\\
K_{ij} &= \mathcal{G}_{ij} + \sum_{r=1}^{i-1} \ell_{ri}\mathcal{G}_{rj}\\
K_{jj} &= \mathcal{G}_{jj} + \sum_{r=1}^{j-1} \ell_{rj}\mathcal{G}_{rj}
\end{aligned}
\tag{7.14}
$$

Equation 7.14 may be solved for $\mathcal{G}_{ij}$ and ℓ_{ij}, giving

$j = 1$

$$\mathcal{G}_{jj} = \mathcal{G}_{11} = K_{11}$$

$j = 2, 3, \ldots, N$

$i = 1, 2, \ldots, j - 1$

$$
\mathcal{G}_{ij} = \begin{cases} K_{ij} & (i = 1)\\ K_{ij} - \sum_{r=1}^{i-1} \ell_{ri}\mathcal{G}_{rj} & (i > 1)\end{cases}
\tag{7.15}
$$

$$\ell_{ij} = \frac{\mathcal{G}_{ij}}{\mathcal{G}_{ii}}$$

$$\mathcal{G}_{jj} = K_{jj} - \sum_{r=1}^{j-1} \ell_{rj}\mathcal{G}_{rj}$$

The coefficients ℓ_{jj} are not included in this algorithm because they are always equal to unity and need not be calculated. We have already noticed that the array $\mathcal{G}$ is generated directly by gaussian elimination, so it is not surprising to see the strong resemblance between the decomposition algorithm of Eq. 7.15 and the elimination algorithm represented by the first expression of Eq. 7.1a. A close inspection will reveal that the calculations are the same in both cases; only the sequencing of calculations is different.

As an aid to understanding the application of the gaussian decomposition algorithm, it will now be used to directly factorize the stiffness matrix **K** of Eq. A into the lower triangular array $\mathcal{L}^T$ and the upper triangular array $\mathcal{G}$.

The equations of Eq. 7.15 are applied, as needed, in the order listed, to determine successive columns of $\mathcal{G}$ and corresponding coefficients of $\mathcal{L}^T$.

First column of $\mathcal{G}$: $j = 1$:

$i = 1$:

$$g_{11} = K_{11} = 2$$

Second column of $\mathcal{G}$: $j = 2$:

$i = 1$:

$$g_{12} = K_{12} = -1$$

$$\ell_{12} = \frac{g_{12}}{g_{11}} = \frac{-1}{2} = -.5$$

$$g_{22} = K_{22} - \sum_{r=1}^{1} \ell_{r2} g_{r2} = 1 - (-.5)(-1) = .5$$

Third column of $\mathcal{G}$: $j = 3$:

$i = 1$:

$$g_{13} = K_{13} = -1$$

$$\ell_{13} = \frac{g_{13}}{g_{11}} = \frac{-1}{2} = -.5$$

$i = 2$:

$$g_{23} = K_{23} - \sum_{r=1}^{1} \ell_{r2} g_{r3} = 0 - (-.5)(-1) = -.5$$

$$\ell_{23} = \frac{g_{23}}{g_{22}} = \frac{-.5}{.5} = -1$$

$$g_{33} = K_{33} - \sum_{r=1}^{2} \ell_{r3} g_{r3} = 1.5 - (-.5)(-1) - (-1)(-.5) = .5$$

Fourth column of $\mathcal{G}$: $j = 4$:

$i = 1$:

$$g_{14} = K_{14} = 0$$

$$\ell_{14} = \frac{g_{14}}{g_{11}} = \frac{0}{2} = 0$$

$i = 2$:

$$g_{24} = K_{24} - \sum_{r=1}^{1} \ell_{r2} g_{r4} = 0 - (-.5) \times 0 = 0$$

$$\ell_{24} = \frac{g_{24}}{g_{22}} = \frac{0}{.5} = 0$$

$i = 3$:

$$g_{34} = K_{34} - \sum_{r=1}^{2} \ell_{r3} g_{r4} = -.5 - (-.5) \times 0 - (-1) \times 0 = -.5$$

$$\ell_{34} = \frac{g_{34}}{g_{33}} = \frac{-.5}{.5} = -1$$

$$g_{44} = K_{44} - \sum_{r=1}^{3} \ell_{r4} g_{r4} = 2 - 0 \times 0 - 0 \times 0 - (-1)(-.5) = 1.5$$

These coefficients obtained by gaussian decomposition are seen to be equivalent to the coefficients in the $\mathcal{L}^T$ and $\mathcal{G}$ arrays developed by gaussian elimination and shown in Eq. F. As we have thus demonstrated that it is possible to directly decompose the stiffness matrix according to the form of Eq. 7.12, it is now appropriate to consider obtaining a solution to the resulting set of equations. Toward this end, Eq. F is repeated as Eq. F′, with $\mathbf{R} = < 4 \ \ -10 \ \ 0 \ \ 0 >$.

$$\begin{bmatrix} 1 & 0 & 0 & 0 \\ -.5 & 1 & 0 & 0 \\ -.5 & -1 & 1 & 0 \\ 0 & 0 & -1 & 1 \end{bmatrix} \begin{bmatrix} 2 & -1 & -1 & 0 \\ 0 & .5 & -.5 & 0 \\ 0 & 0 & .5 & -.5 \\ 0 & 0 & 0 & 1.5 \end{bmatrix} \begin{pmatrix} D_1 \\ D_2 \\ D_3 \\ D_4 \end{pmatrix} = \begin{pmatrix} 4 \\ -10 \\ 0 \\ 0 \end{pmatrix} \tag{F′}$$

Equation 7.11, which is a symbolic representation of Eq. F′, is repeated here for completeness.

$$\mathcal{L}^T \mathcal{G} \mathbf{D} = \mathbf{R} \tag{7.11}$$

A solution to Eq. F′ may be conveniently obtained by introducing a new vector, $\mathbf{V} = \langle V_1 \ V_2 \ V_3 \ V_4 \rangle$, so that

$$\mathcal{G} \mathbf{D} = \mathbf{V} \tag{7.16}$$

or

$$\begin{bmatrix} 2 & -1 & -1 & 0 \\ 0 & .5 & -.5 & 0 \\ 0 & 0 & .5 & -.5 \\ 0 & 0 & 0 & 1.5 \end{bmatrix} \begin{pmatrix} D_1 \\ D_2 \\ D_3 \\ D_4 \end{pmatrix} = \begin{pmatrix} V_1 \\ V_2 \\ V_3 \\ V_4 \end{pmatrix} \tag{G}$$

and

$$\mathcal{L}^T \mathbf{V} = \mathbf{R} \tag{7.17}$$

or

$$\begin{bmatrix} 1 & 0 & 0 & 0 \\ -.5 & 1 & 0 & 0 \\ -.5 & -1 & 1 & 0 \\ 0 & 0 & -1 & 1 \end{bmatrix} \begin{pmatrix} V_1 \\ V_2 \\ V_3 \\ V_4 \end{pmatrix} = \begin{pmatrix} 4 \\ -10 \\ 0 \\ 0 \end{pmatrix} \tag{H}$$

Because of the lower triangular character of $\mathcal{L}^T$, a forward elimination solution to Eq. H simplifies to a forward substitution solution that is perhaps most easily carried out in scalar form. Thus the first of the scalar equations in matrix Eq. H yields

$$V_1 = 4$$

Substituting this result into the second equation gives

$$V_2 = -10 + (.5)\,4 = -8$$

Continuing in this fashion into the third and fourth equations, we get, respectively,

$$V_3 = 0 + (.5)\,4 - 8 = -6$$

and

$$V_4 = 0 - 6 = -6$$

This solution for the vector **V** may be obtained symbolically by multiplying Eq. 7.17 by $(\mathcal{L}^T)^{-1}$, giving

$$\mathbf{V} = [\mathcal{L}^T]^{-1}\mathbf{R} \tag{7.18}$$

Substituting the detailed results into Eq. G, we get

$$\begin{bmatrix} 2 & -1 & -1 & 0 \\ 0 & .5 & -.5 & 0 \\ 0 & 0 & .5 & -.5 \\ 0 & 0 & 0 & 1.5 \end{bmatrix} \begin{pmatrix} D_1 \\ D_2 \\ D_3 \\ D_4 \end{pmatrix} = \begin{pmatrix} 4 \\ -8 \\ -6 \\ -6 \end{pmatrix} \tag{I}$$

A backward substitution process now yields

$$\begin{aligned} D_4 &= \frac{1}{1.5}(-6) = -4 \\ D_3 &= \frac{1}{.5}[-6 + (.5)(-4)] = -16 \\ D_2 &= \frac{1}{.5}[-8 + (.5)(-16)] = -32 \\ D_1 &= \frac{1}{2}(4 - 32 - 16) = -22 \end{aligned} \tag{J}$$

The symbolic representation of this solution is obtained by substituting from Eq. 7.18 into 7.16 and multiplying that result by $\mathcal{G}^{-1}$, which yields

$$\mathbf{D} = \mathcal{G}^{-1}(\mathcal{L}^T)^{-1}\mathbf{R} \tag{7.19}$$

Of course, the inverses of the matrices $\mathcal{G}$ and $\mathcal{L}^T$ are never actually calculated, as may be seen from the detailed solution displayed earlier. However, the concept of matrix inversion allows the symbolic solution to be displayed conveniently.

For computer applications, the matrix $\mathcal{G}$ can be overwritten on the stiffness matrix **K**, and the matrix $\mathcal{L}^T$ need not be stored, since it can be obtained simply and directly from $\mathcal{G}$. And, as was the case with gaussian elimination, gaussian decomposition can be developed to operate row-wise as well as columnwise. Thus gaussian decomposition also lends itself to application along with any of the data handling schemes discussed in Sec. 7.3.

There are other decomposition methods that have this same broad range of application. One of these is a modification of the gaussian decomposition method.

7.4.2 Modified Gaussian Decomposition

In presenting the modified gaussian decomposition method, we will also, coincidentally, broaden the example problem slightly by considering multiple load cases. All the techniques for simultaneous-equation solutions discussed in this chapter are capable of handling multiple load cases; but in the interest of simplicity, we have avoided the subject thus far. You are urged not to misconstrue the consideration of this more general example now as an implication of a broader capability of the modified gaussian method. The more general example is offered now only because the circumstances are considered appropriate.

Rewriting Eq. F of Sec. 7.4.1 with the **R** vector and the **D** vector both expanded to 4 × 2, we have

$$\begin{bmatrix} 1 & 0 & 0 & 0 \\ -.5 & 1 & 0 & 0 \\ -.5 & -1 & 1 & 0 \\ 0 & 0 & -1 & 1 \end{bmatrix} \begin{bmatrix} 2 & -1 & -1 & 0 \\ 0 & .5 & -.5 & 0 \\ 0 & 0 & .5 & -.5 \\ 0 & 0 & 0 & 1.5 \end{bmatrix} \begin{pmatrix} D_{11} & D_{12} \\ D_{21} & D_{22} \\ D_{31} & D_{32} \\ D_{41} & D_{42} \end{pmatrix} = \begin{pmatrix} 4 & 12 \\ -10 & 0 \\ 0 & 6 \\ 0 & 0 \end{pmatrix} \tag{A}$$

Equation A may still be represented symbolically as

$$\mathcal{L}^T\mathcal{G}\mathcal{D} = \mathbf{R} \tag{7.11}$$

The first and second columns of **R** of Eq. A are the load vectors for Cases I and II, respectively, while the first and second columns of **D** are the corresponding nodal displacement vectors.

If we recall the relationships among the coefficients of the $\mathcal{G}$ and $\mathcal{L}^T$ arrays, it seems rather natural to factor a diagonal array out of $\mathcal{G}$, so that Eq. A may be written as

$$\begin{bmatrix} 1 & 0 & 0 & 0 \\ -.5 & 1 & 0 & 0 \\ -.5 & -1 & 1 & 0 \\ 0 & 0 & -1 & 1 \end{bmatrix} \begin{bmatrix} 2 & 0 & 0 & 0 \\ 0 & .5 & 0 & 0 \\ 0 & 0 & .5 & 0 \\ 0 & 0 & 0 & 1.5 \end{bmatrix} \begin{bmatrix} 1 & -.5 & -.5 & 0 \\ 0 & 1 & -1 & 0 \\ 0 & 0 & 1 & -1 \\ 0 & 0 & 0 & 1 \end{bmatrix} \begin{pmatrix} D_{11} & D_{12} \\ D_{21} & D_{22} \\ D_{31} & D_{32} \\ D_{41} & D_{42} \end{pmatrix} = \begin{pmatrix} 4 & 12 \\ -10 & 0 \\ 0 & 6 \\ 0 & 0 \end{pmatrix} \tag{A'}$$

or

$$\mathcal{L}^T \mathcal{D} \mathcal{L} \mathbf{D} = \mathbf{R} \tag{7.20}$$

Comparing Eqs. 4.14a and 7.20, we can obtain a symbolic representation of modified gaussian decomposition:

$$\mathbf{K} = \mathcal{L}^T \mathcal{D} \mathcal{L} \tag{7.21}$$

The modified gaussian decomposition algorithm that will yield the coefficients of Eq. A′ may be obtained in a manner similar to that used to obtain the unmodified gaussian decomposition algorithm of Sec. 7.4.1, giving (Ref. 7.13)

$$j = 1$$

$$d_{jj} = d_{11} = K_{11}$$

$$j = 2, 3, \ldots, N$$

$$i = 1, 2, \ldots, j - 1$$

$$g_{ij} = \begin{cases} K_{ij} & (i = 1) \\ K_{ij} - \sum_{r=1}^{i-1} \ell_{ri} g_{rj} & (i > 1) \end{cases} \tag{7.22}$$

$$\ell_{ij} = \frac{g_{ij}}{d_{ii}}$$

$$d_{jj} = K_{jj} - \sum_{r=1}^{j-1} \ell_{rj} g_{rj}$$

The step in which the coefficient g_{ij} is calculated in this algorithm is shown indented to emphasize that it is an intermediate step and that g_{ij} will not be saved. This algorithm will require no more computer storage space than the unmodified algorithm, because the diagonal coefficients of $\mathcal{D}$ can be stored in place of the diagonal coefficients of $\mathcal{L}$, which are known to be unity and therefore need not be stored. Finally, as before, the coefficients that arise during decomposition (the coefficients in the combined $\mathcal{D}$ and $\mathcal{L}$ array in this case) can be overwritten on **K** as decomposition proceeds.

The application of the modified gaussian decomposition algorithm of Eq. 7.22 is very similar to that of the standard algorithm of Eq. 7.15. Since we have already demonstrated the standard algorithm, we will forgo a demonstration of the modified algorithm. On the other hand, the solutions to the equations generated by using the two decomposition methods are enough different to make it worthwhile to consider the detailed solution of Eq. A′. Of additional interest is the handling of the multiple load and displacement vectors during the solution process.

Once again an array **V** is introduced, this time as

$$\mathcal{D}\mathcal{L}\mathbf{D} = \mathbf{V} \tag{7.23}$$

or

$$\begin{bmatrix} 2 & 0 & 0 & 0 \\ 0 & .5 & 0 & 0 \\ 0 & 0 & .5 & 0 \\ 0 & 0 & 0 & 1.5 \end{bmatrix}\begin{bmatrix} 1 & -.5 & -.5 & 0 \\ 0 & 1 & -1 & 0 \\ 0 & 0 & 1 & -1 \\ 0 & 0 & 0 & 1 \end{bmatrix}\begin{pmatrix} D_{11} & D_{12} \\ D_{21} & D_{22} \\ D_{31} & D_{32} \\ D_{41} & D_{42} \end{pmatrix} = \begin{pmatrix} V_{11} & V_{12} \\ V_{21} & V_{22} \\ V_{31} & V_{32} \\ V_{41} & V_{42} \end{pmatrix} \tag{B}$$

and

$$\mathcal{L}^T\mathbf{V} = \mathbf{R} \tag{7.24}$$

or

$$\begin{bmatrix} 1 & 0 & 0 & 0 \\ -.5 & 1 & 0 & 0 \\ -.5 & -1 & 1 & 0 \\ 0 & 0 & -1 & 1 \end{bmatrix}\begin{pmatrix} V_{11} & V_{12} \\ V_{21} & V_{22} \\ V_{31} & V_{32} \\ V_{41} & V_{42} \end{pmatrix} = \begin{pmatrix} 4 & 12 \\ -10 & 0 \\ 0 & 6 \\ 0 & 0 \end{pmatrix} \tag{C}$$

Because of the multiple load and **V** vectors, it is convenient to solve Eq. C in matrix form, using a series of row operations. Thus adding one half of the first row to the second and third rows gives

$$\begin{bmatrix} 1 & 0 & 0 & 0 \\ 0 & 1 & 0 & 0 \\ 0 & -1 & 1 & 0 \\ 0 & 0 & -1 & 1 \end{bmatrix}\begin{pmatrix} V_{11} & V_{12} \\ V_{21} & V_{22} \\ V_{31} & V_{32} \\ V_{41} & V_{42} \end{pmatrix} = \begin{pmatrix} 4 & 12 \\ -8 & 6 \\ 2 & 12 \\ 0 & 0 \end{pmatrix} \tag{D}$$

Now adding the second row to the third row, we get

$$\begin{bmatrix} 1 & 0 & 0 & 0 \\ 0 & 1 & 0 & 0 \\ 0 & 0 & 1 & 0 \\ 0 & 0 & -1 & 1 \end{bmatrix}\begin{pmatrix} V_{11} & V_{12} \\ V_{21} & V_{22} \\ V_{31} & V_{32} \\ V_{41} & V_{42} \end{pmatrix} = \begin{pmatrix} 4 & 12 \\ -8 & 6 \\ -6 & 18 \\ 0 & 0 \end{pmatrix} \tag{E}$$

Finally, adding the third row to the fourth row yields

$$\begin{bmatrix} 1 & 0 & 0 & 0 \\ 0 & 1 & 0 & 0 \\ 0 & 0 & 1 & 0 \\ 0 & 0 & 0 & 1 \end{bmatrix} \begin{pmatrix} V_{11} & V_{12} \\ V_{21} & V_{22} \\ V_{31} & V_{32} \\ V_{41} & V_{42} \end{pmatrix} = \begin{pmatrix} 4 & 12 \\ -8 & 6 \\ -6 & 18 \\ -6 & 18 \end{pmatrix} \tag{F}$$

or

$$\begin{bmatrix} V_{11} & V_{12} \\ V_{21} & V_{22} \\ V_{31} & V_{32} \\ V_{41} & V_{42} \end{bmatrix} = \begin{bmatrix} 4 & 12 \\ -8 & 6 \\ -6 & 18 \\ -6 & 18 \end{bmatrix} \tag{G}$$

Notice that the first column of **V** matches the solution obtained previously and shown on the right-hand side of Eq. I in Sec. 7.4.1. The solution displayed in Eq. G may be represented symbolically as

$$\mathbf{V} = [\mathcal{L}^T]^{-1}\mathbf{R} \tag{7.25}$$

Substituting from Eq. G into B gives

$$\begin{bmatrix} 2 & 0 & 0 & 0 \\ 0 & .5 & 0 & 0 \\ 0 & 0 & .5 & 0 \\ 0 & 0 & 0 & 1.5 \end{bmatrix} \begin{bmatrix} 1 & -.5 & -.5 & 0 \\ 0 & 1 & -1 & 0 \\ 0 & 0 & 1 & -1 \\ 0 & 0 & 0 & 1 \end{bmatrix} \begin{pmatrix} D_{11} & D_{12} \\ D_{21} & D_{22} \\ D_{31} & D_{32} \\ D_{41} & D_{42} \end{pmatrix} = \begin{pmatrix} 4 & 12 \\ -8 & 6 \\ -6 & 18 \\ -6 & 18 \end{pmatrix} \tag{H}$$

or, once again,

$$\mathcal{D}\mathcal{L}\mathbf{D} = \mathbf{V} \tag{7.23}$$

Since $\mathcal{D}$ is a diagonal array, its inverse is easily determined as a diagonal array with coefficients equal to the reciprocals of the corresponding coefficients in the original array. Multiplying Eq. H by $\mathcal{D}^{-1}$ yields

$$\begin{bmatrix} 1 & -.5 & -.5 & 0 \\ 0 & 1 & -1 & 0 \\ 0 & 0 & 1 & -1 \\ 0 & 0 & 0 & 1 \end{bmatrix} \begin{pmatrix} D_{11} & D_{12} \\ D_{21} & D_{22} \\ D_{31} & D_{32} \\ D_{41} & D_{42} \end{pmatrix} = \begin{bmatrix} .5 & 0 & 0 & 0 \\ 0 & 2 & 0 & 0 \\ 0 & 0 & 2 & 0 \\ 0 & 0 & 0 & 2/3 \end{bmatrix} \begin{pmatrix} 4 & 12 \\ -8 & 6 \\ -6 & 18 \\ -6 & 18 \end{pmatrix} \tag{I}$$

or

$$\mathcal{L}\mathbf{D} = \mathcal{D}^{-1}\mathbf{V} \tag{7.26}$$

Equation I may be simplified to

$$\begin{bmatrix} 1 & -.5 & -.5 & 0 \\ 0 & 1 & -1 & 0 \\ 0 & 0 & 1 & -1 \\ 0 & 0 & 0 & 1 \end{bmatrix} \begin{pmatrix} D_{11} & D_{12} \\ D_{21} & D_{22} \\ D_{31} & D_{32} \\ D_{41} & D_{42} \end{pmatrix} = \begin{pmatrix} 2 & 6 \\ -16 & 12 \\ -12 & 36 \\ -4 & 12 \end{pmatrix} \tag{I'}$$

Backward elimination row operations may now be carried out on Eq. I′. Starting by adding the fourth row to the third, we get

$$\begin{bmatrix} 1 & -.5 & -.5 & 0 \\ 0 & 1 & -1 & 0 \\ 0 & 0 & 1 & 0 \\ 0 & 0 & 0 & 1 \end{bmatrix} \begin{pmatrix} D_{11} & D_{12} \\ D_{21} & D_{22} \\ D_{31} & D_{32} \\ D_{41} & D_{42} \end{pmatrix} = \begin{pmatrix} 2 & 6 \\ -16 & 12 \\ -16 & 48 \\ -4 & 12 \end{pmatrix} \tag{J}$$

Adding the third row to the second row, and one half of the third row to the first, we get

$$\begin{bmatrix} 1 & -.5 & 0 & 0 \\ 0 & 1 & 0 & 0 \\ 0 & 0 & 1 & 0 \\ 0 & 0 & 0 & 1 \end{bmatrix} \begin{pmatrix} D_{11} & D_{12} \\ D_{21} & D_{22} \\ D_{31} & D_{32} \\ D_{41} & D_{42} \end{pmatrix} = \begin{pmatrix} -6 & 30 \\ -32 & 60 \\ -16 & 48 \\ -4 & 12 \end{pmatrix} \tag{K}$$

Finally, adding one half of the second row to the first yields

$$\begin{bmatrix} 1 & 0 & 0 & 0 \\ 0 & 1 & 0 & 0 \\ 0 & 0 & 1 & 0 \\ 0 & 0 & 0 & 1 \end{bmatrix} \begin{pmatrix} D_{11} & D_{12} \\ D_{21} & D_{22} \\ D_{31} & D_{32} \\ D_{41} & D_{42} \end{pmatrix} = \begin{pmatrix} -22 & 60 \\ -32 & 60 \\ -16 & 48 \\ -4 & 12 \end{pmatrix} \tag{L}$$

or

$$\begin{pmatrix} D_{11} & D_{12} \\ D_{21} & D_{22} \\ D_{31} & D_{32} \\ D_{41} & D_{42} \end{pmatrix} = \begin{pmatrix} -22 & 60 \\ -32 & 60 \\ -16 & 48 \\ -4 & 12 \end{pmatrix} \tag{L'}$$

The first column of the **D** array of Eq. L′ represents the solution for the nodal displacements corresponding to the first load vector. This result is seen to agree with the solution obtained by unmodified gaussian decomposition and shown in Eq. J of Sec. 7.4.1. Symbolically, the solution displayed in Eq. L′ may be obtained by operating on Eq. 7.26, giving

$$\mathbf{D} = \mathcal{L}^{-1}\mathcal{D}^{-1}\mathbf{V} \tag{7.27}$$

7.4.3 Choleskian Decomposition

Another widely used method of solving large sets of simultaneous equations is Choleskian decomposition. It is convenient, though not necessary, to base the development of Cholesky's method on modified gaussian decomposition. For this purpose, Eq. A′ of Sec. 7.4.2 is repeated here, but this time, for simplicity, with only a single displacement vector.

$$\begin{bmatrix} 1 & 0 & 0 & 0 \\ -.5 & 1 & 0 & 0 \\ -.5 & -1 & 1 & 0 \\ 0 & 0 & -1 & 1 \end{bmatrix}\begin{bmatrix} 2 & 0 & 0 & 0 \\ 0 & .5 & 0 & 0 \\ 0 & 0 & .5 & 0 \\ 0 & 0 & 0 & 1.5 \end{bmatrix}\begin{bmatrix} 1 & -.5 & -.5 & 0 \\ 0 & 1 & -1 & 0 \\ 0 & 0 & 1 & -1 \\ 0 & 0 & 0 & 1 \end{bmatrix}\begin{pmatrix} D_1 \\ D_2 \\ D_3 \\ D_4 \end{pmatrix} = \begin{pmatrix} 4 \\ -10 \\ 0 \\ 0 \end{pmatrix} \tag{A}$$

It may be recalled that Eq. A represents the stiffness relationships for the example of Fig. 7.12, with the stiffness matrix decomposed into three arrays by modified gaussian decomposition. The symbolic representation of Eq. A is

$$\mathcal{L}^T\mathcal{D}\mathcal{L}\mathbf{D} = \mathbf{R} \tag{7.20}$$

To develop Choleskian decomposition, it is necessary to factor the $\mathcal{D}$ array into two arrays, each of whose diagonal terms is equal to the square root of the corresponding term of the $\mathcal{D}$ array. In the case of Eq. A this gives

$$\begin{bmatrix} 1 & 0 & 0 & 0 \\ -.5 & 1 & 0 & 0 \\ -.5 & -1 & 1 & 0 \\ 0 & 0 & -1 & 1 \end{bmatrix}\begin{bmatrix} 1.414 & 0 & 0 & 0 \\ 0 & .7071 & 0 & 0 \\ 0 & 0 & .7071 & 0 \\ 0 & 0 & 0 & 1.225 \end{bmatrix}\begin{bmatrix} 1.414 & 0 & 0 & 0 \\ 0 & .7071 & 0 & 0 \\ 0 & 0 & .7071 & 0 \\ 0 & 0 & 0 & 1.225 \end{bmatrix}$$

$$\times\begin{bmatrix} 1 & -.5 & -.5 & 0 \\ 0 & 1 & -1 & 0 \\ 0 & 0 & 1 & -1 \\ 0 & 0 & 0 & 1 \end{bmatrix}\begin{pmatrix} D_1 \\ D_2 \\ D_3 \\ D_4 \end{pmatrix} = \begin{pmatrix} 4 \\ -10 \\ 0 \\ 0 \end{pmatrix} \tag{B}$$

or

$$\mathcal{L}^T\mathcal{D}^{1/2}\mathcal{D}^{1/2}\mathcal{L}\mathbf{D} = \mathbf{R} \tag{7.28}$$

Now, carrying out the products $\mathcal{L}^T\mathcal{D}^{1/2}$ and $\mathcal{D}^{1/2}\mathcal{L}$ in detail, we get

$$\begin{bmatrix} 1.414 & 0 & 0 & 0 \\ -.7071 & .7071 & 0 & 0 \\ -.7071 & -.7071 & .7071 & 0 \\ 0 & 0 & -.7071 & 1.225 \end{bmatrix}\begin{bmatrix} 1.414 & -.7071 & -.7071 & 0 \\ 0 & .7071 & -.7071 & 0 \\ 0 & 0 & .7071 & -.7071 \\ 0 & 0 & 0 & 1.225 \end{bmatrix}\begin{pmatrix} D_1 \\ D_2 \\ D_3 \\ D_4 \end{pmatrix} = \begin{pmatrix} 4 \\ -10 \\ 0 \\ 0 \end{pmatrix} \tag{C}$$

or

$$\overline{\mathcal{L}}^T\overline{\mathcal{L}}\mathbf{D} = \mathbf{R} \tag{7.29}$$

where

$$\begin{aligned} \overline{\mathcal{L}}^T &= \mathcal{L}^T\mathcal{D}^{1/2} \\ \overline{\mathcal{L}} &= \mathcal{D}^{1/2}\mathcal{L} \end{aligned} \tag{7.30}$$

Comparing Eqs. 4.14a and 7.29, we find that the symbolic representation of Choleskian decomposition is

$$\mathbf{K} = \overline{\mathcal{L}}^T\overline{\mathcal{L}} \tag{7.31}$$

Using an approach similar to that used in Sec. 7.4.1, we could now determine the Choleskian decomposition algorithm as (Ref. 7.12)

$$\begin{aligned} & j = 1, 2, \ldots, N \\ & \bar{\ell}_{ij} = \frac{K_{ij} - \sum_{r=1}^{i-1} \bar{\ell}_{ri}\bar{\ell}_{rj}}{\bar{\ell}_{ii}} \qquad i = 1, 2, \ldots, j-1 \\ & \bar{\ell}_{jj} = \left(K_{jj} - \sum_{r=1}^{j-1} \bar{\ell}_{rj}^2\right)^{1/2} \end{aligned} \tag{7.32}$$

The algorithm of Eq. 7.32 is written somewhat more compactly than were the previous two decomposition algorithms, but its application is similar to theirs. It is therefore left to you to demonstrate that this algorithm can be used to obtain the coefficients of the decomposed stiffness matrix arrays of Eq. C. We will also forgo a detailed solution of Eq. C because of its similarity to the solution carried out in Sec. 7.4.1.

As in the cases of previously discussed solution algorithms, Choleskian decomposition can be applied either columnwise as indicated in Eq. 7.32 or row-wise. In addition, once again the coefficients generated during decomposition can be overwritten on the stiffness matrix. Thus the Choleskian algorithm has the same broad range of application as the other algorithms already discussed. Cholesky's method does, however, require a small percentage of extra numerical operations in application (Ref. 7.13).

Before we leave the subject of simultaneous equation solution algorithms, it is appropriate to offer one further remark. Because of the very large number of times that the innermost loops of these algorithms are used in the solution of a set of equations, it has been found (Ref. 7.13) that coding these loops in assembler language adds significantly to the efficiency of operation of the algorithms at the expense of making the solution routine machine-dependent.

7.5 ERRORS AND CONDITIONING

There are a variety of causes for solution error, some of which are more under the control of the analyst to avoid or detect and correct than others. The major causes of error that the analyst can avoid by exercising reasonable care are: (1) improper idealization, (2) data input blundering, and (3) data output misinterpretation. Sources of error that may be difficult or impossible to avoid are (1) wide variations in the magnitudes of stiffnesses of the individual members of the structure, (2) static instability of the structure, and (3) very large structures.

7.5.1 Idealization Errors

It is the responsibility of the analyst to develop a mathematical model that properly represents the actual behavior of the structure being studied. For example, the use of a structural analysis program that is based on linear elastic theory to analyze a structure that undergoes very large displacements will produce erroneous results regardless of how good the computer program is or how much care is used in the preparation of data. Thus, as emphasized in Chap. 1, the analyst must possess and exercise a good understanding of structural behavior. The analyst must also make the effort to learn the limitations of any computer program used.

7.5.2 Data Input Blundering

Data input blundering, as another source of solution error, is self-descriptive and needs no further elaboration. It is worth mention, however, that the incidence of this particular type of error will likely diminish markedly in the future as graphical input of data becomes better developed and more widely available.

7.5.3 Data Output Misinterpretation

In the case of discrete-membered structures being studied in this text, data output misinterpretation is mainly a matter of not being critical enough of the results provided by the computer. Again as emphasized in Chap. 1, computer results must be viewed with suspicion. They must be scrutinized to determine their reasonableness. As the analyst, you must continually be asking the rhetorical question, "Does this make sense?" If the program does not provide a self-check on equilibrium, you must do so. Even if the program is self-checking, you will want to make your own independent spot checks. These precautions will not detect all types of errors, but they will detect some. It is important not to become complacent, however, just because a good equilibrium check is obtained. Modeling errors and most data input errors will be transparent to equilibrium checks since the program will often operate on the incorrect data supplied and produce a valid solution to an invalid model. Equilibrium checks will mainly detect program malfunctions and round-off errors. (Round-off errors will be discussed in the next section.) Thus we are left with "reasonableness" as a qualitative check on results. Quantitative checks can be obtained by working the problem with another computer program, preferably on a different computer and/or by another method of analysis. It may also be useful to re-solve the problem using a different sequence of equations. Sometimes exact or approximate spot checks of results carried out independently by hand are also possible. The amount of effort you will need to put into checking results will depend on your experience and the complexity and importance of the structure.

7.5.4 Relatively Large Stiffnesses

It was demonstrated in Sec. 5.4 that if the very large stiffnesses of support members contribute to off-diagonal terms in the stiffness matrix, the resulting set of simultaneous equations may be ill-conditioned. A set of equations is said to be ill-conditioned if small changes in the coefficients result in large changes in the solution. Therefore if certain members other than support members are relatively stiffer than the remaining members of the structure, ill-conditioning can also occur. How serious this ill-conditioning will be, if it does occur, will depend on the magnitude of the relative difference in stiffness, the location and orientation of the relatively stiff members in the structure, and details of the solution procedure. The numerical errors that occur in such a case are truncation and round-off errors. The computer represents the stiffness coefficients as well as forces in the load vector with a finite number of significant figures. Thus the structure stiffness coefficients will be initially approximated by a finite number of significant figures, and the resulting error is referred to as a *truncation* error. Truncation error is the obvious error that occurred in the example solution of Sec. 5.4. A less obvious error that also occurred during that solution process was *round-off* error. Round-off error results from carrying out the arithmetic steps in the solution process with the use of a finite number of significant figures. In this example, truncation error was serious, whereas round-off error was negligible and was, in fact, ignored. In the gaussian elimination solution shown in Eqs. A to H of Sec. 7.2.1, round-off error is seen in the slight loss of symmetry of off-diagonal terms of **K** that occurs during the solution process.

It is seen that accuracy lost due to truncation is irretrievable; but accuracy lost due to round-off may be, at least partially, recovered. A method often suggested for reducing round-off error is known as iteration on residuals. Recalling the symbolic representation of the structure stiffness force-displacement relationships of Eq. 4.14a, we have

$$\mathbf{KD} = \mathbf{R} \tag{4.14a}$$

From our discussion it is clear that the computer will yield only an approximate solution $\mathbf{D}_a$ to this equation. Backsubstituting this approximate solution into Eq. 4.14a will yield a corresponding approximate load vector $\mathbf{R}_a$, or

$$\mathbf{R}_a = \mathbf{KD}_a \tag{7.33}$$

The residual load that has been unaccounted for during the solution process is seen to be

$$\delta\mathbf{R} = \mathbf{R} - \mathbf{R}_a = \mathbf{R} - \mathbf{KD}_a \tag{7.34}$$

If $\delta\mathbf{R}$ is null, equilibrium is exactly satisfied by the solution. But this is a necessary and sufficient check on the accuracy of the solution only if the structure is statically determinate. This limitation holds because there is a unique set of internal forces that can produce static equilibrium in a statically determinate structure, whereas, by definition, there exists an infinite number of sets of internal forces that can produce static equilibrium in a statically indeterminate structure. Thus the fact that $\delta\mathbf{R}$ is small or null will not, in general, be sufficient to ensure an accurate solution.

The detection of errors due to round-off can be made certain by extending the equilibrium check we have described to include a complete check of not only equilibrium but also compatibility for all of the nodes as well as the members of the structure. But this is an expensive and little-used method, since even after detection there still remains the problem of improving accuracy.

Although iteration to reduce the magnitude of the residual forces may improve the solution, there is no guarantee of its effectiveness. Even so, the concept is presented for completeness. After the residual forces are determined, they are substituted as a new load vector into the original stiffness relationships, which are then solved to obtain a correction to the displacement vector. Thus

$$\delta\mathbf{D} = \mathbf{K}^{-1}\delta\mathbf{R} \tag{7.35}$$

and

$$\mathbf{D}_a \longrightarrow \mathbf{D}_a + \delta\mathbf{D} \tag{7.36}$$

The cycle of finding residuals, solving for displacement corrections, and obtaining new displacement approximations is repeated until the residuals are acceptably small. It has been shown (Ref. 7.14) that this process of iteration must be carried out in double-precision arithmetic to yield improved results, and even this is no guarantee of improvement.

Because of the uncertainty of the effectiveness of residuals to assure accuracy or of iteration on residuals to improve accuracy, many analysts prefer the alternative of using double-precision arithmetic to solve the original problem. Comparison of single- and double-precision solutions provides an indicator of ill-conditioning, while the double-precision solution itself provides improved accuracy. For this approach to work, it is necessary not only to solve but also to formulate the problem in double precision; otherwise truncation error will not be reduced. Double-precision representation and arithmetic is, of course, expensive; but it is often the only viable method of obtaining a good solution when ill-conditioning exists.

Other, less commonly used methods of improving solution accuracy have been suggested, and research continues. If you are interested in learning about these methods, see Ref. 7.14.

7.5.5 Static Instability

As explained in Chap. 1, a statically unstable structure is a structure in which the basic unknown forces are unable to satisfy all equilibrium conditions. This situation may be due to the existence of an insufficient number of basic forces or a poor arrangement of basic forces (for example, forces that can be induced only as a result of very large displacements). The beam of Fig. 7.13a is in the former category, whereas the beam of Fig. 7.13b is in the latter. In both cases, the stiffness matrix will be singular—the ultimate case of ill-conditioning—and no solution can be obtained. It may happen that a structure is merely close to being statically unstable, as is the beam of Fig. 7.13c. In this case the stiffness matrix will be almost singular, and the stiffness equations will be ill-conditioned. Very soft members can have an effect similar to that of poorly placed or nonexistent members; this effect occurs because, in the limit, as a member becomes softer and softer, it will disappear altogether. All of the near-instability conditions must be detected and remedied by methods such as those described in the previous section. Fortunately, these conditions do not often actually exist in structures, and their manifestation in a set of equations is then a sign of an input error that can be easily remedied once detected.

7.5.6 Very Large Structures

Very large structures generally are accompanied by very large sets of simultaneous equations. Even if the equations are well-conditioned, the largeness of the set may itself pose an accuracy problem, since round-off errors can have a cumulative effect. If this problem occurs, it may once again be necessary to use double-precision arithmetic.

7.5.7 Conditioning

We have already discussed reasons to suspect ill-conditioning of equations and considered methods to obtain solutions to sets of equations even though they may be ill-conditioned. There are also methods of predicting the quality of the conditioning of a given set of equations. One such method (Ref. 7.14) develops a condition number as the ratio of the maximum and minimum eigenvalues of the stiffness matrix and uses this condition number to obtain an estimate of the number of accurate digits that can be expected in the solution for displacements, given the number of significant figures carried in the computer. Another method (Ref. 7.15) makes use of the determinate of the stiffness matrix, scaled in a special way, to predict equation conditioning. If you are interested, you can obtain the details of these methods from the literature.

A simple method of checking on equation conditioning during solution (Ref. 7.16) is to monitor the ratio of the initial and final values of each diagonal term in the stiffness array during forward elimination (or decomposition). A large ratio indicates that the diagonal term has lost many significant figures during elimination (or decomposition). How large a value this ratio may be permitted to take on before the solution is aborted will depend on the number of significant figures carried in the computer. Once again, this check method will not be universally successful.

One further method of improving the conditioning of a set of equations is to use either partial or complete pivoting (Ref. 7.3) during the elimination (or decomposition) process. Pivoting was discussed and used to produce significant

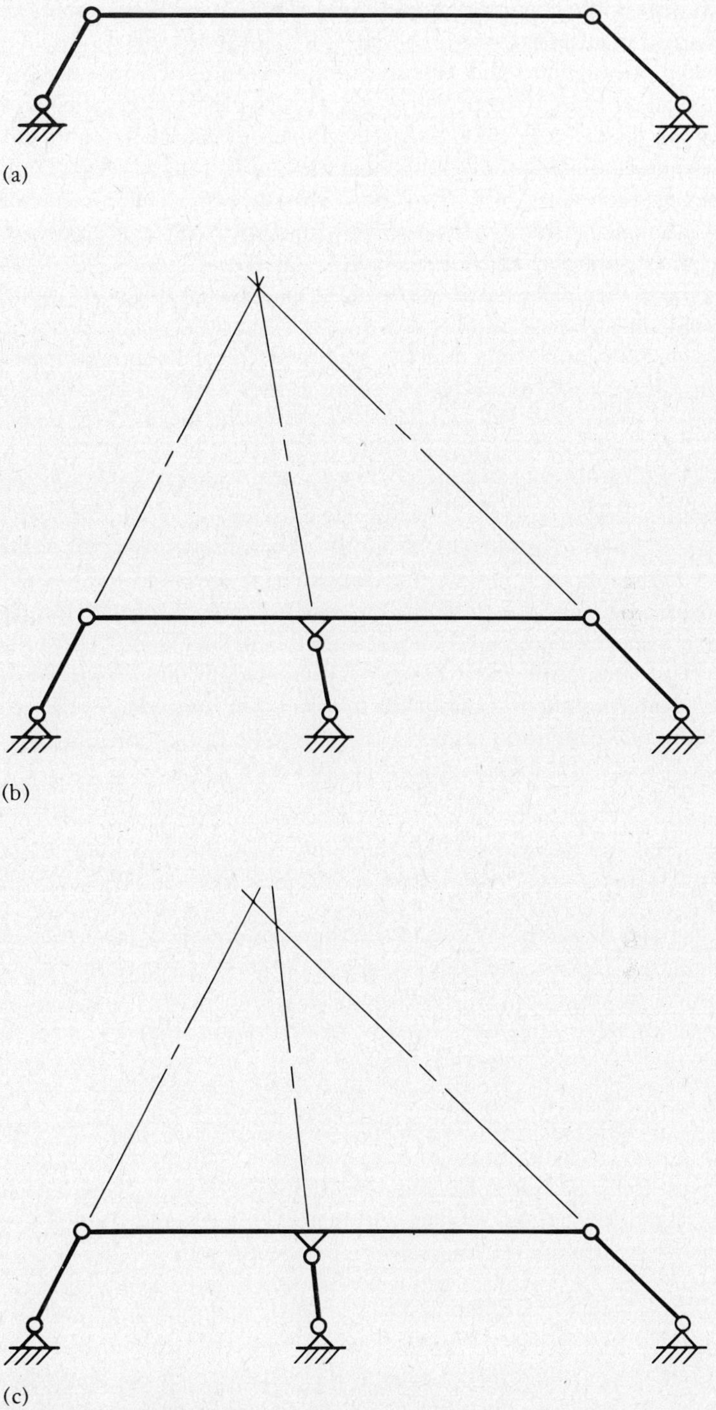

Fig. 7.13 Structural instability: (a) too few basic forces; (b) poorly arranged basic forces; (c) "almost" statically unstable.

improvement in the conditioning of the equilibrium equations generated by the force method of analysis (see Sec. 6.3.3). A similar improvement does not usually occur when pivoting is used in conjunction with the displacement method. Hence its use is not generally considered to be worthwhile, especially because partial pivoting would destroy the symmetry of the stiffness matrix and reduce the efficiency of solution algorithms, and complete pivoting would involve extensive bookkeeping chores.

In summary, it must be stated that, in general, no methods of accuracy assessment or improvement are foolproof. The analyst must therefore remain alert to the possibility of error persisting even subsequent to the application of these methods.

Finally, it is important that the analyst not mistakenly conclude that the solution, even of a problem represented by well-conditioned equations, possesses the accuracy implied by the number of significant figures used in the solution or shown in the output. The accuracy of the output data can never be considered to be better than the accuracy of the input data, which is often only known to two or three significant figures. Why, then, this need to carry out a solution with so many significant figures? The answer is simple. To achieve even, say, three-place accuracy in the output, problems may need to be represented and solved by using many-place accuracy, as is well demonstrated by the example of Sec. 5.4.

In retrospect, the picture of accuracy we have painted appears to be rather dismal. This pessimistic view has been presented to emphasize the need for caution in interpretation of computer solutions. On the positive side, it must be stated that well-written and properly used computer programs will yield excellent solutions to many, if not most, structural problems.

7.6 SUMMARY

Direct methods of solution of sets of simultaneous equations have been discussed in preference to indirect methods because of the general superiority of the former in the modern computational environment. Of the direct methods available, gaussian elimination and other closely related elimination and decomposition methods such as the Choleskian method appear to be the most effective and therefore are the most widely used. Any of these solution algorithms can be developed to be useful with the various data handling schemes that have been devised to provide data storage and manipulation efficiency in dealing with the structural stiffness force-displacement relationships. Among these data handling schemes, band solvers and frontal solvers are, perhaps, most prominent, while hypermatrix, sparse matrix, and substructure techniques are also important. All these techniques are useful for out-of-core as well as in-core solutions.

All solution methods will be subject to errors, and there is no perfect method for detecting and reducing errors. As the analyst, you must therefore be well-educated regarding the behavior of the structure you are analyzing and the basis of the program you are using. You must remain cautious in your acceptance of computer results.

PROBLEMS

7.1 Prove that the backsubstitution accomplished in Sec. 7.2.2 can be carried out by a series of row operations on matrix Eq. H of Sec. 7.2.2.

7.2 Use the direct stiffness method to generate the structure force-displacement relationships for the example of Fig. P7.2. Use the algorithms of Eqs. 7.1 and 7.2 to obtain a solution to this set of simultaneous equations.

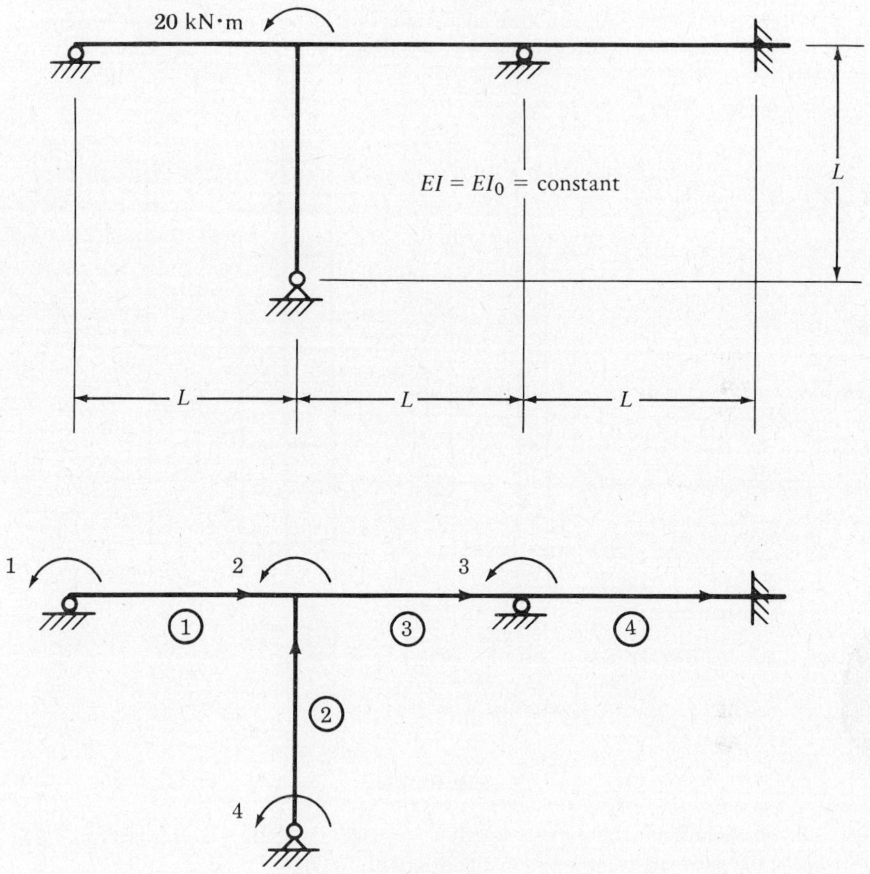

Fig. P7.2

7.3 Develop an appropriate algorithm and use it to build the upper half of **K**, along with **R**, directly in rectangular form for the example of Fig. P7.2. Carry out the elimination process by using the gaussian algorithms of Eq. 7.1b, which are intended to operate directly on the upper half band of **K** in rectangular form. Carry out the backsubstitution process by means of the algorithms of Eq. 7.2a.

7.4 Starting with the reduced structure stiffness relations of Fig. 7.3, carry out the backsubstitution process by means of the algorithms of Eq. 7.2a.

7.5 Having built the structure stiffness relationships in rectangular form as indicated in Prob. 7.3, carry out the elimination process by using the algorithms of Eq. 7.1c, which avoid operations involving zeros.

7.6 Use the frontal method as described in Sec. 7.3.4 to build and solve the stiffness relationships associated with the example of Fig. P7.2.

7.7 The multistory frame of Fig. P7.7 has a uniformly distributed load of intensity q applied downward at each story level. Each girder has a stiffness of $2EI_0$, while each column has a stiffness of EI_0. Noting the plane of symmetry, consider only the left half of the frame. For simplicity it may be assumed that inextensional behavior applies. Obtain the stiffness relationships for the following and then solve:
(a) A typical level 2 substructure condensing out the internal degrees of freedom
(b) A typical level 1 substructure condensing out the internal degrees of freedom
(c) The structure imposing boundary conditions

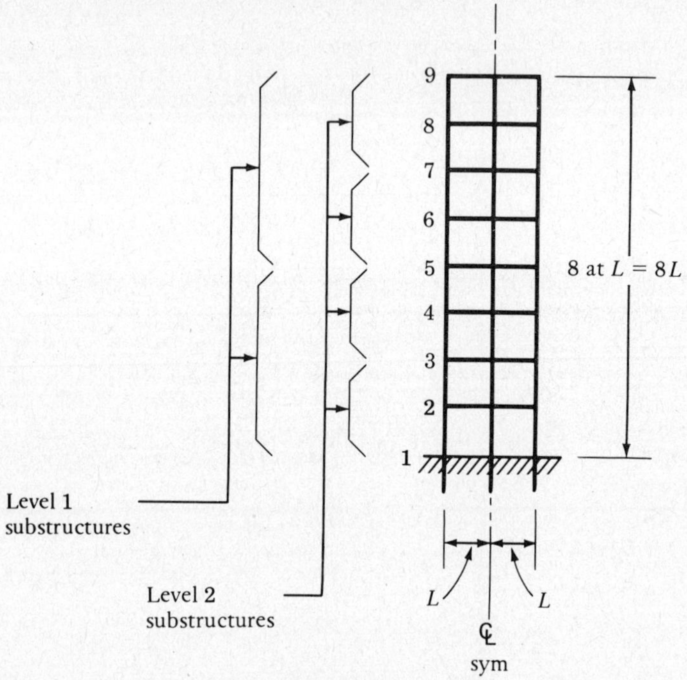

Fig. P7.7

7.8 Obtain a solution to the stiffness equations associated with the example of Prob. 7.2 by using the gaussian decomposition algorithms of Eq. 7.15.

7.9 Obtain a solution to the stiffness equations associated with the example of Prob. 7.2 by using the modified gaussian decomposition algorithms of Eq. 7.22.

7.10 Obtain a solution to the stiffness equations associated with the example of Prob. 7.2 by using the Choleskian decomposition algorithms of Eq. 7.32.

REFERENCES

7.1 Salvadori, M. G., and Baron, M. L. *Numerical Methods in Engineering*. Englewood Cliffs, N.J.: Prentice-Hall, 1952, p. 25.

7.2 Melosh, R. J., and Bamford, R. M. "Efficient Solution of Load-Deflection Equations." *Journal of the Structural Division, ASCE* 95 (1969): 661.

7.3 Meyer, C. "Solution of Linear Equations—State of the Art." *Journal of the Structural Division, ASCE* 99 (1973): 1507–1526.

7.4 Klyuyev, V. V., and Kokovkin-Scherbak, N. L. "On the Minimization of the Number of Arithmetic Operations for the Solution of Linear Algebraic Systems of Equations." *Technical Report CS 24*, Stanford University, Computer Science Department, Stanford, Calif., 1965.

7.5 Kamel, H. A., and McCabe, M. W. "Direct Numerical Solution of Large Sets of Simultaneous Equations." *Computers and Structures* 9 (1978): 113–123.

7.6 Bathe, K.-J., and Wilson, E. L. *Numerical Methods in Finite Element Analysis*. Englewood Cliffs, N.J.: Prentice-Hall, 1976, pp. 209–213.

7.7 Jensen, H. G., and Parks, G. A. Efficient Solutions for Linear Matrix Equations." *Journal of the Structural Division, ASCE* 96 (1970): 49–64.

7.8 Irons, B. M. "A Frontal Solution Program for Finite Element Analysis." *International Journal for Numerical Methods in Engineering* 2 (1970): 5–32.

7.9 Noor, A. K., and Fulton, R. E. "Impact of CDC-STAR-100 Computer on Finite Element Systems." *Journal of the Structural Division, ASCE* 101 (1975): 731–750.

7.10 Przemieniecki, J. S. *Theory of Matrix Structural Analysis*, New York: McGraw-Hill, 1968.

7.11 McCormick, C. W. "Sparse Matrix Operations in NASTRAN." In *Theory and Practice in Finite Element Structural Analysis*, edited by Y. Yamada and R. H. Gallagher. Tokyo: University of Tokyo Press, 1973, pp. 612–631.

7.12 von Fuchs, G.; Roy, J. R.; and Schrem, E. "Hypermatrix Solution of Large Sets of Symmetric Positive-Definite Linear Equations." *Computer Methods in Applied Mechanics and Engineering* 1 (1972): 197–216.

7.13 Wilson, E. L.; Bathe, K.-J.; Doherty, W. P. "Direct Solution of Large Systems of Linear Equations." *Computers and Structures* 4 (1974): 363–372.

7.14 Meyer, C. "Special Problems Related to Linear Equation Solvers." *Journal of the Structural Division, ASCE* 101 (1975): 869–890.

7.15 McGuire, W., and Gallagher, R. H. *Matrix Structural Analysis*. New York: Wiley, 1979, pp. 410–417.

7.16 Cook, R. D. *Concepts and Applications of Finite Element Analysis*. New York: Wiley, 1974, pp. 322–323.

Chapter 8
Space Structures and Gridworks—Displacement Method

8.1 INTRODUCTION

Until now we have presented all matrix structural analysis concepts in a two-dimensional context, confining our attention to planar structures subjected to loads in their own planes. In this brief chapter we will demonstrate that extension of these concepts to three dimensions and to structures subjected to out-of-plane loadings is quite easy. In the interest of brevity and because of its broad appeal, only the direct stiffness method will be discussed.

8.2 MEMBER STIFFNESS RELATIONSHIPS

As in the case of two-dimensional analysis, it is again convenient here to initiate our discussions by considering individual member behavior.

8.2.1 Three-Dimensional Truss Member

Figure 8.1 shows the orientation of a typical space-truss member with reference to a system of rectangular cartesian global coordinates. Also shown in the figure are both the basic and global system of member-end forces. By inspection of this figure, you can readily see that

$$\begin{pmatrix} r_1 \\ r_2 \\ r_3 \\ r_4 \\ r_5 \\ r_6 \end{pmatrix} = \begin{bmatrix} -\ell_1 \\ -\ell_2 \\ -\ell_3 \\ \ell_1 \\ \ell_2 \\ \ell_3 \end{bmatrix} (P) \tag{8.1}$$

where

$$\begin{aligned} \ell_1 &= \cos \alpha_x \\ \ell_2 &= \cos \alpha_y \\ \ell_3 &= \cos \alpha_z \end{aligned} \tag{8.2}$$

and, for example, $\cos \alpha_x = x_{ji}/(x_{ji}^2 + y_{ji}^2 + z_{ji}^2)^{1/2}$. Equation 8.1 may be represented symbolically as

$$\mathbf{r} = \mathbf{TP} \tag{8.1a}$$

Using familiar notation and applying the contragredient law, we can write the matrix relationship between basic and global systems of member-end displacements as

$$(\Delta) = [-\ell_1 \quad -\ell_2 \quad -\ell_3 \quad \ell_1 \quad \ell_2 \quad \ell_3] \begin{pmatrix} d_1 \\ d_2 \\ d_3 \\ d_4 \\ d_5 \\ d_6 \end{pmatrix} \tag{8.3}$$

or

$$\mathbf{\Delta} = \mathbf{T}^T\mathbf{d} \tag{8.3a}$$

The basic force and displacement in a truss member are both seen to remain unchanged in two- or three-dimensional space; thus we may again use the basic stiffness relationship for the truss member developed in Chap. 3. This relationship is repeated here for convenience.

$$P = \frac{EA}{L} \Delta \tag{3.6}$$

or, in matrix form,

$$\mathbf{P} = \mathbf{k}\mathbf{\Delta} \tag{3.6a}$$

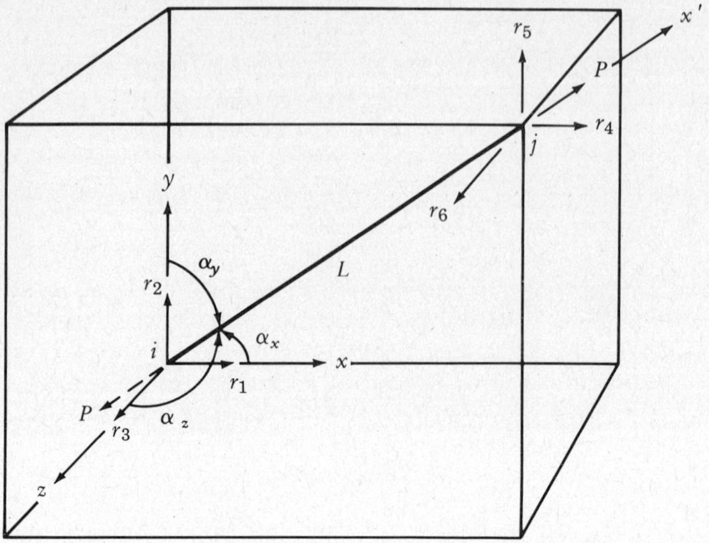

Fig. 8.1 Member-end forces, space truss member.

The global stiffness relationships for a space truss member may now be obtained by using Eq. 8.1, 8.3, and 3.6, so that we get

$$\begin{Bmatrix} r_1 \\ r_2 \\ r_3 \\ r_4 \\ r_5 \\ r_6 \end{Bmatrix} = \begin{bmatrix} -\ell_1 \\ -\ell_2 \\ -\ell_3 \\ \ell_1 \\ \ell_2 \\ \ell_3 \end{bmatrix} \frac{EA}{L} [-\ell_1 \quad -\ell_2 \quad -\ell_3 \quad \ell_1 \quad \ell_2 \quad \ell_3] \begin{Bmatrix} d_1 \\ d_2 \\ d_3 \\ d_4 \\ d_5 \\ d_6 \end{Bmatrix} \tag{8.4}$$

or

$$\mathbf{r} = \mathbf{T}\not{k}\mathbf{T}^T\mathbf{d} \tag{8.4a}$$

Carrying out the matrix products indicated on the right side of Eq. 8.4, we have

$$\begin{Bmatrix} r_1 \\ r_2 \\ r_3 \\ r_4 \\ r_5 \\ r_6 \end{Bmatrix} = \frac{EA}{L} \begin{bmatrix} \ell_1^2 & \ell_1\ell_2 & \ell_1\ell_3 & -\ell_1^2 & -\ell_1\ell_2 & -\ell_1\ell_3 \\ & \ell_2^2 & \ell_2\ell_3 & -\ell_1\ell_2 & -\ell_2^2 & -\ell_2\ell_3 \\ & & \ell_3^2 & -\ell_1\ell_3 & -\ell_2\ell_3 & -\ell_3^2 \\ & & & \ell_1^2 & \ell_1\ell_2 & \ell_1\ell_3 \\ & \text{Symmetric} & & & \ell_2^2 & \ell_2\ell_3 \\ & & & & & \ell_3^2 \end{bmatrix} \begin{Bmatrix} d_1 \\ d_2 \\ d_3 \\ d_4 \\ d_5 \\ d_6 \end{Bmatrix} \tag{8.4b}$$

or

$$\mathbf{r} = \mathbf{k}\mathbf{d} \tag{8.4c}$$

8.2.2 Plane Gridwork Member

As a preliminary step to considering the three-dimensional space frame member, it is useful to review the behavior of the plane frame member and investigate the behavior of the plane gridwork member. The plane frame member that was first displayed in Fig. 2.10 is shown again in Fig. 8.2 with slight changes in notation to accommodate the subsequent discussion of space frames. A comparison of Figs. 2.10 and 8.2 will reveal that the end moments that were previously represented by curved arrows in the xy plane are now represented by double-headed vectors perpendicular to the xy plane. Moment directions are interpreted by using the right-hand rule. The z' axis has been added to Fig. 8.2 to help establish the direction of these moment vectors. Finally, in the basic system of forces of Fig. 8.2a, the subscripts z' and y' have been added to the moment and shear terms, respectively. This addition indicates that the bending of the frame member is taking place about an axis of the cross section that is parallel to the z' axis and that the shear forces induced are in a direction parallel to the y' axis. It is assumed here (as it was tacitly assumed previously) that the axis about which bending is occurring is parallel to a principal axis of the cross section of the member. It is important to realize that, although the z' axis has been added for reference in Fig. 8.2, the frame member and its loading are still assumed to be in the xy plane; that is, the member is still the same plane frame member.

The local stiffness relationships for the plane frame member of constant cross section were obtained in Chap. 4 and displayed in Eqs. 4.6 and 4.6a. These relationships, which include the effect of distributed loading on the member, are repeated here for completeness.

$$\begin{Bmatrix} r'_1 \\ r'_2 \\ r'_3 \\ r'_4 \\ r'_5 \\ r'_6 \end{Bmatrix} = \frac{2EI_{z'}}{L} \begin{bmatrix} A/2I_{z'} & 0 & 0 & -A/2I_{z'} & 0 & 0 \\ & 6/L^2 & 3/L & 0 & -6/L^2 & 3/L \\ & & 2 & 0 & -3/L & 1 \\ & & & A/2I_{z'} & 0 & 0 \\ \text{Symmetric} & & & & 6/L^2 & -3/L \\ & & & & & 2 \end{bmatrix} \begin{Bmatrix} d'_1 \\ d'_2 \\ d'_3 \\ d'_4 \\ d'_5 \\ d'_6 \end{Bmatrix} + \begin{Bmatrix} r'_{F1} \\ r'_{F2} \\ r'_{F3} \\ r'_{F4} \\ r'_{F5} \\ r'_{F6} \end{Bmatrix} \qquad (4.6)$$

$$\mathbf{r}' = \mathbf{k}'\mathbf{d}' + \mathbf{r}'_F \qquad (4.6a)$$

The subscript z' has been added to the moment of inertia term to again indicate that bending is occurring about an axis parallel to z'.

The relationship between global and local sets of member-end forces for this member was obtained in Chap. 2 as

$$\begin{Bmatrix} r_1 \\ r_2 \\ r_3 \\ r_4 \\ r_5 \\ r_6 \end{Bmatrix} = \begin{bmatrix} c & -s & 0 & 0 & 0 & 0 \\ s & c & 0 & 0 & 0 & 0 \\ 0 & 0 & 1 & 0 & 0 & 0 \\ 0 & 0 & 0 & c & -s & 0 \\ 0 & 0 & 0 & s & c & 0 \\ 0 & 0 & 0 & 0 & 0 & 1 \end{bmatrix} \begin{Bmatrix} r'_1 \\ r'_2 \\ r'_3 \\ r'_4 \\ r'_5 \\ r'_6 \end{Bmatrix} \qquad (2.35)$$

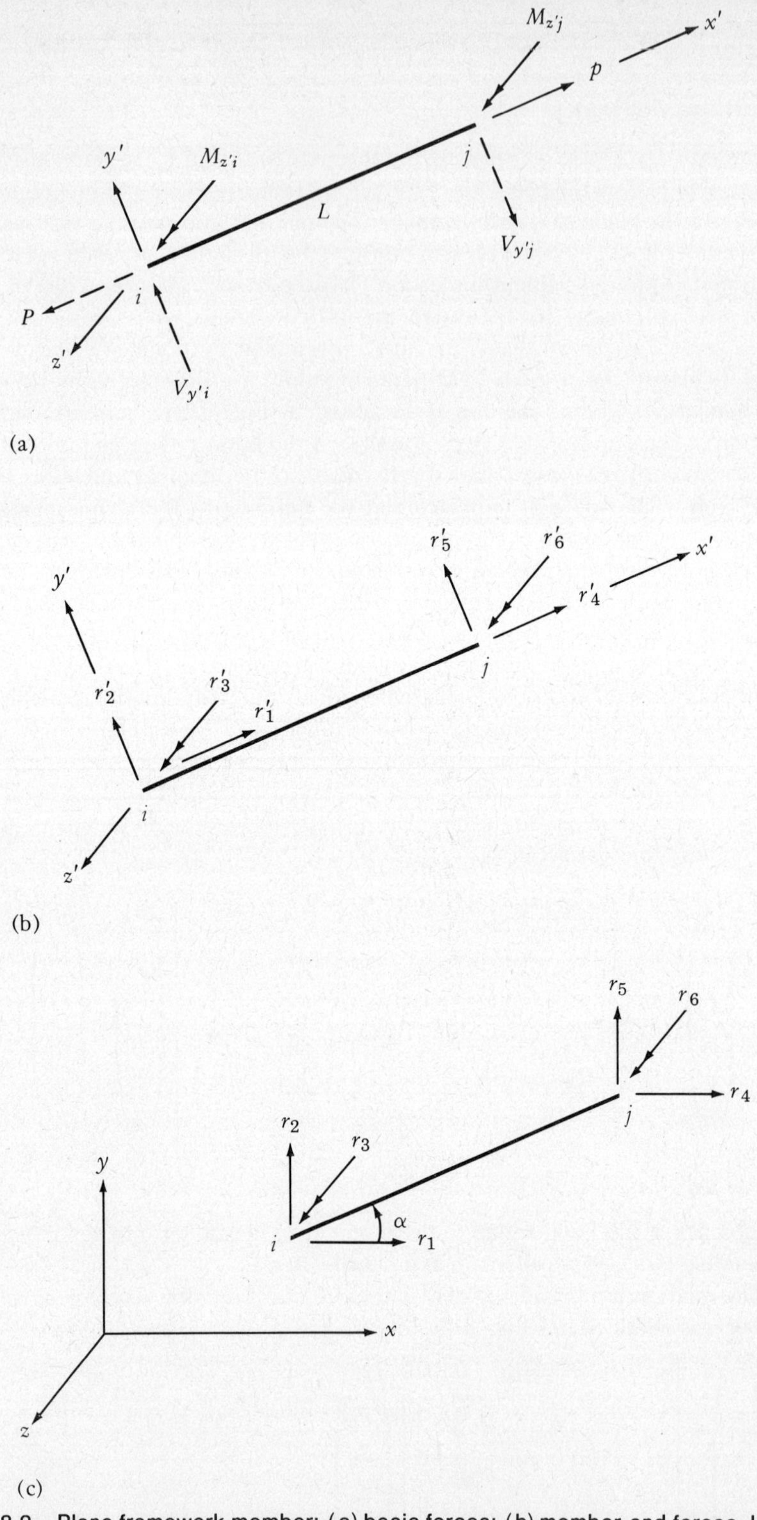

Fig. 8.2 Plane framework member: (a) basic forces; (b) member-end forces, local coordinates; (c) member-end forces, global coordinates.

or

$$\mathbf{r} = \boldsymbol{\alpha}\mathbf{r}' \tag{2.35a}$$

The corresponding relationship of member-end displacements was shown to be

$$\mathbf{d}' = \boldsymbol{\alpha}^T\mathbf{d} \tag{2.36a}$$

Using Eqs. 2.35a, 2.36a, and 4.6a, we determined the global stiffness relationship for the plane frame member symbolically as

$$\mathbf{r} = \boldsymbol{\alpha}\mathbf{k}'\boldsymbol{\alpha}^T\mathbf{d} + \boldsymbol{\alpha}\mathbf{r}'_F \tag{4.8}$$

or

$$\mathbf{r} = \mathbf{k}\mathbf{d} + \mathbf{r}_F \tag{4.8a}$$

As shown in Eq. 4.9, but not repeated here, the detailed global stiffness relationship could now be obtained by appropriate substitution and by carrying out the matrix products indicated in Eq. 4.8.

It is now appropriate to develop the stiffness relationship for the plane gridwork member of Fig. 8.3. Whereas the plane frame member was assumed to be part of a structure that was loaded only in its own plane, the gridwork member is assumed to be part of a structure that is loaded only transversely to its own plane. Thus, comparing Figs. 8.3 and 8.2, we see that the set of forces induced at the ends of the gridwork member differs from that induced at the ends of the plane frame member. (It will be important to our subsequent consideration of the space frame member that the two sets of end forces, taken together, constitute the full set of forces that can be induced at the ends of the member under any orientation and loading condition.) In Fig. 8.3a the subscript y' on the moments indicates that the principal axis of the cross section about which bending of the gridwork member takes place is parallel to the y' axis; and the subscript z' on the shears indicates that the shears induced are in a direction parallel to the z' axis.

From overall equilibrium of the member in Fig. 8.3a, it is seen that

$$V_{z'j} = -V_{z'i} = \frac{M_{y'i} + M_{y'j}}{L} \tag{8.5}$$

Inspection of the force systems of Figs. 8.3a and b reveals the basic to local member-end force relationship for the gridwork member to be

$$\begin{pmatrix} r'_1 \\ r'_2 \\ r'_3 \\ r'_4 \\ r'_5 \\ r'_6 \end{pmatrix} = \begin{bmatrix} -1/L & -1/L & 0 \\ 0 & 0 & -1 \\ 1 & 0 & 0 \\ 1/L & 1/L & 0 \\ 0 & 0 & 1 \\ 0 & 1 & 0 \end{bmatrix} \begin{pmatrix} M_{y'i} \\ M_{y'j} \\ T \end{pmatrix} \tag{8.6}$$

or

$$\mathbf{r}' = \mathbf{L}\mathbf{P} \tag{8.6a}$$

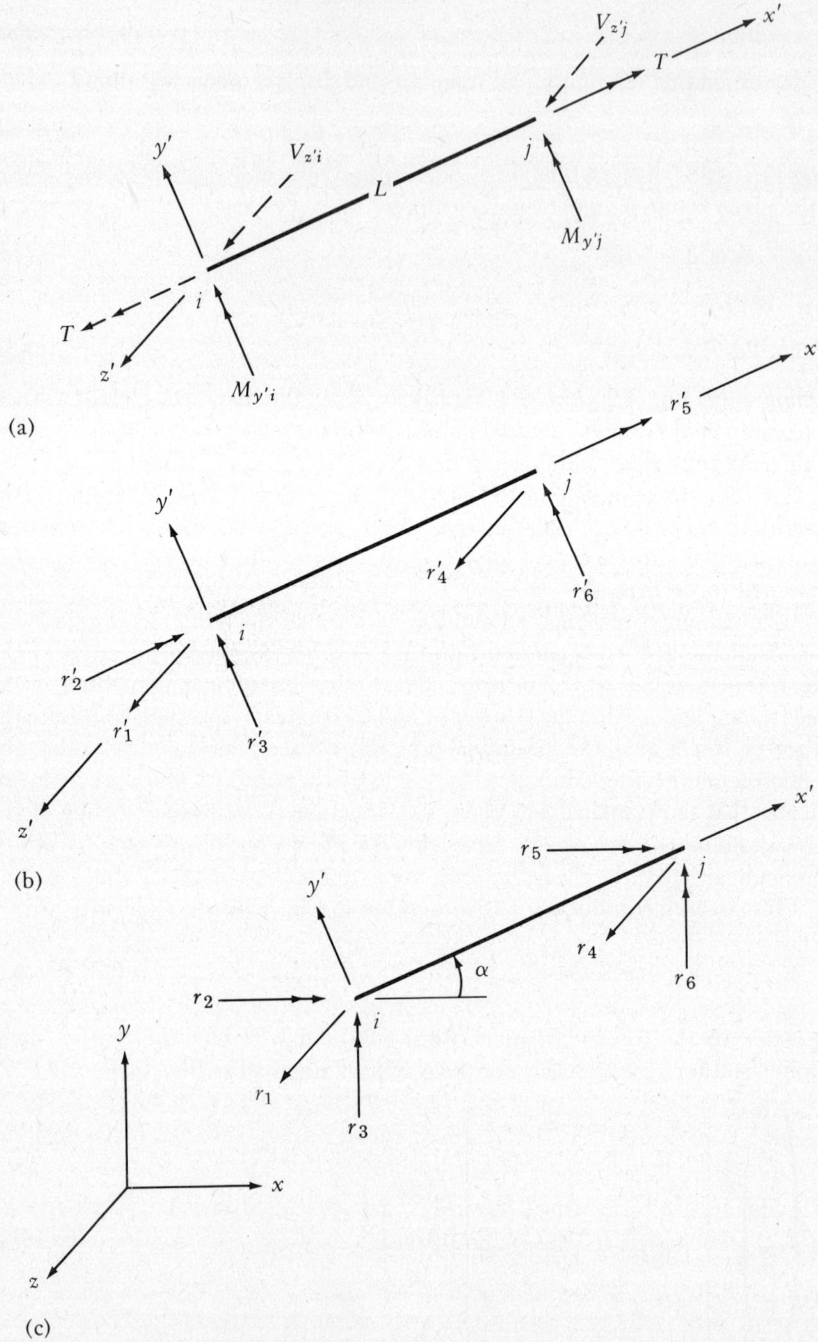

Fig. 8.3 Gridwork member: (a) basic forces; (b) member-end forces, local coordinates; (c) member-end forces, global coordinates.

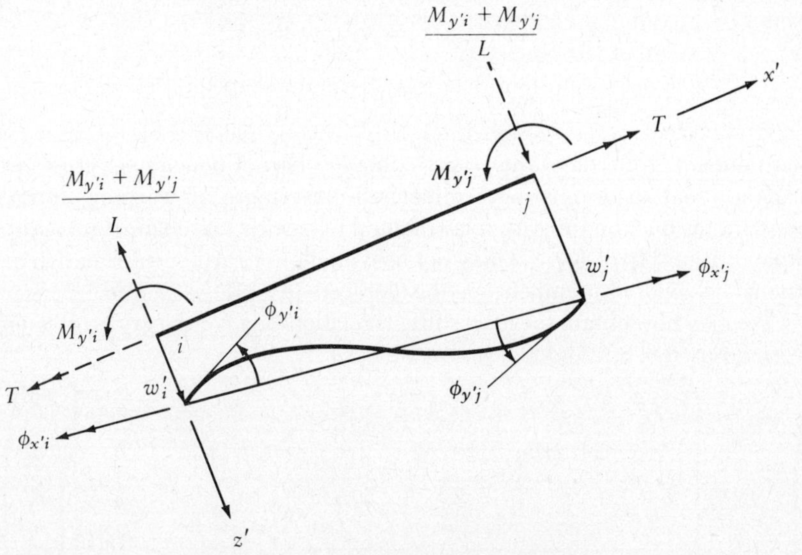

Fig. 8.4 Prismatic gridwork member.

Once again, the contragredient law dictates that the corresponding displacement relationship will be

$$\begin{pmatrix} \phi_{y'i} \\ \phi_{y'j} \\ \phi_{x'} \end{pmatrix} = \begin{bmatrix} -1/L & 0 & 1 & 1/L & 0 & 0 \\ -1/L & 0 & 0 & 1/L & 0 & 1 \\ 0 & -1 & 0 & 0 & 1 & 0 \end{bmatrix} \begin{pmatrix} d'_1 \\ d'_2 \\ d'_3 \\ d'_4 \\ d'_5 \\ d'_6 \end{pmatrix} \tag{8.7}$$

or

$$\mathbf{\Delta} = \mathbf{L}^T \mathbf{d}' \tag{8.7a}$$

where $\phi_{x'} = \phi_{x'i} + \phi_{x'j}$. The terms $\phi_{x'i}$, $\phi_{x'j}$, $\phi_{y'i}$, and $\phi_{y'j}$ are all defined in Fig. 8.4.

The basic stiffness relationships for the gridwork member may be written as

$$\begin{pmatrix} M_{y'i} \\ M_{y'j} \\ T \end{pmatrix} = \frac{2EI_{y'}}{L} \begin{bmatrix} 2 & 1 & 0 \\ 1 & 2 & 0 \\ 0 & 0 & GJ/2EI_{y'} \end{bmatrix} \begin{pmatrix} \phi_{y'i} \\ \phi_{y'j} \\ \phi_{x'} \end{pmatrix} \tag{8.8}$$

where the moment-rotation relationships have been written by analogy with those previously obtained and shown as the first two of scalar equations of matrix Eq. 3.19. The torque-rotation relationship has been taken as

$$T = \frac{GJ}{L} \phi_{x'} \tag{8.9}$$

where G = modulus of rigidity
L = member length
J = geometric constant that depends on the shape of the cross section

For bars of circular cross section, it is relatively easy to show (Ref. 8.1) that J is the polar moment of inertia of the cross section. For bars of noncircular cross section, the determination of J is more difficult. Furthermore, if warping restraint is important (as may be the case for structural bars with I or channel sections), the simple relationship of Eq. 8.9 may not hold at all. For further information on the torsional behavior of members, see the literature (e.g., Refs. 8.1 and 8.2).

We may now obtain the local stiffness relationships for the gridwork member by using Eqs. 8.6, 8.7, and 8.8, so that we have

$$\begin{Bmatrix} r_1' \\ r_2' \\ r_3' \\ r_4' \\ r_5' \\ r_6' \end{Bmatrix} = \frac{2EI_{y'}}{L} \begin{bmatrix} 6/L^2 & 0 & -3/L & -6/L^2 & 0 & -3/L \\ & GJ/2EI_{y'} & 0 & 0 & -GJ/2EI_{y'} & 0 \\ & & 2 & 3/L & 0 & 1 \\ & & & 6/L^2 & 0 & 3/L \\ & \text{Symmetric} & & & GJ/2EI_{y'} & 0 \\ & & & & & 2 \end{bmatrix} \begin{Bmatrix} d_1' \\ d_2' \\ d_3' \\ d_4' \\ d_5' \\ d_6' \end{Bmatrix} \tag{8.10}$$

or

$$\mathbf{r}' = \mathbf{k}'\mathbf{d}' \tag{8.10a}$$

The effect of distributed loading may be accounted for by superposition, as in the case of the plane frame member (see Sec. 4.2.3). Thus we get

$$\begin{Bmatrix} r_1' \\ r_2' \\ r_3' \\ r_4' \\ r_5' \\ r_6' \end{Bmatrix} = \frac{2EI_{y'}}{L} \begin{bmatrix} 6/L^2 & 0 & -3/L & -6/L^2 & 0 & -3/L \\ & GJ/2EI_{y'} & 0 & 0 & -GJ/2EI_{y'} & 0 \\ & & 2 & 3/L & 0 & 1 \\ & & & 6/L^2 & 0 & 3/L \\ & \text{Symmetric} & & & GJ/2EI_{y'} & 0 \\ & & & & & 2 \end{bmatrix} \times \begin{Bmatrix} d_1' \\ d_2' \\ d_3' \\ d_4' \\ d_5' \\ d_6' \end{Bmatrix} + \begin{Bmatrix} r_{F1}' \\ r_{F2}' \\ r_{F3}' \\ r_{F4}' \\ r_{F5}' \\ r_{F6}' \end{Bmatrix} \tag{8.11}$$

or

$$\mathbf{r}' = \mathbf{k}'\mathbf{d}' + \mathbf{r}_F' \tag{8.11a}$$

Comparing Figs. 8.3b and c, we see that

$$\begin{pmatrix} r_1 \\ r_2 \\ r_3 \\ r_4 \\ r_5 \\ r_6 \end{pmatrix} = \begin{bmatrix} 1 & 0 & 0 & 0 & 0 & 0 \\ 0 & c & -s & 0 & 0 & 0 \\ 0 & s & c & 0 & 0 & 0 \\ 0 & 0 & 0 & 1 & 0 & 0 \\ 0 & 0 & 0 & 0 & c & -s \\ 0 & 0 & 0 & 0 & s & c \end{bmatrix} \begin{pmatrix} r_1' \\ r_2' \\ r_3' \\ r_4' \\ r_5' \\ r_6' \end{pmatrix} \tag{8.12}$$

or

$$\mathbf{r} = \boldsymbol{\alpha}\mathbf{r}' \tag{8.12a}$$

and it follows that, by the contragredient law,

$$\mathbf{d}' = \boldsymbol{\alpha}^T\mathbf{d} \tag{8.13}$$

Using Eqs. 8.11a, 8.12a, and 8.13, we then have

$$\mathbf{r} = \boldsymbol{\alpha}\mathbf{k}'\boldsymbol{\alpha}^T\mathbf{d} + \boldsymbol{\alpha}\mathbf{r}_F' \tag{8.14}$$

or

$$\mathbf{r} = \mathbf{k}\mathbf{d} + \mathbf{r}_F \tag{8.14a}$$

where, by definition, the gridwork member global stiffness matrix is

$$\mathbf{k} = \boldsymbol{\alpha}\mathbf{k}'\boldsymbol{\alpha}^T \tag{8.15}$$

Substituting the detailed arrays into Eq. 8.15 and carrying out the indicated matrix products, we get for $\alpha = 0°$,

$$\mathbf{k} = \frac{2EI_{y'}}{L}\begin{bmatrix} 6/L^2 & 0 & -3/L & -6/L^2 & 0 & -3/L \\ & GJ/2EI_{y'} & 0 & 0 & -GJ/2EI_{y'} & 0 \\ & & 2 & 3/L & 0 & 1 \\ & \text{Symmetric} & & 6/L^2 & 0 & 3/L \\ & & & & GJ/2EI_{y'} & 0 \\ & & & & & 2 \end{bmatrix} \tag{8.16}$$

and for $\alpha = 90°$,

$$\mathbf{k} = \frac{2EI_{y'}}{L}\begin{bmatrix} 6/L^2 & 3/L & 0 & -6/L^2 & 3/L & 0 \\ & 2 & 0 & -3/L & 1 & 0 \\ & & GJ/2EI_{y'} & 0 & 0 & -GJ/2EI_{y'} \\ & & & 6/L^2 & -3/L & 0 \\ & \text{Symmetric} & & & 2 & 0 \\ & & & & & GJ/2EI_{y'} \end{bmatrix} \tag{8.17}$$

8.2.3 Three-Dimensional Frame Member

The action of the space frame member of Fig. 8.5 is a composite of the actions of the plane frame member of Fig. 8.2 and the gridwork member of Fig. 8.3. This composite is best seen by considering the behavior of each member with respect to its local coordinate system. Note that the space frame member is subjected to bending in the $x'y'$ and $x'z'$ planes and to twisting about and extension along the x' axis. Thus the local end forces induced at the ends of the space frame member are the sum of the forces induced at the ends of the plane frame and gridwork members, as may be seen by inspecting Figs. 8.2b, 8.3b, and 8.5. It follows that the local stiffness relationships for the space frame member may be obtained directly from the local stiffness relationships of the plane frame and gridwork members. Thus we can get the local space-frame stiffness relationships of Eq. 8.18 by superimposing Eqs. 4.6 and 8.11 with end forces and displacements renumbered according to the following coordinate renumbering scheme:

1F	2F	1G	2G	3G	3F	4F	5F	4G	5G	6G	6F	←	{Framework / Gridwork
1	2	3	4	5	6	7	8	9	10	11	12	←	Space frame

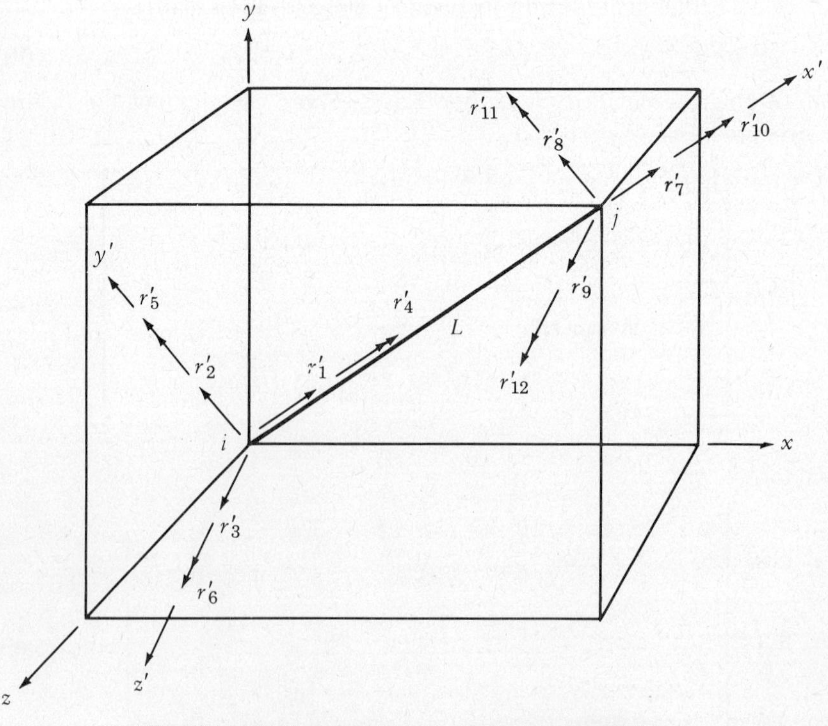

Fig. 8.5 Member-end forces, space frame member.

$$
\begin{Bmatrix} r'_1 \\ r'_2 \\ r'_3 \\ r'_4 \\ r'_5 \\ r'_6 \\ r'_7 \\ r'_8 \\ r'_9 \\ r'_{10} \\ r'_{11} \\ r'_{12} \end{Bmatrix} =
\begin{bmatrix}
\frac{EA}{L} & 0 & 0 & 0 & 0 & 0 & \frac{-EA}{L} & 0 & 0 & 0 & 0 & 0 \\
 & \frac{12EI_{z'}}{L^3} & 0 & 0 & 0 & \frac{6EI_{z'}}{L^2} & 0 & \frac{-12EI_{z'}}{L^3} & 0 & 0 & 0 & \frac{6EI_{z'}}{L^2} \\
 & & \frac{12EI_{y'}}{L^3} & 0 & \frac{-6EI_{y'}}{L^2} & 0 & 0 & 0 & \frac{-12EI_{y'}}{L^3} & 0 & \frac{-6EI_{y'}}{L^2} & 0 \\
 & & & \frac{GJ}{L} & 0 & 0 & 0 & 0 & 0 & \frac{-GJ}{L} & 0 & 0 \\
 & & & & \frac{4EI_{y'}}{L} & 0 & 0 & 0 & \frac{6EI_{y'}}{L^2} & 0 & \frac{2EI_{y'}}{L} & 0 \\
 & & & & & \frac{4EI_{z'}}{L} & 0 & \frac{-6EI_{z'}}{L^2} & 0 & 0 & 0 & \frac{2EI_{z'}}{L} \\
 & & & & & & \frac{EA}{L} & 0 & 0 & 0 & 0 & 0 \\
 & & \text{Symmetric} & & & & & \frac{12EI_{z'}}{L^3} & 0 & 0 & 0 & \frac{-6EI_{z'}}{L^2} \\
 & & & & & & & & \frac{12EI_{y'}}{L^3} & 0 & \frac{6EI_{y'}}{L^2} & 0 \\
 & & & & & & & & & \frac{GJ}{L} & 0 & 0 \\
 & & & & & & & & & & \frac{4EI_{y'}}{L} & 0 \\
 & & & & & & & & & & & \frac{4EI_{z'}}{L}
\end{bmatrix}
\begin{Bmatrix} d'_1 \\ d'_2 \\ d'_3 \\ d'_4 \\ d'_5 \\ d'_6 \\ d'_7 \\ d'_8 \\ d'_9 \\ d'_{10} \\ d'_{11} \\ d'_{12} \end{Bmatrix} +
\begin{Bmatrix} r'_{F1} \\ r'_{F2} \\ r'_{F3} \\ r'_{F4} \\ r'_{F5} \\ r'_{F6} \\ r'_{F7} \\ r'_{F8} \\ r'_{F9} \\ r'_{F10} \\ r'_{F11} \\ r'_{F12} \end{Bmatrix}
\tag{8.18}
$$

Equation 8.18 may be represented in the usual symbolic fashion as

$$\mathbf{r}' = \mathbf{k}'\mathbf{d}' + \mathbf{r}'_F \tag{8.18a}$$

As before, the relationship between the sets of member-end forces in global and local coordinates may be represented symbolically as

$$\mathbf{r} = \boldsymbol{\alpha}\mathbf{r}' \tag{8.19}$$

The corresponding member-end displacement relationship is

$$\mathbf{d}' = \boldsymbol{\alpha}^T\mathbf{d} \tag{8.20}$$

Using Eqs. 8.18a, 8.19, and 8.20, we may write

$$\mathbf{r} = \boldsymbol{\alpha}\mathbf{k}'\boldsymbol{\alpha}^T\mathbf{d} + \boldsymbol{\alpha}\mathbf{r}'_F \tag{8.21}$$

or

$$\mathbf{r} = \mathbf{k}\mathbf{d} + \mathbf{r}_F \tag{8.21a}$$

where Eq. 8.21a is the usual symbolic representation of the member global stiffness relationships. The $\boldsymbol{\alpha}$ matrix may be shown to be

$$\boldsymbol{\alpha} = \begin{bmatrix} \boldsymbol{\alpha}^\circ & \boldsymbol{\Theta} & \boldsymbol{\Theta} & \boldsymbol{\Theta} \\ \boldsymbol{\Theta} & \boldsymbol{\alpha}^\circ & \boldsymbol{\Theta} & \boldsymbol{\Theta} \\ \boldsymbol{\Theta} & \boldsymbol{\Theta} & \boldsymbol{\alpha}^\circ & \boldsymbol{\Theta} \\ \boldsymbol{\Theta} & \boldsymbol{\Theta} & \boldsymbol{\Theta} & \boldsymbol{\alpha}^\circ \end{bmatrix} \tag{8.22}$$

where

$$\boldsymbol{\alpha}^\circ = \begin{bmatrix} \ell_1 & m_1 & n_1 \\ \ell_2 & m_2 & n_2 \\ \ell_3 & m_3 & n_3 \end{bmatrix} \tag{8.23}$$

and where

ℓ_1, ℓ_2, ℓ_3 = direction cosines of x' relative to x, y, z (see Eq. 8.2)

m_1, m_2, m_3 = direction cosines of y' relative to x, y, z

n_1, n_2, n_3 = direction cosines of z' relative to x, y, z

8.3 SPACE TRUSS EXAMPLE

The space truss of Fig. 8.6 is a steel structure with members of thin-walled, closed, rectangular tubular section. A reference cross section for these members is shown in Fig. 8.6c, while relative values of cross sectional properties are stated in Fig. 8.6a.

The yz plane is seen to be a plane of symmetry for this space truss structure. Thus, by properly accounting for symmetry, we may obtain a solution for this example by considering only the half structure on the $+x$ side of the yz plane, as

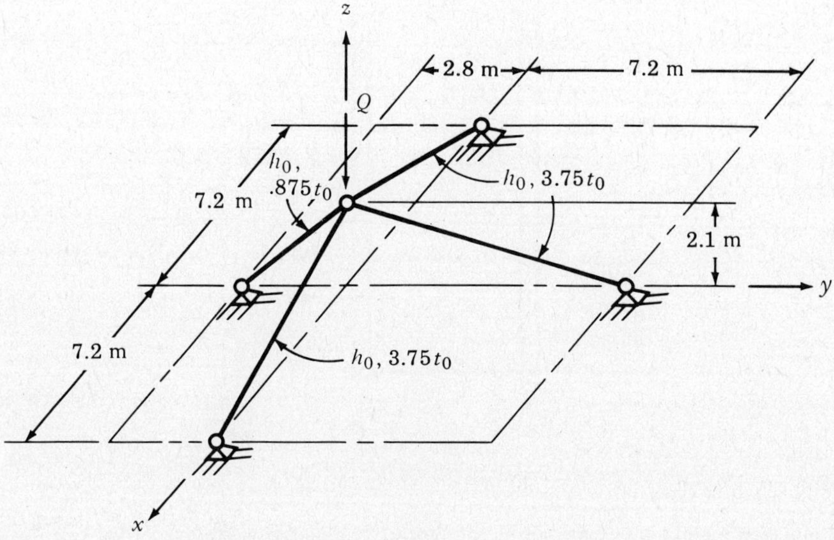

(a)

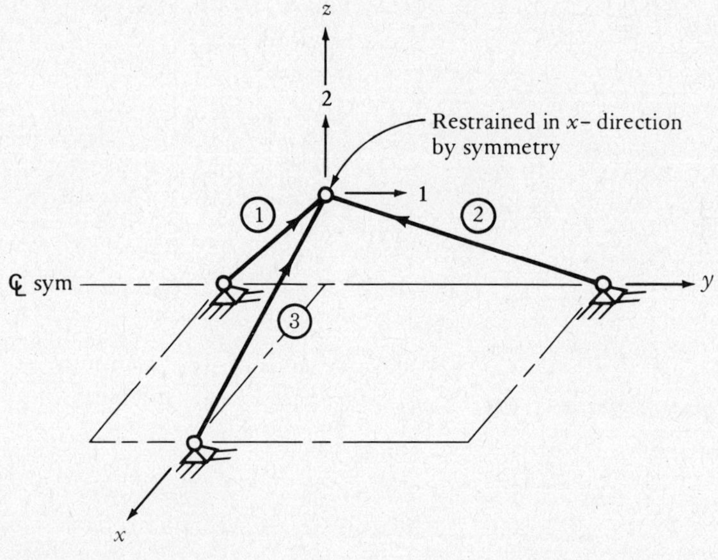

(b)

Fig. 8.6 Space truss structure: (a) geometry, properties, and loading; (b) model numbering and member directions.

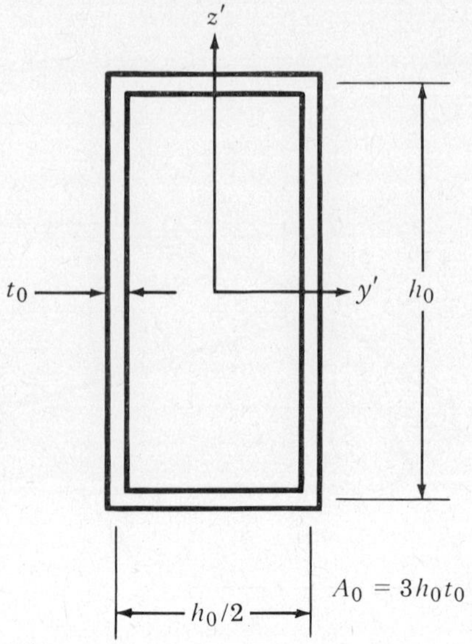

Fig. 8.6 (continued) Space truss structure: (c) member reference cross section.

shown in Fig. 8.6b. Using this approach, we can obtain the individual member stiffness arrays by appropriately specializing the array of Eq. 8.4b as follows:

Member 1

$$\ell_1 = \frac{0}{(0^2 + 2.8^2 + 2.1^2)^{1/2}} = \frac{0}{3.5} = 0$$

$$\ell_2 = \frac{2.8}{3.5} = .8$$

$$\ell_3 = \frac{2.1}{3.5} = .6$$

$$A = \frac{1}{2}3h_0(.875t_0) = .4375A_0$$

$$\frac{EA}{L} = \frac{E(.4375A_0)}{3.5} = \frac{EA_0}{8}$$

$$\mathbf{k}^{(1)} = \frac{EA_0}{8}\begin{bmatrix} 0 & 0 & 0 & 0 & 0 & 0 \\ & .64 & .48 & 0 & -.64 & -.48 \\ & & .36 & 0 & -.48 & -.36 \\ & & & 0 & 0 & 0 \\ \text{Symmetric} & & & & .64 & .48 \\ & & & & & .36 \end{bmatrix} = \frac{EA_0}{4}\begin{bmatrix} 0 & 0 & 0 & 0 & 0 & 0 \\ & .32 & .24 & 0 & -.32 & -.24 \\ & & .18 & 0 & -.24 & -.18 \\ & & & 0 & 0 & 0 \\ \text{Symmetric} & & & & .32 & .24 \\ & & & & & .18 \end{bmatrix}$$

Member 2

$$\ell_1 = \frac{0}{(0^2 + 7.2^2 + 2.1^2)^{1/2}} = \frac{0}{7.5} = 0$$

$$\ell_2 = \frac{-7.2}{7.5} = -.96$$

$$\ell_3 = \frac{2.1}{7.5} = .28$$

$$A = \frac{1}{2}\, 3h_0(3.75t_0) = 1.875A_0$$

$$\frac{EA}{L} = \frac{E\, 1.875A_0}{7.5} = \frac{EA_0}{4}$$

$$\mathbf{k}^{(2)} = \frac{EA_0}{4}\begin{bmatrix} 0 & 0 & 0 & 0 & 0 & 0 \\ & .9216 & -.2688 & 0 & -.9216 & .2688 \\ & & .0784 & 0 & .2688 & -.0784 \\ & & & 0 & 0 & 0 \\ \text{Symmetric} & & & & .9216 & -.2688 \\ & & & & & .0784 \end{bmatrix}$$

Member 3

$$\ell_1 = \frac{-7.2}{7.5} = -.96$$

$$\ell_2 = \frac{0}{7.5} = 0$$

$$\ell_3 = \frac{2.1}{7.5} = .28$$

$$A = 3h_0\,(3.75t_0) = 3.75A_0$$

$$\frac{EA}{L} = \frac{E\, 3.75A_0}{7.5} = \frac{EA_0}{2}$$

$$\mathbf{k}^{(3)} = \frac{EA_0}{4}\begin{bmatrix} 1.8432 & 0 & -.5376 & -1.8432 & 0 & .5376 \\ & 0 & 0 & 0 & 0 & 0 \\ & & .1568 & .5376 & 0 & -.1568 \\ & & & 1.8432 & 0 & -.5376 \\ & \text{Symmetric} & & & 0 & 0 \\ & & & & & .1568 \end{bmatrix}$$

For convenience, a common factor of $EA_0/4$ has been kept as a multiplier of each member stiffness array.

The unconstrained coordinate codes for these members are

Member number ↓	1	2	3	4	5	6 ← Coordinate number
1	0	0	0	0	1	2
2	0	0	0	0	1	2
3	0	0	0	0	1	2

Using these codes, we can now build the structure stiffness array for the example truss from the individual member stiffness arrays as

$$\mathbf{K} = \frac{EA_0}{4}\begin{bmatrix} .32 + .9216 + 0 & .24 - .2688 + 0 \\ \text{Symmetric} & .18 + .0784 + .1568 \end{bmatrix}$$

$$= \frac{EA_0}{4}\begin{bmatrix} 1.2416 & -.0288 \\ -.0288 & .4152 \end{bmatrix}$$

Given the loading of Fig. 8.6a, the structure stiffness relationships may be written as

$$\begin{pmatrix} 0 \\ -.5Q \end{pmatrix} = \frac{EA_0}{4}\begin{bmatrix} 1.2416 & -.0288 \\ -.0288 & .4152 \end{bmatrix}\begin{pmatrix} D_1 \\ D_2 \end{pmatrix} \tag{A}$$

Solving Eq. A, we get

$$\begin{pmatrix} D_1 \\ D_2 \end{pmatrix} = \frac{4}{EA_0}\begin{bmatrix} .8067 & .05596 \\ .05596 & 2.412 \end{bmatrix}\begin{pmatrix} 0 \\ -.5Q \end{pmatrix} = \frac{4Q}{EA_0}\begin{pmatrix} -.02798 \\ -1.206 \end{pmatrix} \tag{B}$$

Making use of these results along with the individual member stiffness arrays and the member codes, we can now obtain the member-end forces in global coordinates:

$$\begin{aligned} \mathbf{r}^{(1)} &= \mathbf{k}^{(1)}\mathbf{d}^{(1)} = \langle\ 0 \quad .2984 \quad .2238 \quad 0 \quad -.2984 \quad -.2238\rangle\, Q \\ \mathbf{r}^{(2)} &= \mathbf{k}^{(2)}\mathbf{d}^{(2)} = \langle\ 0 \quad -.2984 \quad .0870 \quad 0 \quad .2984 \quad -.0870\rangle\, Q \\ \mathbf{r}^{(3)} &= \mathbf{k}^{(3)}\mathbf{d}^{(3)} = \langle -.6483 \quad 0 \quad .1891 \quad .6483 \quad 0 \quad -.1891\rangle\, Q \end{aligned} \tag{C}$$

A check of these force results will reveal that they satisfy all equilibrium requirements.

8.4 GRIDWORK EXAMPLE

The gridwork of Fig. 8.7 is seen to be a flattened version of the space truss of Fig. 8.6, except that the gridwork has rotationally rigid connections between members and between members and supports. Again the yz plane is a plane of symmetry, and only the half structure of Fig. 8.7b need be analyzed. As in the case of the space truss, it is again assumed that the members of the gridwork are closed steel

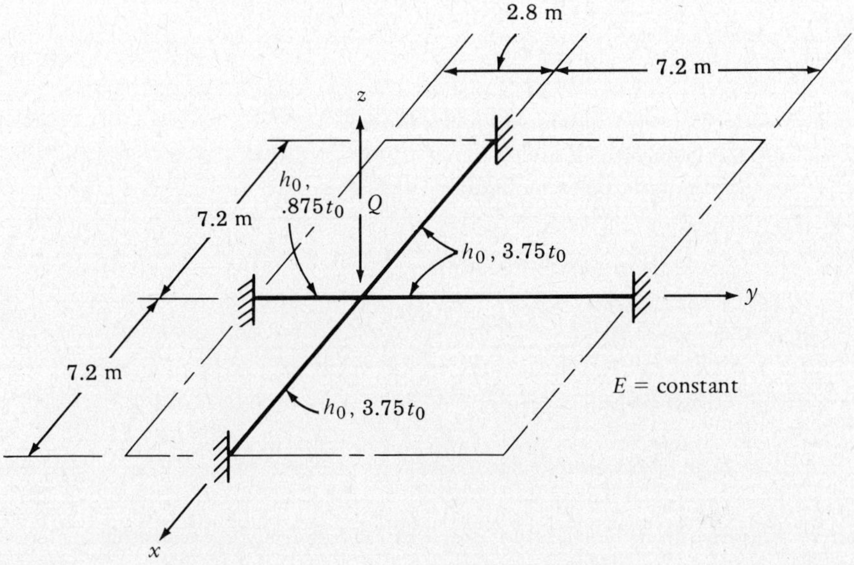

(a)

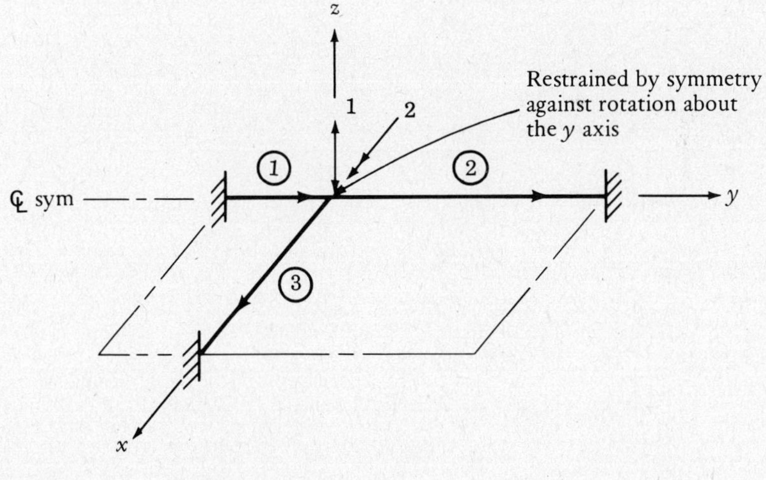

(b)

Fig. 8.7 Gridwork structure: (a) geometry, properties, and loading; (b) model numbering and coordinate directions.

box-sections of constant thickness, as shown in Fig. 8.6c. It follows (Ref. 8.2) that if warping restraint is neglected, the member torsional geometric constant J in Eq. 8.10 is

$$J = \frac{4a^2t}{p} \tag{8.24}$$

where a = area of box inside median line of wall
t = thickness of wall
p = perimeter around median line of wall

Thus from Fig. 8.6c,

$$J_0 = \frac{4\left(\frac{h_0}{2}h_0\right)^2 t_0}{3h_0} = \frac{h_0^3 t_0}{3} \tag{A}$$

and, using familiar formulas, we get

$$I_0 \cong 2(.5h_0)t_0\frac{h_0^2}{4} + \frac{1}{12} \times .2t_0h_0^3 = \frac{5h_0^3t_0}{12} \tag{B}$$

where Eq. B will be considered exact, assuming $t_0 \lll h_0$. Considering these section property formulas and specializing Eqs. 8.16 and 8.17, we can obtain the individual member stiffness arrays as follows:

Member 1

$\alpha = 90°$

$$I_{y'} = \frac{1}{2} \times \frac{5}{12} \times h_0^3(.875t_0) = .4375I_0$$

$$\frac{2EI_{y'}}{L} = \frac{2E(.4375I_0)}{2.8} = \frac{EI_0}{3.2}$$

$$J = \frac{1}{2} \times \frac{1}{3}h_0^3(.875t_0) = .4375J_0$$

Taking $\nu = .25$ gives

$$\frac{GJ}{2EI_{y'}} = \frac{E}{2(1 + .25)} \cdot \frac{(.4375J_0)}{2E(.4375I_0)} = \frac{J_0}{5I_0} = \frac{.8}{5} = .16$$

$$\mathbf{k}^{(1)} = \frac{EI_0}{3.2}\begin{bmatrix} .7653 & 1.071 & 0 & -.7653 & 1.071 & 0 \\ & 2 & 0 & -1.071 & 1 & 0 \\ & & .16 & 0 & 0 & .16 \\ & & & .7653 & -1.071 & 0 \\ \text{Symmetric} & & & & 2 & 0 \\ & & & & & .16 \end{bmatrix}$$

or

$$\mathbf{k}^{(1)} = \frac{EI_0}{1.92}\begin{bmatrix} .4592 & .6428 & 0 & -.4592 & .6428 & 0 \\ & 1.2 & 0 & -.6428 & .6 & 0 \\ & & .096 & 0 & 0 & -.096 \\ & & & .4592 & -.6428 & 0 \\ \text{Symmetric} & & & & 1.2 & 0 \\ & & & & & .096 \end{bmatrix}$$

Member 2

$$\alpha = 90°$$

$$I_{y'} = \frac{1}{2} \times \frac{5}{12} h_0^3(3.75t_0) = 1.875I_0$$

$$\frac{2EI_{y'}}{L} = \frac{2E(1.875I_0)}{7.2} = \frac{EI_0}{1.92}$$

$$J = \frac{1}{2} \times \frac{1}{3} h_0^3(3.75t_0) = 1.875J_0$$

Taking $\nu = .25$ gives

$$\frac{GJ}{2EI_{y'}} = \frac{E}{2(1 + .25)} \cdot \frac{(1.875J_0)}{2E(1.875I_0)} = .16$$

$$\mathbf{k}^{(2)} = \frac{EI_0}{1.92}\begin{bmatrix} .1157 & .4167 & 0 & -.1157 & .4157 & 0 \\ & 2 & 0 & -.4167 & 1 & 0 \\ & & .16 & 0 & 0 & -.16 \\ & & & .1157 & -.4167 & 0 \\ \text{Symmetric} & & & & 2 & 0 \\ & & & & & .16 \end{bmatrix}$$

Member 3

$$\alpha = 0°$$

$$I_{y'} = \frac{5}{12} h_0^3(3.75t_0) = 3.75I_0$$

$$\frac{2EI_{y'}}{L} = \frac{2E(3.75I_0)}{7.2} = \frac{EI_0}{.96}$$

$$J = \frac{1}{3} h_0^3(3.75t_0) = 3.75J_0$$

Taking $\nu = .25$ gives

$$\frac{GJ}{2EI_{y'}} = \frac{E}{2(1+.25)} \cdot \frac{(3.75J_0)}{2E(3.75I_0)} = .16$$

$$\mathbf{k}^{(3)} = \frac{EI_0}{.96}\begin{bmatrix} .1157 & 0 & -.4167 & -.1157 & 0 & -.4167 \\ & .16 & 0 & 0 & -.16 & 0 \\ & & 2 & .4167 & 0 & 1 \\ & & & .1157 & 0 & .4167 \\ & \text{Symmetric} & & & .16 & 0 \\ & & & & & 2 \end{bmatrix}$$

or

$$\mathbf{k}^{(3)} = \frac{EI_0}{1.92}\begin{bmatrix} .2315 & .32 & -.8333 & -.2315 & 0 & -.8333 \\ & .32 & 0 & 0 & -.32 & 0 \\ & & 4 & .8333 & 0 & 2 \\ & & & .2315 & 0 & .8333 \\ & \text{Symmetric} & & & .32 & 0 \\ & & & & & 4 \end{bmatrix}$$

The unconstrained coordinate codes for the three members described may be constructed by inspection of Figs. 8.3c and 8.7b, giving

Member number ↓	1	2	3	4	5	6 ← Member coordinate number
1	0	0	0	1	2	0
2	1	2	0	0	0	0
3	1	2	0	0	0	0

Using these codes, we can obtain the stiffness array for the example gridwork:

$$\mathbf{K} = \frac{EI_0}{1.92}\begin{bmatrix} .4592 + .1157 + .2315 & -.6428 + .4167 + 0 \\ \text{Symmetric} & 1.2 + 2 + .32 \end{bmatrix}$$

$$= \frac{EI_0}{1.92}\begin{bmatrix} .8064 & -.2261 \\ -.2261 & 3.52 \end{bmatrix}$$

Given the loading of Fig. 8.7a, we may now write the gridwork stiffness relationships as

$$\begin{pmatrix} -.5Q \\ 0 \end{pmatrix} = \frac{EI_0}{1.92}\begin{bmatrix} .8064 & -.2261 \\ -.2261 & 3.52 \end{bmatrix}\begin{pmatrix} D_1 \\ D_2 \end{pmatrix} \tag{C}$$

Solving Eq. C gives

$$\begin{pmatrix} D_1 \\ D_2 \end{pmatrix} = \frac{1.92}{EI_0}\begin{bmatrix} 1.263 & .08112 \\ .08112 & .2893 \end{bmatrix}\begin{pmatrix} -.5Q \\ 0 \end{pmatrix} = \frac{1.92Q}{EI_0}\begin{pmatrix} -.6315 \\ -.04056 \end{pmatrix} \tag{D}$$

Backsubstituting these results into the individual member stiffness relationships, we get

$$\begin{aligned} \mathbf{r}^{(1)} &= \langle\ .2639 \quad .3816 \quad 0 \quad -.2639 \quad .3475 \quad 0\ \rangle Q \\ \mathbf{r}^{(2)} &= \langle -.08997 \quad -.3443 \quad 0 \quad .08997 \quad -.3031 \quad 0\ \rangle Q \\ \mathbf{r}^{(3)} &= \langle -.1462 \quad -.01298 \quad .5262 \quad .1462 \quad .01298 \quad .5262 \rangle Q \end{aligned} \tag{E}$$

A close inspection of these results will reveal that they satisfy all equilibrium requirements.

Consider now the same gridwork subjected to a uniformly distributed line loading of q kN/m intensity applied vertically downward along the full length of each member. This loading will induce the following fixed-end forces in the members:

$$\begin{aligned} \mathbf{r}_F^{(1)} &= \langle\ .7 \quad .3267 \quad 0 \quad .7 \quad -.3267 \quad 0\ \rangle q \\ \mathbf{r}_F^{(2)} &= \langle 1.8 \quad 2.16 \quad 0 \quad 1.8 \quad -2.16 \quad 0\ \rangle q \\ \mathbf{r}_F^{(3)} &= \langle 3.6 \quad 0 \quad -4.32 \quad 3.6 \quad 0 \quad 4.32 \rangle q \end{aligned} \tag{F}$$

Using the unconstrained codes listed previously for this example gives

$$\mathbf{R}_F = \langle .7 + 1.8 + 3.6 \quad -.3267 + 2.16 + 0 \rangle q$$

or

$$\mathbf{R}_F = \langle 6.1 \quad 1.833 \rangle q \tag{G}$$

Replacing the load vector in Eq. D with $-\mathbf{R}_F$ from Eq. G and carrying out the matrix product yields

$$\mathbf{D} = \frac{1.92q}{EI_0}\langle -7.853 \quad -1.025 \rangle \tag{H}$$

Letting $24.4q = Q$ for purposes of comparison with the previous loading results, we get

$$\mathbf{D} = \frac{1.92Q}{EI_0}\langle -.3218 \quad -.04200 \rangle \tag{I}$$

As expected, the vertical displacement is smaller than for the case of the concentrated load because of the effect of spreading the concentrated loading over the structure.

Backsubstituting from Eq. H into the member stiffness relationships (including the member fixed-end force vectors), we get

$$\begin{aligned}
\mathbf{r}^{(1)} &= \langle 3.647 \quad 4.760 \quad 0 \quad -2.247 \quad 3.491 \quad 0\ \rangle\, q \\
\mathbf{r}^{(2)} &= \langle .4643 \quad -3.162 \quad 0 \quad 3.136 \quad -6.450 \quad 0\ \rangle\, q \\
\mathbf{r}^{(3)} &= \langle 1.782 \quad -.3280 \quad 2.224 \quad 5.418 \quad -.3280 \quad 10.86 \rangle\, q
\end{aligned} \tag{J}$$

Again these results satisfy equilibrium.

8.5 SPACE FRAME EXAMPLE

The space frame example of Fig. 8.8 is seen to be yet another variation of the space truss of Fig. 8.6. It may therefore be thought of either as the space truss with rotationally rigid connections or as the gridwork of Fig. 8.7 with the central node raised 2.1 m above the xy plane. The loading is seen to be identical with the second loading considered on the gridwork. We can obtain the individual member global stiffness arrays by specializing the local stiffness array of Eq. 8.18 and then transforming this result to global coordinates by using a properly specialized $\boldsymbol{\alpha}$ array as defined in Eqs. 8.22 and 8.23. The familiar transformation required may be observed by comparing Eqs. 8.21 and 8.21a, from which it is seen that

$$\mathbf{k} = \boldsymbol{\alpha}\mathbf{k}'\boldsymbol{\alpha}^T \tag{8.25}$$

The reference cross section for the members of the space frame is that shown in Fig. 8.6c. Most of the section properties for the reference section were calculated in Secs. 8.3 and 8.4 as

$$A_0 = 3h_0t_0 \tag{A}$$

$$I_0 = (= I_{y'0}) = \frac{5h_0^3t_0}{12} \tag{B}$$

$$J_0 = \frac{h_0^3t_0}{3} \tag{C}$$

Additionally, we now need

$$I_{z'0} = 2h_0t_0\frac{h_0^2}{16} + 2 \times \frac{1}{12}t_0\frac{h_0^3}{8} = \frac{7h_0^3t_0}{48} \tag{D}$$

With a view toward subsequently factoring a multiple of $2EI_{y'}/L$ from each member stiffness array, we may also write

$$\frac{I_{z'0}}{I_0} = .35 \tag{E}$$

Taking $h_0 = .6$ m gives

$$\frac{A_0}{2I_0} = 10 \tag{F}$$

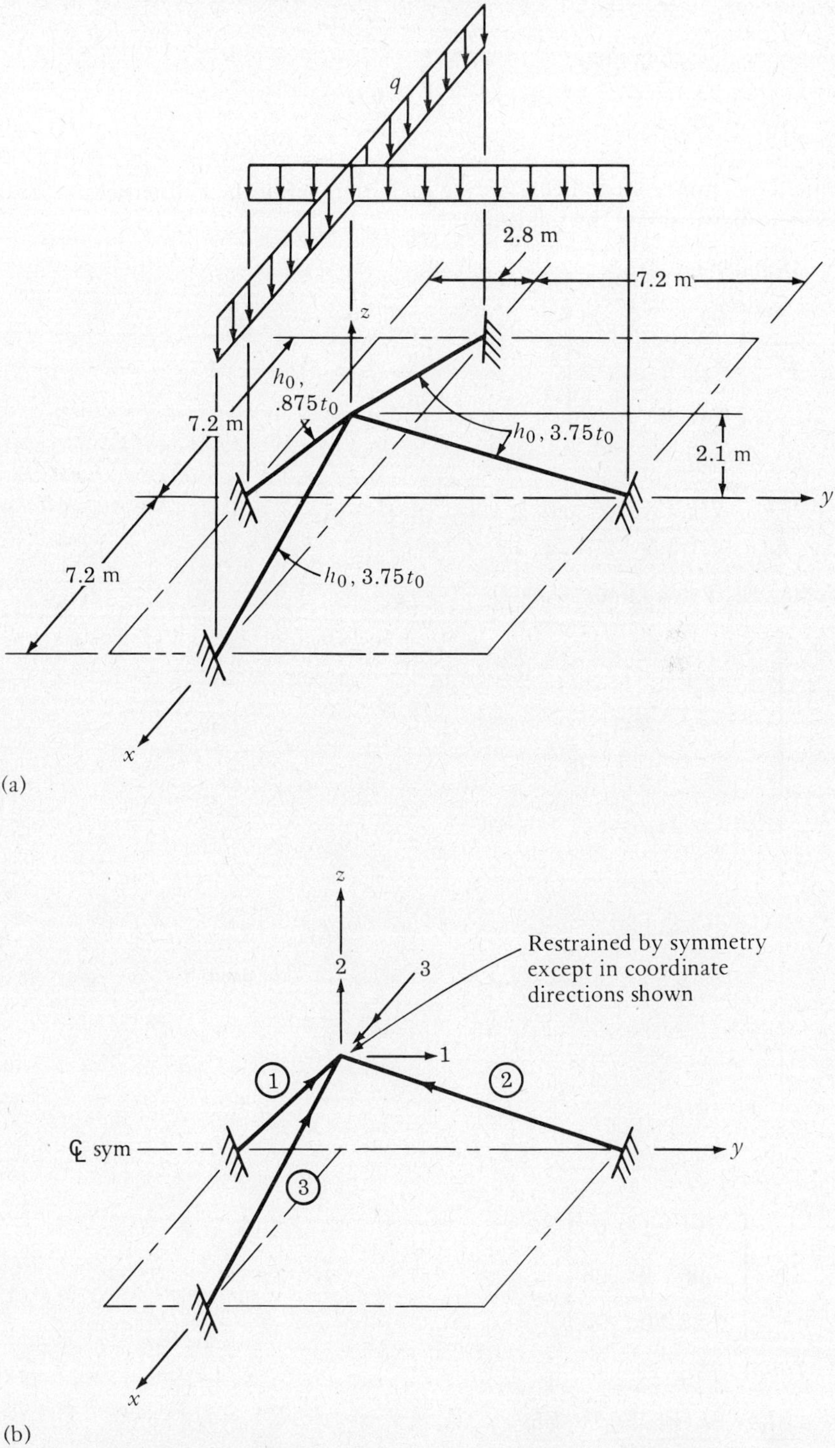

Fig. 8.8 Space frame structure: (a) geometry, properties, and loading; (b) model numbering and member directions.

and taking $\nu = .25$ gives

$$\frac{GJ_0}{2EI_0} = .16 \tag{G}$$

Using these properties and considering each member of the half-structure shown in Fig. 8.8b, we obtain the following:

Member 1

$$\boldsymbol{\alpha}^{\circ} = \begin{bmatrix} 0 & -1 & 0 \\ .8 & 0 & -.6 \\ .6 & 0 & .8 \end{bmatrix}$$

$$I_{y'} = .4375 I_0$$

$$\frac{2EI_{y'}}{L} = \frac{2E(.4375I_0)}{3.5} = \frac{EI_0}{4}$$

Keeping $EI_0/2$ as a factor, we get

$$\mathbf{k}'^{(1)} = \frac{EI_0}{2}\begin{bmatrix}
5 & \cdot & \cdot & \cdot & \cdot & \cdot & -5 & \cdot & \cdot & \cdot & \cdot & \cdot \\
 & .08571 & \cdot & \cdot & \cdot & .15 & \cdot & -.08571 & \cdot & \cdot & \cdot & .15 \\
 & & .2449 & \cdot & -.4286 & \cdot & \cdot & \cdot & -.2449 & \cdot & -.4286 & \cdot \\
 & & & .08 & \cdot & \cdot & \cdot & \cdot & \cdot & -.08 & \cdot & \cdot \\
 & & & & 1 & \cdot & \cdot & \cdot & .4286 & \cdot & .5 & \cdot \\
 & & & & & .35 & \cdot & -.15 & \cdot & \cdot & \cdot & .175 \\
 & & & & & & 5 & \cdot & \cdot & \cdot & \cdot & \cdot \\
 & & \text{Symmetric} & & & & & .08571 & \cdot & \cdot & \cdot & -.15 \\
 & & & & & & & & .2449 & \cdot & .4286 & \cdot \\
 & & & & & & & & & .08 & \cdot & \cdot \\
 & & & & & & & & & & 1 & 0 \\
 & & & & & & & & & & & .35
\end{bmatrix}$$

Member 2

$$\boldsymbol{\alpha}^{\circ} = \begin{bmatrix} 0 & 1 & 0 \\ -.96 & 0 & .28 \\ .28 & 0 & .96 \end{bmatrix}$$

$$I_{y'} = 1.875 I_0$$

$$\frac{2EI_{y'}}{L} = \frac{2E(1.875I_0)}{7.5} = \frac{EI_0}{2}$$

Keeping $EI_0/2$ as a factor, we get

$$\mathbf{k}'^{(2)} = \frac{EI_0}{2}\begin{bmatrix} 10 & \cdot & \cdot & \cdot & \cdot & \cdot & -10 & \cdot & \cdot & \cdot & \cdot & \cdot \\ & .03733 & \cdot & \cdot & \cdot & .14 & \cdot & -.03733 & \cdot & \cdot & \cdot & .14 \\ & & .1067 & \cdot & -.4 & \cdot & \cdot & \cdot & -.1067 & \cdot & -.4 & \cdot \\ & & & .16 & \cdot & \cdot & \cdot & \cdot & \cdot & -.16 & \cdot & \cdot \\ & & & & 2 & \cdot & \cdot & \cdot & .4 & \cdot & 1 & \cdot \\ & & & & & .7 & \cdot & -.14 & \cdot & \cdot & \cdot & .35 \\ & & & & & & 10 & \cdot & \cdot & \cdot & \cdot & \cdot \\ & \text{Symmetric} & & & & & & .03733 & \cdot & \cdot & \cdot & -.14 \\ & & & & & & & & .1067 & \cdot & .4 & \cdot \\ & & & & & & & & & .16 & \cdot & \cdot \\ & & & & & & & & & & 2 & \cdot \\ & & & & & & & & & & & .7 \end{bmatrix}$$

Member 3

$$\alpha^{\circ} = \begin{bmatrix} -.96 & 0 & .28 \\ 0 & -1 & 0 \\ .28 & 0 & .96 \end{bmatrix}$$

$$I_{y'} = 3.75 I_0$$

$$\frac{2EI_{y'}}{L} = \frac{2E(3.75 I_0)}{7.5} = EI_0$$

Keeping $EI_0/2$ as a factor again, we get

$$\mathbf{k}'^{(3)} = 2\mathbf{k}'^{(2)}$$

These three member local stiffness arrays have been transformed to global coordinates according to Eq. 8.25 by using the SMIS3 program, as shown in the printout below.

****** *Computer Output:*

```
START
REMARK          1

TRANSFORMS STIFFNESS MATRIX FOR MEMBER 1 FROM LOCAL TO GLOBAL COORDINATES

LOAD      ALPH           3    3
   3 ROWS,     3 COLUMNS
PRINT     ALPH           1
```

****** *Computer Output (continued):*

```
THIS IS ALPH, SUBARRAY OF ALPHA FOR MEMBER 1


             1            2            3
 1           0-1.0000E+00            0
 2  8.0000E-01            0-6.0000E-01
 3  6.0000E-01            0 8.0000E-01


FORM        ALPHA          12     12
   12 ROWS,     12 COLUMNS
STOSM       ALPHA     ALPH            1      1
 ROW NUMBER   1,   COLUMN NUMBER   1
STOSM       ALPHA     ALPH            4      4
 ROW NUMBER   4,   COLUMN NUMBER   4
STOSM       ALPHA     ALPH            7      7
 ROW NUMBER   7,   COLUMN NUMBER   7
STOSM       ALPHA     ALPH           10     10
 ROW NUMBER  10,   COLUMN NUMBER  10
LOAD        KA              4     12
    4 ROWS,     12 COLUMNS
PRINT       KA              1

THIS IS KA, SUBARRAY OF THE LOCAL STIFFNESS ARRAY FOR MEMBER 1


             1            2            3            4            5            6
 1  5.0000E+00            0            0            0            0            0
 2           0 8.5710E-02            0            0            0 1.5000E-01
 3           0            0 2.4490E-01            0-4.2860E-01            0
 4           0            0            0 8.0000E-02            0            0
             7            8            9           10           11           12
 1 -5.0000E+00            0            0            0            0            0
 2           0-8.5710E-02            0            0            0 1.5000E-01
 3           0            0-2.4490E-01            0-4.2860E-01            0
 4           0            0            0-8.0000E-02            0            0

LOAD        KB              2     12
    2 ROWS,     12 COLUMNS
PRINT       KB              1

THIS IS KB, SUBARRAY OF THE LOCAL STIFFNESS ARRAY FOR MEMBER 1


             1            2            3            4            5            6
 1           0            0-4.2860E-01            0 1.0000E+00            0
 2           0 1.5000E-01            0            0            0 3.5000E-01
             7            8            9           10           11           12
 1           0            0 4.2860E-01            0 5.0000E-01            0
 2           0-1.5000E-01            0            0            0 1.7500E-01

LOAD        KC              2     12
    2 ROWS,     12 COLUMNS
PRINT       KC              1

THIS IS KC, SUBARRAY OF THE LOCAL STIFFNESS ARRAY FOR MEMBER 1


             1            2            3            4            5            6
 1           0            0-4.2860E-01            0 5.0000E-01            0
 2           0 1.5000E-01            0            0            0 1.7500E-01
             7            8            9           10           11           12
 1           0            0 4.2860E-01            0 1.0000E+00            0
 2           0-1.5000E-01            0            0            0 3.5000E-01
```

****** *Computer Output (continued):*

```
FORM      K               12    12
   12 ROWS,     12 COLUMNS
STOSM     K          KA              1     1
 ROW NUMBER   1,  COLUMN NUMBER   1
SCALE     KA            -1.00000
 SCALAR= -1.0000000E+00
STOSM     K          KA              7     1
 ROW NUMBER   7,  COLUMN NUMBER   1
STOSM     K          KB              5     1
 ROW NUMBER   5,  COLUMN NUMBER   1
STOSM     K          KC             11     1
 ROW NUMBER  11,  COLUMN NUMBER   1
MULT      ALPHA      K          KTEM
DELETE    K
TRANS     ALPHA      ALPHAT
DELETE    ALPHA
MULT      KTEM       ALPHAT     K
PRINT     K               1

THIS IS GLOBAL K FOR MEMBER 1 WITH EI(0)/2 FACTORED OUT

            1           2           3           4           5           6
  1  8.5710E-02           0           0           0 9.0000E-02-1.2000E-01
  2           0 3.2882E+00 2.2824E+00-2.5716E-01           0           0
  3           0 2.2824E+00 1.9567E+00 3.4288E-01           0           0
  4           0-2.5716E-01 3.4288E-01 1.0000E+00           0           0
  5  9.0000E-02           0           0           0 1.7720E-01-1.2960E-01
  6 -1.2000E-01           0           0           0-1.2960E-01 2.5280E-01
  7 -8.5710E-02           0           0           0-9.0000E-02 1.2000E-01
  8           0-3.2882E+00-2.2824E+00 2.5716E-01           0           0
  9           0-2.2824E+00-1.9567E+00-3.4288E-01           0           0
 10           0-2.5716E-01 3.4288E-01 5.0000E-01           0           0
 11  9.0000E-02           0           0           0 1.1800E-02-1.2240E-01
 12 -1.2000E-01           0           0           0-1.2240E-01 8.3200E-02
            7           8           9          10          11          12
  1 -8.5710E-02           0           0           0 9.0000E-02-1.2000E-01
  2           0-3.2882E+00-2.2824E+00-2.5716E-01           0           0
  3           0-2.2824E+00-1.9567E+00 3.4288E-01           0           0
  4           0 2.5716E-01-3.4288E-01 5.0000E-01           0           0
  5 -9.0000E-02           0           0           0 1.1800E-02-1.2240E-01
  6  1.2000E-01           0           0           0-1.2240E-01 8.3200E-02
  7  8.5710E-02           0           0           0-9.0000E-02 1.2000E-01
  8           0 3.2882E+00 2.2824E+00 2.5716E-01           0           0
  9           0 2.2824E+00 1.9567E+00-3.4288E-01           0           0
 10           0 2.5716E-01-3.4288E-01 1.0000E+00           0           0
 11 -9.0000E-02           0           0           0 1.7720E-01-1.2960E-01
 12  1.2000E-01           0           0           0-1.2960E-01 2.5280E-01

STOP
```

****** *Computer Output:*

```
START
REMARK          1

TRANSFORMS STIFFNESS MATRIX FOR MEMBER 2 FROM LOCAL TO GLOBAL COORDINATES

LOAD      ALPH            3     3
    3 ROWS,      3 COLUMNS
PRINT     ALPH            1
```

****** *Computer Output (continued):*

```
THIS IS ALPH, SUBARRAY OF ALPHA FOR MEMBER 2


            1            2            3
 1            0 1.0000E+00            0
 2 -9.6000E-01            0 2.8000E-01
 3  2.8000E-01            0 9.6000E-01


FORM      ALPHA          12    12
   12 ROWS,      12 COLUMNS
STOSM     ALPHA     ALPH                 1     1
 ROW NUMBER   1,   COLUMN NUMBER   1
STOSM     ALPHA     ALPH                 4     4
 ROW NUMBER   4,   COLUMN NUMBER   4
STOSM     ALPHA     ALPH                 7     7
 ROW NUMBER   7,   COLUMN NUMBER   7
STOSM     ALPHA     ALPH                10    10
 ROW NUMBER  10,   COLUMN NUMBER  10
LOAD      KA              4     12
    4 ROWS,      12 COLUMNS
PRINT     KA              1


THIS IS KA, SUBARRAY OF THE LOCAL STIFFNESS ARRAY FOR MEMBER 2


            1            2            3            4            5            6
 1  1.0000E+01            0            0            0            0            0
 2            0 3.7330E-02            0            0            0 1.4000E-01
 3            0            0 1.0670E-01            0-4.0000E-01            0
 4            0            0            0 1.6000E-01            0            0
            7            8            9           10           11           12
 1 -1.0000E+01            0            0            0            0            0
 2            0-3.7330E-02            0            0            0 1.4000E-01
 3            0            0-1.0670E-01            0-4.0000E-01            0
 4            0            0            0-1.6000E-01            0            0


LOAD      KB              2     12
    2 ROWS,      12 COLUMNS
PRINT     KB              1


THIS IS KB, SUBARRAY OF THE LOCAL STIFFNESS ARRAY FOR MEMBER 2


            1            2            3            4            5            6
 1            0            0-4.0000E-01            0 2.0000E+00            0
 2            0 1.4000E-01            0            0            0 7.0000E-01
            7            8            9           10           11           12
 1            0            0 4.0000E-01            0 1.0000E+00            0
 2            0-1.4000E-01            0            0            0 3.5000E-01


LOAD      KC              2     12
    2 ROWS,      12 COLUMNS
PRINT     KC              1


THIS IS KC, SUBARRAY OF THE LOCAL STIFFNESS ARRAY FOR MEMBER 2


            1            2            3            4            5            6
 1            0            0-4.0000E-01            0 1.0000E+00            0
 2            0 1.4000E-01            0            0            0 3.5000E-01
            7            8            9           10           11           12
 1            0            0 4.0000E-01            0 2.0000E+00            0
 2            0-1.4000E-01            0            0            0 7.0000E-01
```

****** *Computer Output (continued):*

```
FORM       K              12     12
   12 ROWS,     12 COLUMNS
STOSM      K          KA             1      1
 ROW NUMBER   1,  COLUMN NUMBER   1
SCALE      KA            -1.00000
 SCALAR= -1.0000000E+00
STOSM      K          KA             7      1
 ROW NUMBER   7,  COLUMN NUMBER   1
STOSM      K          KB             5      1
 ROW NUMBER   5,  COLUMN NUMBER   1
STOSM      K          KC            11      1
 ROW NUMBER  11,  COLUMN NUMBER   1
MULT       ALPHA      K        KTEM
DELETE     K
TRANS      ALPHA      ALPHAT
DELETE     ALPHA
MULT       KTEM       ALPHAT     K
PRINT      K               1

THIS IS GLOBAL K FOR MEMBER 2 WITH EI(0)/2 FACTORED OUT


           1          2          3          4          5          6
 1  3.7330E-02          0          0          0 3.9200E-02 1.3440E-01
 2           0 9.2244E+00-2.6593E+00-1.1200E-01          0          0
 3           0-2.6593E+00 8.8233E-01-3.8400E-01          0          0
 4           0-1.1200E-01-3.8400E-01 2.0000E+00          0          0
 5  3.9200E-02          0          0          0 2.0234E-01 1.4515E-01
 6  1.3440E-01          0          0          0 1.4515E-01 6.5766E-01
 7 -3.7330E-02          0          0          0-3.9200E-02-1.3440E-01
 8           0-9.2244E+00 2.6593E+00 1.1200E-01          0          0
 9           0 2.6593E+00-8.8233E-01 3.8400E-01          0          0
10           0-1.1200E-01-3.8400E-01 1.0000E+00          0          0
11  3.9200E-02          0          0          0-1.2002E-01 1.3709E-01
12  1.3440E-01          0          0          0 1.3709E-01 3.1002E-01
           7          8          9         10         11         12
 1 -3.7330E-02          0          0          0 3.9200E-02 1.3440E-01
 2           0-9.2244E+00 2.6593E+00-1.1200E-01          0          0
 3           0 2.6593E+00-8.8233E-01-3.8400E-01          0          0
 4           0 1.1200E-01 3.8400E-01 1.0000E+00          0          0
 5 -3.9200E-02          0          0          0-1.2002E-01 1.3709E-01
 6 -1.3440E-01          0          0          0 1.3709E-01 3.1002E-01
 7  3.7330E-02          0          0          0-3.9200E-02-1.3440E-01
 8           0 9.2244E+00-2.6593E+00 1.1200E-01          0          0
 9           0-2.6593E+00 8.8233E-01 3.8400E-01          0          0
10           0 1.1200E-01 3.8400E-01 2.0000E+00          0          0
11 -3.9200E-02          0          0          0 2.0234E-01 1.4515E-01
12 -1.3440E-01          0          0          0 1.4515E-01 6.5766E-01

STOP
```

****** *Computer Output:*

```
START
REMARK          1

TRANSFORMS STIFFNESS MATRIX FOR MEMBER 3 FROM LOCAL TO GLOBAL COORDINATES


LOAD       ALPH              3      3
    3 ROWS,     3 COLUMNS
PRINT      ALPH              1
```

****** *Computer Output (continued):*

```
THIS IS ALPH, SUBARRAY OF ALPHA FOR MEMBER 3


              1            2            3
 1 -9.6000E-01            0 2.8000E-01
 2            0-1.0000E+00            0
 3  2.8000E-01            0 9.6000E-01


FORM       ALPHA         12    12
   12 ROWS,    12 COLUMNS
STOSM      ALPHA      ALPH             1      1
 ROW NUMBER   1,  COLUMN NUMBER   1
STOSM      ALPHA      ALPH             4      4
 ROW NUMBER   4,  COLUMN NUMBER   4
STOSM      ALPHA      ALPH             7      7
 ROW NUMBER   7,  COLUMN NUMBER   7
STOSM      ALPHA      ALPH            10     10
 ROW NUMBER  10,  COLUMN NUMBER  10
LOAD       KA             4    12
    4 ROWS,    12 COLUMNS
PRINT      KA             1


THIS IS KA, SUBARRAY OF THE LOCAL STIFFNESS ARRAY FOR MEMBER 3


              1            2            3            4            5            6
 1  1.0000E+01            0            0            0            0            0
 2            0 3.7330E-02            0            0            0 1.4000E-01
 3            0            0 1.0670E-01            0-4.0000E-01            0
 4            0            0            0 1.6000E-01            0            0
              7            8            9           10           11           12
 1 -1.0000E+01            0            0            0            0            0
 2            0-3.7330E-02            0            0            0 1.4000E-01
 3            0            0-1.0670E-01            0-4.0000E-01            0
 4            0            0            0-1.6000E-01            0            0


LOAD       KB             2    12
    2 ROWS,    12 COLUMNS
PRINT      KB             1


THIS IS KB, SUBARRAY OF THE LOCAL STIFFNESS ARRAY FOR MEMBER 3


              1            2            3            4            5            6
 1            0            0-4.0000E-01            0 2.0000E+00            0
 2            0 1.4000E-01            0            0            0 7.0000E-01
              7            8            9           10           11           12
 1            0            0 4.0000E-01            0 1.0000E+00            0
 2            0-1.4000E-01            0            0            0 3.5000E-01

LOAD       KC             2    12
    2 ROWS,    12 COLUMNS
PRINT      KC             1


THIS IS KC, SUBARRAY OF THE LOCAL STIFFNESS ARRAY FOR MEMBER 3


              1            2            3            4            5            6
 1            0            0-4.0000E-01            0 1.0000E+00            0
 2            0 1.4000E-01            0            0            0 3.5000E-01
              7            8            9           10           11           12
 1            0            0 4.0000E-01            0 2.0000E+00            0
 2            0-1.4000E-01            0            0            0 7.0000E-01
```

****** *Computer Output (continued):*

```
FORM       K                 12      12
  12 ROWS,      12 COLUMNS
STOSM      K           KA                  1       1
 ROW NUMBER   1,  COLUMN NUMBER    1
SCALE      KA              -1.00000
 SCALAR= -1.0000000E+00
STOSM      K           KA                  7       1
 ROW NUMBER   7,  COLUMN NUMBER    1
STOSM      K           KB                  5       1
 ROW NUMBER   5,  COLUMN NUMBER    1
STOSM      K           KC                 11       1
 ROW NUMBER  11,  COLUMN NUMBER    1
SCALE      K                2.00000
 SCALAR=  2.0000000E+00
MULT       ALPHA       K              KTEM
DELETE     K
TRANS      ALPHA       ALPHAT
DELETE     ALPHA
MULT       KTEM        ALPHAT     K
PRINT      K                 1

THIS IS GLOBAL K FOR MEMBER 3 WITH EI(0)/2 FACTORED OUT

             1              2              3              4              5              6
 1  1.8449E+01              0-5.3186E+00              0 2.2400E-01              0
 2              0 7.4660E-02              0-7.8400E-02              0-2.6880E-01
 3 -5.3186E+00              0 1.7647E+00              0 7.6800E-01              0
 4              0-7.8400E-02              0 4.0467E-01              0 2.9030E-01
 5  2.2400E-01              0 7.6800E-01              0 4.0000E+00              0
 6              0-2.6880E-01              0 2.9030E-01              0 1.3153E+00
 7 -1.8449E+01              0 5.3186E+00              0-2.2400E-01              0
 8              0-7.4660E-02              0 7.8400E-02              0 2.6880E-01
 9  5.3186E+00              0-1.7647E+00              0-7.6800E-01              0
10              0-7.8400E-02              0-2.4003E-01              0 2.7418E-01
11  2.2400E-01              0 7.6800E-01              0 2.0000E+00              0
12              0-2.6880E-01              0 2.7418E-01              0 6.2003E-01
             7              8              9             10             11             12
 1 -1.8449E+01              0 5.3186E+00              0 2.2400E-01              0
 2              0-7.4660E-02              0-7.8400E-02              0-2.6880E-01
 3  5.3186E+00              0-1.7647E+00              0 7.6800E-01              0
 4              0 7.8400E-02              0-2.4003E-01              0 2.7418E-01
 5 -2.2400E-01              0-7.6800E-01              0 2.0000E+00              0
 6              0 2.6880E-01              0 2.7418E-01              0 6.2003E-01
 7  1.8449E+01              0-5.3186E+00              0-2.2400E-01              0
 8              0 7.4660E-02              0 7.8400E-02              0 2.6880E-01
 9 -5.3186E+00              0 1.7647E+00              0-7.6800E-01              0
10              0 7.8400E-02              0 4.0467E-01              0 2.9030E-01
11 -2.2400E-01              0-7.6800E-01              0 4.0000E+00              0
12              0 2.6880E-01              0 2.9030E-01              0 1.3153E+00

STOP
```

The member global fixed-end forces may be obtained directly by inspection of Fig. 8.8:

$$
\begin{aligned}
\mathbf{r}_F^{(1)} &= \langle 0 \quad 0 \quad .7 \quad .3267 \quad 0 \quad 0 \quad 0 \quad 0 \quad .7 \quad -.3267 \quad 0 \quad 0 \rangle\, q \\
\mathbf{r}_F^{(2)} &= \langle 0 \quad 0 \quad 1.8 \quad -2.16 \quad 0 \quad 0 \quad 0 \quad 0 \quad 1.8 \quad 2.16 \quad 0 \quad 0 \rangle\, q \qquad \text{(H)} \\
\mathbf{r}_F^{(3)} &= \langle 0 \quad 0 \quad 3.6 \quad 0 \quad 4.32 \quad 0 \quad 0 \quad 0 \quad 3.6 \quad 0 \quad -4.32 \quad 0 \rangle\, q
\end{aligned}
$$

The space-frame structure stiffness relationships may now be built by using the following unconstrained coordinate code:

	1	2	3	4	5	6	7	8	9	10	11	12
1	0	0	0	0	0	0	0	1	2	3	0	0
2	0	0	0	0	0	0	0	1	2	3	0	0
3	0	0	0	0	0	0	0	1	2	3	0	0

$$\begin{pmatrix} R_1 \\ R_2 \\ R_3 \end{pmatrix} = \frac{EI_0}{2}\begin{bmatrix} 3.288 + 9.224 + .07466 & 2.282 - 2.659 + 0 & .2572 + .1120 + .07840 \\ & 1.957 + .8823 + 1.765 & -.3429 + .3840 + 0 \\ \text{Symmetric} & & 1 + 2 + .4047 \end{bmatrix}$$

$$\times \begin{pmatrix} D_1 \\ D_2 \\ D_3 \end{pmatrix} + \begin{pmatrix} 0 + 0 + 0 \\ .7 + 1.8 + 3.6 \\ -.3267 + 2.16 + 0 \end{pmatrix} q \tag{I}$$

or

$$\frac{EI_0}{2}\begin{bmatrix} 12.59 & -.377 & .4476 \\ & 4.604 & .0411 \\ \text{Symmetric} & & 3.405 \end{bmatrix}\begin{pmatrix} D_1 \\ D_2 \\ D_3 \end{pmatrix}\begin{pmatrix} 0 \\ -6.1 \\ -1.833 \end{pmatrix} q \tag{J}$$

Solving, we get

$$\mathbf{D} = \frac{2q}{EI_0}\langle -.02111 \quad -1.322 \quad -.5197 \rangle \tag{K}$$

Backsubstituting these results into the member stiffness relationships, we could find the member-end forces and verify that they satisfy equilibrium requirements. In the interest of brevity, we will forgo that step here. It is interesting, however, as a check on our work to consider the space frame of Fig. 8.8 subjected to the loading on the space truss of Fig. 8.6. This gives the load vector

$$\mathbf{R} = \langle 0 \quad -.5 \quad 0 \rangle Q \tag{L}$$

Solving, again using this load vector, gives

$$\mathbf{D} = \frac{2Q}{EI_0}\langle -.003323 \quad -.1089 \quad .001751 \rangle \tag{M}$$

or, since $A_0 = 20I_0$,

$$\mathbf{D} = \frac{4Q}{EA_0}\langle -.03323 \quad -1.089 \quad .01751 \rangle \tag{N}$$

Comparing the result of Eq. N with that of Eq. B in Sec. 8.3, we see that the effect of rigid connections on displacements is small in this case, as would be expected.

8.6 SUMMARY

The extension of the direct stiffness method to three-dimensional structures and gridwork structures has proved to be conceptually quite straightforward. Although we made no attempt to implement in three-dimensional analysis all the special features of the direct stiffness method described in Chaps. 4 and 5, it is clear that this also would be possible. We could furthermore similarly extend the force method of analysis as discussed in Chap. 6 to three dimensions. Thus the discussions of this text that have been confined mainly to two-dimensional examples should provide a useful basis for the study of more general examples. You will find that this is true not only for three-dimensional analysis of discrete-membered structures, as demonstrated in this chapter, but also for a wide range of numerical analysis examples as mentioned in Chap. 1.

PROBLEMS

8.1 Solve for the nodal displacements and member-end forces for the space truss of Fig. P8.1. Solve by hand using the direct stiffness method and considering symmetry.

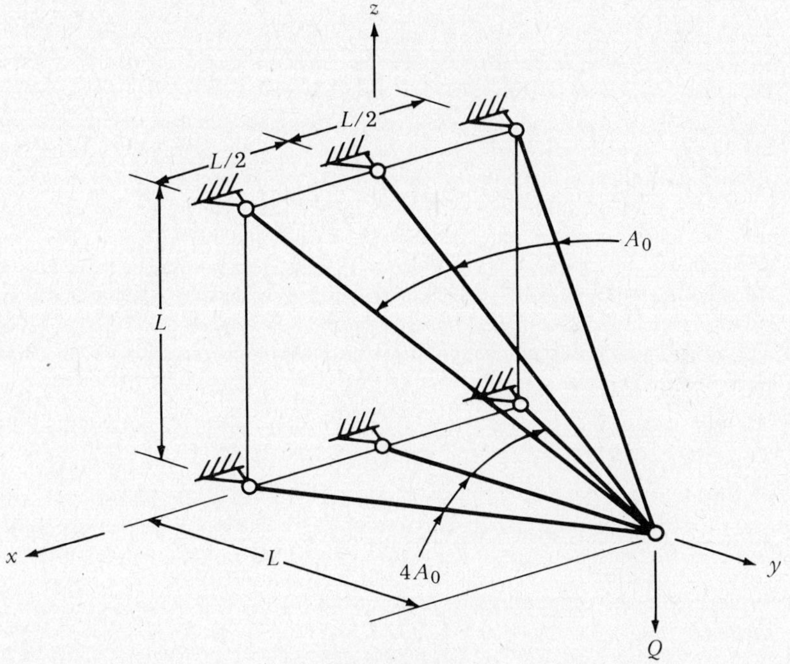

E = constant

Fig. P8.1

8.2 Solve Prob. 8.1 using an appropriate structural analysis program.

8.3 Solve for the nodal displacements and member-end forces for the gridwork of Fig. P8.3. Solve by hand using the direct stiffness method and considering both planes of symmetry. It is given that $I_{beams} = I_0$, $I_{girder} = 4I_0$, $J_{beams} = J_0$, $J_{girder} = 4J_0$, $I_0/J_0 = 3$, $E =$ constant, and $\nu = .25$. All connections are totally rigid. Warping restraint effects are negligible. A uniformly distributed load of intensity q is applied vertically downward on the deck supported by the gridwork. The one-way decking is simply supported on the beams.

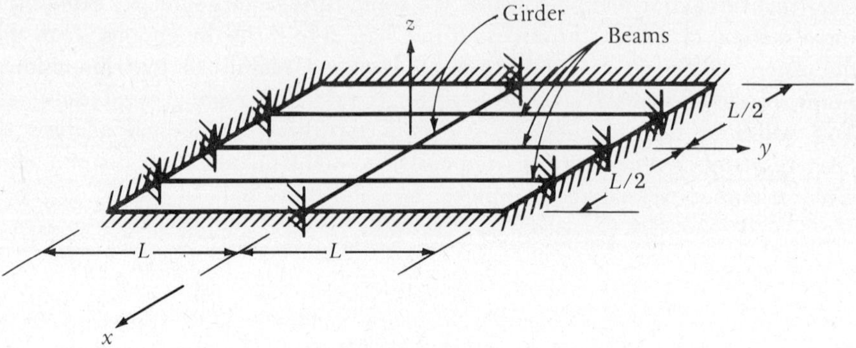

$L = 10$ m

Fig. P8.3

8.4 Solve Prob. 8.3 using an appropriate structural analysis program.

8.5 The space framework of Fig. P8.5 is a slight simplification of the roof structure for Mackey Arena at Purdue University. Solve for the nodal displacements and member-end forces for this space framework by using an appropriate structural analysis program. Consider the roof to be subjected to the dead weight of the roof structure plus a live load of 30 psf uniformly distributed over the horizontal projected area of the roof. The roof deck weight may be taken to be 5 psf, and the deck may be assumed to be simply supported on adjacent purlins (W 10s). The purlins, in turn, are simply supported at their ends. All other connections are rigid except that the tension ring rests on a slipping surface. The proposed structural model shown in Fig. P8.5 takes into account structural and loading symmetry and, in the process, neglects the diagonal braces for this loading.

The SMIS3 program has been written to allow certain subroutines to be conveniently replaced by user-generated subroutines. Problems 8.6 and 8.7 make use of this feature. To successfully accomplish these replacements, it will be necessary to study the appropriate sections of the documentation presented in Sec. A2.6 of App. 2.

8.6 Write FORTRAN subroutines to perform the following functions for a space truss member:

(a) Generate the basic stiffness array $\mathscr{k}$. Replace subroutine TRBKF described in part E2 of Sec. A2.6.

(b) Generate the basic-local transformation array **L**. Replace subroutine TDTL described in part E4 of Sec. A2.6.

(c) Generate the local-global transformation array $\boldsymbol{\alpha}$. Replace subroutine TDTA described in part E3 of Sec. A2.6.

(d) Transform the basic stiffness array $\mathscr{k}$ into the global stiffness array **k**. Replace subroutine FORMKT described in part E5 of Sec. A2.6.

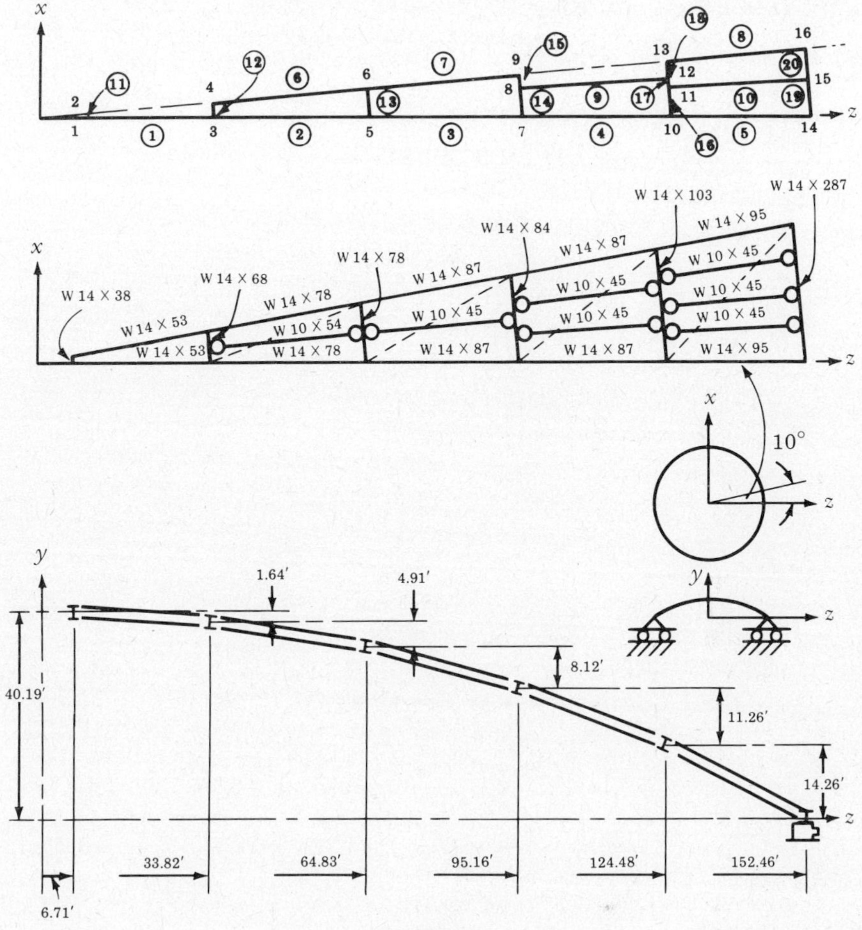

Fig. P8.5

Check the functioning of these replacement subroutines by working Prob. 8.1.

8.7 Write FORTRAN subroutines to perform the following functions for a space frame member:

(a) Generate the basic stiffness array $\mathscr{k}$. Replace subroutine TDFBKF described in part F2 of Sec. A2.6.

(b) Generate the basic-local transformation array **L**. Replace subroutine TDFL described in part F4 of Sec. A2.6.

(c) Generate the local-global transformation array $\boldsymbol{\alpha}$. Replace subroutine TDFA described in part F3 of Sec. A2.6.

(d) Transform the basic stiffness array $\mathscr{k}$ into the global stiffness array **k**. Replace subroutine FORMKT described in part F5 of Sec. A2.6.

Check the functioning of these replacement subroutines by working the example of Fig. 8.8.

REFERENCES

8.1 Boresi, Arthur P., et al. *Advanced Mechanics of Materials*. New York: Wiley, 1978, pp. 209–257.

8.2 Flugge, W. *Handbook of Engineering Mechanics*. New York: McGraw-Hill, 1962, pp. 36-12 to 36-13.

Appendix 1
Matrix Algebra Review

A1.1 INTRODUCTION

The idealized structural system of Fig. A1.1a will be used as a vehicle for demonstrating, and thereby reviewing, the matrix algebra manipulations commonly encountered in matrix structural analysis. This approach allows you not only to review matrix algebra topics but also to observe how they relate to structural analysis concepts. The structural analysis example allows a fairly logical ordering of matrix algebra concepts, but if you wish to review a specific topic, you may want to consult the summary of topics at the end of this appendix to determine where the topic appears.

A1.2 STRUCTURAL EXAMPLE

The masses of the system of Fig. A1.1a are subjected to forces R_1, R_2, and R_3, which are assumed to be gradually applied, causing the system to distort statically and the masses to move from their initial positions to some new positions at which equilibrium is again satisfied. The distances between the initial and final equilibrium positions of the masses are defined as D_1, D_2, and D_3, respectively. These displacements may be determined by considering the conditions of equilibrium displayed on the free-body diagrams of Fig. A1.1b.

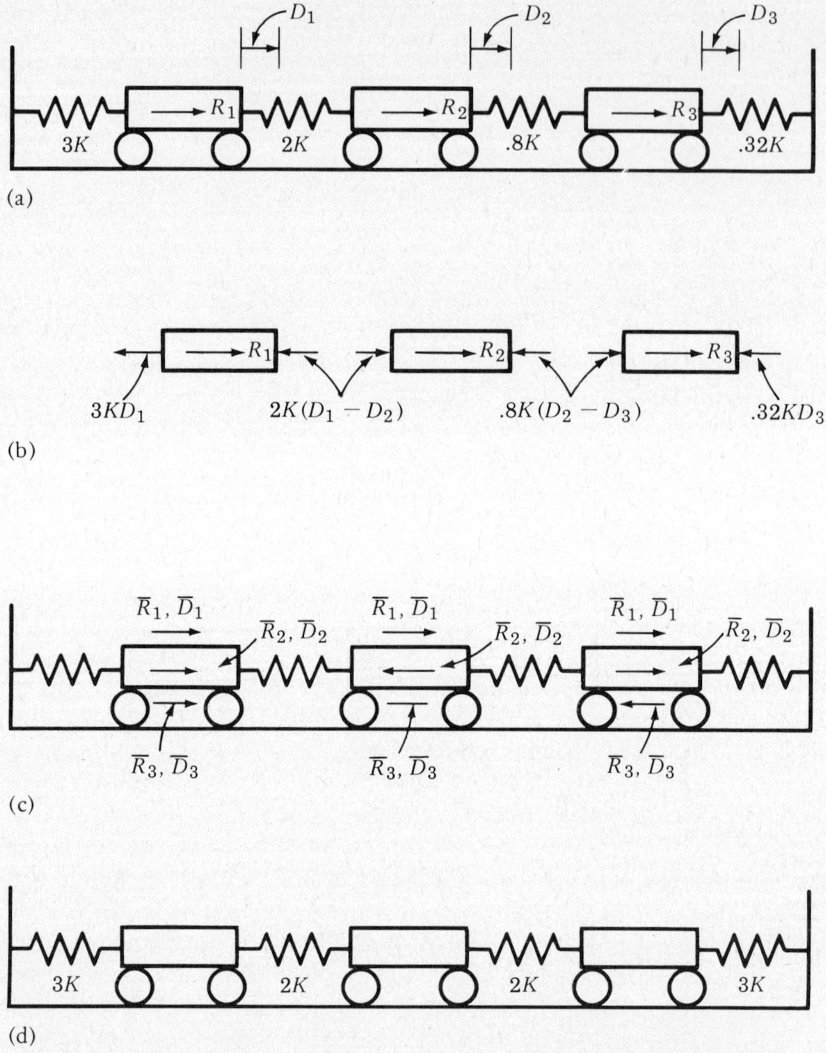

Fig. A1.1 Spring-mass systems: (a) loading and properties; (b) free body diagrams; (c) alternate coordinate system; (d) alternate properties.

Summing the forces horizontally on each free body gives, respectively,

$$
\begin{aligned}
&-3KD_1 - 2K(D_1 - D_2) + R_1 = 0 \\
&2K(D_1 - D_2) - .8K(D_2 - D_3) + R_2 = 0 \qquad \text{(A1.1)} \\
&-.8K(D_2 - D_3) - .32KD_3 + R_3 = 0
\end{aligned}
$$

where the coefficients, $3K$, $2K$, and so on, represent the stiffnesses of the springs (which in turn may be thought of as idealized structural members). The stiffness coefficients have units of force per unit displacement. The product of the spring stiffness times the relative displacement of the ends of the spring yields the force in (and at each end of) the spring. The magnitudes of the stiffness coefficients depend on the physical properties of the members represented by the springs. Rearranging, we may write Eq. A1.1 as follows:

$$\begin{aligned} 5KD_1 - 2KD_2 - R_1 &= 0 \\ -2KD_1 + 2.8KD_2 - .8KD_3 - R_2 &= 0 \\ -.8KD_2 + 1.32KD_3 - R_3 &= 0 \end{aligned} \tag{A1.1a}$$

A cursory examination of Eq. A1.1a reveals that there are three distinct groups of terms present: unknown displacements, known applied forces, and given coefficients. It is convenient to be able to isolate each of these types of terms so that they may be manipulated together in their respective groups. Matrix algebra provides the ideal mechanism for accomplishing this. Using matrix algebra, we can gather the different types of terms into groups called arrays and then set up rules to manipulate the arrays in meaningful ways.

Rewriting Eq. A1.1a in matrix notation gives

$$K\begin{bmatrix} 5 & -2 & 0 \\ -2 & 2.8 & -.8 \\ 0 & -.8 & 1.32 \end{bmatrix}\begin{pmatrix} D_1 \\ D_2 \\ D_3 \end{pmatrix} = \begin{pmatrix} R_1 \\ R_2 \\ R_3 \end{pmatrix} \tag{A1.1b}$$

Any of these arrays may be correctly termed matrices. In addition, the one-dimensional arrays may be referred to as vectors. In more detail, the two-dimensional array of coefficients would be called a 3×3 square matrix and the other two arrays would be called 3×1 column vectors. Now, by defining rules for matrix manipulation, we may prove that matrix Eq. A1.1b exactly represents the scalar equations of Eq. A1.1a. First, the rule for multiplying an array by a scalar (each term of the array is multiplied by the scalar) may be used, giving

$$\begin{bmatrix} -5K & -2K & 0 \\ -2K & 2.8K & -.8K \\ 0 & -.8K & 1.32K \end{bmatrix}\begin{pmatrix} D_1 \\ D_2 \\ D_3 \end{pmatrix} = \begin{pmatrix} R_1 \\ R_2 \\ R_3 \end{pmatrix} \tag{A1.1c}$$

Matrix Eq. A1.1c may be represented symbolically as

$$\begin{bmatrix} K_{11} & K_{12} & K_{13} \\ K_{21} & K_{22} & K_{23} \\ K_{31} & K_{32} & K_{33} \end{bmatrix}\begin{pmatrix} D_1 \\ D_2 \\ D_3 \end{pmatrix} = \begin{pmatrix} R_1 \\ R_2 \\ R_3 \end{pmatrix} \tag{A1.1d}$$

or, more compactly, as

$$\mathbf{KD} = \mathbf{R} \tag{A1.1e}$$

Coefficients are seen to be located in arrays by subscripting. The **K** array of Eq. A1.1d indicates that the first and second subscripts of a coefficient are equal to its row and column numbers, respectively. These symbolic representations will be useful for future reference.

Next, the rule for multiplying two arrays (a term in the product array equals the inner product of the vector composed of the terms in the corresponding row of the first array and the vector composed of the terms in the corresponding column of the second array) may be invoked, giving

$$\begin{pmatrix} 5KD_1 - 2KD_2 \\ -2KD_1 + 2.8KD_2 - .8KD_3 \\ -.8KD_2 + 1.32KD_3 \end{pmatrix} = \begin{pmatrix} R_1 \\ R_2 \\ R_3 \end{pmatrix} \tag{A1.1f}$$

Notice, in passing, that the rule for multiplication of arrays does not follow a commutative law. In fact, if the array order were reversed in Eq. A1.1c, a product array would not even be defined. Thus the concepts of premultiplication and postmultiplication are needed. For this example, it may be stated that the square matrix premultiplies the column vector or the column vector postmultiplies the square matrix. It will be shown later, however, that multiplication of arrays does follow a distributive law. Finally, using the rule for addition of two arrays (a term in the sum array equals the sum of the corresponding terms in the two arrays being summed) and adding the negative of the right-hand side vector to both sides of Eq. A1.1f, we get

$$\begin{pmatrix} 5KD_1 - 2KD_2 \\ -2KD_1 + 2.8KD_2 - .8KD_3 \\ -.8KD_2 + 1.32KD_3 \end{pmatrix} + \begin{pmatrix} -R_1 \\ -R_2 \\ -R_3 \end{pmatrix} = \begin{pmatrix} R_1 \\ R_2 \\ R_3 \end{pmatrix} + \begin{pmatrix} -R_1 \\ -R_2 \\ -R_3 \end{pmatrix} \tag{A1.1g}$$

or

$$\begin{pmatrix} 5KD_1 - 2KD_2 - R_1 \\ -2KD_1 + 2.8KD_2 - .8KD_3 - R_2 \\ -.8KD_2 + 1.32KD_3 - R_3 \end{pmatrix} = \begin{pmatrix} 0 \\ 0 \\ 0 \end{pmatrix} \tag{A1.1h}$$

Notice that we have, in effect, carried out matrix subtraction by addition of a negative matrix. Thus matrix subtraction is included in matrix addition. Brief consideration of the matrix addition process will reveal that this process is both commutative and distributive in nature. It could also be shown in a more general context that the sequence of multiplication and then addition is the same as the sequence of addition and then multiplication. The demonstration of these truths is left to you.

It is not surprising of course, to find that the vector of zeros in Eq. A1.1h is called a null vector. Now, imposing on that equation the rather self-evident truth that for any two arrays to be equal, they must be equal on a term-by-term basis, we obtain the following:

$$\begin{aligned} 5KD_1 - 2KD_2 - R_1 &= 0 \\ -2KD_1 + 2.8KD_2 - .8KD_3 - R_2 &= 0 \\ -.8KD_2 + 1.32KD_3 - R_3 &= 0 \end{aligned} \tag{A1.1i}$$

which is seen to identically check Eq. A1.1a, demonstrating that the rules for manipulating arrays as defined do work properly.

There are, of course, additional matrix manipulation techniques. Just as there was a subtraction technique closely associated with the addition process, one would expect there to be a division technique closely associated with the multiplication process. It turns out, however, that there is no such thing as matrix division, but rather a sort of synthetic division known as matrix inversion. Matrix inversion is best demonstrated in the context of carrying out a solution of Eq. A1.1a.

The need to solve a set of simultaneous equations is fundamentally important to structural analysis. Thus it is appropriate here to demonstrate the most efficient method of solving a general set of simultaneous equations—gaussian elimination. Gaussian elimination will be demonstrated in solving the set of equations that arose from the example problem without any attempt being made at this point to develop a general algorithm. Those interested in a more general development of this method should see Chap. 7.

Rearranging Eq. A1.1a gives

$$\begin{aligned} 5KD_1 - 2KD_2 &= R_1 \\ -2KD_1 + 2.8KD_2 - .8KD_3 &= R_2 \\ -.8KD_2 + 1.32KD_3 &= R_3 \end{aligned} \tag{A1.1j}$$

Dividing the first equation of Eq. A1.1j by the coefficient of the first term in that equation gives

$$D_1 - .4D_2 = \frac{.2R_1}{K} \tag{A1.2}$$

This equation may now be used to eliminate the unknown D_1 from the remaining equations; in this case only the second equation is involved. Multiplying Eq. A1.2 by the negative of the coefficient of D_1 in the second equation of Eq. A1.1j and adding the result to the second equation of Eq. A1.1j gives

$$2KD_2 - .8KD_3 = R_2 + .4R_1 \tag{A1.3}$$

Dividing Eq. A1.3 by the coefficient of the unknown D_2 in that equation, we get

$$D_2 - .4D_3 = \frac{.5R_2}{K} + \frac{.2R_1}{K} \tag{A1.3a}$$

Multiplying this equation by the negative of the coefficient of the unknown D_2 in the third equation of Eq. A1.1j and adding the result to the third equation of Eq. A1.1j, we can eliminate D_2 from that equation, so that we have

$$KD_3 = R_3 + .4R_2 + .16R_1 \tag{A1.4}$$

Dividing this result by the coefficient of D_3 and rearranging gives

$$D_3 = \frac{.16R_1}{K} + \frac{.4R_2}{K} + \frac{R_3}{K} \tag{A1.4a}$$

Backsubstituting this result into Eq. A1.3a gives

$$D_2 = \frac{.264R_1}{K} + \frac{.66R_2}{K} + \frac{.4R_3}{K} \tag{A1.5}$$

Now, backsubstituting this result into Eq. A1.2, we get

$$D_1 = \frac{.3056R_1}{K} + \frac{.264R_2}{K} + \frac{.16R_3}{K} \tag{A1.6}$$

Rewriting Eqs. A1.4a, A1.5, and A1.6 in matrix form gives

$$\begin{pmatrix} D_1 \\ D_2 \\ D_3 \end{pmatrix} = \begin{bmatrix} .3056/K & .264/K & .16/K \\ .264/K & .66/K & .4/K \\ .16/K & .4/K & 1/K \end{bmatrix} \begin{pmatrix} R_1 \\ R_2 \\ R_3 \end{pmatrix} \tag{A1.7}$$

or, symbolically,

$$\mathbf{D} = \mathbf{K}^{-1}\mathbf{R} \tag{A1.7a}$$

The matrix $\mathbf{K}^{-1}$ is defined as the inverse of the matrix $\mathbf{K}$. (The matrix $\mathbf{K}$ is also the inverse of the matrix $\mathbf{K}^{-1}$.) The product of a matrix and its inverse has an interesting result. Consider the product $\mathbf{K}^{-1}\mathbf{K}$.

$$\begin{bmatrix} .3056/K & .264/K & .16/K \\ .264/K & .66/K & .4/K \\ .16/K & .4/K & 1/K \end{bmatrix} \begin{bmatrix} 5K & -2K & 0 \\ -2K & 2.8K & -.8K \\ 0 & -.8K & 1.32K \end{bmatrix} = \begin{bmatrix} 1 & 0 & 0 \\ 0 & 1 & 0 \\ 0 & 0 & 1 \end{bmatrix} \tag{A1.8}$$

or

$$\mathbf{K}^{-1}\mathbf{K} = \mathbf{I} \tag{A1.8a}$$

where a matrix that has unity as a value for each term of what is known as the main diagonal and that has zeros as values for all of the off-diagonal terms is called an identity matrix. A property of an identity matrix is that when it multiplies another matrix, the resulting product matrix is the same as the matrix that was multiplied by the identity matrix. Thus premultiplying both sides of Eq. A1.1e by $\mathbf{K}^{-1}$ gives

$$\mathbf{K}^{-1}\mathbf{K}\mathbf{D} = \mathbf{K}^{-1}\mathbf{R} \tag{A1.9}$$

or

$$\mathbf{I}\mathbf{D} = \mathbf{K}^{-1}\mathbf{R} \tag{A1.9a}$$

or

$$\mathbf{D} = \mathbf{K}^{-1}\mathbf{R} \tag{A1.9b}$$

which checks Eq. A1.7a and which, if carried out in detail, would exactly check Eq. A1.7.

You have probably noticed that the terms in the $\mathbf{K}$ matrix at corresponding positions on either side of the main diagonal are equal to each other; that is, $K_{ij} =$

K_{ji}. Matrices for which this is true are referred to as symmetric matrices. Another way of stating that a matrix is symmetric is by saying that it is equal to its transpose or

$$\mathbf{K} = \mathbf{K}^T \tag{A1.10}$$

The transpose of a matrix is defined as the matrix formed by interchanging corresponding rows and columns of the original matrix.

It should also be noticed that the matrix $\mathbf{K}^{-1}$ is symmetric, demonstrating the general rule that the inverse of a symmetric matrix is also a symmetric matrix. It is, of course, no accident that the **K** matrix, and hence also its inverse, has turned out to be symmetric. There is a solid physical reason behind these symmetries, as discussed in Chap. 3. There is also a consistent and systematic treatment of the order of equations and unknowns that has led to this result. Note that the equilibrium equations were listed in exactly the same order as were the unknowns. If this practice had not been followed, the **K** matrix and its inverse would not have been symmetric.

A1.3 STRUCTURAL EXAMPLE—ALTERNATIVE COORDINATE SYSTEM

It is often useful to transform the statement of a problem from one coordinate system to another. Suppose, for example, that it is of interest to write the statement of the spring-mass problem in the alternate coordinate system of Fig. A1.1c. The displacements and forces in this coordinate system are generalized in nature. For example, the "displacement" $\overline{D}_1$ is actually the amplitude of a displacement mode that consists of equal displacements of each mass, positive to the right. Similarly, $\overline{D}_2$ and $\overline{D}_3$ are amplitudes of two other modes of displacement of the system, as indicated in Fig. A1.1c. It should be noticed that these displacement modes are linearly independent of one another; thus it will be possible to represent any displaced configuration of the system in terms of these modes. This representation may be accomplished by inspection of Figs. A1.1a and c, the result being

$$\begin{aligned} D_1 &= \overline{D}_1 + \overline{D}_2 + \overline{D}_3 \\ D_2 &= \overline{D}_1 - \overline{D}_2 \\ D_3 &= \overline{D}_1 + \overline{D}_2 - \overline{D}_3 \end{aligned} \tag{A1.11}$$

which may be written in matrix form as

$$\begin{pmatrix} D_1 \\ D_2 \\ D_3 \end{pmatrix} = \begin{bmatrix} 1 & 1 & 1 \\ 1 & -1 & 0 \\ 1 & 1 & -1 \end{bmatrix} \begin{pmatrix} \overline{D}_1 \\ \overline{D}_2 \\ \overline{D}_3 \end{pmatrix} \tag{A1.11a}$$

or

$$\mathbf{D} = \mathbf{E}\overline{\mathbf{D}} \tag{A1.11b}$$

The **E** matrix is seen to transform the $\overline{\mathbf{D}}$ displacement vector into the **D** displacement vector. Thus the **E** matrix is referred to as a transformation matrix.

(The symbol **E** is used here because transformation matrices that may be developed by simple modifications of an identify matrix are sometimes called elementary matrices.)

The relationship between the corresponding force vectors **R** and $\overline{\mathbf{R}}$ is perhaps not quite so obvious as was the relationship between the displacement vectors **D** and $\overline{\mathbf{D}}$. To establish the relationship between force vectors, it is useful to consider the work done by each set of forces as the system undergoes a virtual displacement. Let the virtual displacement of the system be represented as

$$\delta\mathbf{D} = \langle \delta D_1 \quad \delta D_2 \quad \delta D_3 \rangle^{*} \tag{A1.12}$$

where the symbol δ is called a variational operator that indicates a small, arbitrary, imaginary change in the parameter being operated upon (displacements in this case). The work that the forces of the **R** vector do as a result of the preceding displacements is

$$\delta W = R_1\,\delta D_1 + R_2\,\delta D_2 + R_3\,\delta D_3 \tag{A1.13}$$

or

$$\delta W = \delta\mathbf{D}^T\,\mathbf{R} \tag{A1.13a}$$

The work done by the forces of the $\overline{\mathbf{R}}$ vector must of course be the same as that done by the forces of the **R** vector; thus

$$\delta W = \overline{R}_1\,\delta\overline{D}_1 + \overline{R}_2\,\delta\overline{D}_2 + \overline{R}_3\,\delta\overline{D}_3 \tag{A1.14}$$

or

$$\delta W = \delta\overline{\mathbf{D}}^T\,\overline{\mathbf{R}} \tag{A1.14a}$$

Comparing Eqs. A1.13a and A1.14a, we may write

$$\delta\mathbf{D}^T\mathbf{R} = \delta\overline{\mathbf{D}}^T\overline{\mathbf{R}} \tag{A1.15}$$

But it is seen from Eq. A1.11b that

$$\delta\mathbf{D} = \mathbf{E}\,\delta\overline{\mathbf{D}} \tag{A1.16}$$

Since it can be shown that the transpose of the product of two arrays is equal to the reverse order product of the transposes of the arrays, we may write

$$\delta\mathbf{D}^T = \delta\overline{\mathbf{D}}^T\,\mathbf{E}^T \tag{A1.17}$$

Substituting from Eq. A1.17 into A1.15 gives

$$\delta\overline{\mathbf{D}}^T\,\mathbf{E}^T\mathbf{R} = \delta\overline{\mathbf{D}}^T\,\overline{\mathbf{R}} \tag{A1.18}$$

or

$$\delta\overline{\mathbf{D}}^T\,(\mathbf{E}^T\mathbf{R} - \overline{\mathbf{R}}) = 0 \tag{A1.18a}$$

The only way that Eq. A1.18a can be satisfied for arbitrary $\delta\overline{\mathbf{D}}^T$ is that

$$\mathbf{E}^T\mathbf{R} - \overline{\mathbf{R}} = \Theta \tag{A1.19}$$

*The angle brackets, $\langle\ \ \rangle$, indicate that the array is in reality a column vector.

where Θ represents a null vector. Thus

$$\overline{\mathbf{R}} = \mathbf{E}^T\mathbf{R} \tag{A1.20}$$

which may be written in detail as

$$\begin{pmatrix} \overline{R}_1 \\ \overline{R}_2 \\ \overline{R}_3 \end{pmatrix} = \begin{bmatrix} 1 & 1 & 1 \\ 1 & -1 & 1 \\ 1 & 0 & -1 \end{bmatrix} \begin{pmatrix} R_1 \\ R_2 \\ R_3 \end{pmatrix} \tag{A1.20a}$$

The relationships between the forces of the **R** vector and those of the $\overline{\mathbf{R}}$ vector have thus been established. It is interesting and important to notice that the transformation matrix relating the $\overline{\mathbf{R}}$ vector to the **R** vector is exactly the transpose of the transformation matrix relating the **D** vector to the $\overline{\mathbf{D}}$ vector. These two transformations are said to follow a contragredient law. It will be seen that forces and displacements undergoing linear transformations invariably follow a contragredient law. This behavior is important because, if we are aware of it and if we have determined a linear displacement transformation relationship, we will automatically know the corresponding transformation of forces too, and vice versa.

It is now possible to transform the statement of the problem from the **D**, **R** system to the $\overline{\mathbf{D}}$, $\overline{\mathbf{R}}$ system of coordinates. Substituting from Eq. A1.11a into A1.1c and then substituting the result into Eq. A1.20a gives

$$\begin{bmatrix} 1 & 1 & 1 \\ 1 & -1 & 1 \\ 1 & 0 & -1 \end{bmatrix} \begin{bmatrix} 5K & -2K & 0 \\ -2K & 2.8K & -.8K \\ 0 & -.8K & 1.32K \end{bmatrix} \begin{bmatrix} 1 & 1 & 1 \\ 1 & -1 & 0 \\ 1 & 1 & -1 \end{bmatrix} \begin{pmatrix} \overline{D}_1 \\ \overline{D}_2 \\ \overline{D}_3 \end{pmatrix} = \begin{pmatrix} \overline{R}_1 \\ \overline{R}_2 \\ \overline{R}_3 \end{pmatrix} \tag{A1.21}$$

or

$$\begin{bmatrix} 3K & 0 & .52K \\ 7K & -5.6K & 2.12K \\ 5K & -1.2K & -1.32K \end{bmatrix} \begin{bmatrix} 1 & 1 & 1 \\ 1 & -1 & 0 \\ 1 & 1 & -1 \end{bmatrix} \begin{pmatrix} \overline{D}_1 \\ \overline{D}_2 \\ \overline{D}_3 \end{pmatrix} = \begin{pmatrix} \overline{R}_1 \\ \overline{R}_2 \\ \overline{R}_3 \end{pmatrix} \tag{A1.21a}$$

or

$$\begin{bmatrix} 3.52K & 3.52K & 2.48K \\ 3.52K & 14.72K & 4.88K \\ 2.48K & 4.88K & 6.32K \end{bmatrix} \begin{pmatrix} \overline{D}_1 \\ \overline{D}_2 \\ \overline{D}_3 \end{pmatrix} = \begin{pmatrix} \overline{R}_1 \\ \overline{R}_2 \\ \overline{R}_3 \end{pmatrix} \tag{A1.21b}$$

or

$$\overline{\mathbf{K}}\overline{\mathbf{D}} = \overline{\mathbf{R}} \tag{A1.21c}$$

where

$$\overline{\mathbf{K}} = \mathbf{E}^T\mathbf{K}\mathbf{E} \tag{A1.22}$$

A transformation of the form of Eq. A1.22 is called a symmetric transformation. The detailed $\overline{\mathbf{K}}$ matrix shown in Eq. A1.21b demonstrates that a symmetric transformation of a symmetric matrix produces a symmetric matrix.

A1.4 ALTERNATE STRUCTURAL EXAMPLE

In the case of the spring-mass system of Fig. A1.1a there is no advantage to transforming the statement of the problem to the particular alternate set of coordinates chosen; that is, Eq. A1.1c is, if anything, simpler than its transformation, Eq. A1.21b. Consider, however, the alternate spring-mass system of Fig. A1.1d. This system possesses physical symmetry about its centerline. Writing the equilibrium equations for this system in terms of **R**, **D** coordinates gives

$$\begin{bmatrix} 5K & -2K & 0 \\ -2K & 4K & -2K \\ 0 & -2K & 5K \end{bmatrix} \begin{pmatrix} D_1 \\ D_2 \\ D_3 \end{pmatrix} = \begin{pmatrix} R_1 \\ R_2 \\ R_3 \end{pmatrix} \tag{A1.23}$$

(Notice, in passing, the evidence of physical symmetry in the symmetry of the coefficient matrix about its minor diagonal.) Transformation of Eq. A1.23 to the $\overline{\mathbf{R}}$, $\overline{\mathbf{D}}$ coordinate system gives

$$\left[\begin{array}{cc|c} 6K & 6K & 0 \\ 6K & 22K & 0 \\ \hline 0 & 0 & 10K \end{array}\right] \begin{pmatrix} \overline{D}_1 \\ \overline{D}_2 \\ \hline \overline{D}_3 \end{pmatrix} = \begin{pmatrix} \overline{R}_1 \\ \overline{R}_2 \\ \hline \overline{R}_3 \end{pmatrix} \tag{A1.24}$$

Equation A1.24 is seen to possess a significant advantage over Eq. A1.23 in that the modes of behavior represented in the $\overline{\mathbf{R}}$, $\overline{\mathbf{D}}$ coordinates have become partially uncoupled into symmetric and antisymmetric sets. This uncoupling is represented by the partitioning of Eq. A1.24 as shown. This equation may be rewritten as

$$\begin{gathered} \begin{bmatrix} 6K & 6K \\ 6K & 22K \end{bmatrix} \begin{pmatrix} \overline{D}_1 \\ \overline{D}_2 \end{pmatrix} = \begin{pmatrix} \overline{R}_1 \\ \overline{R}_2 \end{pmatrix} \\ [10K](\overline{D}_3) = (\overline{R}_3) \end{gathered} \tag{A1.24a}$$

Thus two sets of equations, each roughly half the size of the original set, may be solved in preference to the original set. This procedure can be shown to be a significant advantage, which was made possible by the fact that the structure was symmetric and that its response was broken into symmetric and antisymmetric components. (See Sec. 4.6.)

A1.5 SUMMARY

The purpose of this appendix is to call your attention to matrix algebra concepts that will be used in matrix structural analysis. If you are unfamiliar with any of these concepts, you will want to review them more thoroughly by consulting the literature. (See, for example, Ref. 2.1 or A1.1.) Although a simple structural

Table A1.1 SUMMARY OF MATRIX ALGEBRA TOPICS

TOPICS	EQUATIONS
Array	
Matrix	
Vector	
Multiplication of an array by a scalar	(A.1b) and (A1.1c)
Array row and column numbering	(A1.1d)
Symbolic representation of arrays	(A1.1e)
Matrix multiplication	(A1.1c) and (A1.1f)
Matrix addition & subtraction	(A1.1f), (A1.1g), and (A1.1h)
Null matrix	(A1.1h)
Matrix inversion	(A1.1c) and (A1.1j) to (A1.9b)
Gaussian elimination	(A1.1j) to (A1.6)
Identity matrix	(A1.8) and (A1.8a)
Main diagonal	
Symmetric matrix	
Transpose of a matrix	(A1.11a) and (A1.20a)
Transformation matrix	(A1.11a) and (A1.11b)
Elementary matrix	(A1.11a) and (A1.11b)
< > notation	(A1.12)
Symmetric transformation	(A1.22)
Minor diagonal	
Matrix partitioning	(A1.24)

system has been used as a vehicle for this discussion and a number of matrix structural analysis concepts have naturally arisen incidental to the discussion, it is not the purpose of this appendix to consider these topics for their own sakes. That is, if some of the structural analysis concepts mentioned are not totally clear, there is no need for you to be concerned, as these topics will be thoroughly covered in the main body of this work. To help you identify the matrix algebra review topics of this appendix, they are summarized in Table A1.1 in the sequence encountered in the text.

REFERENCE

A1.1 Ayres, Frank, Jr. *Theory and Problems of Matrices*. New York: Schaum Publishing Company, 1962.

Appendix 2
SMIS3: An Advanced Symbolic Matrix Interpretive System

A2.1 INTRODUCTION

The Advanced Symbolic Matrix Interpretive System, SMIS3, provides the user with a broad range of capabilities. In addition to matrix manipulation, the user can analyze continuous beams or two- or three-dimensional truss or frame structures under static loading. SMIS3 is an extended form of SMIS2 (Ref. A2.1),* which is itself a modified form of SMIS (Ref. A2.2). Users of SMIS3 are not required to write their own subroutines to perform the operations provided. They can solve most of the usual matrix-oriented problems without any knowledge of the FORTRAN language. Of more importance, however, is the fact that SMIS3 has been written in such a way that replacement subroutines can be conveniently furnished by the users to carry out various operations. Thus, the users can gain the experience of programming major matrix manipulation and matrix structural analysis operations without having to be burdened with input/output and the many other difficulties associated with setting up an independent program. They can therefore keep all of their attention focused on the logical steps in the

*Portions of App. 2 are modifications or reproductions of portions of Ref. A2.1 and are used with the generous permission of A. Fred Schkade.

operations. Thus the users have a significant learning device that allows integration of the theory and the translation of that theory into a functioning program.

A "normal" program run consists of executing the operations designated on the operation cards (data deck) sequentially, in the order in which they are encountered. If the user has properly provided replacement subroutines, they will be executed during the run instead of the program's subroutines.

All matrices are designated by symbols that are selected by the user. SMIS3 is particularly easy to use because all data input is unformatted.

A2.2 RESTRICTIONS ON USE

A2.2.1 Computer System

As outlined in this manual, SMIS3 was first implemented only on the CDC 6500 computer at Purdue University. SMIS3 may be used on many other computers if a copy of the 3000-card FORTRAN deck is obtained. *There are, however, some parameters in the program that are machine dependent and will have to be changed to fit the computer being used.*

A2.2.2 Matrices

The size of any matrix is limited only by the availability of sufficient core storage. No more than 100 matrices may be retained in core at any one time. The total core storage available for *all* matrices is variable at execution time through manipulation of the field length parameter on the first control card. Matrix storage capability on tapes or disk files is practically unlimited.

Each matrix is referred to by its name—any sequence of from one to six keypunch characters may be used to label a matrix. If an attempt is made to load or generate a matrix with the same name as a previously defined matrix, the previous matrix is destroyed when the operation is executed.

A2.3 PROCEDURES FOR USE

A2.3.1 Control Cards

The form of the matrix storage in SMIS3 allows the user to reserve storage space at execution time by simply specifying the proper parameter on the job card. Since execution priority on most computers is strongly dependent on the amount of core storage requested, this feature allows the user to obtain a higher execution priority for runs that require few matrix storage locations, or, conversely, to increase the storage size for very large problems.

There is an internal check in SMIS3 to guard against the attempt to store more matrix elements than there are storage locations available. Any such attempt will result in the printout of an error message. Any succeeding operation and data cards will be ignored unless another START card is encountered (see part A1 in Sec. A2.4).

A2.3.2 Program Input

The input deck is composed of *operation cards* and *data cards*. An operation card designates the operation to be performed, along with its "calling sequence." The calling sequence contains the names of matrices to be used in the operations, the sizes of the matrices, and/or other necessary parameters. One or more data cards are needed following certain operation cards (e.g., LOAD) to specify quantities too large or too numerous to fit on the operation cards themselves. All operation cards and data cards may be punched in free format; only the ordering of parameters in each operation must be consistent with that shown in the operation definition. In addition, each parameter must be separated from its neighbors by a blank. Data for matrices must be punched row-wise. For example, the matrix

$$\begin{bmatrix} 11.2 & 1.0 & 0.0 \\ 2.1 & 3.0 & 4.2 \\ 0.0 & 0.0 & 5.0 \end{bmatrix}$$

may be typed as

11.2 1 0 2.1 3 4.2 2∘0 5

More than one data card may, of course, be used if desired. Each coefficient must be separated by at least one blank from its neighbors.

A2.4 OPERATION CODE WORDS

This section contains the definitions of the SMIS3 operation code words or commands and their calling sequences.

A. Control

1. START

   This command, which causes the elimination of all matrices currently in core storage, must appear as the first card of any data deck. If an error is encountered in the input deck, all cards following the card that caused the error will be read and printed, but they *will not be executed* unless another START card is encountered. When a new START card is found, execution is resumed from that point after all existing matrices are deleted from memory. It is to the user's advantage, therefore, that essentially independent input data segments always begin with a new START card.

2. STOP

   This command causes a normal exit from execution. When the STOP card is encountered, the job is terminated. A STOP card must be used *only* at the end of a series of input data segments, when no more operation cards are to be read.

   When the STOP card is encountered, the program will print out a table of the storage used during the run. This printout can be used in determining the minimum field length that will be required by future runs.

3. TIME

   This command causes a printout of the total memory (T.M.) time that has elapsed since the program began executing.

4. REWIND N1

   N1 = number of the tape unit to be rewound (1 or 2)

   This command causes the specified tape unit or disk file to be rewound to its load point. Only tapes numbered 1 and 2 are available for use. Any attempt to use another tape will result in an error message.

5. BACKSP N1

   N1 = number of the tape unit to be backspaced (1 or 2)

   This command causes the specified tape unit or disk file to be backspaced one record. Only tapes numbered 1 and 2 may be used.

B. Input

1. LOAD A N1 N2

   A = name of the matrix to be loaded
   N1 = number of rows in the matrix
   N2 = number of columns in the matrix

   This command causes the specified matrix to be read from the data cards following the operation card. The elements of the matrix must be punched row-wise on the data cards.

2. READ A N1

   A = name of the matrix to be read from tape
   N1 = number of the tape unit to be used (1 or 2)

   This command causes the specified matrix to be read from the specified binary tape or disk file. Only tapes numbered 1 and 2 may be used. The matrix must previously have been written on this tape by using the WRITE command, and the tape must be rewound or backspaced to the beginning of the matrix information before the matrix can be read.

C. Output

1. PRINT A N1

   A = name of the matrix to be printed
   N1 = number of cards following the PRINT operation card to be printed as a heading. *Caution:* A zero must be punched for N1 if no heading is desired.

   This command causes the specified matrix to be printed out, along with a user-supplied heading if desired. Normally, the matrix will be printed out six columns at a time, with each row and column labeled with its number. If a heading is to be printed above the matrix printout, the desired heading should be punched as data cards following the PRINT operation card, and the number of cards to be printed as a heading should be punched in the N1 field on the PRINT operation card.

2. REMARK N1

   N1 = number of cards following the operation card to be printed as remarks

   This command causes the specified number of data cards following the REMARK operation card to be printed. This command may be used, for example, to identify various steps in an extended sequence of calculations.
3. WRITE A N1

   A = name of the matrix to be written on tape
   N1 = number of the tape unit to be used (1 or 2)

   This command causes the specified matrix to be written on the specified binary tape unit or disk file. Only tapes numbered 1 and 2 may be used. Two records are written, the first containing the number of rows and columns in the matrix and the second containing the elements of the matrix. To backspace the tape to the beginning of a matrix that has just been written, it is therefore necessary to have *two* BACKSP commands before the tape read is attempted.
4. PUNCH A

   A = name of the matrix to be punched

   This command causes the punching of a deck of cards in the form required by the LOAD operation. The first card punched is a LOAD operation card containing the specified name of the matrix and its size. After the LOAD operation card has been punched, the matrix is punched in the proper format for loading. Use of the PUNCH command allows the user to transfer matrices in toto from one program to another without having to manually punch a single card.

D. Matrix Generation and Deletion

1. FORM A N1 N2

   A = name to be given to the matrix
   N1 = number of rows in the matrix
   N2 = number of columns in the matrix

   This command causes the formation of a null matrix of specified size with the specified name.
2. DUPL A B

   A = name of the matrix to be duplicated
   B = name to be given to the duplicate matrix

   This command causes the creation of a new matrix identical with the specified matrix.
3. DELETE A

   A = name of the matrix to be deleted from core

This command causes the specified matrix and all associated reference information to be erased from core storage. Such an operation may be useful when core storage space is at a premium.

E. Operations on Elements of Matrices

Note: After each of the following operations, the original matrix no longer exists in core unless a duplicate was made before the operation was requested.

1. SCALE A S1

A = name of the matrix to be scaled
S1 = value of the scalar by which each element in the matrix is to be multiplied

This command causes each element in the specified matrix to be replaced by the designated scalar multiple of that element. *See introductory note to this section.*

2. SQREL A

A = name of the matrix whose elements are to be replaced by their square roots

This command causes each element in the specified matrix to be replaced by its (positive) square root. *See introductory note to this section.*

3. LOG A

A = name of the matrix whose elements are to be replaced by their natural logarithms

This command causes each element in the specified matrix to be replaced by its natural logarithm. *See introductory note to this section.*

4. INVEL A

A = name of the matrix whose elements are to be replaced by their reciprocals

This command causes each element in the specified matrix to be replaced by its reciprocal (inverse). *See introductory note to this section.*

Note: This operation is *not* the same as inversion of a matrix.

5. MSCALE A B

A = name of the matrix to be scaled
B = name of the matrix whose first (1,1) element is to be used as the scalar multiplier

This command causes each element in the specified matrix to be replaced by a scalar multiple of that element, where the scalar multiplier is taken to be the first (1,1) element of the second specified matrix. *See introductory note to this section.*

F. Matrix Algebra

Note: After some of the following operations, the original matrix no longer exists in core unless a duplicate was made before the operation was requested.

1. ADD A B

   A = name of the matrix to be replaced by the sum of the matrices A and B
   B = name of the matrix to be added to A

   This command causes the replacement of the first specified matrix by the sum of the two specified matrices. *See introductory note to this section.*

2. SUB A B

   A = name of the matrix to be replaced by the difference between matrices A and B
   B = name of the matrix to be subtracted from A

   This command causes the replacement of the first specified matrix by the difference between the first specified matrix and the second specified matrix. *See introductory note to this section.*

3. TRANS A B

   A = name of the matrix whose transpose is desired
   B = name to be given to the transposed matrix

   This command causes the creation of the second specified matrix as the transpose of the first specified matrix.

4. MULT A B C

   A = name of the premultiplier matrix
   B = name of the postmultiplier matrix
   C = name to be given to the product matrix

   This command causes the first specified matrix to premultiply the second specified matrix with the result being stored in the third specified matrix.

5. INVERT A

   A = the name of the matrix to be replaced by its inverse

   This command causes the specified matrix to be replaced by its matrix inverse. *See introductory note to this section.*

6. SYMINV A

   A = the name of the symmetric matrix to be replaced by its inverse

   This command causes the specified symmetric matrix to be replaced by its matrix inverse. It can be used *only* to invert symmetric matrices. SYMINV, however, is about twice as fast as INVERT. *See introductory note to this section.*

7. SOLVE A B C

   A = name of the coefficient matrix
   B = name of the matrix of knowns
   C = name to be given to the solution matrix

   This command causes a solution of the set of simultaneous equations expressed in matrix form as AC = B. The number of rows in A is the same as that in B and C. A is a square array.

G. Operations Involving Submatrices

1. RMVSM A B N1 N2 N3 N4

   A = name of the larger matrix from which the submatrix is to be copied
   B = name to be given to the new submatrix
   N1 = row number in the larger matrix of the first (1,1) element of the new submatrix
   N2 = column number in the larger matrix of the first (1,1) element of the new submatrix
   N3 = number of rows in the new submatrix
   N4 = number of columns in the new submatrix

   This command causes the creation of the specified matrix as a duplicate of the specified submatrix.

2. STOSM A B N1 N2

   A = name of the larger matrix whose submatrix is to be replaced
   B = name of the submatrix to be placed in the larger matrix
   N1 = number of the row in the larger matrix into which the first (1,1) element of the submatrix is to be placed
   N2 = number of the column in the larger matrix into which the first (1,1) element of the submatrix is to be placed

   This command causes the replacement by the specified submatrix of the specified elements in the larger matrix.

3. ADDSM A B N1 N2

   A = name of the larger matrix to which the submatrix is to be added
   B = name of the submatrix to be added into the larger matrix
   N1 = number of the row in the larger matrix to which the first (1,1) element of the submatrix is to be added
   N2 = number of the column in the larger matrix to which the first (1,1) element of the submatrix is to be added

   This command causes the addition of the specified submatrix to the specified elements in the larger matrix.

4. RMVDG A B

   A = name of the square matrix whose major diagonal is to be copied
   B = name to be given to the new row matrix

   This command causes the major diagonal of the specified square matrix to be duplicated as the specified row matrix.

5. STODG A B

   A = name of the square matrix whose diagonal elements are to be replaced
   B = name of the row matrix whose elements are to replace the major diagonal of the larger matrix

   This command causes the replacement of the major diagonal elements of the specified square matrix by the elements of the specified row matrix.

H. Special Operations

1. ENVEL A B N1 S1

A = name of the matrix whose envelope values are to be determined
B = name to be given to the column matrix whose elements, corresponding row to row, are the envelope values
N1 = number of cards following the ENVEL operation card to be printed as a heading
S1 = scalar to be used as the interval between columns of the envelope values

This command causes the generation of the specified column matrix as the envelope values (elements of maximum absolute value in each row) of the specified matrix. In addition, each envelope value is printed out, preceded by its column number (the number of the column in which the element of maximum absolute value was found) times a specified scalar interval value.

2. FUNGN A B N1 S1

A = name of the matrix that defines the input function in terms of line segments; the size of this matrix is $n \times 2$
B = name of the matrix to be generated as the function defined by interpolation among the y values at equal x intervals of the input function
N1 = number of y values to be generated; the first value will be y_1 of the input function
S1 = x interval to be used

This command causes the generation of the specified row matrix "function" as the specified number of y values at specified x intervals. This function is determined by a linear interpolation with a "function" defined by the end points of successive line segments $(x_1, y_1), \ldots, (x_n, y_n)$—a discretized approximation to a curve—stored as a specified $n \times 2$ matrix. The first y value of the generated function will be at x_1, with succeeding elements evaluated at $x_1 + dx$, $x_1 + 2dx$, and so on.

Note: To guard against the possibility of inaccurate extrapolation beyond the limits of the input function, any attempt to evaluate the generated function at an abscissa greater than the value of x_n will result in the printout of a caution message to that effect.

I. Special Operations for Structural Analysis

The term *element* as used here is synonymous with the term *member* as used throughout this book.

1. BEAMEL K1 NL RF1

K1 = name to be given to the 4×4 stiffness matrix
NL = number of separate loads on the element
RF1 = name to be given to the 4×1 vector of fixed-end forces in global coordinates; RF1 to be specified *only* if NL $\neq 0$.

One data card is required following this operation card as follows:

L E I $\bar{A}$ G

L = member length
E = modulus of elasticity
I = cross-sectional moment of inertia
$\bar{A}$ = effective shear area of the cross section (= zero if shear distortions are neglected)
G = shear modulus (= zero if shear distortions are neglected)

For NL $\neq$ 0, NL additional data cards, each of which is of one of the following types, are required:

Type F:
LT C D WC WD

LT = F indicates an applied force loading normal to the element
C = dimension to beginning point of load (see Fig. A2.1)
D = dimension to ending point of load (see Fig. A2.1)
WC = intensity of loading at C
WD = intensity of loading at D

Note: If C = D and WC = WD, load is interpreted as a concentrated force of magnitude WC applied at C. If C = D and WC = −WD, load is interpreted as a concentrated moment of magnitude WC, counterclockwise positive, applied at C.

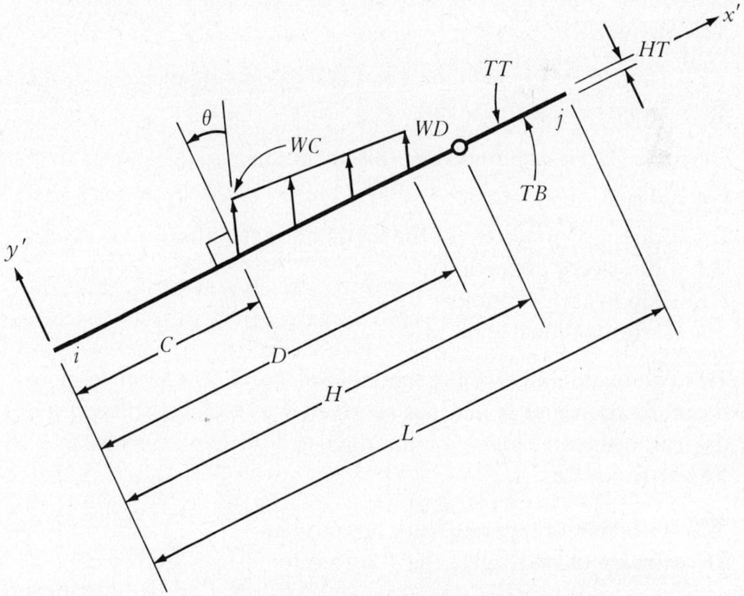

Fig. A2.1 Element load parameters.

Type T:
LT ΔT HT β
LT = T indicates a thermal loading
ΔT = temperature difference between the bottom and the top of the element; positive if the bottom is warmer than the top
HT = height of the element
β = coefficient of thermal expansion

Type R:
LT RL1 RL2 RL3 RL4
LT = R indicates a fixed-end force vector is to be read in
RL1, RL2, RL3, RL4 = fixed-end forces in local coordinates

This command causes the generation of the specified 4 × 4 global stiffness matrix and 4 × 1 global fixed-end force vector for the beam element. The forms of this stiffness matrix and fixed-end force vector are discussed in Sec. A2.5.

1a. BEAMLD NL RF1

NL = number of separate loads on the element
RF1 = name to be given to the 4 × 1 vector of fixed-end forces

NL data cards are required following this operation card; each is to be of type F, T, or R as described under BEAMEL.

This command causes the generation of the specified global fixed-end force vector for a beam element. This command may be used repeatedly to create alternate fixed-end force vectors for an element only immediately subsequent to that element's creation by a BEAMEL command.

2. TRUSEL A1 K1

A1 = name to be given to the 4 × 4 matrix relating global and local forces
K1 = name to be given to the 4 × 4 stiffness matrix

One data card is required after this operation card as follows:

DX DY E $\bar{A}$
DX = directed projection of the member on the x axis
DY = directed y projection
E = modulus of elasticity
$\bar{A}$ = cross-sectional area

This command causes the generation of the 4 × 4 local to global force transformation matrix and the specified 4 × 4 global stiffness matrix for the truss element. This element is discussed further in Sec. A2.5.

2a. TRUSLD NL RF1

NL = number of separate loads on the element
RF1 = name to be given to the 4 × 1 vector of fixed-end forces

NL data cards are required following this operation card; each is to be of one of the following types:

Type F:
LT C D WC WD
LT = F indicates an applied force loading parallel to the element
C = dimension to beginning point of load (see Fig. A2.1)
D = dimension to ending point of load (see Fig. A2.1)
WC = intensity of loading at C
WD = intensity of loading at D

Note: If C = D, load is interpreted as a concentrated force of magnitude WC applied at C.

Type I:
LT ΔL
LT = I indicates an initial strain loading
ΔL = initial change in length of the bar (increase is positive)

Type R:
LT RL1 RL2 RL3 RL4
LT = R indicates a fixed-end force vector is to be read in
RL1, RL2, RL3, RL4 = fixed-end forces in local coordinates

This command causes the generation of the specified global fixed-end force vector for a truss element. This command may be used repeatedly to create alternate fixed-end force vectors for an element only immediately subsequent to that element's creation by a TRUSEL command.

3. FRAMEL A1 K1 NL RF1

A1 = name to be given to the 6×6 matrix relating global and local forces
K1 = name to be given to the 6×6 stiffness matrix
NL = number of separate loads on the element
RF1 = name to be given to the 6×1 vector of fixed-end forces in global coordinates, to be input *only* if NL $\neq$ 0.

One data card is required after this operation card as follows:

DX DY E I $\bar{\text{A}}$
DX = directed projection of the member on the x axis
DY = directed y projection
E = modulus of elasticity
I = cross-sectional moment of inertia
$\bar{\text{A}}$ = cross-sectional area

For NL $\neq$ 0, NL additional data cards, each of which is of one of the following types, are required:

Type F:
LT C D WC WD Θ
LT = F indicates an applied force loading at some angle Θ to the element normal
C = dimension to beginning point of load (see Fig. A2.1)

D = dimension to ending point of load (see Fig. A2.1)
WC = intensity of loading at C
WD = intensity of loading at D
Θ = angle (in degrees) measured counterclockwise from the direction of the load to the y' direction, as shown in Fig. A2.1

Note: If C = D and WC = WD, load is interpreted as a concentrated force of magnitude WC applied at C. If C = D and WC = −WD, load is interpreted as a concentrated moment of magnitude WC, counterclockwise positive, applied at C. (In the case of the moment, the angle Θ should not be input.)

Type T:
LT TT TB HT β
LT = T indicates a thermal loading
TT = temperature change at the top of the element
TB = temperature change at the bottom of the element
HT = height of the element
β = coefficient of thermal expansion

Type R:
LT RL1 RL2 RL3 RL4 RL5 RL6
LT = R indicates a fixed-end force vector is to be read in
RL1, . . . , RL6 = fixed-end forces in local coordinates

This command causes the generation of the specified 6 × 6 local to global force transformation matrix, the specified 6 × 6 global stiffness matrix, and the specified 6 × 1 global fixed-end force vector for the frame element. This element is discussed further in Sec. A2.5.

3a. FRAMLD NL RF1

NL = number of separate loads on the element
RF1 = name to be given to the 6 × 1 vector of fixed-end forces

NL data cards are required following this operation card; each is to be of type F, T, or R as described under FRAMEL.

This command causes the generation of the specified global fixed-end force vector for a frame element. This command may be used repeatedly to create alternate fixed-end force vectors for an element only immediately subsequent to that element's creation by a FRAMEL command.

4. SUPTEL K1

K1 = name given to the 1 × 1 stiffness matrix

One data card is required following this operation card as follows:

K11
K11 = stiffness of this element

This command causes the generation of the specified 1 × 1 global stiffness matrix for the support element.

5. 3DTREL A1 K1

   A1 = name to be given to the 6 × 6 matrix relating global and local forces
   K1 = name to be given to the 6 × 6 stiffness matrix
   One data card is required following this operation card as follows:

   DX DY DZ E Ā
   DX = directed projection of the member on the x axis
   DY = directed y projection
   DZ = directed z projection
   E = modulus of elasticity
   Ā = cross-sectional area

   This command causes the generation of the specified 6 × 6 three-dimensional (3D) truss-element force transformation and global stiffness matrices. The forms of these matrices are discussed in Sec. A2.5.

6. 3DFREL A1 K1

   A1 = name to be given to the 12 × 12 matrix relating global and local forces
   K1 = name to be given to the 12 × 12 stiffness matrix

   One data card is required following this operation card as follows:

   DX DY DZ G E Ā I_y I_z J α
   DX = directed projection of the member on the x axis
   DY = directed y projection
   DZ = directed z projection
   G = shear modulus
   E = modulus of elasticity
   Ā = cross-sectional area
   I_y = cross-sectional moment of inertia about element y' axis
   I_z = cross-sectional moment of inertia about element z' axis
   J = cross-sectional torsional geometric constant
   α = angle of twist of the element about its x' axis. The angle is measured ($0 \le \alpha \le 90°$) in a plane normal to the x' axis from the projection of the z' axis on the xz plane to the z' axis (right-hand rule). If the x axis is parallel to the y axis, α is the angle measured ($0° \le \alpha \le 90°$) from the z axis to the z' axis (right-hand rule) (see Fig. A2.7).

   This command causes the generation of the specified 12 × 12 three-dimensional frame-element force transformation and global stiffness matrices. The forms of these matrices are discussed in Sec. A2.5.

7. HINGED H1 H2

   H1 = dimension to hinge number 1 (see Fig. A2.1)
   H2 = dimension to hinge number 2 (if H1 = H2, only one hinge is assumed)

   This command is used to modify a frame or beam element by inserting one or two hinges in the element. This command may be used only immediately preceding the command that generates the element to be modified by this command. *Caution:* If two hinges are inserted, the member must not be loaded.

8. SKEWED ALPHI ALPHJ

ALPHI = angle, measured in degrees, counterclockwise from the global x axis to the skewed $\hat{x}$ axis, end i
ALPHJ = angle at end j

This command is used to transform the stiffness and fixed-end force arrays for a frame or truss element to a coordinate system that may be skewed from the global system of coordinates at either end or both ends of the element. This command may be used only immediately preceding the command that generates the element stiffness relationships to be transformed by this command.

9. OFFSET DXI DYI DXJ DYJ

DXI = directed x projection of offset, end i
DYI = directed y projection of offset, end i
DXJ = directed x projection of offset, end j
DYJ = directed y projection of offset, end j

This command may be used to modify a frame element by adding rigid offsets to the ends of the element. This command may be used only immediately preceding the FRAMEL command that generates the frame element to be modified by this command.

10. ADDSTF K NE

K = the name of the structure stiffness matrix to which element stiffness coefficients are to be added
NE = number of elements whose stiffness coefficients are to be added to the structure stiffness matrix

NE data cards, each of the following form, are required following this operation card:

K1 C1 N1 N2 N3 N4 . . . . .
K1 = name of the element stiffness matrix
C1 = name to be given to the element coordinate code
N1, N2 . . . . = coordinate code for the element

The coordinate code consists of the identification numbers of the structure degrees of freedom to which the element degrees of freedom correspond. The order given for the affected structure degrees of freedom must correspond to the element direction established in generating the element stiffness matrix—hence the importance of using *directed* projections in generating the element stiffnesses.

Consider, for example, the frame element shown in Fig. A2.2. The structure degrees of freedom corresponding to the element degrees of freedom are numbered in the figure. The proper coordinate code for this

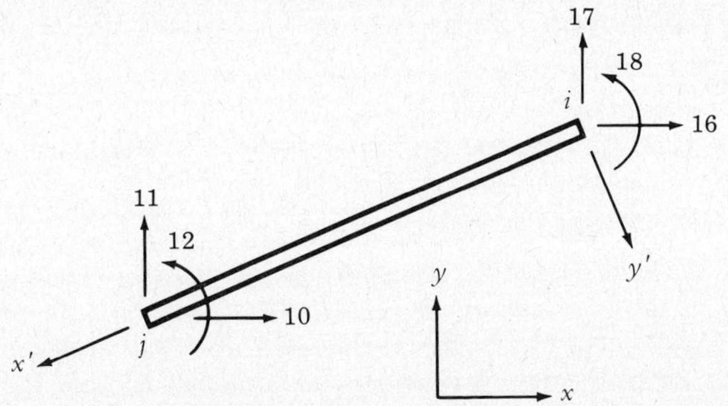

Fig. A2.2 Structure coordinates for a typical frame member.

data card is

16 17 18 10 11 12

If the element being considered is, say, a truss element, the coordinate code would contain only four numbers.

This command causes coefficients of the specified element stiffness matrix to be added to coefficients of the specified structure stiffness matrix with the use of the direct stiffness method. A structure stiffness matrix of the proper size must have been defined, for example with a FORM command, before this operation can be used.

11. ADDFEF R NE

R = name of the structure nodal force vector from which element fixed-end forces are to be subtracted

NE = number of elements whose fixed-end forces are to be subtracted from the structure nodal force vector

NE data cards, each of the following form, are required following this operation card:

RF1 C1

RF1 = name of the element fixed-end force vector

C1 = name of the element coordinate code previously established by ADDSTF

This command causes forces of the specified element fixed-end force vectors to be subtracted from forces of the specified structure nodal force vector with the use of the coordinate code established by the ADDSTF command. Because of this, it is necessary that the ADDFEF command (if required) be used after the ADDSTF command and its data cards. It is also, of course, necessary that a structure nodal force vector must have been defined, for example with a FORM command, before this operation can be used.

12. BOUNDC K R NB

K = name of the *completed* structure stiffness matrix
R = name of the vector of applied loads—the RHS vector in the equation system—on the structure; this vector must be formed beforehand
NB = number of displacement boundary conditions to be input—that is, the number of boundary-condition cards to be read

This command causes the specified structure boundary conditions to be imposed on the completed structure stiffness equation system. One data card is required following the operation card for each boundary condition to be imposed. Each of the data cards should be of the following form:

I D_0
I = identification number of the structure degree of freedom to which this boundary condition applies
D_0 = value of the initial displacement in the direction of the specified degree of freedom

Note: As an alternative to using the BOUNDC command, zero displacement constraints may be imposed by inserting zeros at appropriate locations in the coordinate code supplied with the ADDSTF command. Thus constrained structure degrees of freedom are not numbered, and stiffness equations corresponding to these degrees of freedom are not generated.

13. BEAMFC D NE

D = name of the vector of structure nodal displacements
NE = number of beam elements whose end forces are to be calculated

NE data cards, each of the following form, are required following this operation card:

K1 C1 RF1
K1 = name of the element stiffness matrix
C1 = name of the element coordinate code
RF1 = name of the element fixed-end force vector. *Note: If this vector does not exist or is null, supply a zero instead of a vector name here.*

This command causes the calculation and printing of the element-end forces in local coordinates. This force output will be identified by element code name and listed in the order that element data cards are supplied below this command. (If it is desired to be able to identify and further manipulate these element-end force results, the parameter NE should be input as a negative number. It will then be further necessary to supply a name [e.g., RL1] at the end of each of the NE data cards for each element-end local force vector. This procedure will be required, for example, on those rare occasions when it is desired to superpose the force responses of two or more structure load cases within the computer.)

14. TRUSFC D NE

D = name of the vector of structure nodal displacements
NE = number of truss elements whose end forces are to be calculated

NE data cards, each of the following form, are required following this operation card:

A1 K1 C1 RF1
A1 = name of the element local to global force transformation matrix
K1 = name of the element stiffness matrix
C1 = name of the element coordinate code
RF1 = name of the element fixed-end force vector *Note: If this vector does not exist or is null, supply a zero instead of a vector name here.*

See paragraph at the end of part I13, please.

15. FRAMFC D NE

D = name of the vector of structure nodal displacements
NE = number of frame elements whose end forces are to be calculated

NE data cards, each of the form described above under part I14, are required following this operation card.

See paragraph at the end of part I13, please.

16. SUPTFC D NE

D = name of the vector of structure nodal displacements
NE = number of support elements whose end forces are to be calculated

NE data cards, each of the following form, are required following this operation card:

K1 C1

K1 = name of the element stiffness matrix
C1 = name of the element coordinate code

See paragraph at the end of part I13, please.

17. MEMFCE D NE

D = name of the vector of structure nodal displacements
NE = number of elements in the structure

This command causes the calculation and printing of the element-end forces in local coordinates. Forces will be calculated for all elements, and output will be identified by element code name. The MEMFCE command accesses the required element arrays automatically and is therefore more efficient for the user. On the other hand, the functioning of the BEAMFC, TRUSFC, and FRAMFC commands are easier for the user to understand.

The steps for using the analysis commands of SMIS3 to successfully carry out a direct stiffness solution of a structural problem are as follows:

(a) The structure stiffness and equivalent nodal load arrays must be initialized using the FORM and/or LOAD commands.
(b) Element stiffness matrices (and fixed-end force vectors, if applicable) must be generated by using one of the element commands. If the structure contains elements with identical physical properties and orientations, an element command need only be used once for the entire set of such elements. If loadings vary from element to element within a set of identical elements, the element load commands may be used to generate an element fixed-end force vector for each different loading. Element load commands must always be used in sequence immediately following the corresponding element command.
(c) The element stiffnesses and fixed-end forces thus generated may be added to the structure stiffness and equivalent nodal load arrays, respectively, by use of the ADDSTF and ADDFEF commands.
(d) Structure boundary conditions may be enforced either by use of the BOUNDC command or by use of an appropriate unconstrained coordinate code.
(e) The resulting set of structural stiffness equations must be solved for the nodal displacements.
(f) Either the BEAMFC, TRUSFC, FRAMFC, or SUPTFC or the MEMFCE command may now be used to obtain the desired element end forces.

18. INITLZ KUU RU D0 NN NF NC NL NE

KUU = name to be given to the structure stiffness submatrix associated with unconstrained degrees of freedom
RU = the name to be given to the structure nodal load subvector associated with unconstrained degrees of freedom
D0 = name to be given to the structure initial displacement vector
NN = number of structure nodes
NF = number of degrees of freedom per structure node
NC = number of constraints
NL = number of nodal loads
NE = number of elements

Three groups of data cards are required following this operation card.

Group 1: NC data cards of the following form:

N CC DIS
N = structure node number
CC = constrained coordinate direction (1, 2 = x, y for truss; 1, 2 = y, θ for beam; 1, 2, 3 = x, y, θ for frame)
DIS = magnitude of imposed displacement. (Imposed zero displacements must be input as well as those that are nonzero.)

Group 2: NL data cards of the following form:

N LC LD

N = structure node number

LC = coordinate direction (1, 2 = x, y for truss; 1, 2 = y, θ for beam; 1, 2, 3 = x, y, θ for frame)

LD = magnitude of applied force

Group 3: NE data cards of the following form

C1 NI NJ

C1 = name to be given to the element coordinate code generated internally

NI = structure node number corresponding to ith end of element

NJ = structure node number corresponding to jth end of element

This command forms a null structure stiffness matrix of size (NN·NF-NC) × (NN·NF-NC) and generates a structure nodal load vector of size (NN·NF-NC) × 1. It also makes use of a constraint list to generate a structure initial nodal displacement vector of size (NN·NF) × 1. The constraint list is also used along with a nodal code to generate a coordinate code for each element of the structure.

Note: When use is made of this command, the usual SMIS3 stiffness method command sequence, as described on page 464, may be followed with two exceptions: The INITLZ command is to be used in place of the FORM and LOAD commands of step (a), and the BLDSTF command (described next) is to be used instead of the ADDSTF and ADDFEF commands of step (c). Caution: The INITLZ command can be used only when all members in the structure are of one type.

19. BLDSTF KUU RU D0 NE

KUU = name of the structure stiffness submatrix associated with unconstrained degrees of freedom

D0 = name of the structure initial displacement vector

RU = name of the structure nodal load subvector associated with unconstrained degrees of freedom

NE = number of elements contributing to the structure stiffness relationships

NE data cards, each of the following form, are required after this operation card:

K1 RF1 C1

K1 = name of the element stiffness matrix

RF1 = name of the element fixed-end force vector.

(*Note:* If fixed-end forces are generated in an element owing to initial displacement of an adjacent node but a fixed-end force vector has not previously been named, it must be named here. If the element is not subjected to fixed-end forces from any source, a zero must be input here.)

C1 = name of the element coordinate code

This command builds the indicated structure stiffness submatrix and structure load subvector according to the procedure outlined in Sec. 4.5.3. This command is to be used only in conjunction with the INITLZ command described in part I18.

A2.5 SMIS3 STRUCTURAL ELEMENTS AND STIFFNESS MATRICES

In SMIS3 the element stiffness equations are set up in a cartesian coordinate system fixed to the element and defined in relation to user-designated ends i and j. The coordinate system origin is located at end i, and the x' coordinate is positive in the direction of end j. The degrees of freedom appearing in the stiffness equations are arranged so that those at end i are listed first and those at end j last. The form of the stiffness equations in local coordinates for each type of element will be discussed in detail later in this section.

The transformation from local to structure coordinates is determined by SMIS3 from the input values of the *directed* projections of the element along each of the global coordinate directions. The directed projections are defined as follows:

$$DX = x_j - x_i$$

$$DY = y_j - y_i$$

$$DZ = z_j - z_i$$

Negative values of DX, DY, and DZ are possible and *must* be so used when they occur.

A2.5.1 BEAMEL Element

The BEAMEL element is shown in Fig. A2.3. The stiffness matrix for this element, including the effect of shear distortions, is

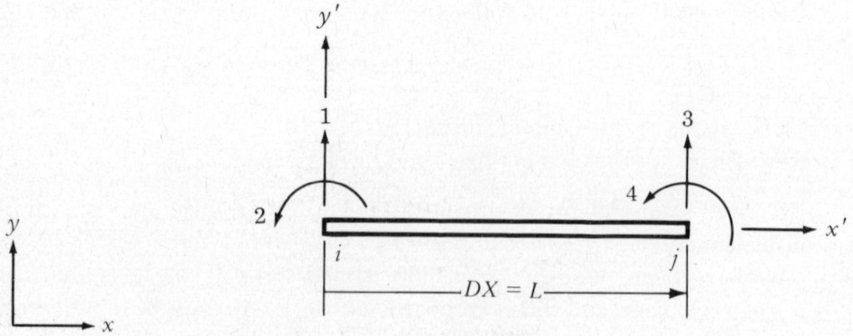

Fig. A2.3 BEAMEL element.

$$\mathbf{k} = \frac{2EI}{L(1+2\beta)}\begin{bmatrix} 6/L^2 & 3/L & -6/L^2 & 3/L \\ 3/L & 2+\beta & -3/L & 1-\beta \\ -6/L^2 & -3/L & 6/L^2 & -3/L \\ 3/L & 1-\beta & -3/L & 2+\beta \end{bmatrix}$$

where $\beta = 6EI/\bar{A}GL^2$.

If desired, it is possible to ignore shear distortions and calculate the element stiffness matrix as

$$\mathbf{k} = \frac{2EI}{L}\begin{bmatrix} 6/L^2 & 3/L & -6/L^2 & 3/L \\ 3/L & 2 & -3/L & 1 \\ -6/L^2 & -3/L & 6/L^2 & -3/L \\ 3/L & 1 & -3/L & 2 \end{bmatrix}$$

by punching zeros for the parameters $\bar{A}$ and G on the data card following the BEAMEL operation card.

The fixed-end force vector is of the form

$$\mathbf{r}_F = \langle r_{F1} \; r_{F2} \; r_{F3} \; r_{F4} \rangle$$

where the nature of these four forces depends on the load case. Since the local and global coordinates coincide in direction in the case of the BEAMEL element, no transformation need be carried out. The element end forces calculated by the BEAMFC command correspond to the coordinate numbers and directions shown in Fig. A2.3.

A2.5.2 TRUSEL Element

The basic stiffness matrix for the TRUSEL element of Fig. A2.4 is

$$k = \left[\frac{\bar{A}E}{L}\right]$$

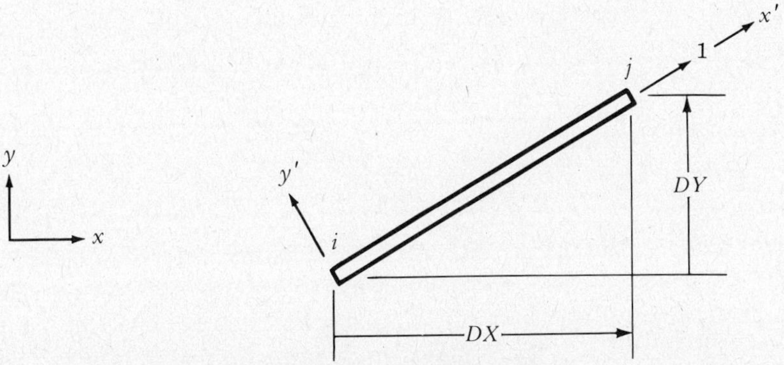

Fig. A2.4 TRUSEL element.

The matrix for the transformation from basic to global coordinates is

$$\mathbf{T} = \frac{1}{L}[-DX \ \ -DY \ \ DX \ \ DY]$$

The single element end force calculated by the TRUSFC command is the axial force (positive if tensile).

A2.5.3 FRAMEL Element

The FRAMEL element is shown in Fig. A2.5. Its stiffness matrix in local coordinates for the case where an internal hinge exists is

$$\mathbf{k}' = \begin{bmatrix} \frac{\bar{A}E}{L} & 0 & 0 & 0 & 0 & 0 \\ & \frac{3EI}{a^3+b^3} & \frac{3EIa}{a^3+b^3} & 0 & \frac{-3EI}{a^3+b^3} & \frac{3EIb}{a^3+b^3} \\ & & \frac{3EIa^2}{a^3+b^3} & 0 & \frac{-3EIa}{a^3+b^3} & \frac{3EIab}{a^3+b^3} \\ & & & \frac{-\bar{A}E}{L} & 0 & 0 \\ & \text{Symmetric} & & & \frac{3EI}{a^3+b^3} & \frac{-3EIb}{a^3+b^3} \\ & & & & & \frac{3EIb^2}{a^3+b^3} \end{bmatrix}$$

where a = H and b = L-H.

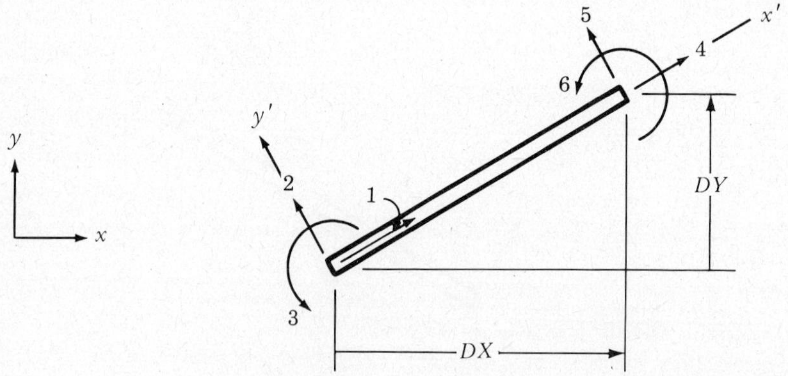

Fig. A2.5 FRAMEL element.

The local coordinate element stiffness matrix for the case where no hinge exists is

$$
\mathbf{k}' = \frac{1}{L}\begin{bmatrix}
\overline{A}E & 0 & 0 & -\overline{A}E & 0 & 0 \\
 & \dfrac{12EI}{L^2} & \dfrac{6EI}{L} & 0 & -\dfrac{12EI}{L^2} & \dfrac{6EI}{L} \\
 & & 4EI & 0 & -\dfrac{6EI}{L} & 2EI \\
 & & & -\overline{A}E & 0 & 0 \\
 & \text{Symmetric} & & & \dfrac{12EI}{L^2} & -\dfrac{6EI}{L} \\
 & & & & & 4EI
\end{bmatrix}
$$

The element fixed-end force vector is of the form

$$\mathbf{r}'_F = \langle \mathbf{r}'_{F1}\ \ \mathbf{r}'_{F2}\ \ \mathbf{r}'_{F3}\ \ \mathbf{r}'_{F4}\ \ \mathbf{r}'_{F5}\ \ \mathbf{r}'_{F6} \rangle$$

where the nature of these six forces depends on the load case. The matrix for the transformation from local to global coordinates is

$$
\alpha = \frac{1}{L}\begin{bmatrix}
DX & DY & 0 & 0 & 0 & 0 \\
-DY & DX & 0 & 0 & 0 & 0 \\
0 & 0 & L & 0 & 0 & 0 \\
0 & 0 & 0 & DX & DY & 0 \\
0 & 0 & 0 & -DY & DX & 0 \\
0 & 0 & 0 & 0 & 0 & L
\end{bmatrix}
$$

The element end forces calculated by the FRAMFC command correspond to the coordinate numbers and directions shown in Fig. A2.5.

A2.5.4 3DTREL Element

The basic stiffness matrix for the 3DTREL element of Fig. A2.6 is

$$\mathscr{k} = \left[\frac{\overline{AE}}{L}\right]$$

No distributed loading is allowed on the 3DTREL element. The matrix for the transformation from basic to global coordinates is

$$\alpha = \left(\frac{1}{L}\right)[-DX\ \ -DY\ \ -DZ\ \ DX\ \ DY\ \ DZ]$$

The single element end force calculated by the TROSFC command is the axial force in the member (positive if tensile).

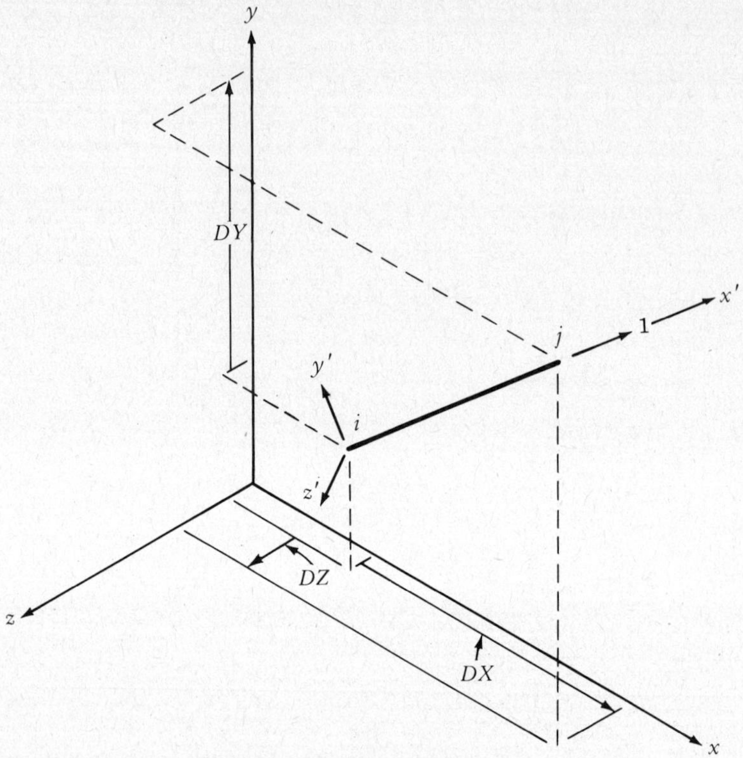

Fig. A2.6 3DTREL element.

A2.5.5 3DFREL Element (Ref. A2.3)

The element stiffness matrix in local coordinates for the 3DFREL element of Fig. A2.7 is

$$
\mathbf{k}' = \begin{bmatrix}
k_1 & 0 & 0 & 0 & 0 & 0 & -k_1 & 0 & 0 & 0 & 0 & 0 \\
 & k_3 & 0 & 0 & 0 & k_5 & 0 & -k_3 & 0 & 0 & 0 & k_5 \\
 & & k_4 & 0 & -k_6 & 0 & 0 & 0 & -k_4 & 0 & -k_6 & 0 \\
 & & & k_2 & 0 & 0 & 0 & 0 & 0 & -k_2 & 0 & 0 \\
 & & & & 2k_8 & 0 & 0 & 0 & k_6 & 0 & k_8 & 0 \\
 & & & & & 2k_7 & 0 & -k_6 & 0 & 0 & 0 & k_7 \\
 & & & & & & k_1 & 0 & 0 & 0 & 0 & 0 \\
 & & & & & & & k_3 & 0 & 0 & 0 & -k_5 \\
 & & \text{Symmetric} & & & & & & k_4 & 0 & k_6 & 0 \\
 & & & & & & & & & k_2 & 0 & 0 \\
 & & & & & & & & & & 2k_8 & 0 \\
 & & & & & & & & & & & 2k_7
\end{bmatrix}
$$

where

$$k_1 = \frac{AE}{L} \quad k_3 = \frac{12EI_z}{L^3} \quad k_5 = \frac{6EI_z}{L^2} \quad k_7 = \frac{2EI_z}{L}$$

$$k_2 = \frac{GJ}{L} \quad k_4 = \frac{12EI_y}{L^3} \quad k_6 = \frac{6EI_y}{L^2} \quad k_8 = \frac{2EI_y}{L}$$

No distributed loading is allowed on the 3DFREL element.

The matrix for the transformation from local to global coordinates is

$$\mathbf{L} = \begin{bmatrix} \mathbf{L}^* & \Theta & \Theta & \Theta \\ \Theta & \mathbf{L}^* & \Theta & \Theta \\ \Theta & \Theta & \mathbf{L}^* & \Theta \\ \Theta & \Theta & \Theta & \mathbf{L}^* \end{bmatrix} \qquad \text{where } \mathbf{L}^* = \begin{bmatrix} L_1 & L_2 & L_3 \\ L_4 & L_5 & L_6 \\ L_7 & L_8 & L_9 \end{bmatrix}$$

with

$$L_1 = \frac{DX}{L} \qquad L_4 = \frac{-L_1L_2 \cos\alpha - L_3 \sin\alpha}{L_{10}} \qquad L_7 = \frac{L_1L_2 \sin\alpha - L_3 \cos\alpha}{L_{10}}$$

$$L_2 = \frac{DY}{L} \qquad L_5 = L_{10} \cos\alpha \qquad L_8 = -L_{10} \sin\alpha$$

$$L_3 = \frac{DZ}{L} \qquad L_6 = \frac{-L_2L_3 \cos\alpha + L_1 \sin\alpha}{L_{10}} \qquad L_9 = \frac{L_2L_3 \sin\alpha + L_1 \cos\alpha}{L_{10}}$$

$$L_{10} = (L_1^2 + L_3^2)^{1/2} \qquad \alpha = \text{orientation angle}$$

For vertical members, the transformation submatrix is defined by

$$\mathbf{L}^* = \begin{bmatrix} 0 & L_2 & 0 \\ -L_2 \cos\alpha & 0 & \sin\alpha \\ L_2 \sin\alpha & 0 & \cos\alpha \end{bmatrix}$$

The element end forces calculated by the FRAMFC command are in directions shown in Fig. A2.7 (positive as indicated).

A2.6 SUBROUTINE DOCUMENTATION

The SMIS3 program has been written so that certain commands each access a hierarchy of subroutines. This hierarchy consists of three types of subroutines referred to as bookkeeping, calling, and working subroutines. The initial subroutine called is the bookkeeping subroutine, and its functions are to locate matrices that have already been generated and to store matrices that will be generated by this command and used later. These matrices, along with information on their sizes, are then transferred to the calling subroutine. This subroutine reads in additional data, if required, and then calls one or more working subroutines. With the information provided to them, the working subroutines perform the actual calculations and generate the appropriate matrices. In some instances there is no need for a calling subroutine, and the bookkeeping subroutine accesses the working subroutines directly.

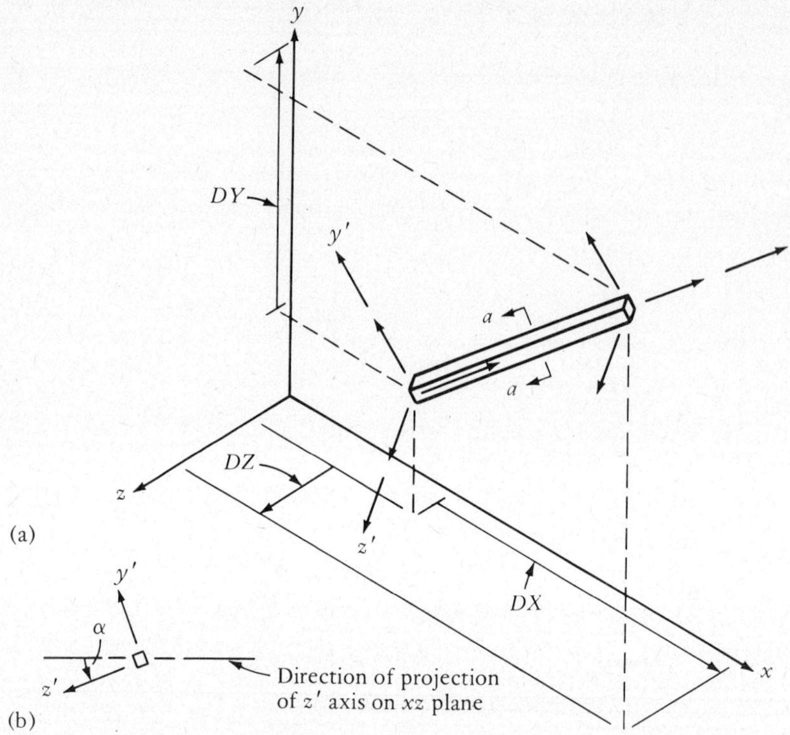

Fig. A2.7 3DFREL element: (a) element; (b) section *a-a*.

The program has been written in this manner to facilitate replacement of important subroutines by user-generated subroutines for instructional purposes. Table A2.1 lists commands and their related bookkeeping, calling, and working subroutines. The commands have been documented in Sec. A2.4. Individual calling and working subroutines are documented in the remainder of this section.

The documentation gives the subroutine names and parameter lists whose formats must be followed exactly. In some subroutines, common blocks are shown, and again the format given must be followed exactly. Matrix declarations are provided only for those matrices transferred through the parameter list. Other matrices, to be used only within the subroutine, may also need to be declared. Unless otherwise declared, the standard FORTRAN method of declaring variables as real or integer by default is employed.

It is possible to use subroutines developed earlier in the later subroutines. The matrix manipulation subroutines are especially helpful in this regard. Moreover, some subroutines developed for one type of member can easily be modified to work for other types of members.

A. Matrix Manipulation Subroutines

1. SUBROUTINE ADSB(A,B,NR,NC,XX)
   REAL A(NR,NC),B(NR,NC)
   A = first matrix to be added
   B = second matrix to be added (or subtracted)

Table A2.1 COMMANDS AND RELATED SUBROUTINES

SMIS3 COMMAND	BOOKKEEPING SUBROUTINE	CALLING SUBROUTINE	WORKING SUBROUTINES
ADD	ADDSUB		ADSB
SUB	ADDSUB		ADSB
TRANS	TRANS		TRAN1
MULT	MULT		MULT1
BEAMEL	BEAMEL	BEAMKT	BMBKF,BMBKFH,BMT,BMTH, FORMK
BEAMLD	FEFORC	FEFOR	BMFEF,BMFEFH,BMFEFT
TRUSEL	TRUSEL	TRAKT	TRBKF,TRA,TRSA,TRL, FORMKT
TRUSLD	FEFORC	FEFOR	TRFEF,TRFEFI,FORMRF
FRAMEL	FRAMEL	FRAKT	FRBKF,FRBKFH,FRA,FRSA, FRL,FRLH, FORMKT
FRAMLD	FEFORC	FEFOR	FRFEF,FRFEFH,FRFEFT, FORMRF
3DTREL	TDTREL	TDTAKT	TRBKF,TDTA,TDTL,FORMKT
3DFREL	TDFREL	TDFAKT	TDFBKF,TDFA,TDFL,FORMKT
ADDSTF	ADDSTF		ADDST
ADDFEF	ADDFEF		ADDFE
BOUNDC	BOUNDC		BOUDC
INITLZ	INITLZ		INITLD,INITLR,INITLC, REARRD
BLDSTF	BLDSTF		ADDST,FEFINT,ADDFE
BEAMFC	BEAMFC		BEAMF
TRUSFC	FRTRFC		FRTRF
FRAMFC	FRTRFC		FRTRF

NR = number of rows in matrices A and B
NC = number of columns in matrices A and B
XX = indicator of whether to perform addition or subtraction

This subroutine adds matrix B to or subtracts matrix B from matrix A and stores the result in matrix A. If XX equals 1.0, the subroutine adds; if XX equals −1.0, the subroutine subtracts.

2. SUBROUTINE TRAN1 (A,B,NR,NC)
REAL A(NR,NC),B(NC,NR)

A = matrix whose transpose is to be determined
B = transpose of matrix A
NR, NC = matrix dimensions

This subroutine determines the transpose of matrix A and stores the result in matrix B.

3. SUBROUTINE MULT1 (A,B,C,NRA,NCA,NCB)
REAL A(NRA,NCA),B(NCA,NCB),C(NRA,NCB)

A = premultiplying matrix
B = postmultiplying matrix
C = product matrix of matrices A and B
NRA, NCA, NCB = matrix dimensions

This subroutine carries out the product of matrix A and matrix B and stores the result in matrix C.

B. Subroutines Associated with BEAMEL

1. SUBROUTINE BEAMKT (K,T,BKF,NC)
COMMON / SPCSIN / INF,INH,INO,INS
COMMON / SPCSVL / H1,H2,DXI,DYI,DXJ,DYJ,THETAI,THETAJ
COMMON / FEFDTA / A,E,I,L
REAL K(4,4),T(4,NC),BKF(NC,NC),I,L

K = global stiffness matrix
T = basic to global transformation matrix
BKF = basic stiffness or flexibility matrix
NC = matrix dimension

This calling subroutine first reads* in the required beam element material and geometric properties as described in part I1 of Sec. A2.4 and then calls, as appropriate, the working subroutines described in the next three parts (B2 to B4). The parameters contained in the two common blocks, SPCSIN and SPCSVL, have already been set or read in if required. See Table A2.2 for a description and discussion of the applicability of these parameters to a beam member. The relevant parameters in the common block FEFDTA are passed to subroutine FEFOR for possible use in calculating member fixed-end forces. NC is set ahead of this subroutine to 2 if no hinge is present or to 1 if there is an internal hinge.

2. SUBROUTINE BMBKF (BKF,L,E,I,A,G,INF)
REAL BKF(2,2),L,I

BKF = basic stiffness or flexibility matrix
L = member length
E = modulus of elasticity
I = cross-sectional moment of inertia
A = effective shear area of the cross section
G = shear modulus
INF = stiffness or flexibility method indicator (see Table A2.2)

This subroutine generates the basic stiffness or flexibility array, depending on the value of the indicator INF for a beam member. If A or G is input as zero, shear effects are neglected.

2a. SUBROUTINE BMBKFH (BKF,L,E,I,A,G,H,INF)
REAL BKF(1,1),L,I

Definitions of the parameters in the list are the same as those given in part B2, with the addition of the following:

H = dimension to internal hinge (see Fig. A2.1)

*The free-format input is accomplished by using subroutines RR and RI, for Read Real and Read Integer, respectively. For example, to read in the material and geometric properties for a beam member would simply require five lines, each in standard FORTRAN, of the form CALL RR(XX), where XX is the real parameter to be read in.

Table A2.2 Parameters of COMMON Blocks SPCSIN and SPCSVL

PARAMETER	DESCRIPTION
INF	Indicates whether the program is to be used for the stiffness method (INF=0) or for the flexibility method (INF=1). Applicable to all types of members. (Flexibility method capability currently not complete.)
INH	Indicates whether there is (INH=1) or is not (INH=0) a hinge present in the member. Applicable only to beam and frame members. See part I7 of Sec. A2.4 for a discussion of the special case values of H1 and H2 associated with hinges.
INO	Indicates whether there is (INO=1) or is not (INO=0) a rigid offset present. Applicable only to the frame member. See part I9 of Sec. A2.4 for a discussion of the special case values of DXI, DYI, DXJ, DYJ, associated with a rigid offset.
INS	Indicates whether a member rests on a skewed support (INS=1) or on a globally oriented support (INS=0). Applicable only to frame and truss members. See part I8 of Sec. A2.4 for a discussion of the special case values of THETAI and THETAJ associated with a skewed support.

This subroutine performs the same function as the subroutine described in part B2, except that the effect of an internal hinge is included.

3. SUBROUTINE BMT (T,L)
REAL T(4,2),L

T = basic to global transformation matrix
L = member length

This subroutine generates the T array for a beam member.

3a. SUBROUTINE BMTH (T,L,H)
REAL T(4,1),L

T = basic to global transformation matrix
L = member length
H = dimension to internal hinge (see Fig. A2.1)

This subroutine generates the T array for a beam member with an internal hinge.

4. SUBROUTINE FORMK (T,BKF,K,NR,NC)
REAL T(NR,NC),BKF(NC,NC),K(NR,NR)

T = basic to global transformation matrix
BKF = basic stiffness matrix
K = global stiffness matrix

This subroutine generates the K matrix using the matrices T and BKF and the matrix dimensions NR and NC, as provided.

5. SUBROUTINE FEFOR (RFG,NR,NL,INEL)
COMMON / SPCSIN / INF,INH,INO,INS
COMMON / SPCSVL / H1,H2,DXI,DXJ,DYI,DYJ,THETAI,THETAJ
COMMON / FEFDTA / A,E,I,L

COMMON / ALPH / ALPHA
REAL RFG(NR), ALPHA(6,6),I,L

RFG = member global fixed-end force vector
NR = size of vector RFG
NL = number of loads
INEL = indicator of element type (2 for truss, 4 for beam, 6 for frame)

Definition of parameters in COMMON / FEFDTA / and COMMON / ALPH:

A = cross-sectional area (not used for BEAMEL)
E = modulus of elasticity
I = cross-sectional moment of inertia
L = member length
ALPHA = local to global transformation matrix (not used for BEAMEL)

For each load this calling subroutine reads* in the load type and the load data. The subroutine then calls the appropriate working subroutine described in parts B6 to B7 to calculate the global fixed-end force vector, superposing the effects of each load.

6. SUBROUTINE BMFEF (L,C,D,WC,WD,RFG,CV)
REAL RFG(4,1),L

L = member length
C = dimension to beginning point of load (see Fig. A2.1)
D = dimension to ending point of load (see Fig. A2.1)
WC = intensity of loading at C
WD = intensity of loading at D
RFG = member global fixed-end force vector
CV = parameter needed only when considering effect of shear distortions on RFG due to a concentrated couple

This subroutine generates the RFG vector for a force loaded beam member. See part I1 of Sec. A2.4 for limitations on the types of loads that can be applied to the beam member.

6a. SUBROUTINE BMFEFH (L,C,D,H,WC,WD,RFG)
REAL RFG(4,1),L

Definitions of the parameters in the list are the same as those given in part B6, with the addition of the following:

H = dimension to internal hinge (see Fig. A2.1)

This subroutine performs the same function as the subroutine described in part B6, except that the effect of an internal hinge is included.

7. SUBROUTINE BMFEFT (E,I,BETA,DT,HT,RFG)
REAL RFG(4,1),I

E = modulus of elasticity
I = cross-sectional moment of inertia

*See footnote on page 474.

BETA = coefficient of thermal expansion
DT = change in temperature over member height
RFG = member global fixed-end force vector

This subroutine generates the RFG vector for a thermally loaded beam member.

C. Subroutines Associated with TRUSEL

1. SUBROUTINE TRAKT (ALPHA,K,T,BKF)
COMMON / SPCSIN / INF,INH,INO,INS
COMMON / SPCSVL / H1,H2,DXI,DYI,DXJ,DYJ,THETAI,THETAJ
COMMON / FEFDTA / A,E,I,L
REAL ALPHA(4,4),K(4,4),T(4,1),BKF(1,1)

ALPHA = local to global transformation matrix
K = global stiffness matrix
T = basic to global transformation matrix
BKF = basic stiffness or flexibility matrix

This calling subroutine first reads* in the truss element material and geometric properties as described in part I2 of Sec. A2.4. After calculating the member length, the subroutine calls, as appropriate, the subroutines described next (parts C2 to C5). The parameters contained in the two common blocks SPCSIN and SPCSVL have already been set or read in as required. See Table A2.2 for a description and discussion of the applicability of these parameters to a truss member. The relevant parameters in the common block FEFDTA are passed to subroutine FEFOR for possible use in calculating member fixed-end forces.

2. SUBROUTINE TRBKF (BKF,E,A,L,INF)
REAL BKF(1,1),L

BKF = basic stiffness or flexibility matrix
E = modulus of elasticity
A = cross-sectional area
L = member length
INF = stiffness or flexibility method indicator (see Table A2.2)

This subroutine generates the basic stiffness or flexibility array, depending on the value of the indicator INF, for a truss member.

3. SUBROUTINE TRA (DX,DY,L,ALPHA)
REAL ALPHA(4,4),L

DX = directed projection of the member on the x axis
DY = directed y projection
L = member length
ALPHA = local to global transformation matrix

This subroutine generates the ALPHA matrix for a truss member.

*See footnote on page 474.

3a. SUBROUTINE TRSA (DX,DY,THETAI,THETAJ,ALPHA)
REAL ALPHA(4,4)
Definitions of the parameter list are the same as those in part C3, with the addition of the following:

THETAI = angle, measured in degrees, counterclockwise from the global x axis to the skewed x axis, end i
THETAJ = angle at end j

This subroutine performs the same function as the subroutine described in part C3, except that the truss member can now rest on skewed supports.

4. SUBROUTINE TRL (LL)
REAL LL(4,1)

LL = basic to local transformation matrix

This subroutine generates the LL matrix for a truss member.

5. SUBROUTINE FORMKT (ALPHA,LL,BKF,T,K,NR,NC,INF)
REAL ALPHA(NR,NR),LL(NR,NC),BKF(NC,NC),T(NR,NC),K(NR,NR)

ALPHA = local to global transformation matrix
LL = basic to local transformation matrix
BKF = basic stiffness or flexibility matrix
T = basic to global transformation matrix
K = global stiffness matrix
NR,NC = matrix dimensions
INF = stiffness or flexibility method indicator (see Table A2.2)

This subroutine generates the T matrix using the matrices ALPHA and LL and the matrix dimensions NR and NC, as provided. Then, if the stiffness method is being used, the K matrix is generated using the T and BKF matrices along with the matrix dimensions.

6. SUBROUTINE FEFOR (RFG,NR,NL,INEL)
COMMON / SPCSIN / INF,INH,INO,INS /
COMMON / SPCSVL / H1,H2,DXI,DYI,DXJ,DYJ,THETAI,THETAJ
COMMON / FEFDTA / A,E,I,L
COMMON / ALPH / ALPHA
REAL RFG(RN),ALPHA(6,6),I,L
The definitions of the parameters in the list and the function of this subroutine are the same as those described in part B5.

7. SUBROUTINE TRFEF (L,C,D,WC,WD,RF)
REAL RF(4,1),L
L = member length
C = dimension to beginning point of load (see Fig. A2.1)
D = dimension to ending point of load (see Fig. A2.1)
WC = intensity of loading at C
WD = intensity of loading at D
RF = local fixed-end force vector

This subroutine generates the RF vector for an axially loaded truss member. See part I2a of Sec. A2.4 for limitations on the types of loads that can be applied to a truss member.

8. SUBROUTINE TRFEFI (A,E,L,DL,RF)
   REAL RF(4,1),L

   A = cross-sectional area
   E = modulus of elasticity
   L = member initial length
   DL = change in member length (increase is positive)
   RF = local fixed-end force vector

   This subroutine generates the RF vector for an initially strained truss member.

9. SUBROUTINE FORMRF(ALPHA,RF,RFG,NR)
   REAL ALPHA(NR,NR),RF(NR,1),RFG(NR,1)

   ALPHA = local to global transformation matrix
   RF = local fixed-end force vector
   RFG = global fixed-end force vector
   NR = matrix dimension

   This subroutine generates the RFG vector using the matrices RF and ALPHA and the matrix dimension NR, as provided.

D. Subroutines Associated with FRAMEL

1. SUBROUTINE FRAKT(ALPHA,K,T,BKF,NC)
   COMMON / SPCSIN / INF,INH,INO,INS
   COMMON / SPCSVL / H1,H2,DXI,DYI,DXJ,DYJ,THETAI,THETAJ
   COMMON / FEFDTA / A,E,I,L
   REAL ALPHA(6,6),K(6,6),T(6,NC),BKF(NC,NC)

   ALPHA = local to global transformation matrix
   K = global stiffness matrix
   T = basic to global transformation matrix
   BKF = basic stiffness or flexibility matrix

   This calling subroutine first reads* in the frame member material and geometric properties as described in part I3 of Sec. A2.4. After calculating the member length, this subroutine calls, as appropriate, the subroutines described next (parts D2 to D5). The parameters contained in the two common blocks SPCSIN and SPCSVL have already been set or read in as required. See Table A2.2 for a description and discussion of the applicability of these parameters to a frame member. The relevant parameters in the common block FEFDTA are passed to subroutine FEFOR for possible use in calculating member fixed-end forces. NC is set ahead of this subroutine to 3 if no hinge is present or to 2 if there is an internal hinge.

*See footnote on page 474.

2. SUBROUTINE FRBKF(BKF,E,I,A,L,INF)
REAL BKF(3,3),I,L

BKF = basic stiffness or flexibility array
E = modulus of elasticity
I = cross-sectional moment of inertia
A = cross-sectional area
L = member length
INF = stiffness or flexibility method indicator (see Table A2.2)

This subroutine generates the basic stiffness or flexibility array, depending on the value of INF, for a frame member.

2a. SUBROUTINE FRBKFH(BKF,E,I,A,L,H,INF)
REAL BKF(2,2),I,L

Definitions of the parameters in the list are the same as those given in part D2 except for the addition of the following:

H = dimension to internal hinge (see Fig. A2.1)

This subroutine performs the same function as the subroutine described in part D2, except that the effect of an internal hinge is included.

3. SUBROUTINE FRA(DX,DY,L,ALPHA)
REAL ALPHA(6,6),L

DX = directed projection of the member on the x axis
DY = directed y projection
L = member length
ALPHA = local to global transformation matrix

This subroutine generates the ALPHA matrix for a frame member.

3a. SUBROUTINE FRSA (DX,DY,THETAI,THETAJ,ALPHA)
REAL ALPHA(6,6)

Definitions of the parameters in the list are the same as those given in part D3, with the addition of the following:

THETAI = angle, measured in degrees, counterclockwise from the global x axis to the skewed x axis, end i
THETAJ = angle at end j

This subroutine performs the same function as the subroutines described in part D3, except that the frame member can now rest on skewed supports.

4. SUBROUTINE FRL(LL,L)
REAL LL(6,3),L

LL = basic to local transformation matrix
L = member length

This subroutine generates the LL matrix for a frame member.

4a. SUBROUTINE FRLH(LL,L,H)
REAL LL(6,2),L

Definitions of the parameters in the list are the same as those given in part D4, with the addition of the following:

H = dimension to internal hinge (see Fig. A2.1)

This subroutine performs the same function as the subroutine described in part D4, except that the effect of an internal hinge is included.

5. SUBROUTINE FORMKT(ALPHA,LL,BKF,T,K,NR,NC,INF)
REAL ALPHA(NR,NR),LL(NR,NC),BKF(NC,NC),T(NR,NC),K(NR,NR)

The definitions of the parameters in the list and the function of this subroutine are identical to those described in part C5.

6. SUBROUTINE FEFOR (RFG,NR,NL,INEL)
COMMON / SPCSIN / INF,INH,INO,INS /
COMMON / SPCSVL / H1,H2,DXI,DYI,DXJ,DYJ,THETAI,THETAJ
COMMON / FEFDTA / A,E,I,L
COMMON / ALPH / ALPHA
REAL RFG(NR),ALPHA(6,6),I,L

The definitions of the parameters in the list and the function of this subroutine are the same as those described in part B5.

7. SUBROUTINE FRFEF(L,C,D,WC,WD,THETA,RF)
REAL RF(6,1),L

L = member length
C = dimension to beginning point of load (see Fig. A2.1)
D = dimension to ending point of load (see Fig. A2.1)
WC = intensity of load at C
WD = intensity of load at D
THETA = angle, in degrees, measured counterclockwise from the direction of the load to the y' direction (see Fig. A2.1)
RF = local fixed-end force vector

This subroutine generates the RF vector for a force loaded frame member. See part I3 of Sec. A2.4 for limitations on the types of loads that can be applied to a frame member.

7a. SUBROUTINE FRFEFH(L,C,D,H,WC,WD,THETA,RF)
REAL RF(6,1),L

Definitions of the parameters in the list are the same as those given in part D7, with the addition of the following:

H = dimension to internal hinge (see Fig. A2.1)

This subroutine performs the same function as the subroutine described in part D7, except that the effect of an internal hinge is included.

8. SUBROUTINE FRFEFT(A,E,I,BETA,TB,TT,HT,RF)
   REAL RF(6,1),I

   A = cross-sectional area
   E = modulus of elasticity
   I = cross-sectional moment of inertia
   BETA = coefficient of thermal expansion
   TT = temperature change at the top of the member
   TB = temperature change at the bottom of the member
   HT = height of the member
   RF = local fixed-end force vector

   This subroutine generates the RF vector for a thermally loaded frame member.
9. FORMRF(ALPHA,RF,RFG,NR)
   REAL ALPHA(NR,NR),RF(NR,1),RFG(NR,1)
   The definitions of the parameters in the list and the function of this subroutine are identical to those described in part C9.

E. Subroutines Associated with 3DTREL
SUBROUTINE TDTAKT(ALPHA,K,T,BKF)
COMMON / SPCSIN / INF,INH,INO,INS
REAL ALPHA(6,6),K(6,6),T(6,1),BKF(1,1)

ALPHA = local to global transformation matrix
K = global stiffness matrix
T = basic to global transformation matrix
BKF = basic stiffness or flexibility matrix

This calling subroutine first reads* in the 3D-truss element material and geometric properties as described in part I5 of Sec. A2.4. After calculating the member length, this subroutine calls the subroutines described next (parts E2 to E5). The parameters in the common block have already been read in or set. See Table A2.2 for a description and discussion of the applicability of these parameters to a 3D-truss member.

2. SUBROUTINE TRBKF(BKF,E,A,L,INF)
   REAL BKF(1,1),L
   The definitions of the parameters in the list and the function of this subroutine are identical to those described in part C2.
3. SUBROUTINE TDTA(DX,DY,DZ,L,ALPHA)
   REAL ALPHA(6,6),L

   DX = directed projection of the member on the x axis
   DY = directed y projection
   DZ = directed z projection
   L = member length
   ALPHA = local to global transformation matrix

   This subroutine generates the ALPHA matrix for a 3D-truss member.

*See footnote on page 474.

4. SUBROUTINE TDTL(LL)
   REAL LL(6,1)

   LL = basic to local transformation matrix

   This subroutine generates the LL matrix for a 3D-truss member.
5. SUBROUTINE FORMKT(ALPHA,LL,BKF,T,K,NR,NC,INF)
   REAL ALPHA(NR,NR),LL(NR,NC),BKF(NC,NC),T(NR,NC),K(NR,NR)
   The definitions of the parameters in the list and the function of this subroutine are identical to those described in part C5.

F. Subroutines Associated with 3DFREL
1. SUBROUTINE TDFAKT(ALPHA,K,T,BKF)
   COMMON / SPCSIN / INF,INH,INO,INS
   REAL ALPHA(12,12),K(12,12),T(12,6),BKF(6,6)

   ALPHA = local to global transformation matrix
   K = global stiffness matrix
   T = basic to global transformation matrix
   BKF = basic stiffness or flexibility matrix

   This calling subroutine first reads* in the 3D-frame member material and geometric properties as described in part I6 of Sec. A2.4. After calculating the member length, this subroutine calls the subroutines described next (parts F2 to F5). The parameters in the common block have already been read in or set. See Table A2.2 for a description and discussion of the applicability of these parameters to a 3D-frame member.
2. SUBROUTINE TDFBKF(BKF,G,E,IY,IZ,J,A,L,INF)
   REAL BKF(6,6),IY,IZ,J,L

   BKF = basic stiffness or flexibility array
   G = shear modulus
   E = modulus of elasticity
   IY = cross-sectional moment of inertia about member y' axis
   IZ = cross-sectional moment of inertia about member z' axis
   J = cross-sectional torsional geometric constant
   A = cross-sectional area
   L = member length
   INF = stiffness or flexibility method indicator (see Table A2.2)

   This subroutine generates the basic stiffness or flexibility array, depending on the value of INF, for a 3D-frame member.
3. SUBROUTINE TDFA(DX,DY,DZ,L,XAL,ALPHA)
   REAL ALPHA(12,12),L

   DX = directed projection of the member on the x axis
   DY = directed y projection
   DZ = directed z projection

*See footnote on page 474.

L = member length
XAL = the angle of twist of the element about its x'-axis (see part I6 of Sec. A2.4)
ALPHA = local to global transformation matrix

This subroutine generates the ALPHA matrix for a 3D-frame member.

4. SUBROUTINE TDFL(LL,L)
   REAL LL(12,6),L

   LL = basic to local transformation matrix
   L = member length

   This subroutine generates the LL matrix for a 3D-frame member.

5. SUBROUTINE FORMKT(ALPHA,L,T,BKF,K,NR,NC,INF)
   The definitions of the parameters in the list and the function of this subroutine are identical to those described in part C5.

G. Subroutines to Build the Structure Stiffness and Load Matrices

1. SUBROUTINE ADDST(SK,NRS,K,NR,CODE)
   REAL SK(NRS,NRS),K(NR,NR)
   INTEGER CODE(NR)

   SK = structure global stiffness matrix
   NRS = number of rows and columns in SK
   K = member global stiffness matrix
   NR = number of rows and columns in K
   CODE = member coordinate code

   This subroutine uses the coordinate code to appropriately add coefficients from the matrix K to coefficients in the matrix SK.

2. ADDFE(R,NRS,RFG,NR,CODE)
   REAL R(NRS,1),RFG(NR,1)
   INTEGER CODE(NR)

   R = structure equivalent nodal load vector
   NRS = size of R vector
   RFG = member global fixed-end force vector
   NR = size of RFG
   CODE = member coordinate code

   This subroutine uses the coordinate code to appropriately subtract elements of the vector RFG from elements of the vector R.

3. SUBROUTINE BOUDC(SK,R,NR,NB)
   REAL SK(NR,NR),R(NR,1)

   SK = structure global stiffness matrix
   R = structure equivalent nodal load vector
   NR = matrix dimension
   NB = number of boundary conditions

This subroutine reads in the initial displacement associated with each boundary condition and then appropriately modifies the completed SK and R matrices.

4. SUBROUTINE INITLD(D0,ND,NC,NF,VRC)
   REAL D0(ND)
   INTEGER VRC(ND)

   D0 = structure initial displacement vector
   ND = number of structure degrees of freedom
   NC = number of constraints
   NF = number of degrees of freedom per structure node
   VRC = vector of rearranged coordinates

   This subroutine reads* in the node number, coordinate direction, and imposed displacement for each constrained structure degree of freedom. The value of the imposed displacement is then appropriately placed in the D0 vector. Finally, the vector VRC is generated.

5. SUBROUTINE INITLR(RU,NRR,NL,NF,VRC,ND)
   REAL RU(NRR)
   INTEGER VRC(ND)

   RU = vector of structure loads in unconstrained coordinate directions
   NRR = size of the RU vector
   NL = number of nodal loads
   NF = number of degrees of freedom per structure node
   VRC = vector of rearranged coordinates
   ND = size of the VRC vector

   This subroutine reads* in the node number, coordinate direction, and magnitude of the load for each applied nodal load. Using the VRC vector, the subroutine then generates the RU vector.

6. SUBROUTINE INITLC(CODE,NC,NI,NJ,NF,VRC,ND)
   INTEGER CODE(NC),VRC(ND)

   CODE = member unconstrained coordinate code
   NC = size of CODE vector
   NI = structure node number at *i* end of member
   NJ = structure node number at *j* end of member
   NF = number of degrees of freedom per structure node
   VRC = vector of rearranged coordinates
   ND = size of VRC vector

   This subroutine generates the unconstrained coordinate code vector CODE for a member between structure nodes NI and NJ.

*See footnote on page 474.

7. SUBROUTINE REARRD (D0,ND,VRC,TEM)
   REAL D0(ND),TEM(ND)
   INTEGER VRC(ND)

   D0 = structure initial displacement vector
   ND = size of vectors, D0, VRC, and TEM
   VRC = vector of rearranged coordinates
   TEM = temporary storage vector

   This subroutine rearranges the D0 vector, making use of the information contained in the VRC vector.
8. SUBROUTINE FEFINT(RFG,NRR,K,D0,ND,CODE)
   REAL RFG(NRR),K(NRR,NRR),D0(ND)
   INTEGER CODE(NRR)

   RFG = member global fixed-end force vector
   NRR = matrix dimension
   K = member global stiffness matrix
   D0 = rearranged structure initial displacement vector
   ND = size of D0 vector
   CODE = member unconstrained coordinate code

   This subroutine superimposes the effects of initial displacements on the RFG vector.

H. Subroutines to Determine the Final Local Force Vectors
1. SUBROUTINE BEAMF(RL,K,D,NRD,RFG,CODE)
   REAL D(NRD),RL(4),K(4,4),RFG(4)
   INTEGER CODE(4)

   RL = member local force vector
   K = member global stiffness matrix
   D = structure nodal displacement vector
   NRD = size of D vector
   RFG = member global fixed-end force vector
   CODE = member coordinate code

   This subroutine generates the RL vector for a beam.
2. SUBROUTINE FRTRF(RL,NRR,ALPHA,K,D,NRD,RFG,CODE)
   REAL RL(NRR),ALPHA(NRR,NRR),K(NRR,NRR),D(NRD),RFG(NRR)
   INTEGER CODE(NRR)

   RL = member local force vector
   NRR = matrix dimension
   ALPHA = local to global transformation matrix
   K = member global stiffness matrix
   D = structure nodal displacement vector
   NRD = size of D vector

°See footnote on page 474.

RFG = member global fixed-end force vector
CODE = member coordinate code

This subroutine generates the RL vector for a truss, frame, 3D-truss, or 3D-frame member.

REFERENCES

A2.1 Schkade, A. Fred. *SMIS2 User's Manual.* Austin: University of Texas, Department of Aerospace Engineering, 1969.

A2.2 Wilson, Edward P. *SMIS, Symbolic Matrix Interpretive System.* Report No. UCSESM 73-3. Berkeley: University of California, Department of Civil Engineering, n.d.

A2.3 Weaver, William, Jr. *Computer Programs for Structural Analysis.* Princeton, N.J.: Van Nostrand, 1967, pp. 19–31.

Answers to Problems

2.2 $\overline{\mathbf{P}} = \langle 20 \quad -80 \quad 0 \quad -20 \quad -60 \quad -20 \quad 84.85 \quad 28.29 \quad 60 \quad 80 \quad 0 \quad -80 \rangle$

2.3 $\overline{\mathbf{P}} = \langle -46.88 \quad -46.88 \quad 53.12 \quad -50 \quad 53.12 \quad 0 \quad -25 \quad -25 \rangle$

2.5 $\overline{\mathbf{P}} = \langle 54 \quad 36 \quad 54 \quad -60.38 \quad -52.33 \quad -52.33 \quad -60.38 \quad -16.10 \quad 18 \quad 18 \quad -16.10 \quad 0 \quad -36 \quad -36 \rangle$

2.7 $\overline{\mathbf{P}} = \langle 0 \quad -18 \quad 18 \quad 0 \quad -33.75 \quad -56.25 \rangle$

2.9 $\overline{\mathbf{P}} = \langle 0 \quad 0 \quad 13.33 \quad -80 \quad 0 \quad 0 \quad 0 \quad 13.33 \quad -13.33 \rangle$

2.11 $\overline{\mathbf{P}} = \langle 0 \quad -40 \quad 34.64 \quad 40 \quad 0 \quad 6 \quad 34.64 \quad 10 \quad 10 \rangle$

2.13 $\mathbf{D} = \langle 0 \quad 0 \quad 5.616 \times 10^{-3} \quad 0 \quad 5.616 \times 10^{-3} \quad 0 \quad 0 \quad 0 \quad 5.616 \times 10^{-3} \quad 0 \quad 5.616 \times 10^{-3} \quad 0 \rangle$

4.1 $\mathbf{D} = \dfrac{4}{EA_0} \langle 6 \times 10^{-4} \quad 8 \times 10^{-4} \quad 309.7 \quad 20 \quad 506.3 \quad 20 \quad 0 \quad -8 \times 10^{-4} \quad 249.7 \quad -80 \quad 486.3 \quad -100 \rangle$

$P^{(1)} = 20$ kN
$P^{(2)} = -80$ kN
$P^{(7)} = 84.85$ kN

4.3 $\mathbf{D} = \dfrac{4}{EA_0} \langle 2.8 \times 10^{-4} \quad 8 \times 10^{-4} \quad 142.4 \quad 51.98 \quad 276.7 \quad 64.88 \quad 3.2 \times 10^{-4} \quad -8 \times 10^{-4} \quad 127.3 \quad -48.03 \quad 269.6 \quad -55.12\rangle$

$P^{(3)} = 12.90$ kN
$P^{(6)} = -7.10$ kN
$P^{(10)} = -18.24$ kN

4.5 $\mathbf{D} = \dfrac{2}{EI_0} \langle -4.977 \times 10^{-10} \quad -2.22 \times 10^{-9} \quad 6.63 \times 10^{-10} \quad .02488 \quad -.2221 \quad -6.648 \quad 4.977 \times 10^{-10} \quad -7.787 \times 10^{-10} \quad 6.713\rangle$

$\mathbf{r}'^{(1)} = \langle 22.21 \quad -4.98 \quad -6.63 \quad -22.21 \quad 4.98 \quad -13.28\rangle$
$\mathbf{r}'^{(2)} = \langle 4.98 \quad 22.21 \quad 13.28 \quad -4.98 \quad 7.79 \quad 0\rangle$

4.7 $\mathbf{D} = \dfrac{1}{EI_0} \langle 0 \quad 0 \quad 0 \quad 30 \quad 42 \quad 18\rangle$

$\mathbf{r}^{(1)} = \langle 41.62 \quad 63 \quad 30.38 \quad -18\rangle$
$\mathbf{r}^{(2)} = \langle 18 \quad 18 \quad 0 \quad 0\rangle$

4.9 $\mathbf{D} = \dfrac{4}{EA_0} \langle 142.4 \quad 51.98 \quad 276.7 \quad 64.88 \quad 127.3 \quad -48.03 \quad 269.6 \quad -55.12\rangle$

4.10 (a) Same as Problem 4.5
(b) $\mathbf{D} = \langle 1.10 \times 10^{-5} \quad -1.51 \times 10^{-2} \quad -2.93 \times 10^{-3} \quad 7.47 \times 10^{-3}\rangle$
$\mathbf{r}'^{(1)} = \langle 21.46 \quad -3.29 \quad -4.39 \quad -21.46 \quad 3.29 \quad -8.78\rangle$
$\mathbf{r}'^{(2)} = \langle 3.29 \quad 21.46 \quad 8.78 \quad -3.29 \quad 8.54 \quad 0\rangle$

4.12 $\mathbf{D} = \langle .01104 \quad -0.04381 \quad -5.4 \times 10^{-4} \quad -1.855 \times 10^{-4} \quad 0 \quad -.1362 \quad 0 \quad .04381 \quad -5.4 \times 10^{-4} \quad 1.855 \times 10^{-4} \quad -.01104\rangle$
$\mathbf{r}'^{(1)} = \mathbf{r}'^{(4)} = \langle 450 \quad -299.5 \quad 0 \quad -450 \quad 299.5 \quad -1797\rangle$
$\mathbf{r}'^{(2)} = \mathbf{r}'^{(3)} = \langle 426.4 \quad 332.2 \quad 1797 \quad -284.1 \quad 94.70 \quad 456.3\rangle$

4.13 Symmetric case:
$\mathbf{D} = \langle 4.034 \times 10^{-6} \quad -4.007 \times 10^{-5} \quad -8.984 \times 10^{-4} \quad -1.118 \times 10^{-4}\rangle$
$\mathbf{r}'^{(1)} = \langle 10.02 \quad -2.017 \quad -2.686 \quad -10.02 \quad 2.017 \quad -5.381\rangle$
Antisymmetric case:
$\mathbf{D} = \langle 2.358 \times 10^{-3} \quad -4.8 \times 10^{-5} \quad -1.772 \times 10^{-3} \quad 2.356 \times 10^{-3} \quad 1.784 \times 10^{-3} \quad -1.775 \times 10^{-3}\rangle$
$\mathbf{r}'^{(1)} = \langle 12 \quad -1.334 \quad -.01074 \quad -12 \quad 1.334 \quad -5.327\rangle$
Total:
$\mathbf{D} = \langle 2.362 \times 10^{-3} \quad -8.807 \times 10^{-5} \quad -2.671 \times 10^{-3} \quad 2.356 \times 10^{-3} \quad -1.118 \times 10^{-4} \quad 1.784 \times 10^{-3} \quad -1.775 \times 10^{-3} \quad 2.354 \times 10^{-3} \quad 7.919 \times 10^{-6} \quad -8.738 \times 10^{-4}\rangle$
$\mathbf{r}'^{(1)} = \langle 22.02 \quad -3.351 \quad -2.697 \quad -22.02 \quad 3.351 \quad -10.07\rangle$

5.1 $\mathbf{D} = 10^{-3} \langle -3.31 \quad -21.48 \quad -8.13 \quad -4.96 \quad -3.31 \quad -5.53 \quad -20\rangle$

$\overline{\mathbf{P}} = \dfrac{EA_0}{3} \langle 0 \quad .003446 \quad 0 \quad .003309 \quad .00193 \quad -.003446 \quad -.00383\rangle$

5.2 $\mathbf{D} = \langle -.01498 \quad -.0000146 \quad .00125 \quad -.02 \quad -.00312\rangle$

$\mathbf{r}'^{(1)} = \dfrac{EI_0}{2} \times 10^{-3} \langle 1.456 \quad -4.680 \quad -9.985 \quad -1.456 \quad 4.680 \quad -8.737\rangle$

$\mathbf{r}'^{(2)} = \dfrac{EI_0}{2} \times 10^{-3} \langle 4.680 \quad 1.457 \quad 8.737 \quad -4.680 \quad -1.456 \quad 0\rangle$

5.3 $\mathbf{D} = 10^{-3}\langle 2.51 \quad .66 \quad 2.84 \quad .59 \quad 3.10 \quad .66 \quad 2.77 \quad .59\rangle$

$$\overline{\mathbf{P}} = \frac{EA_0}{4} \times 10^{-5}\langle 65.66 \quad 65.66 \quad -6.80 \quad -6.80 \quad 58.86 \quad -6.80 \quad -92.86 \quad 9.615 \quad -92.86 \quad 9.615\rangle$$

5.5 $D_1 = 4.597 \times 10^{-4}$
$\mathbf{r}^{(1)} = \langle 136.95 \quad 0 \quad -136.95 \quad 6573.8\rangle$

5.7 $\mathbf{r}_F = qL\langle .4312 \quad .05289L \quad .5688 \quad -.1217L\rangle$

5.9 Inextensional:

$$\mathbf{D} = \frac{2}{EI_0}\langle -6.667 \quad 6.667\rangle$$

$r_3'^{(2)} = 13.333$
Extensional:

$$\mathbf{D} = \frac{2}{EI_0}\langle -6.648 \quad 6.713\rangle$$

$r_3'^{(2)} = 13.278$

5.12 $\mathbf{D} = 10^{-3}\langle -6.31 \quad -.462 \quad -5.67 \quad 0 \quad -.924 \quad 6.31 \quad -.462 \quad -5.4\rangle$
$\mathbf{r}'^{(1)} = \langle 55.42 \quad 20.45 \quad 0 \quad -55.42 \quad 29.56 \quad -13.65\rangle$
$\mathbf{r}'^{(2)} = \langle 55.42 \quad 29.56 \quad 13.65 \quad -55.42 \quad 20.45 \quad 0\rangle$
$P^{(3)} = 62.81$
$P^{(4)} = -59.11$
$P^{(5)} = 62.81$

5.15 $\mathbf{D} = 10^{-3}\langle -12.06 \quad -13.88 \quad -1.70 \quad -12.43 \quad .51 \quad .79 \quad -13.10 \quad .69 \quad 2.76 \; 9.09\rangle$
$\mathbf{r}'^{(1)} = \langle 56.35 \quad 9.48 \quad 5.72 \quad -65.35 \quad 8.52 \quad -4.11\rangle$

5.16 $$\mathbf{D} = -\frac{q}{EI_0}\langle .1174 \quad .0221\rangle$$

$\mathbf{r}'^{(1)} = q\langle 1.25 \quad 0 \quad -.2604 \quad -1.25 \quad -1.25 \quad .5208\rangle$

6.1 $\overline{\mathbf{P}} = \langle 0 \quad -36 \quad 36 \quad 0 \quad 4.5 \quad -22.5\rangle$

$$\mathbf{D} = \frac{1}{3EI_0}\langle 0 \quad 72 \quad 0 \quad -144 \quad -432 \quad -252\rangle$$

6.4 Case I:
$\overline{\mathbf{P}}_I = \langle 48 \quad -48 \quad 3 \quad -3 \quad 0 \quad 0 \quad 0\rangle$
$\mathbf{R}_F = \langle 36 \quad 48 \quad 45 \quad -45 \quad 9 \quad -3\rangle$
Case II:
$\overline{\mathbf{P}}_{II} = \langle 15 \quad 30 \quad 15 \quad 3 \quad -41.625 \quad -63 \quad -48.375\rangle$
Total:
$\overline{\mathbf{P}} = \langle 63 \quad -18 \quad 18 \quad 0 \quad -41.625 \quad -63 \quad -48.375\rangle$

$$\mathbf{D} = \frac{1}{3EI_0}\langle 0 \quad 0 \quad 0 \quad 90 \quad 126 \quad 54\rangle$$

6.7 $\overline{\mathbf{P}} = \langle 56.7 \quad -30.6 \quad 30.6 \quad 0 \quad -39.3 \quad -56.7 \quad -57.04 \quad 6.3\rangle$

$$\mathbf{D} = \frac{1}{3EI_0}\langle 0 \quad 0 \quad 0 \quad 52.2 \quad 0 \quad -21.6\rangle$$

7.2 $\mathbf{D} = \langle -2.105 \quad 4.210 \quad -1.053 \quad -2.105 \rangle$

7.7 Joint rotation at fourth level $= -.003515 \dfrac{qL^3}{EI_0}$

Joint rotation at top level $= -.005422 \dfrac{qL^3}{EI_0}$

8.1 $\mathbf{D} = \dfrac{QL}{EA_0} \langle -.1029 \quad -1.160 \rangle$

$\overline{\mathbf{P}} = Q \langle .2643 \quad -.2057 \quad .4697 \quad -.3292 \rangle$

8.3 $\mathbf{D} = \dfrac{q}{EI_0} \langle -61.71 \quad -34.79 \quad -18.47 \rangle$

$\mathbf{r}^{(1)} = q \langle 5.880 \quad 8.566 \quad 0 \quad 6.620 \quad -12.27 \quad 0 \rangle$

Index

83 84 85 86 87 9 8 7 6 5 4 3 2 1

NOTATION

MDIF — maximum difference in structure nodal numbers associated with any one member of a structure

n_1, n_2, n_3 — direction cosines of z' axis relative to x, y, and z axes, respectively

N — total number of degrees of freedom of a structure

N_D — number of degrees of freedom per structure node

p — intensity of distributed external force, y direction

P — member basic axial force

$\mathbf{P}$ — vector of member basic forces

$\overline{\mathbf{P}}$ — vector of basic forces for collection of all members and supports

$\mathbf{P}_F$ — vector of member basic fixed-end forces

$\overline{\mathbf{P}}_F$ — vector of basic fixed-end forces for collection of all members and supports

P_c — translational constraint basic force

P_{cF} — constraint basic fixed-end force

$\overline{\mathbf{P}}_Q$ — subarray of the vector, $\overline{\mathbf{P}}$, representing a vector of redundant basic forces ($\overline{\mathbf{P}}_Q \equiv \mathbf{Q}$)

$\overline{\mathbf{P}}_R$ — subarray of the vector, $\overline{\mathbf{P}}$, representing a vector of nonredundant basic forces

q — intensity of distributed external force, x direction

Q — magnitude of concentrated external force

$\mathbf{Q}$ — vector of redundant forces that are applied external to the cut-back structure

$\mathbf{r}$ — vector of member-end forces, global coordinates

$\mathbf{r}'$ — vector of member-end forces, local coordinates

$\hat{\mathbf{r}}$ — vector of member-end forces, skew coordinates

$\overline{\mathbf{r}}$ — vector of member-end forces for the collection of all members and supports, global coordinates

$\mathbf{r}_E$, $\mathbf{r}_I$ — vectors of substructure forces associated with external and internal degrees of freedom, respectively

$\mathbf{r}_F$ — vector of member fixed-end forces, global coordinates

$\mathbf{r}'_F$ — vector of member fixed-end forces, local coordinates

$\overline{\mathbf{r}}_F$ — vector of member fixed-end forces for the collection of all members and supports, global coordinates

$\mathbf{r}_{FE}$, $\mathbf{r}_{FI}$ — vectors of substructure fixed-end forces associated with external and internal degrees of freedom, respectively

$\mathbf{r}_{F0}$ — vector of member fixed-end forces due to initial strains, global coordinates

(*Continued on next page*)

$\mathbf{r}'_{F0}$	vector of member fixed-end forces due to initial strains, local coordinates
r_c	constraint force, global coordinates
$\mathbf{R}$	vector of structure nodal forces, global coordinates
$\hat{\mathbf{R}}$	vector of structure nodal forces, generalized coordinates or mixed coordinates
$\mathbf{R}_A$, $\mathbf{R}_S$	respective antisymmetric and symmetric components of the $\mathbf{R}$ vector for a symmetric structure
$\mathbf{R}_c$, $\mathbf{R}_u$	subvectors of the $\mathbf{R}$ vector associated with constrained and unconstrained degrees of freedom, respectively
$\mathbf{R}_F$	vector of structure nodal fixed-node forces, global coordinates
$\mathbf{R}_{F0}$	vector of structure nodal fixed-node forces due to initial strains, global coordinates
$\mathbf{R}_{FA}$, $\mathbf{R}_{FS}$	respective antisymmetric and symmetric components of the $\mathbf{R}_F$ vector for a symmetric structure
$\mathbf{R}_{Fc}$	a subvector of the $\mathbf{R}_F$ vector associated with constrained degrees of freedom
$\mathbf{R}_{FL}$, $\mathbf{R}_{FR}$	respective subvectors of the $\mathbf{R}_F$ vector for the left and right sides of a symmetric structure
$\mathbf{R}_L$, $\mathbf{R}_R$	respective subvectors of the $\mathbf{R}$ vector for the left and right sides of a symmetric structure
s	sine α
s_i, s_j	sine α_i and sine α_j, respectively
S	member internal axial force
t	temperature
$\mathbf{t}$	thermal load vector
$\mathbf{T}$	array of member force influence coefficients relating $\mathbf{r}$ and $\mathbf{P}$
$\mathbf{T}^T$	array of member displacement influence coefficients relating $\boldsymbol{\Delta}$ and $\mathbf{d}$
$\overline{\mathbf{T}}$	array of force influence coefficients for the collection of all members and supports ($\overline{\mathbf{r}} = \overline{\mathbf{T}}\,\overline{\mathbf{P}}$)
$\overline{\mathbf{T}}^T$	array of displacement influence coefficients for the collection of all members and supports ($\overline{\boldsymbol{\Delta}} = \overline{\mathbf{T}}^T\overline{\mathbf{d}}$)
U	potential strain energy
U_A	axial strain energy
U_B	bending strain energy
U_V	shearing strain energy
V	member internal shearing force
$\mathbf{V}$	intermediate solution array generated during forward elimination
V_A^c	shearing force at end A of a conjugate beam
V_i	member-end shear, end i
V_j	member-end shear, end j